Chemical
Oceanography

VOLUME 3
2ND EDITION

Chemical Oceanography

Edited by
J. P. RILEY

Department of Oceanography,
The University of Liverpool, England

and

G. SKIRROW

Department of Inorganic, Physical and Industrial Chemistry,
The University of Liverpool, England

VOLUME 3

2ND EDITION

1975

ACADEMIC PRESS

LONDON   NEW YORK   SAN FRANCISCO

A Subsidiary of Harcourt Brace Jovanovich, Publishers

ACADEMIC PRESS INC. (LONDON) LTD.
24/28 Oval Road,
London NW1

United States Edition published by
ACADEMIC PRESS INC.
111 Fifth Avenue
New York, New York 10003

Library of Congress Catalog Card Number: 74–5679
ISBN: 0–12–588603–9

Printed in Great Britain by
PAGE BROS (NORWICH) LTD
NORWICH

Contributors to Volume 3

J. D. BURTON, *Department of Oceanography, The University, Southampton, England*

W. D. DEUSER, *Woods Hole Oceanographic Institution, Woods Hole, Massachusetts 02543, U.S.A.*

J. W. R. DUTTON, *Fisheries Radiobiological Laboratory, Lowestoft, Suffolk, England.*

E. D. GOLDBERG, *Scripps Institution of Oceanography, La Jolla, California 92037, U.S.A.*

N. T. MITCHELL, *Fisheries Radiobiological Laboratory, Lowestoft, Suffolk, England*

J. P. RILEY, *Department of Oceanography, The University of Liverpool, England*

D. E. ROBERTSON, *Battelle Pacific Northwest Laboratory, Richland, Washington 99352, U.S.A.*

P. J. LE B. WILLIAMS, *Department of Oceanography, The University, Southampton, England*

Preface to the Second Edition

Rapid progress has occurred in all branches of Chemical Oceanography since the publication of the first edition of this book a decade ago. Particularly noteworthy has been the tendency to treat the subject in a much more quantitative fashion; this has become possible because of our much improved understanding of the physical chemistry of sea water systems in terms of ionic and molecular theories. For these reasons chapters dealing with sea water as an electrolyte system, with speciation and with aspects of colloid chemistry are now to be considered as essential in any up-to-date treatment of the subject. Fields of research which were little more than embryonic only ten years ago, for example sea surface chemistry, have now expanded so much that they merit separate consideration. Since the previous edition, there has arisen a general awareness of the potential threat to the sea caused by man's activities, in particular its use as a "rubbish bin" and a receptacle for toxic wastes. Although it was inevitable that there should be some over-reaction to this, there is real cause for concern. Clearly, it is desirable to have available reasoned discussions of this topic and also an examination of the role of the sea as a potential source of raw materials in view of the imminent exhaustion of many high grade ores; these subjects are treated in the second, third and fourth volumes.

Most branches of marine chemistry make use of analytical techniques; the number and range of these has increased dramatically over recent years. Consequently, it has been necessary to expand greatly and restructure the sections dealing with analytical methodology. These developments are extending increasingly into the very important and rapidly developing area of organic chemistry.

Rapid advances which have taken place in geochemistry, particularly those that have stemmed from the Deep Sea Drilling Project, have made it necessary to devote a whole volume to topics in sedimentary geochemistry.

Both the range and accuracy of the physical constants available have increased since the first edition and a selection of tabulated values of these constants are to be found at the end of each of the first four volumes.

No attempt has been made to discuss Physical Oceanography except where a grasp of the physical concepts is necessary for a better understanding of the chemistry. For a treatment of the physical processes occurring in the sea the reader is referred to the numerous excellent texts now available on physical oceanography. Likewise, since the distribution of salinity in the sea is of greater relevance to the physical oceanographer and is well discussed in these texts, it will not be considered in the present volumes.

This series is not intended to serve as a practical handbook of Marine Chemistry, and if practical details are required the original references given in the text should be consulted. In passing, it should be mentioned that, although those practical aspects of sea water chemistry which are of interest to biologists are reasonably adequately covered in the "Manual of Sea Water Analysis" by Strickland and Parsons, there is an urgent need for a more general laboratory manual.

The editors are most grateful to the various authors for their helpful co-operation which has greatly facilitated the preparation of this book. They would particularly like to thank Messrs R. F. C. Mantoura and A. Dickson for their willing assistance with the arduous task of proof reading; without their aid many errors would have escaped detection. They would also like to acknowledge the courtesy of the various copyright holders, both authors and publishers, for permission to use tables, figures, and photographs. In conclusion, they wish to thank Academic Press, and in particular Mr. E. A. S. Cotton, for their efficiency and ready co-operation which has much lightened the task of preparing this book for publication.

Liverpool   J. P. RILEY
November, 1974   G. SKIRROW

CONTENTS

CONTRIBUTORS TO VOLUME 3 v
PREFACE vii
CONTENTS OF VOLUMES 1, 2 AND 4 xiv
SYMBOLS AND UNITS USED IN THE TEXT xv

Chapter 16 by W. G. DEUSER
Reducing Environments

16.1. Introduction 1
16.2. Oxygen-deficient Conditions in the Open Ocean.................. 2
16.3. The Development of Anoxic Conditions in Restricted Basins ... 12
16.4. Time Scale of Water Renewal in Anoxic Basins 22
16.5. Selected Aspects of the Decomposition of Organic Matter in
      Marine Anoxic Environments 29
16.6. Acknowledgements 34
References 35

Chapter 17 by E. D. GOLDBERG
Marine Pollution

17.1. Introduction 39
17.2. Transport Processes 42
17.3. Radioactivity 48
17.4. Petroleum 53
17.5. Metals 60
17.6. Synthetic Organic Chemicals 70
17.7. Marine Litter 82
17.8. Control of Anthropogenic Inputs to the Environment 84
References 86

Chapter 18 by J. D. BURTON
Radioactive Nuclides in the Marine Environment

18.1. Introduction 91
18.2. Long-lived Nuclides of Elements with Stable Isotopes 93
18.3. Uranium, Thorium and Actino-uranium Series.................... 97
18.4. Nuclides Produced by Cosmic Radiation 139
18.5. Artificially Produced Nuclides 157
References 177

Chapter 19 *by* J. P. RILEY
With contributions by D. E. ROBERTSON, J. W. R. DUTTON *and* N. T.
MITCHELL, P. J. le B. WILLIAMS

Analytical Chemistry of Sea Water

19.1.   Introduction   193
19.2.   Analytical Error   197
19.3.   Water Sampling   199
19.4.   Filtration of Water Samples   211
19.5.   Storage of Sea Water Samples   218
19.6.   Determination of Salinity and Chlorinity   223
19.7.   Determination of Total Cations by Ion Exchange   225
19.8.   Determination of Major Components   225
19.9.   Determination of Dissolved Gases.........................   253
19.10. Determination of Minor Inorganic Constituents.............   269
19.11. Determination of Micronutrient Elements...................   408
19.12. Determination of the Radioelements, Uranium-238, Uranium-
        235 and Thorium and their Decay Products..................   431
19.13. Determination of Radionuclides other than those of the
        Uranium and Thorium Series................................   437
19.14. Determination of Organic Components   443
References   477

Appendix *compiled by* J. P. RILEY

Table of Physical and Chemical Constants Relevant to Marine Chemistry

TABLE 1.   Some physical properties of pure water   515
TABLE 2.   Concentrations of the major ions in sea water of various
           salinities   516
TABLE 3.   Preparation of artificial sea water   517
TABLE 4.   Collected conversion factors   518
TABLE 5.   Table for conversion of weights of nitrogen, phosphorus
           and silicon expressed in terms of µg into µg-at   518
TABLE 6.   Solubility of oxygen in sea water ($cm^3\ l^{-1}$)   519
TABLE 7.   Solubility of nitrogen in sea water ($cm^3\ l^{-1}$) ..   521
TABLE 8.   Solubility of argon in sea water ($cm^3\ l^{-1}$)   522
TABLE 9.   Literature citations for solubilities of other gases in sea
           water   522

TABLE 10.   Density of artificial sea water as a function of temperature
            and chlorinity.................................... 523
TABLE 11.   Expansibility of artificial sea water as a function of tem-
            perature and chlorinity........................... 524
TABLE 12.   Isothermal compressibility of sea water at 1 atm as a
            function of salinity and temperature 524
TABLE 13.   Observed values for the change in the specific volume of
            sea water at various pressures and salinities 525
TABLE 14.   Specific gravity and percentage volume reduction of sea
            water under pressure.............................. 526
TABLE 15.   Percentage reduction in volume of sea water under a
            pressure of 1000 db at various temperatures and salinities 526
TABLE 16.   Thermal expansion of sea water under pressure 527
TABLE 17a. Velocity of sound in sea water...................... 527
      17b. Effect of salinity on sound velocity.................. 528
TABLE 18.   Specific heat of sea water at constant pressure at various
            salinities and temperatures.......................... 529
TABLE 19.   Relative partial equivalent heat capacity of sea salt....... 530
TABLE 20.   Thermal conductivity of sea water at various temperatures
            and pressures 530
TABLE 21.   Freezing point of sea water at atmospheric and higher
            pressures....................................... 531
TABLE 22.   Boiling point elevation of sea water at various tempera-
            tures 531
TABLE 23.   Osmotic pressure and vapour pressure depression of sea
            water at 25°C 532
TABLE 24.   Surface tension of clean sea water at various salinities and
            temperatures 533
TABLE 25.   Viscosity coefficient of sea water as a function of tem-
            perature and chlorinity 534
TABLE 26.   Relative viscosity of Standard Sea Water at various
            temperatures and pressures 535
TABLE 27.   Specific conductivity of sea water 536
TABLE 28.   Effect of pressure on the conductivity of sea water 537
TABLE 29.   Conductivity ratio of sea water relative to sea water of
            salinity 35·000 ‰ 538
TABLE 30.   Corrections to be applied to conductivity ratios measured
            at temperatures differing from 20°C to correct them to
            ratios at 20°C.................................... 538
TABLE 31.   Light absorption of typical sea waters 539

TABLE 32. Differences between the extinctions of sea waters and pure water 539

FIG. 1. Absorption curves for ocean water and pure water 540

TABLE 33. Refractive index differences for sea water ($\lambda = 589\cdot3$ nm) at various temperatures and salinities 541

TABLE 34. Refractive index differences for sea water of salinity $35\cdot00\%_0$ at various temperatures and wavelengths 542

TABLE 35. Absolute refractive index of sea water ($S = 35\cdot00\%_0$) as a function of temperature pressure and wavelength 544

TABLE 36. Velocity of light ($\lambda = 589\cdot3$ nm) in sea water at 1 atm. 545

References 546

Subject Index... 549

Contents of Volume 1

Chapter 1 *by* K. F. BOWDEN
Oceanic and Estuarine Mixing Processes
Chapter 2 *by* M. WHITFIELD
Sea Water as an Electrolyte Solution
Chapter 3 *by* WERNER STUMM and PHYLLIS A. BRAUNER
Chemical Speciation
Chapter 4 *by* GEORGE A. PARKS
Adsorption in the Marine Environment
Chapter 5 *by* FRED T. MACKENZIE
Sedimentary Cycling and the Evolution of Sea Water
Chapter 6 *by* T. R. S. WILSON
Salinity and the Major Elements of Sea Water
Chapter 7 *by* PETER G. BREWER
Minor Elements in Sea Water
Chapter 8 *by* DANA A. KESTER
Dissolved Gases other than CO_2
Tables of Chemical and Physical Constants
Subject Index

Contents of Volume 2

Chapter 9 *by* GEOFFREY SKIRROW
The Dissolved Gases—Carbon Dioxide
Chapter 10 *by* PETER S. LISS
Chemistry of the Sea Surface Microlayer
Chapter 11 *by* C. P. SPENCER
The Micronutrient Elements
Chapter 12 *by* P. LE B. WILLIAMS
Biological and Chemical Aspects of Dissolved Organic Material in Sea Water
Chapter 13 *by* T. R. PARSONS
Particulate Organic Carbon in the Sea
Chapter 14 *by* G. E. FOGG
Primary Productivity
Chapter 15 *by* G. GRASSHOFF
The Hydrochemistry of Landlocked Basins and Fjords
Tables of Chemical and Physical Constants
Subject Index

Contents of Volume 4

Chapter 20 *by* M. WHITFIELD
The Electroanalytical Chemistry of Sea Water
Chapter 21 *by* W. F. MCILHENNY
Extraction of Economic Inorganic Materials from Sea Water
Chapter 22 *by* E. BOOTH
Seaweeds in Industry
Chapter 23 *by* HEBER W. YOUNGKEN JR. and YUZURA SHIMIZU
Marine Drugs: Chemical and Pharmacological Aspects
Table of Chemical and Physical Constants
Subject Index

Symbols and units used in the text

A list of the more important symbols used in the text is given below. It is not exhaustive and inevitably there is some duplication of usage since some symbols have different accepted usages in two or more disciplines. The generally accepted symbols have been altered only when there is a possibility of ambiguity.

Concentration. There are several systems in common use for expressing concentration. The more important of these are the molarity scale (g molecules l^{-1} of solution = mol l^{-1}) usually designated by c_i, the molality scale (g molecules kg^{-1} of solvent* = mol kg^{-1}) designated by m_i, and the mole fraction scale usually denoted by x_i, which is of more fundamental significance in physical chemistry. In each instance the subscript i indicates the solute species; when i is an ion the charge is not included in the subscript unless confusion is likely to arise. Some other means of indicating the concentration are also to be found in the text, these include: g or mg kg^{-1} of solution (for major components), µg or ng l^{-1} or kg^{-1} of solution (for trace elements and nutrients) and µg-at l^{-1} of solution (for nutrients). Factors for conversion of µg to µg-at are to be found in Appendix Tables 4 and 5.

Activity. When an activity or activity coefficient is associated with a species the symbols a_i and γ_i are used respectively regardless of the method of expressing concentration, where the subscript i has the significance indicated above. Further qualifying symbols may be added as superscripts and/or subscripts as circumstances demand. It is important to realize that the numerical values of the activity and activity coefficient depend on the standard state chosen. It should also be noted that since activity is a relative quantity it is dimensionless.

UNITS

Where practicable SI units (and the associated notations) have been adopted in the text except where their usage goes contrary to established oceanographic practice.

* A common practice is to regard sea water as the solvent for minor constituents.

LENGTH

Å	= Ångstrom unit	$= 10^{-10}$ m
nm	= nanometre	$= 10^{-9}$ m
μm	= micrometre	$= 10^{-6}$ m
mm	= millimetre	$= 10^{-3}$ m
cm	= centimetre	$= 10^{-2}$ m
m	= metre	
km	= kilometre	$= 10^{3}$ m
mi	= nautical mile (6080 ft)	$= 1{\cdot}85$ km

WEIGHT

pg	= picogram	$= 10^{-12}$ g
ng	= nanogram	$= 10^{-9}$ g
μg	= microgram	$= 10^{-6}$ g
mg	= milligram	$= 10^{-3}$ g
g	= gram	
kg	= kilogram	$= 10^{3}$ g
ton	= metric ton	$= 10^{6}$ g

VOLUME

μl	= microlitre	$= 10^{-6}$ l
ml	= millilitre	$= 10^{-3}$ l
l	= litre	
dm^3	= litre	

CONCENTRATION

ppm	= parts per million ($\mu g\,g^{-1}$ or $mg\,l^{-1}$)
ppb	= parts per billion ($ng\,g^{-1}$ or $\mu g\,l^{-1}$)
μg-at l^{-1}	= μg atoms $l^{-1} = $ (μg/atomic weight) l^{-1}

ELECTRICAL

V	= volt
A	= ampere
Ω	= ohm

TIME

s = second
min = minute
h = hour
d = day
yr = year

ENERGY AND FORCE

J = Joule = 0·2390 cal
N = Newton = 10^5 dynes
W = Watt

LIGHT FLUX

klux = kilolux

RADIOACTIVITY

pCi picocurie (10^{-12}Ci)
Ci curie
MCi megacurie

GENERAL SYMBOLS

A absorbance = optical density ($\log I_0/I$)
A activity of a radioactive material
AOU apparent oxygen utilization
c_i molar concentration of species i
Cl chlorinity (g kg^{-1} = ‰)
D_e coefficient of eddy diffusion
D_m coefficient of molecular diffusion (see p. 127)
dpm radioactive disintegrations per minute
E the E.M.F. of a cell
E^o standard potential
F Faraday equivalent of electric charge

h	relative humidity
I	attenuated light intensity
I_o	initial light intensity
K	equilibrium constant
P	primary production mg C m^{-2} per unit time
P_G	partial pressure of gas, G, in solution
p_G	partial pressure of gas, G, in the atmosphere
p_o	saturated vapour pressure of pure water
p_s	saturated vapour pressure of sea water or an aqueous solution
S	salinity (g kg^{-1} = ‰)
T	temperature in K
t	temperature in °C
t	age of a deposit
u	component of velocity of water in x-direction
V	volume
v, w	components of velocity of water in y and z directions respectively

Greek Symbols

γ_i	activity coefficient of species i
Δ	Change of (as in ΔG)
$\Delta^{14}C$	see p. 144
$\delta^{13}C$	permillage enrichment of ^{13}C relative to a given standard (see p. 144)
$\delta^{14}C$	permillage enrichment of ^{14}C relative to a given standard (see p. 144)
δN	nitrate deficiency (see p. 7)
ε	molar absorptivity
η	viscosity
λ	wavelength
λ	radio-nuclide decay constant
ρ	density
τ	residence time of an element in sea water
σ	standard deviation

Chapter 16

Reducing Environments*

W. G. DEUSER

Woods Hole Oceanographic Institution
Woods Hole, Massachusetts 02543, U.S.A.

SECTION 16.1 INTRODUCTION 1

16.2 OXYGEN-DEFICIENT CONDITIONS IN THE OPEN OCEAN......................... 2

16.3 THE DEVELOPMENT OF ANOXIC CONDITIONS IN RESTRICTED BASINS 12

16.4 TIME SCALE OF WATER RENEWAL IN ANOXIC BASINS 22

16.5 SELECTED ASPECTS OF THE DECOMPOSITION OF ORGANIC MATTER IN MARINE
ANOXIC ENVIRONMENTS 29

16.6 ACKNOWLEDGEMENTS 34

REFERENCES 35

16.1. INTRODUCTION

This chapter examines those aspects of the chemistry of reducing environments which have been the subject of extensive investigations in the past several years. It is assumed that the reader has some familiarity with the basic physical and chemical characteristics of anoxic basins and fjords (see e.g. Richards, 1965 and Chapter 15). The aspects to be considered are:

(1) The development of anoxic conditions in oxygen-deficient parts of the open ocean.

(2) The circumstances which lead to and accompany anoxic conditions in restricted basins.

(3) The time scale of water renewal in anoxic basins.

(4) Some selected aspects of the decomposition of organic matter in anoxic environments. Although this division is arbitrary, it provides a convenient basis on which much of the research effort of the past years can be discussed.

---

* Contribution No. 3314 from the Woods Hole Oceanographic Institution.

1

16.2. OXYGEN-DEFICIENT CONDITIONS IN THE OPEN OCEAN

Reducing conditions can arise in natural waters only where the consumption rate of oxygen exceeds that of supply. In the marine environment, the consumption of oxygen is intimately linked to the oxidation of organic matter (both living and dead), and the dissolved oxygen which is used in this oxidation is derived from the atmosphere and photosynthesis. The supply of oxygen to waters below the photic zone is thus entirely dependent on advection and diffusion from surface waters in which high concentrations of dissolved oxygen are maintained by physicochemical and biological processes. The oxidation of most of the organic matter introduced into or formed in the sea takes place in the photic zone and the rate of oxidation decreases sharply at greater depth. If this were not so, most of the deep parts of the oceans would be anoxic. The changing balance between oxygen supply and consumption in the open ocean typically leads to a characteristic dissolved oxygen profile exhibiting a minimum at intermediate depth below which the concentration rises again. In the deep water in parts of the Atlantic, for example, the oxygen concentration may exceed $6 \, ml \, l^{-1}$. It was long believed that complete exhaustion of the oxygen content could occur only in enclosed basins where a combination of physical barriers and density stratification of the water severely limits advection of oxygen to the deep water. Typical examples of such situations are the Black Sea, the Cariaco Trench and several fjords on the coasts of Norway and British Columbia. The first report of the absence of measurable quantities of oxygen from open ocean waters was made by Schmidt (1925) for a station in the Gulf of Panama. The water depth at that station was 3140 m, and between 400 and 500 m there was "practically no oxygen at all". Since then, areas of considerable size, both north and south of the equator in the eastern Pacific and in the northern Indian Ocean (Fig. 16.1), have been found where, in the intermediate waters, oxygen is present in very small amounts ($< 20 \, \mu g\text{-}at \, l^{-1}$) or even below the limits of detection. All of these areas are close to regions of upwelling, and consequently of high primary productivity. It has been suggested that other parts of the oceans, such as portions of the eastern subtropical Atlantic may have similar oxygen-deficient layers. However, there are good arguments against the occurrence of such water in the Atlantic. The deep Atlantic is more efficiently flushed than are either the Pacific or the Indian Ocean, and as Redfield (1958) pointed out, in the Atlantic the formation of organic matter is limited by the low phosphate concentration to a level which could not deplete the available oxygen even if complete oxidation occurred.

When the supply of free oxygen in a parcel of water nears exhaustion, the reduction of nitrate to nitrite is the next source of oxygen for the continued

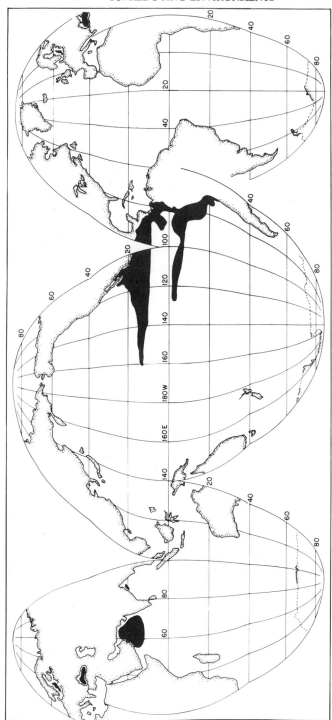

FIG. 16.1. Extent of oxygen-deficient ($<20\ \mu g$-at l^{-1}) intermediate or deep waters. (Goode's homolosine equal-area projection; copyright by the University of Chicago, Department of Geography). Principal data sources: Caspers (1957); Fonselius (1969); Ivanenkov et al. (1964); Reid (1965).

oxidation of organic matter. This process, the first step in denitrification, is the thermodynamically favoured next step in the oxidation of organic matter. It is used as an energy source by bacteria which, apparently, are active only at very low concentrations of oxygen or in its complete absence. Brandhorst (1958, 1959) first discovered a nitrite maximum in the oxygen-depleted water of the eastern tropical Pacific at a depth varying from 200 to 500 m. The nitrite concentration attained levels as high as 2·5 µg-atoms N l^{-1}. The term secondary nitrite maximum is applied to distinguish it from a shallower (primary) maximum which is widely found near the thermocline and which is attributed to the oxidation of organically derived ammonia or to extracellular production of nitrite by phytoplankton during the assimilation of nitrate nitrogen (Vaccaro and Ryther, 1960). Brandhorst showed that below the thermocline significant nitrite concentrations were found only in association with oxygen concentrations of 0·1 ml l^{-1} or less. He concluded from this correlation that the secondary nitrite maximum had a different origin from that associated with the thermocline, namely nitrate reduction coupled to the oxidation of organic matter.

Wooster et al. (1965) found a secondary nitrite maximum in the Pacific south of the equator, between 10° and 24°S, forming a lobe off the coast of Peru with a maximum offshore extent of about 1,200 km. They found a very clear association of the secondary maximum with a water mass of very narrowly defined temperature, salinity and thermosteric anomaly, and concluded from this that circulation plays an important role in the distribution of the nitrite. However, their data provided no basis for attributing the origin of the nitrite either to ammonia oxidation or to nitrate reduction.

The first direct evidence that denitrification leads to the production of molecular nitrogen in sea water was provided by Goering and Dugdale (1966) who incubated waters taken from various depths in Darwin Bay (Isla Genovesa in the Galapagos Islands). The Bay, which has a shallow sill, was found to be anoxic from 175 m to the bottom at about 240 m. Samples collected at 50 m intervals from depths 25 m to 225 m were incubated in the dark at *in situ* temperatures after having been spiked with ^{15}N-labeled KNO_3. The formation of molecular nitrogen enriched in ^{15}N in the deep samples indicated that N_2 is the end result of denitrification. However, these authors were unable to detect the formation of free nitrogen by the same method in waters from the Peru Current which contained very little oxygen (0·11 to 0·37 ml l^{-1}) and varying amounts of nitrite (0·05 to 3·23 µg-at l^{-1}). Thomas (1966) re-examined the nitrite accumulation in the eastern tropical North Pacific. He measured the nitrate and nitrite concentrations as a function of depth at several stations. He also calculated a hypothetical nitrate concentration based on the concentrations of phosphate and the average molar NO_3^-/PO_4^{3-}

ratio of 11·9 ± 1·6 found in that area below the oxygen minimum depth. His calculated nitrate profile showed a significant excess over the measured one in the depth range at which the secondary nitrite peak occurred. However, the difference between the calculated and measured nitrate concentrations was up to six times as large as the measured nitrite concentrations at the depth of the nitrite maximum. Thomas attributed this difference to the formation during denitrification of N_2O or N_2, although direct evidence was lacking.

Goering (1968) reported the formation of both nitrite and free nitrogen during incubation experiments with water from the low-oxygen layer off the west coast of Mexico. He concluded that all the N_2 formed originated by bacterially mediated denitrification and that neither inorganic nitrite reduction nor ammonia oxidation were involved. Although he calculated rates of denitrification from his results, he did not consider them to be typical of the natural *in situ* rate because of the accelerating effect resulting from the proliferation of bacteria at the glass surfaces in the incubation experiments. The question of oceanic denitrification rates assumes importance when attempts are made to balance the global nitrogen budget (see below). Unless nitrogen fixation is important, much of the molecular nitrogen produced during denitrification would be lost to the atmosphere. Other important problems are the identification of limiting oxygen concentrations, below which nitrate reduction can be initiated, and whether or not nitrite results primarily from nitrate reduction or ammonia oxidation.

Fiadeiro and Strickland (1968) studied the secondary nitrite maximum off the coast of Peru in detail. From plots of NO_3^--nitrogen versus apparent oxygen utilization they identified the depth zones at which nitrate reduction occurred. These zones coincided with those in which the oxygen concentration was less than $0.2 \, ml \, l^{-1}$. A remarkable, but inescapable conclusion arising from their work was that nitrate was reduced in water containing less than $0.5 \, mg \, l^{-1}$ of organic carbon of which probably less than 20% was readily reactive material. This conclusion is consistent with the observation by Carlucci and Schubert (1969) that bacteria isolated from this area can reduce nitrate in sea water in the presence of little or no added organic matter. However, it should be pointed out that although it is possible to determine organic carbon in sea water with acceptable *precision*, the *accuracy* of the determination is questionable. There is recent evidence that the true concentration of organic carbon in the deep sea may well be higher than previously reported values. This uncertainty arises from doubts about the completeness of the oxidation in the techniques used for the determination of organic carbon. Furthermore, there is a possibility that volatile organic compounds, which are not determined by classical methods, may be present in metabolically significant amounts in deep waters. Recent suggestions of "deep metabolism" (Craig, 1971) point toward the possible existence of more oxidizable organic carbon in the deep sea than has heretofore been recognized.

Culture experiments by Carlucci and Schubert (1969) indicated that bio-chemical oxidation of organic matter by nitrate can take place in the oxygen-poor water at the depth of the secondary nitrite maximum off Peru. No ammonia was detected, nor did the sum of nitrate- and nitrite-nitrogen at the end of the experiments equal the amount initially present. On the basis of cell counts and an assumed average weight and nitrogen content for the bacteria, Carlucci and Schubert calculated that all of the nitrogen deficit could not have been accounted for by assimilation. They inferred that gaseous end products could have been produced, although this could not be proved, and that denitrification might have occurred. Circumstantial evidence was thus accumulating that nitrate reduction is at least partly responsible for the high secondary nitrite concentrations in the eastern Pacific.

An alternative explanation was considered by Carlucci and McNally (1969). They observed that the oxidation of ammonia by nitrifying bacteria in cultures occurred even at oxygen contents of less than 0.1 ml l^{-1}. In fact, in some of their experiments, more nitrite was produced at very low oxygen concentrations than at intermediate ones. They concluded that nitrification may occur in waters where there is a secondary nitrite maximum and extremely low oxygen concentrations, such as those of the eastern Pacific. However, they also pointed out that denitrification is a more reasonable explanation in all those instances for which an increase in nitrite concentration is accompanied by a decrease in nitrate concentration.

Goering and Cline (1970) followed the processes accompanying the oxidation of organic matter in a sample of sea water taken from the oxygen minimum layer of the northeastern tropical Pacific. Their results indicated clearly that after the rapid consumption of the small amount of oxygen, nitrate was reduced to nitrite, but further reduction of nitrite commenced only after the concentration of nitrate had been reduced from the initial value of 29 to about 12 μg-at l^{-1}. Thereafter, the reduction of nitrate continued to completion at about the same rate, and nitrite reduction occurred, at first simultaneously with nitrate reduction, and subsequently to completion at an undiminished rate. It thus appears that the rate of denitrification is independent of nitrate concentration. These authors also monitored the concentrations of other nitrogen compounds, and concluded that neither ammonia nor amino nitrogen are significant end products of nitrate reduction. Neither nitrous oxide nor hydrogen sulphide could be detected. They attributed the absence of hydrogen sulphide to the consumption of all the readily degradable organic matter during the depletion of nitrate and nitrite. It has been suggested that the lack of sulphate reduction in the eastern Pacific may also be the result of the absence of sulphate reducing bacteria. Since they are obligate anaerobes, they may not have been able to arrive in the oxygen depleted water in a viable state and in sufficient numbers for colonization. To some extent this argument is inconsistent with the known abundance of sulphate reducing bacteria in sediments and the fact that the layers of low-oxygen water are in contact with the sediment along the continental margin of Central and South America. Furthermore, it is generally held by microbiologists that once suitable conditions for a certain bacterial process exist in a natural situation, the process is sure to start before long.

The widespread occurrence of nitrite below the primary maximum in the central Pacific Ocean was recently reported by Wada and Hattori (1972). The concentrations were generally found to be less than 0.024 µg-at l^{-1} except in very oxygen deficient waters (< 0.1 ml l^{-1}) in which higher concentrations were encountered. When samples of the latter waters were incubated at *in situ* temperatures nitrite concentrations invariably increased. Addition of nitrate accelerated nitrite production whereas addition of ammonia suppressed it. These results suggest that the reduction of nitrate to nitrite is commonplace in the ocean wherever oxygen is in short supply.

By making use of a new colorimetric method for the determination of low oxygen concentrations (Broenkow and Cline, 1969), Cline and Richards (1972) found that oxygen concentrations in the oxygen minimum zone of the eastern tropical North Pacific (i.e. off the coasts of Mexico and Central America) are at least an order of magnitude lower than had previously been reported on the basis of Winkler oxygen titrations. Comparison of the two methods showed that at low oxygen concentrations the Winkler method consistently gave values which were higher by 8 to 10 µg-at l^{-1} (about 0.1 ml l^{-1}) than the colorimetric values. They found widespread nearly anoxic conditions between 200 and 800 m off the coast of southern Mexico and often could not detect oxygen even with the sensitivity available with the new method, i.e. 1 µg-at l^{-1}. It is thus quite possible that, in fact, these waters contain no oxygen at all. This water mass may extend more than 750 km from the shore.

Cline and Richards estimated the amount of nitrate reduction from the expression for the nitrate anomaly δN

$$\delta N = [NO_3^-]_{exp} - [NO_3^-]_{obs} - [NO_2^-]_{obs}$$

where $[NO_3^-]_{exp}$ is the concentration of nitrate expected from both oxidative and preformed sources, i.e. the expected concentration before the onset of nitrate reduction, and $[NO_3^-]_{obs}$ and $[NO_2^-]_{obs}$ are the measured concentrations of nitrate and nitrite, respectively. $[NO_3^-]_{exp}$ was estimated from the relationships between apparent oxygen utilization (AOU) and nitrate and phosphate which were independently established for waters in or near the survey area. In the upper 150 m the change in AOU relative to the changes of both phosphate and nitrate corresponded very closely to the theoretical values derived by Redfield *et al.* (1963). The measured ratio $\Delta AOU/\Delta PO_4^{3-}$ was 263 as compared with the theoretical value of 276, and $\Delta AOU/\Delta NO_3^-$ was 16·8 (theoretical value 17·2). Significantly different relationships were found for the water below 900 m. These can be attributed to differences in the rates of release of phosphate and nitrate during the decomposition of organic matter; similar results have been found for other water masses and have also been noted in laboratory experiments (see e.g. Grill and Richards, 1964). From the linear relationship which existed between preformed phosphate and salinity in the depth range 150 and 900 m (the mixing interval between the salinity maximum and minimum), Cline and Richards concluded that the ratio $\Delta AOU/\Delta PO_4^{3-}$ was constant over this interval and assumed the same to be true for $\Delta AOU/\Delta NO_3^-$. Any deviations from the expected value for the latter ratio should, on this basis, be attributable to changes in preformed nitrate concentration and/or nitrate reduction. They computed nitrate anomalies and compared them to the distribution of nitrite. The maximum of the nitrate anomaly was found at a rather constant depth of 500 m throughout the section whereas the nitrite maximum generally lay about 100 m higher. The ratio between the nitrate anomaly and the nitrite concentration was about 10 throughout much of the depth interval. Little horizontal change was observed up to distances of ca 600 km from shore beyond which both the nitrate anomaly and the nitrite concentration decreased with a concomitant increase in oxygen content. The dependence of both the secondary nitrite concentration maximum and the nitrate anomaly on very low oxygen concentrations was very clearly established (Fig. 16.2). Nitrite in excess of 0·1 μg-at l^{-1} was found almost exclusively wherever the oxygen content was less than 2 μg-at l^{-1} whereas lower nitrite concentrations were present at all oxygen concentrations at least up to 9 μg-at l^{-1}, including very low ones. By contrast, very low values of the nitrate anomaly were not found in association with the lowest oxygen concentrations. This may indicate that nitrite may not persist very long, and that it is subject to further reduction, presumably to free nitrogen. Ammonia was found to be roughly inversely related to nitrite, having a minimum concentration of zero or close to zero at the depth of the nitrite maximum; N_2O could not be detected at all even with a sensitivity of 0·5 μmol l^{-1}. Thus, although present evidence indicates that N_2 is the final product of nitrate reduction, a direct measurement of a nitrogen anomaly using better analytical methods is obviously desirable. Judging by the size of the nitrate anomaly the nitrogen excess should reach a level of about

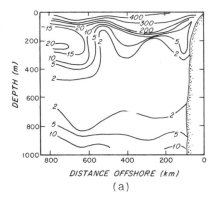 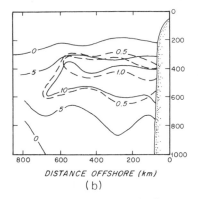

FIG. 16.2. (a) Oxygen concentrations and (b) nitrate anomaly (————) and dissolved nitrite (— — — —), all in μg-at l^{-1}, in the eastern tropical North Pacific along a SW–NE profile off the Gulf of Tehuantepec. (Cline and Richards, 1972).

$0.15 \, ml \, l^{-1}$—not much lower than that measured by Richards and Benson (1961) for the Cariaco Trench.

Richards (1971) and Cline and Richards (1972) have tried to view denitrification in the Pacific in the perspective of the oceanic nitrogen budget. Various estimates of this budget (see e.g. Emery *et al.*, 1955; Eriksson, 1959; Holland, 1970) have led to the conclusion that the annual supply of combined nitrogen to the oceans *via* rivers and the atmosphere exceeds the loss to the sediment by 10^{13} to 10^{14} g. Thus, if a steady state is to be maintained, a return mechanism, presumably denitrification and reflux of free nitrogen to the atmosphere, must be operating to the extent indicated by the imbalance.

Measurements of *in situ* dentrification rates in the open ocean are difficult if not impossible. Goering and Dugdale (1966) and Goering (1968) measured *in vitro* rates at *in situ* temperatures, but they themselves questioned whether these rates apply under natural conditions. Richards and Broenkow (1971) obtained interesting data on the rate of nitrate reduction in Darwin Bay (Galapagos Islands) by observing a change from practically no nitrite to a peak of 8 μg-at l^{-1} over a two-month period. However, the estimation of oceanic denitrification rates even from these field observations involves the assumption that the conditions which apply in this shallow insular environment also apply to a deep-sea situation. Nevertheless Cline and Richards (1972) estimated an upper limit to the denitrification in the eastern tropical North Pacific and arrived at a figure for this area alone of $2.3 \times 10^{14} \, g \, yr^{-1}$. Even if this estimate is too high by an order of magnitude, it appears that denitrification in the oceans is of a magnitude which is comparable to the excess of introduced combined nitrogen over its loss to the sediments.

Although much of the recent research on nearly or fully reducing conditions in the open ocean has been done in the eastern Pacific, significant findings have also been made for the Indian Ocean. The occurrence of reducing conditions in the Arabian Sea (and possibly other parts of the Indian Ocean) might be expected because of the existence of coastal upwelling and high primary productivity similar to those off Central and South America. Ivanenkov and Rosanov (1961) detected small amounts of hydrogen sulphide in the intermediate waters of the Arabian Sea and the Bay of Bengal. Nejman (1961) found extremely low oxygen concentrations in the Arabian Sea and reported occasional massive fish kills in the area. If these were related to the disappearance of oxygen or to upwelling of anoxic water, they may indicate short-term, possibly seasonal, fluctuations in the extent to which reducing conditions are attained in at least part of the affected volume of water. The problem was studied in greater detail during the International Indian Ocean Expedition. Ivanenkov *et al.* (1964) found a large volume of water between about 125 and 1,250 m in the Arabian Sea with oxygen concentrations below 20 μg-at l^{-1}, and in the northeastern part of that sea detected no oxygen at all between 250 m and the bottom at a depth of 800 to 1,000 m. Hydrogen sulphide was again found in the intermediate waters of that region, and Rozanov and Bykova (1964) found a secondary nitrite maximum of up to 5 μg-at N l^{-1} between 150 and 1,500 m. It is thus becoming increasingly apparent that truly reducing conditions can occur in the open ocean outside restricted basins, the controlling factor being the balance between the supply of organic matter and of oxygen to the water below the thermocline. The delicacy of this balance is indicated by the size of the nitrate anomaly (relative to the remaining nitrate) in the Pacific and by the appearance of hydrogen sulphide in the Indian Ocean. No long-term observations of either area exist to indicate whether any or all of these areas are at a steady state or whether at present they are in progressive (or regressive) phases. The fact that in the Arabian Sea hydrogen sulphide is found in the presence of a significant nitrate remainder suggests that only a very minor increase in productivity, a very slight change in circulation in the Pacific, or a slight rise in the surface temperature of the oceans (which would decrease the solubility of oxygen) might lead to the appearance of hydrogen sulphide. It is possible that climatic changes during glacial–interglacial cycles could have brought about such changes in the past. It seems likely that a slight deterioration of the oxygen balance in the tropical Pacific would significantly increase the volume of low-oxygen and anoxic water. The question arises whether this would affect the composition of the atmosphere. This problem is important since both oxygen and nitrogen are involved. This type of control acting over geological periods of time had been proposed by Redfield (1958). Increased denitrification would accelerate the reflux of free nitrogen to the atmosphere and increased burial of reduced carbon and sulphides would increase the net photosynthetic oxygen production. Burial of organic matter represents withdrawal of reduced carbon from the ocean-atmosphere system and leaves an excess of oxygen unless the increased burial rate is balanced by an increased oxidation rate of fossil reduced carbon on land. Reduction of sulphate and the fixation of sulphides in the sediment are also linked to photosynthetic production *via* the oxidation of organic matter. The two processes and their effect on net oxygen production are indicated in the following scheme in which CH_2O represents organic matter:

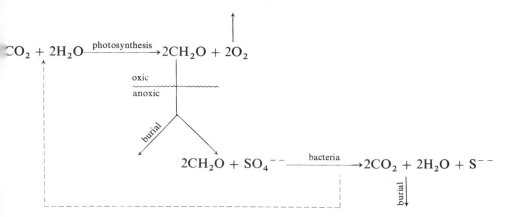

Each step in the scheme is, of course, only partly efficient and whether or not denitrification and sulphate reduction would noticeably alter the atmosphere would depend not only on the extent of reducing conditions in the ocean but also on their duration. The most recent estimates of the nitrogen budget outlined above suggest that the present annual reflux of nitrogen resulting from oceanic dentrification is 10^{-7} to 10^{-8} of the atmospheric reservoir. The present annual net oxygen production is also about 10^{-7} to 10^{-8} of the atmospheric reservoir (Broecker, 1970). It is thus apparent that to achieve, for example, a 1% increase of either the nitrogen or the oxygen content of the atmosphere, the production of either would have to be maintained at ten times the present level for 10^4 to 10^5 years. It should be recognized that this estimate takes no account of any buffering effect; any such effect would of course tend to reduce the magnitude of the change. Only oceanic anoxic conditions on a quasi-global scale might significantly alter the nitrogen or oxygen contents of the atmosphere within geologically short time spans.

16.3. The Development of Anoxic Conditions in Restricted Basins (see also Chapter 15)

Anoxic conditions found in basins having only shallow connections with the open sea may be transient or approximately steady-state with a time scale of thousands of years. The rate at which the transition from aerated to anoxic water proceeds depends mainly on the dimensions of both the basin and its connection with the open sea, on the density structure of the water column and on the rate of supply of organic matter. The latter includes organic matter formed in the basin and introduced from the outside. Examples of inter-mittently anoxic bodies of water are fjords and small bays, whereas the prime example of a long-term, and for practical purposes steady-state, anoxic system is the Black Sea. A list of known marine anoxic basins is given in Table 16.1. The Black Sea is not only the largest such basin, but has also been studied in more detail and for a longer period of time than any other. Con-siderable effort has been expended in attempts to reconstruct the history of the evolution of anoxic conditions in the Black Sea, largely on the basis of studies of sediments and pore waters and through rough models derived from such studies. (For recent treatments of the subject see, e.g. Kvasov, 1968; Manheim and Chan, 1974; Deuser, 1974.) The picture that has emerged from these studies is that the present period of anoxic conditions in the Black Sea began about 1,500 to 2,000 years after the first influx of saline Mediterranean water into the basin. This event was caused by the rise in sea level which followed the last glacial maximum. The influx, which slowly and somewhat irregularly increased with time, brought about a density stratification which impeded the supply of oxygen to the deep water. The widespread onset of deposition of sapropelic sediments, extremely rich in organic matter, over a great depth range and areal extent within less than 1,000 years suggests that the spread of anoxic bottom conditions in the basin, and presumably, the rise in the water column of the interface between oxic and anoxic water was quite rapid. Detailed analyses of the sapropelic sediments indicate that the salinity increase in the basin and, by inference, the degree of reducing conditions did not progress in a smooth fashion, but were subject to numerous short-term fluctuations (Fig. 16.3).

Very detailed observations of another large basin, the Baltic, have been made almost continuously since the late nineteenth century. The Baltic, like the Black Sea, can be considered to be a large estuary with an excess of fresh-water influx and precipitation over evaporation. The salt balance in estuaries is maintained by continuous or periodic influxes of sea water over the seaward sill. The size of this influx depends on the sill geometry and on the freshwater excess of the estuary. For the Baltic the precipitation appears to be just

TABLE 16.1

Anoxic Marine Basins (after Broenkow, 1969, with corrections and additions)

Name and Location Maximum depth (m)	Sill depth (m)	Maximum hydrogen sulfide (μg-atoms l^{-1})	Bottom salinity (‰)	Bottom temperature (°C)	Reference
Black Sea 2243	40	300	22·3	9·0	Caspers, 1957
Bolstadtfjord, Norway 141	1	47	21·1	4·1	Strøm, 1936
Bornholm Basin, Baltic Sea 105	45	variable	up to >17 (variable)		Fonselius, 1969
Cariaco Trench, Caribbean Shelf of Venezuela 1390	150	25	36·2	17	Richards and Vaccaro, 1956
Darwin Bay, Isla Genovesa, Galapagos Islands 220	20	present	34·9	19·4	Richards and Broenkow, 1971
Drammensfjord, Norway 117	8	215	29·9	5·3	Strøm, 1936
Fårö Deep, Baltic Sea 205	140	variable	12·0	5·5	Fonselius, 1969
Fjellangervåg, Norway 75	3	260	30·3	5·5	Strøm, 1936
Framvaren, Norway 175	2	1800	24·6	8·8	Strøm, 1936
Frierfjord, Norway 93	11	25	34·2	6·1	Strøm, 1936

TABLE 16.1 (contd.)

Anoxic Marine Basins (after Broenkow, 1969, with corrections and additions)

Name and Location / Maximum depth (m)	Sill depth (m)	Maximum hydrogen sulfide (µg-atoms l⁻¹)	Bottom salinity (‰)	Bottom temperature (°C)	Reference
Golfo Dulce, Costa Rica / 200	50	9	34·8	15·6	Richards et al., 1971
Gotland Deep, Baltic Sea / 249	60	variable	13	5·9	Fonselius, 1969
Gulf of Cariaco, Venezuela / 90	54	33	36·6	21	Richards, 1960; Kato, 1961
Hellefjord, Norway / 75	9	1800	33·3	5·9	Strøm, 1936
Helvigfjord, Norway / 36	3	290	31·2	7·3	Strøm, 1936
Indre Lyngdalsfjord, Norway / 116	5	300	32·3	7·2	Strøm, 1936
Isefjaerfjord, Norway / 27	1	350	31·1	6·0	Strøm, 1936
Kaoe Bay, Halmahera Island, Indonesia / 491	50	13	34.5	28	Van Riel, 1943
Karlsö Deep, Baltic Sea / 112	101	variable	10		Fonselius, 1969
Lake Nitinat, British Columbia, Canada / 205	4	350	31·2	10	Richards et al., 1965
Lake Suigetsu, Japan / 34	2	3000	16	15	Shigematsu et al., 1961

Location						Reference
Lake Varna, Bulgaria	18		present	14	8	Caspers, 1957
Landsort Deep, Baltic Sea	459	138	variable	11	4·9	Fonselius, 1969
Lenefjord, Norway	240	3	175	32·4	5·3	Strøm, 1936
Lygrefjord, Norway	181	6	3	33·8	6·7	Strøm, 1936
Nordåsvatn, Norway	87	3	340	31·0	6·3	Strøm, 1936
Norrköping Deep, Baltic Sea	205	101	variable	≥10		Fonselius, 1969
Northern Central Basin, Baltic Sea	219	115	variable	11·2	4·9	Fonselius, 1969
Saanich Inlet, British Columbia, Canada	236	65	>20	31·4	9·0	Herlinveaux, 1962; Richards, 1965
Skjoldafjord, Norway	109	3	10	26·8	8·3	Strøm, 1936
Søndeledspoll, Norway	17	2	240	33·1	6·8	Strøm, 1936
Tofino Inlet, British Columbia, Canada	96	66	400	29·6	13	Coote, 1964
Vestrhusfjord, Norway	55	2	440	32·4	5·1	Strøm, 1936

B

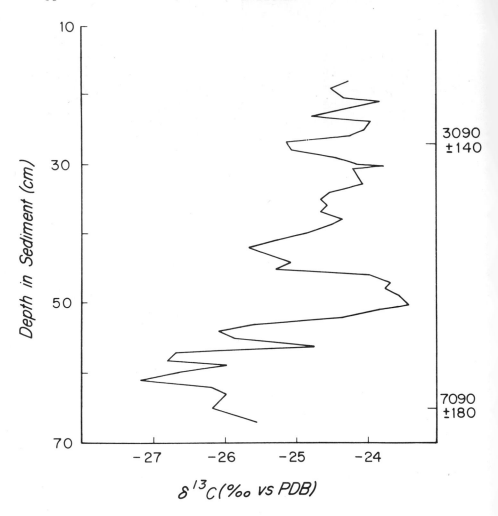

Fig. 16.3. Variation of $\delta^{13}C$* of organic carbon in Black Sea sediments deposited 7000 to 3000 years ago. During that period the Black Sea changed from a fresh or brackish lake, probably aerated throughout, to an anoxic basin with salinities approximating those of today. The changes in $\delta^{13}C$ reflect the fluctuations and long-term trend in the local temperature-salinity regime of the period (Deuser, 1972).

*$\delta^{13}C = 1000 \left[\dfrac{R_{sample}}{R_{PDB}} - 1 \right]$ where R = $^{13}C/^{12}C$ and PDB is the Peedee belemnite isotope standard.

balanced by the evaporation so that the water surplus equals the river discharge into the basin. Brogmus (1952) computed the water balance and found that the influx of saline water from the North Sea through the Danish sounds equalled the river discharge, both being 471 km^3 y^{-1}. Since the outflow which forms the Baltic Current through the Belts and the Öresund must be as large as the sum of influx and river discharge, it amounts to 942 km^3 y^{-1}, or just twice the volume of the influx. From the mean values of salinity of the surface, deep and inflowing waters Fonselius (1969) calculated that the transfer of water from the bottom to the top layer in the Baltic was 1,250 km^3 and that from the top to the bottom layer was ca 750 km^3. Of course, these values are only rough estimates since they assume no changes in either the water balance or in the salinities of the different reservoirs.

In a series of detailed studies Fonselius (1962, 1967, 1969) has presented much information on a number of progressive changes in the hydrography of the Baltic since the latter part of the nineteenth century. The discharge of the major rivers entering the Baltic has shown an overall decrease since records have been kept (Fig. 16.4), and the salinity at a number of stations shows an equally marked increase (Fig. 16.5). It is thus quite apparent that the water and salt balances of the basin must have undergone changes within the last hundred years. The Baltic is clearly not a steady-state system. It is particularly interesting in the present context that during this same time span the deep basins of the Baltic have made the transition from weakly oxic to intermittently anoxic conditions, and if the trend continues, they will progress further to permanent anoxicity. A good example of the decrease in oxygen content of the Baltic, even at moderate depth, is shown in Fig. 16.6. The station is located near the centre of the northern part of the Baltic proper, about half-way between Stockholm and the Gulf of Riga.

Although there is probably an interconnection between the decrease in supply of fresh water to the Baltic, the increase in salinity and the decrease in oxygen content of the basin, other factors are also involved and the complete chain of conditions and events leading to the present situation is likely to be complex. At a depth of 60 to 70 m the primary halocline separates the oxygenated surface water (about 7‰ salinity) from the more saline deep water (about 11‰ salinity). This is the depth to which winter convection, caused by the cooling of fresher surface water, can bring about the downward mixing of oxygenated water and upward transport of more saline and nutrient-rich deeper water. The deep water below the primary halocline is not affected by seasonal mixing, and its oxygen supply depends on downward diffusion through the halocline and on occasional influxes of oxygenated high-salinity water from the North Sea. Under special circumstances, unusually saline water from the Kattegat may pass over the sills, flow along the bottom and depending on its quantity, fill one or several of the deepest basins in the Baltic. Since such water is denser than the Baltic deep water, it is separated from it by a pycnocline which is called the secondary halocline in

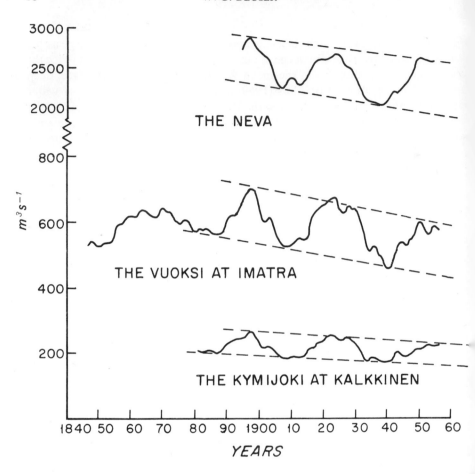

Fig. 16.4. Ten-year running means of runoff values (m³ s⁻¹) for three rivers of the Baltic catchment area showing the long-term decrease in runoff since the beginning of the twentieth century. (Fonselius, 1969).

distinction from the primary one at 60 to 70 m. The depth of the secondary halocline depends on the bottom topography and the volume of the saline influx, and may vary between 70 and 400 m. It is not a permanent feature, but may gradually disappear or be replaced by a new one. A volume of Baltic deep water identical to the volume of inflowing water has to be displaced upwards. Since the deep water is particularly rich in nutrients, this upward transport leads to an exceptional fertilization of the water above the primary halocline, and the additional nutrients which enter the nutrient reservoir are available

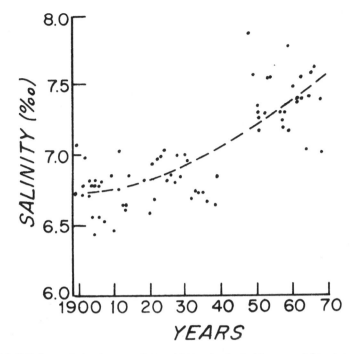

FIG. 16.5. Salinity variation between 250 and 300 m in the Baltic, west of Åland, from 1898 to 1968. (Fonselius, 1969).

for fertilization of the surface water by winter convection. The influx of saline water thus has a dual effect on the long-term oxygen balance of the deep water; the increased salt content of the deep water increases the stability of the water column and the increased fertilization of the surface water leads to an increase in the supply of organic matter. Another factor which may contribute to an accelerating oxygen depletion in the deep water is the 1°C increase in the temperature of the deep water observed during the last 50 years (Fonselius, 1969). An increase in temperature is likely to lead to an increase in the oxidation rate and, thereby, to accelerated oxygen depletion.

If the influx of highly saline water into the Baltic is large enough, the water becomes progressively depleted in oxygen along its generally northward path which follows the deepest parts of the basin. One deep after another may be filled with the new water while the older water is displaced upwards and northwards. A very illustrative series of sections showing the changes in the distributions of oxygen, hydrogen sulphide and phosphate phosphorus along a profile extending through much of the Baltic proper and including several

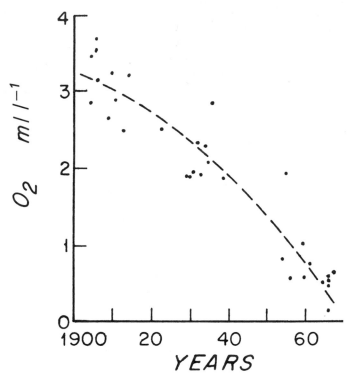

FIG. 16.6. Decrease of oxygen content at 100 m in the Northern Central Basin of the Baltic in this century. (Fonselius, 1969).

of its deeps is given by Fonselius (1970). The sections cover the period from September, 1968 to January, 1970 and demonstrate very clearly the nature and extent of the chemical changes which may occur in the water column at any one place over such a relatively short time span. The original oxygen content of the inflowing dense water is rapidly depleted or consumed by reaction with hydrogen sulphide which may already be present in the deep basins towards which the new water flows. The sulphide may be oxidized to elemental sulphur or higher oxidation forms. Experimental work on this reaction by Cline and Richards (1969) suggests that all oxidation stages, $S_2O_3^{2-}$, SO_3^{2-} and SO_4^{2-} as well as polythionates and polysulphides may form. The formation of elemental sulphur could not be detected, but its formation in the Baltic deeps has been suggested by Fonselius (1969). The precipitation of elemental sulphur on top of anaerobic shallow-water sediments which subsequently come into contact with oxygenated water has been observed in the Baltic by

Hallberg *et al.* (1972) and is presumed to take place in a similar fashion in the deeper parts of the basin although no direct evidence has yet been presented.

Fonselius (1967) concluded that phosphorus is the limiting nutrient in the production of organic matter in the Baltic, and in his follow-up study (Fonselius, 1969) he suggested that the phosphorus requirement of the organisms in the Baltic is only about one-half of that of normal open ocean organisms. If this is so, any increase in the supply of phosphorus to the Baltic might result in a corresponding increase in the total primary production and this will lead to a proportional increase in the supply or organic matter to the deep water. Evidence for increasing primary production has indeed been found (Fonselius, 1970). There are at least two ways by which the supply of phosphorus to the surface waters has increased during the last few decades: the increasing discharge of sewage rich in phosphate and organic matter into the surface water of the Baltic, and the release of phosphate from the sediments below the anoxic waters. Since the volume of, and bottom area covered by, anoxic water have been increasing in the recent past, so also has the amount of phosphate released to the deep water and eventually mixed upward into the surface water. Fonselius (1969) showed that phosphate-phosphorus dissolved in the deep water of several deeps in the Baltic increased from about 1 μg-at l^{-1} in the mid 1950's to 3 or 4 μg-at l^{-1} in the late 1960's.

The release of phosphate from sediments in stagnant basins of the Baltic appears to coincide with the first appearance of the sulphide and ceases whenever an influx of oxygenated water occurs. Olausson (1967) suggested that the lower pH in the Baltic basins (commonly around 7) is primarily responsible for the increased dissolution of phosphate as compared to that in the Cariaco Trench, for example. The experimental work of Hallberg *et al.* (1972) suggests, however, that Eh may be an equal or more important factor in the dissolution of that phosphate which had accumulated in the sediments prior to the recent hydrographic changes in the Baltic. The result is a phosphate content of the deep water which is enormously high in comparison to that which could have originated from the decomposition of plankton. Those concentrations can be calculated from the apparent oxygen utilization (AOU) in these waters and the relationships derived by Redfield *et al.* (1963).

In contrast, in both the Black Sea and the Cariaco Trench the pH in the deep water is 7·5 or higher and, perhaps more importantly, Eh has been negative for prolonged periods; agreement between the phosphate content of the water in these basins and the values calculated from AOU suggests that phosphate is no longer being released from the sediment. It is thus apparent that in the Baltic, even without the additional phosphate input through sewage discharge, the surface waters will be increasingly fertilized with phosphate brought up from the deep water. This, in turn, will trigger increased primary production and eventually increase the oxygen demand in the deep water, thus reinforcing the extension of anoxic deep water and the concomitant liberation of more phosphate. Unless drastic changes in the water and salt balances of the Baltic occur in the near future, so as to reverse the trends of the last 100 years, there appears little hope for averting further deterioration of the oxygen balance in the deep water.

The foregoing may serve as an illustration of both the delicacy of the oxygen balance in a basin which has a density stratification in its water column and the rapidity with which conditions may change once a critical stage has been reached. The most interesting aspect of the changes we are witnessing in the Baltic is the feedback mechanism which exists between the deep and surface water through the release at depth of phosphorus (the limiting nutrient at the surface); once started the process becomes self-accelerating.

The present conditions in the Baltic appear to have been triggered by the decreasing freshwater excess and, therefore, the increasing salinity. Other causes may have similar long-term effects. The effective cross–section of the straits connecting an estuarine basin to the open ocean may vary in response to changes in worldwide sea level or to uplift or subsidence of the land near the straits. Pronounced sea level changes were common during the Pleistocene as a result of waxing and waning of continental and polar ice sheets. Depression and relaxation of the lithosphere both above and below sea level have also occurred as a result of changing overburdens of ice and water. Mörner (1969) has presented evidence that, throughout post-glacial times, sea level has been rising with respect to the land surface along the coast of southwest Sweden, south of the mouth of the Viskan River, and thus presumably also in the areas of the Belts and Öresund. All of these straits are quite shallow and relatively minor uplift or subsidence may cause a significant change in cross section. The Baltic basin has gone through several oscillations between fresher and more brackish stages during the last 15,000 years. The latest trend of relatively long duration was a gradual decrease in salinity over the last 4,000 to 5,000 years. Events witnessed during the last 100 years imply that this trend may now have been reversed.

16.4. TIME SCALE OF WATER RENEWAL IN ANOXIC BASINS

The renewal of water in anoxic basins is intimately related to, though not to be equated with, their redox and nutrient balances. The time scale of the renewal may vary from the order of weeks or months as for small, intermittently anoxic, systems to more than a thousand years as with the Black Sea. In relatively shallow depressions, over which water is constantly moving, marginal oxygen concentrations may be maintained over long periods of time, even in the absence of large scale advection, by downward diffusion (or small scale advection) alone. This phenomenon is exemplified in the Santa Barbara Basin off southern California. Sholkovitz and Gieskes (1971) studied the basin in detail and found that after a flushing with oxygenated water, an event induced by upwelling in the area, an apparently steady-state oxygen concentration of 0·05 to 0·1 ml l^{-1} was reached after two to four months. Their calculations suggest that, at that concentration, the consumption equals the downward diffusion, and complete anoxia is apparently never reached. However, no long-term study of the basin's hydrography has yet been completed, and it is not quite certain whether or not the occasional flushings do play a part in the maintenance of the low oxygen content.

More relevant to the present discussion are those basins which reach truly anoxic conditions, but which are flushed with oxygenated water at certain intervals of time and then return to an anoxic state. Flushing may be brought about by either seasonal changes in the hydrographic conditions or quite irregularly, usually by a coincidence of exceptional deviations from the hydrographical and meteorological norms.

A typical example of a seasonally anoxic basin is Saanich Inlet, a fjord on the south-east side of Vancouver Island, British Columbia (Richards, 1965) which Anderson and Devol (1973) studied for several years. A strong pycnocline isolates the deep water and apparently every year, after flushing, this water runs the entire course from oxygen depletion through nitrate and nitrite reduction to sulphate reduction. Thus, since no nitrate is left in the deep water before flushing, Anderson and Devol were able to combine the average nitrate concentration after flushing, the volume of the deep water and the nitrate concentration in the flushing water in order to calculate the volume of the flushing water. They found that on three different occasions the volume of flushing water varied from 0·23 to 0·6 km^3, corresponding to 18 to 46% of the deep water of the inlet. The flushing is caused by a pronounced annual cycle in density variations (Fig. 16.7) of the water at the sill. This cycle results from an annually recurring upwelling of denser water off the Washington and Oregon coasts. The available data indicate that only during the months of August and September is the density of the water at the sill sufficiently high to make possible the displacement of deep water within the basin. It appears that the flushing water enters the channel connecting the inlet to the open sea as bolases, several of which may form during any flushing season. The new water descends into the basin and seeks its own density level which may either be at the very bottom or at some intermediate depth. Current measurements made by Herlinveaux (1968) and their own observations led Anderson and Devol to estimate that it takes about 12 days for 0·55 km^3 of flushing water (equal to

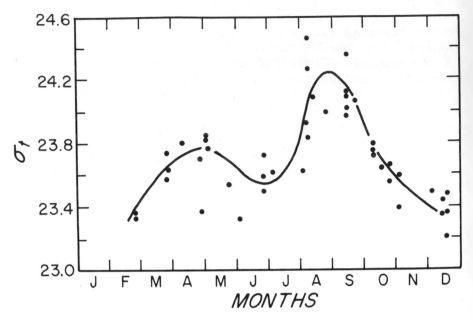

FIG. 16.7. Annual density cycle of the bottom water on the inner sill of Saanich Inlet. Data points cover the period 1961 to 1970. (Anderson and Devol, 1973).

42 % of the deep water) to enter the inlet, most of the net flow occurring on the flood stages of the tide. After flushing of the basin, the density of the deep water slowly decreases as salt diffuses into the overlying water. In this way the basin slowly returns to pre-flush conditions.

The deep water in larger basins generally has a longer residence time. Some of the basins in the Baltic appear at present to be flushed on a time scale of one to several years (Fig. 16.8). However, as indicated earlier, there is reason to believe that, as the oxygen balance of this sea deteriorates, even irregular flushing with aerated water may cease for some of those deeps. Also, if the salinity in the Baltic continues to increase, intrusions of water of even higher salinity will be necessary if a flushing of the deeper parts is to occur. Eventually a point may be reached at which flushing becomes impossible. The deeps will then undergo the transition to quasi-steady state basins in which water renewal is not accomplished by discrete events, but rather by the much slower processes of turbulent mixing and diffusion.

The Cariaco Trench off Venezuela is a basin in which advection of outside water into the deep part has never been documented and may well be completely absent. The hydrography and chemistry were first studied by Richards and Vaccaro (1956) and Richards (1960). Their measurements and those made

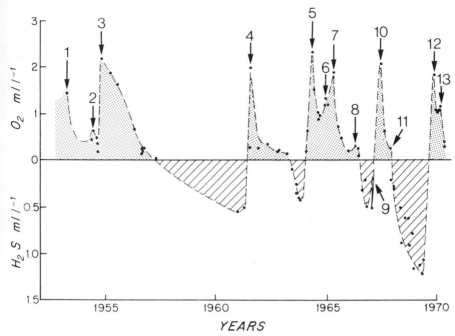

FIG. 16.8. Oscillations between oxic and anoxic conditions at 225 to 240 m in the eastern Gotland Basin. Thirteen successive oxygen maxima could be identified over a 19-year period. (Fonselius and Rattanasen, 1970).

on subsequent visits to the area have shown that the deep part of the trench, from about 500 m to the bottom (greatest depth 1400 m) is practically iso-thermal and isohaline, σ_t is almost constant below 200 m, except for the occasional occurrence of a maximum around 400 m (Okuda et al., 1969). The stability of the water column (in contrast to that of anoxic basins of higher latitudes) is a result of the lower temperature of the deep water rather than of its higher salinity. In spite of the fact that observations cover only about a fifteen-year period, there are some indications of changes in the basin's hydrography. Although Richards and Vaccaro found the oxic-anoxic water interface to be at a depth of 375 m, hydrogen sulphide was detected between 250 and 300 m in 1970 (Deuser, 1973). Moreover, there appears to be a slight but consistent increase in the temperature of the deep water throughout the period of observation (Brewer, in press). Thus, it appears that even this relatively large basin is not in a steady state.

A number of attempts have been made to estimate the age of the deep water in the Trench. Richards and Vaccaro estimated a minimum of 111 years on the basis of a very simple model in which the integrated oxygen and the sulphate consumption in the basin were compared with the primary productivity of the surface layer. They also derived an upper age limit of 2000 years on the assumption that a catastrophic event filled the trench with water having the uniform salinity and temperature now found below 650 m and by calculating the time required by molecular conduction to raise the temperate of that water between sill depth (about 150 m) and 650 m to its present value. Wolgemuth et al. (1971) measured the ^{14}C content of the Trench water and calculated a probable age of less than 200 years for the deep water. Their measurements of ^{137}Cs suggested a residence time of 100–600 years. More recently, Fanning and Pilson (1972) used a vertical diffusion-advection model and the estimated silica flux from the sediment to derive a maximum residence time for the water between 300 and 1000 m of 800 years. Accordingly, the water below 1000 m should have an even longer residence time. However, their estimate of the silica flux may be too low since it was determined on pore water samples squeezed from the sediment. The flux at the sediment surface may well be much higher than that moving within the sediment if a significant portion of the silica is dissolved shortly after arrival on the bottom and before it has a chance to be buried. The estimated residence time can thus be no more than an upper limit. Moreover, the estimate depends on the assumption of a steady state; this may not be valid in the light of the recent evidence.

Deuser (1973) made another attempt to estimate the residence time of the deep water in the Cariaco Trench. The estimate was based on the accumulation of oxidized organic carbon in the water column and an estimate of the probable range of the oxidation rate in the anoxic water. Although the range of values obtained in this manner is 22 to 570 years, the most likely value lies near 100 years.

Most of the evidence on the Cariaco Trench thus indicates that the deep water is renewed on a time scale of the order of a hundred years. It is not clear at this point how the renewal is accomplished. The uniformity of the deep water is perhaps brought about by a system of convecting cells. The cause of the temperature increase is poorly understood. Both the geothermal heat flow and chemical energy released during sulphate reduction and methane formation appear inadequate to account for it. Although advection in the upper part of the anoxic water has been inferred by Fanning and Pilson (1972) there is no observational evidence in support of it. The temperature rise does not appear to be coupled to a change in salinity, at least not to a measurable extent. However, if convection cells do exist, the renewal of the deep water could be accomplished by entrainment of, and mixing with, shallow water at the top and descending parts of the cells. Another possibility (Brewer, in press) is that the recently observed temperature increase is a permanent feature of the basin and periodically leads to a complete overturn of the water column because of breakdown of the density stratification.

The Black Sea is the largest marine anoxic basin, and not surprisingly also has the longest renewal time of its deep water. A number of different approaches to the estimation of this time have been made. Murray (1900) calculated that it would take about 2500 years to renew the water below a thin surface layer. He based this estimate on the rate of water exchange at the Bosporus. Sverdrup *et al.* (1942) confirmed this estimate by making use of more recent data by Merz (Möller, 1928) on the inflow and outflow volumes. However it is not clear how much of the water which flows in through the straits is able to penetrate into the deeper parts of the Black Sea before it becomes entrained in the outflowing current which transports about twice as much water as the inflowing current. Östlund (1969) used the appropriate salinities to calculate that one part of Mediterranean water mixes with four parts of Black Sea surface water to form Black Sea deep water. The volume of this mixture is 950 km^3 per year as compared to the annual input through the Bosporus of about 200 km^3. Östlund made use of this larger annual replacement figure to calculate an average residence time of about 475 years. However, in view of the direct loss to the outflow by entrainment the input of Mediterranean water into the mixture and, proportionately, the volume of the mixture, must be smaller and hence the residence time longer than the above estimate. Radiocarbon measurements on deep water by Östlund (1969) yielded "formal ages" of 935 years for the water between 300 and 700 m and of about 2000 years for the water below 2000 m. Although these are probably the best data at present available, they also only give a lower limit to the actual "age" of the water. The continuous oxidation of more recent organic matter in the deep, and especially the bottom water, serves to impart a higher radiocarbon activity to the deep water than it would have had in the event of mere decay of the activity it obtained by equilibration with the atmosphere when last at the surface. Measurements of the $C^{13}/^{12}C$ ratio suggest that the contribution of oxidized organic matter to the total dissolved inorganic carbon in the deep water is in excess of 20% (Deuser, 1970).

Another lower limit for the residence time may be calculated from the total accumulation of H_2S in the basin and the estimated production rate. Sorokin (1964) measured the rate of sulphate reduction at the sediment surface in different parts of the basin. By weighting his figures according to the extent of the areas represented by his measurements, a total H_2S production of about 10^7 kg day^{-1} may be derived. Taking the volume of the anoxic water as 480,000 km^3 and the mean H_2S content to be 0·25 mM l^{-1}, the minimum buildup time of the reservoir is calculated to be about 1000 years. Of course, this estimate is low because account has been taken neither of the loss of H_2S

at the upper boundary of the anoxic water nor of the fixation of sulphide in the sediment. Further uncertainties are introduced by the assumption that the production rate in the past was the same as that as present. Nevertheless, it may serve as a supporting evidence that the residence time of the deep water in the Black Sea is almost certainly more than 1000 years, and quite possibly even more than twice this value for water near the bottom.

The foregoing discussion of the renewal times of anoxic water in a variety of basins makes it clear that the most important parameters controlling the water exchange are the size and shape of the basin, the relative amounts and densities of inflowing and outflowing water, and the vertical density distribution in the basin. Geographical location and attendant meteorological conditions may also play an important role through their effect on the local current regime and possibly on upwelling of deeper water.

A few remarks on the significance of radiocarbon dating of anoxic waters seem in order. By their very nature anoxic basins accumulate end products of organic decomposition in their water. An important end product is carbon dioxide, and the total dissolved inorganic carbon content is thus always higher in the anoxic zone than in the overlying oxygenated water. The extent of the enrichment is dependent on the turnover rate, as well as on the supply of organic matter and the oxidation rate in the basin. Since the supply and decomposition of organic matter are continuous processes, a constant addition of carbon with a high ^{14}C activity takes place, and the resultant decay of the activity will be a different function of time than it would be in a closed system. It is often assumed that a parcel of water, after leaving the mixed layer where equilibration with the atmosphere occurred, behaves as a closed system. This assumption was made by Östlund when he calculated the "formal ^{14}C ages" for Black Sea waters (see above). However, for deep water, ^{13}C measurements indicate that roughly 1 mM of the 4.3 mM of ΣCO_2 is derived from the oxidation of organic carbon. Fig. 16.9 shows three curves indicating changes in the ^{14}C activity in the Black Sea over a 2000-year. period for the following three different hypothetical circumstances: (1) decay in a closed system, i.e. there is no addition of any kind of carbon, (2) a total of 1 mM of CO_2 per litre is added at a constant rate over 2000 years, the added carbon being assumed to have the same ^{14}C "age" as that in the closed system, (3) a total of 1 mM of CO_2 per litre is added at a constant rate over 2000 years, the added carbon being assumed to have zero age at the time of its addition. The different trends of the three curves are obvious. It is important to realize that curves (1) and (2) correspond to the same specific activity and would lead to the same "formal age" since all the added carbon is of the same specific activity. Curve (3) corresponds to a higher specific activity since "new" carbon is added continuously. This curve would indicate a younger

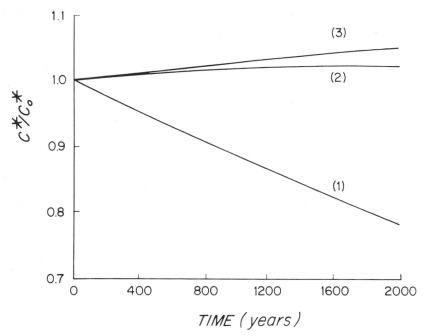

FIG. 16.9. Hypothetical examples of the change in ^{14}C activity of ΣCO_2 with time. (1) closed system: radioactive decay only, (2) continuous addition of carbon with same specific activity as original carbon, (3) continuous addition of carbon of instantaneous "zero-age" activity.

"formal age" than that calculated from either of curves (1) or (2). Curve (3), in fact, comes closest to representing the conditions in the Black Sea, although it is based on the assumption that all the presently observed biogenic carbon has been added over the last 2000 years. The "realistic age" calculated from curve (3) would be in excess of 2000 years. However, it should be clear from an examination of the three curves that, if no account is taken of continued addition of "young" carbon, a "formal age" calculated from the specific activity of anoxic water is only a lower limit of the "true age" of that water. It should be noted that the curve representing the change with time of the radiocarbon concentration in an anoxic basin can assume a variety of shapes depending on the rate of addition of biogenic carbon to the total-CO_2 pool.

16.5. SELECTED ASPECTS OF THE DECOMPOSITION OF ORGANIC MATTER IN MARINE ANOXIC ENVIRONMENTS

It is generally observed that sediments below anoxic water contain a higher concentration of organic matter than do those below aerated water. It is also quite generally true that anoxic waters are enriched in inorganic carbon and the principal nutrients, nitrogen and phosphorus. Hydrogen sulphide is often present in large quantities, but only where nitrate and nitrite have been nearly, or completely, reduced. Where nitrate and nitrite are absent, ammonia may be present in relatively high concentrations. Anoxic waters have been found to contain enhanced amounts of dissolved organic matter only in isolated instances, although significantly higher concentrations of light hydrocarbons are common.

The relatively high content of organic matter in the sediments has often been attributed to lower rates of decomposition in the absence of oxygenated bottom waters. However, the evidence for this contention was examined by Richards (1970) and found to be inconclusive. He suggested that the enhanced preservation might equally well be a consequence of higher sedimentation rates of organic matter in anoxic systems. On this basis the elevated content of organic matter in the sediments is not a consequence of the anoxic conditions *per se* but rather of the same special circumstances, such as restricted circulation and/or higher productivity, which produced the anoxic conditions in the first place. The few lines of evidence we have at this point appear to uphold this view.

Smith and Teal (1973) found oxygen consumption rates of about $0.5 \text{ ml m}^{-2} \text{ h}^{-1}$ at the sediment surface on the continental slope south of New England at a depth of 1850 m and an ambient temperature of about $4°C$. This translates to the oxidation of roughly 1.8 g of carbon $\text{m}^{-2} \text{ yr}^{-1}$. By comparison, Deuser (1971) estimated the oxidation rate in the anoxic zone of the Black Sea to be $10 \text{ g C m}^{-2} \text{ yr}^{-1}$ and calculated a lower limit of $2.3 \text{ g C m}^{-2} \text{ yr}^{-1}$ for the deep water of the Cariaco Trench (Deuser, 1973). As it appears that most of the oxidative activity in anoxic waters is taking place at the sediment surface, the estimates for the anoxic systems and Smith and Teal's measurements may be compared. Even when the temperature differences between these environments are taken into account (the depths are not greatly different), it appears that there is no reduction in the rate of oxidation in the anoxic settings. The final answer to this question must be deferred, however, until more oxidation rate measurements for both oxic and anoxic environments of comparable depths, temperatures and inputs of organic matter are available.

In most instances for which the dissolved organic carbon in anoxic waters has been determined, no significant enrichment with respect to the overlying oxygenated waters has been found, although more or less pronounced minima at the depth of the O_2-H_2S interface were almost always found. Richards and Devol (1970) suggested that this feature may be the result of the adsorption of organic matter onto the continuously forming colloidal particles of manganese, which owe their formation to the oxidation and precipitation of Mn(II) originating from below (see Spencer and Brewer, 1971). This adsorbed material would be removed in the filtration step preceding any determination of dissolved organic matter.

Below the O_2–H_2S interface there is a clear correlation between the organic carbon and the H_2S concentration only in one set of data from the Black Sea (Deuser, 1971), whereas data obtained by Richards and his collaborators and by various Russian workers show a very weak correlation or none at all. The apparent discrepancies in the various sets of data appear, at least in part, to be due to different analytical techniques. It is necessary to caution against putting too much faith in any determination of dissolved organic carbon until independent comparisons of the available methods have been made. Most methods are subject to uncertainties arising from contamination, loss of volatiles, or possibly incomplete oxidation of refractory compounds (see Chapter 14). It is difficult to conceive of processes which could produce the irregular concentration profiles reported by some workers for water columns which either are well-mixed over considerable depth ranges (e.g. the Cariaco Trench) or show very regular distribution of inorganic solutes (e.g. the Black Sea). The problem of the distribution of dissolved organic matter in anoxic systems clearly warrants further study.

The outstanding feature of the qualitative make-up of the dissolved organic matter is the conspicuous occurrence of light hydrocarbons, especially methane, ethane and ethylene, and propane. Atkinson and Richards (1967) studied the occurrence of methane, and Linnenbom and Swinnerton (1968) determined the distributions of light hydrocarbons in the Black Sea, Cariaco Trench and Lake Nitinat. In almost all instances the shape of the methane profile closely resembles that for sulphide, implying that their places of origin and their subsequent distribution through the system are closely similar. Maximum methane concentrations reported by Linnenbom and Swinnerton are 0.15 ml l^{-1} for the Cariaco Trench, 0.17 ml l^{-1} for the Black Sea and 1.6 ml l^{-1} for Lake Nitinat which has a maximum depth of only 205 m.

The formation of methane in anoxic waters can proceed in either of two principal ways, (1) the reduction of carbon dioxide, which effectively corresponds to

$$CO_2 + 4H_2 = CH_4 + 2H_2O$$

or (2) the fermentation of organic substances, especially carbohydrates, which can be considered as a disproportionation of one molecule into a reduced and an oxidized molecule; for example, the fermentation of acetic acid which is equivalent to

$$CH_3COOH = CH_4 + CO_2$$

Both processes are bacterially mediated, the bacteria utilizing the liberated energy. The reduction of CO_2 is thought to proceed only at redox potentials approaching -400 mV, whereas the upper limit for the fermentation process is less well known but may be considerably higher. Richards (1965) suggested on the basis of thermodynamic considerations that the reduction of carbon dioxide should be unimportant until all sulphate has been reduced because this latter process yields more energy. Consequently, methane present in anoxic waters still containing sulphate should arise from fermentation, or be formed in the sediment in which micro-environments depleted in sulphate might favour carbon dioxide reduction. However, it is not clear whether bacterial populations, which are always mixtures of diverse types having specific capabilities, obey or even come close to obeying thermodynamic laws.

The formation of methane appears to take place predominantly near the sediment surface rather than in the water column. Some of the reasons for this are the lower Eh, the higher substrate concentrations and the greater bacterial populations found near the sediment surface. In addition, there is a possibility of the creation and maintenance of microenvironments which deviate from those of the water column and which are more favourable for specific bacterial activities. The distribution of this gas in the water column is thus principally a consequence of diffusion, advection, convection, and oxidation at the oxic-anoxic water interface, another process which is mediated by bacteria.

Higher hydrocarbons, at least up to the butanes, are apparently always associated with the occurrence of methane in anoxic waters, although in much smaller amounts. The methane/ethane ratio, for example, may be in excess of 60,000, although ratios smaller than 500 have also been reported (Linnenbom and Swinnerton, 1968). Above the O_2–H_2S interface the ratio seems to drop drastically, perhaps as a result of the more rapid oxidation of methane by bacteria. Concentrations of the saturated hydrocarbons below the interface are 2 to 4 orders of magnitude higher than those above the interface; no such increase is observed for the unsaturated compounds. The problem of the source in anoxic waters of the hydrocarbons with more than one carbon atom has not been investigated to any great extent and it has only been surmised that, as yet, unidentified microorganisms may be involved.

An interesting aspect of the occurrence of methane in anoxic waters is the possible existence of methane hydrates: solid methane hydrates have been prepared under certain temperature–pressure conditions in laboratory experiments (for a review of the subject, see Hand *et al.*, 1974). Fig. 16.10

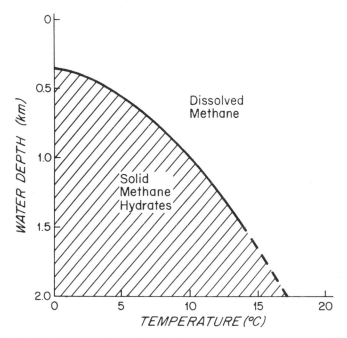

FIG. 16.10. Stability field of solid methane hydrate in terms of water depth and temperature. (After Hand *et al.*, 1974).

shows the stability regime of methane hydrates under marine conditions in terms of water depth and temperature. The position of the curve may be shifted upward by the presence of minor amounts of higher hydrocarbons. For example, 1 % propane may reduce the depth required for hydrate stability by about 200 m at a given temperature. Comparison of the figure with the maximum depths and bottom temperatures of the anoxic basins listed in Table 16.1 shows that only in the Black Sea are the conditions near the sediment–water interface within the stability field of methane hydrates. Their existence there has not yet been established, mainly because of the lack of the special equipment required for bringing them to the surface under conditions within their stability field. It is possible that the presence of hydrates provides an explanation for the relatively low concentration of methane in Black Sea deep water. Although the deep water in the Black Sea appears to be at least ten times as old as that in the Cariaco Trench, for example, methane concentrations near the bottom are 0·17 and 0·15 ml l^{-1}, respectively, in the two basins. Methane produced in the upper layers of the sediment in the Black Sea may be partly retained in the hydrated form rather than being present in a form able to diffuse out and upwards in the water column. Thus, the reduced upward flux of dissolved methane should produce a smaller concentration gradient in the water column and lead to lower deep-water concentrations than would prevail in the absence of hydrate formation. However, the supply of organic matter and water depth also play a major role in the build-up of methane concentrations as indicated by a concentration of $1.6 \, \text{ml} \, l^{-1}$ at 200 m in Lake Nitinat (Linnenbom and Swinnerton, 1968).

Although the presence of hydrocarbons has been abundantly established, relatively little is known about the qualitative make-up of the other organic substances present in anoxic waters. Adams and Richards (1968) reported an increase in the carbon content of dissolved organic matter in Lake Nitinat across the oxic–anoxic interface. They also reported the absence of polar organic constituents which they had found in the overlying oxygenated water. Non-volatile hydrocarbon-type mercaptans were tentatively identified in the anoxic water and sterols or sterol esters; choline-containing lipids, and ninhydrin-positive compounds were found in both oxic and anoxic waters.

A considerable amount of attention has been paid in recent years to the testing with field data of the stoichiometric model for the regeneration of carbon and nutrients as proposed by Redfield et al. (1963) and developed further by Richards (1965). Most of this work was done by Richards and his associates (see e.g. Richards et al., 1965; Richards and Broenkow, 1971; Richards et al., 1971) and in many instances the model was found to agree reasonably well with what was observed, despite some local deviations of the plankton composition from the theoretical norm. This work has significantly refined our understanding of nutrient regeneration and of the chemistry of individual anoxic basins. Recently some aspects of the model have been questioned by Brewer and Murray (1973) who applied the now widely used vertical advection–diffusion model to carbon, nitrogen and phosphorus in the Black Sea. Their results imply consumption of all three elements in the transition zone between the oxic and anoxic zones, quite in contrast to the production rates inferred from the stoichiometric oxidation model. Obviously both models are based on a number of assumptions which may not hold in any given natural situation, and the shortcomings of both approaches need to be elucidated by additional work.

16.6. Acknowledgements

I am indebted to numerous colleagues for providing me with reprints and preprints of their publications. Special thanks are due to Dr. Stig, H. Fonselius for his aid in securing recent information on the Baltic and for helpful comments on the manuscript. Support by the U.S. Office of Naval Research under Contract N00014-66-C0241 is gratefully acknowledged.

REFERENCES

Adams, D. D. and Richards, F. A. (1968). *Deep-Sea Res.* **15**, 471.
Anderson, J. J. and Devol, A. H. (1973). *Estuar. Coast. Mar. Sci.* **1**, 1.
Atkinson, L. P. and Richards, F. A. (1967). *Deep-Sea Res.* **14**, 673.
Brandhorst, W. (1958). *Nature, Lond.* **182**, 679.
Brandhorst, W. (1959). *J. Cons. Int. Explor. Mer.* **25**, 3.
Brewer, P. G. *in press.*
Brewer, P. G. and Murray, J. W. (1973). *Deep-Sea Res.* **20**, 803.
Broecker, W. S. (1970). *Science, N.Y.* **168**, 1537.
Broenkow, W. W. (1969). Unpublished Thesis, University of Washington, Seattle, 207 pp.
Broenkow, W. W. and Cline, J. D. (1969). *Limnol. Oceanogr.* **14**, 450.
Brogmus, W. (1952). *Kiel. Meeresf.* **9**, 15.
Carlucci, A. F. and McNally, P. M. (1969). *Limnol. Oceanogr.* **14**, 736.
Carlucci, A. F. and Schubert, H. R. (1969). *Limnol. Oceanogr.* **14**, 187.
Caspers, H. (1957). *In* "Treatise on Marine Ecology and Paleoecology" (J. W. Hedgpeth, ed.), Vol. 1, pp. 801–889. Memoir 67, Geological Society of America, New York.
Cline, J. D. and Richards, F. A. (1969). *Envir. Sci. Tech.* **3**, 838.
Cline, J. D. and Richards, F. A. (1972). *Limnol. Oceanogr.* **17**, 885.
Coote, A. R. (1964). Unpublished M.S. Thesis, University of British Columbia, Vancouver, 74 pp.
Craig, H. (1971). *J. Geophys. Res.* **76**, 5078.
Deuser, W. G. (1970). *Science, N.Y.* **168**, 1575.
Deuser, W. G. (1971). *Deep-Sea Res.* **18**, 995.
Deuser, W. G. (1972). *J. Geophys. Res.* **77**, 1071.
Deuser, W. G. (1973). *Nature, Lond.* **242**, 601.
Deuser, W. G. (1974). *In* "The Black Sea—Geology, Chemistry, and Biology" (E. T. Degens and D. A. Ross, eds.), pp. 133–136. Memoir 20, American Association of Petroleum Geologists, Tulsa, Oklahoma.
Emery, K. O., Orr, W. L. and Rittenberg, S. C. (1955). *In* "Essays in the Natural Sciences in Honor of Captain Allan Hancock" pp. 299–309. University of Southern California Press, Los Angeles.
Eriksson, E. (1959). *In* "The Atmosphere and the Sea in Motion" (Rossby Memorial Volume) (B. Bolin, ed.), pp. 147–157. Rockefeller Institute Press, New York.
Fanning, K. A. and Pilson, M. E. Q. (1972). *Deep-Sea Res.* **19**, 847.
Fiadeiro, M. and Strickland, J. D. H. (1968). *J. Mar. Res.* **26**, 187.
Fonselius, S. H. (1962). *Fish. Bd. Sweden, Ser. Hydrogr. Rept.* **13**, 42 pp.
Fonselius, S. H. (1967). *Fish. Bd. Sweden, Ser. Hydrogr. Rept.* **20**, 31 pp.
Fonselius, S. H. (1969). *Fish. Bd. Sweden, Ser. Hydrogr. Rept.* **23**, 97 pp.
Fonselius, S. H. (1970). *Tellus,* **22**, 533.
Fonselius, S. H. (1971). *Meddelande från Havsfiskelaboratoriet, Lysekil,* No. 134, 3 pp.
Fonselius, S. H. and Rattanasen, C. (1970). *Meddelande från Havsfiskelaboratoriet, Lysekil,* No. 90, 6 pp.
Goering, J. J. (1968). *Deep-Sea Res.* **15**, 157.
Goering, J. J. and Cline, J. D. (1970). *Limnol. Oceanogr.* **15**, 306.
Goering, J. J. and Dugdale, R. C. (1966). *Science, N.Y.* **154**, 505.

Grill, E. V. and Richards, F. A. (1964). *J. Mar. Res.* **22**, 51.

Hallberg, R. O., Bågander, L. E., Engvall, A.-E. and Schippel, F. A. (1972). *Ambio*, **1**, 71.

Hand, J. H., Katz, D. L. and Verma, V. K. (1974). *In* "Natural Gases in Marine Sediments (I. R. Kaplan, ed.), pp. 179–194, Plenum Press, New York.

Herlinveaux, R. H. (1968). *Fish. Res. Bd. Can. Tech. Rep.* **99**, 34 pp.

Holland, H. D. (1973). *Proc. Symp. Hydrogeochem. Biogeochem.* Tokyo, 1970, Vol. 1, p. 68.

Ivanenkov, V. N. and Rozanov, A. G. (1961). *Okeanologiya*, **1**, 443.

Ivanenkov, V. N., Vintovkin, V. R. and Šackov, K. Z. (1964). *Trudy Inst. Okeanol.* **64**, 115.

Kato, K. (1961). *Bol. Inst. Oceanogr. Univ. Oriente*, **1**, 49.

Kvasov, D. D. (1968). *In* "Doklady na ezhegodnykh chteniyakh pamyati L. S. Berga, VIII–XIV, 1960–1966" pp. 65–80. Izdatel'stvo Nauka Leningradskoye Otdel, Leningrad.

Linnenbom, V. J. and Swinnerton, J. W. (1968). Paper presented at Caribbean Symposium, Curaçao, November, 1968.

Manheim, F. T. and Chan, K. M. (1974). *In* "The Black Sea—Geology, Chemistry and Biology" (E. T. Degens and D. A. Ross, eds.), pp. 155–180. Memoir 20, American Association of Petroleum Geologists, Tulsa, Oklahoma.

Möller, L. (1928). *Veröff. Inst. Meeresk. Berlin Univ., Neue Folge A*, **18**, 1.

Mörner, N.-A. (1969). *Sveriges Geologiska Undersökning, Ser. C, No. 640, Årsbok 63*, *No. 3*, 1.

Murray, J. (1900). *Scott. Geogr. Mag.* **16**, 673.

Nejman, V. G. (1961). *Okeanol. Issled. No. 4*, 62.

Okuda, T., Gamboa, B. R. and García, A. J. (1969). *Bol. Inst. Oceanogr. Univ. Oriente*, **8**, 21.

Olausson, E. (1967). *Progr. Oceanogr.* **4**, 245.

Östlund, H. G. (1969). *Univ. Miami Inst. Mar. Sci. Tech. Rept.* **ML 69167**, 27 pp.

Redfield, A. C. (1958). *Amer. Scient.* **46**, 204.

Redfield, A. C., Ketchum, B. H. and Richards, F. A. (1963). *In* "The Sea" (M. N. Hill, ed.), Vol. 2, pp. 26–77. Interscience, New York.

Reid, J. L., Jr. (1965). "Intermediate Waters of the Pacific", 85 pp. Johns Hopkins Press, Baltimore.

Richards, F. A. (1960). *Deep-Sea Res.* **7**, 163.

Richards, F. A. (1965). *In* "Chemical Oceanography" (J. P. Riley and G. Skirrow, eds.), Vol. 1, pp. 611–645. Academic Press, London.

Richards, F. A. (1970). *In* "Organic Matter in Natural Waters" (D. W. Hood, ed.), pp. 399–411. *Univ. Alaska Inst. Mar. Sci. Occas. Pub. 1*.

Richards, F. A. (1971). *Mer (Bull. Soc. franco-jap. d'océanogr.)* **9**, 68.

Richards, F. A. and Benson, B. B. (1961). *Deep-Sea Res.* **7**, 254.

Richards, F. A. and Broenkow, W. W. (1971). *Limnol. Oceanogr.* **16**, 758.

Richards, F. A. and Devol, A. H. (1973). *Proc. Symp. Hydrogeochem. Biogeochem.*, Tokyo, 1970.

Richards, F. A. and Vaccaro, R. F. (1956). *Deep-Sea Res.* **3**, 214.

Richards, F. A., Cline, J. D., Broenkow, W. W. and Atkinson, L. P. (1965). *Limnol. Oceanogr.* **10**, *Suppl.*, R185.

Richards, F. A., Anderson, J. J. and Cline, J. D. (1971). *Limnol. Oceanogr.* **16**, 43.

Rozanov, A. G. and Bykova, V. S. (1964). *Trudy Inst. Okeanol.* **64**, 94.

Schmidt, J. (1925). *Science, N.Y.* **61**, 592.

Shigematsu, T., Tabushi, M., Nishikawa, Y., Muroga, T. and Matsunaga, Y. (1961). *Bull. Inst. Chem. Res. Kyoto Univ.* **39**, 43.

Sholkovitz, E. R. and Gieskes, J. M. (1971). *Limnol. Oceanogr.* **16**, 479.

Smith, K. L., Jr. and Teal, J. M. (1973). *Science, N.Y.* **179**, 282.

Sorokin, J. I. (1964). *J. Cons. Int. Explor. Mer.* **29**, 41.

Spencer, D. W. and Brewer, P. G. (1971). *J. Geophys. Res.* **76**, 5877.

Strøm, K. M. (1936). *Skr. Norske Vidensk. Akad., Oslo, Mat.-Naturv. Kl.* **7**, 1.

Sverdrup, H. U., Johnson, M. V. and Fleming, R. H. (1942). "The Oceans", 1087 pp., Prentice Hall, Englewood Cliffs, N. J.

Thomas, W. H. (1966). *Deep-Sea Res.* **13**, 1109.

Vaccaro, R. F. and Ryther, J. H. (1960). *J. Cons. Int. Explor. Mer.* **25**, 260.

Van Riel, P. M. (1943). "The Snellius Expedition in the Eastern Part of the Netherlands East-Indies 1929–1930" Vol. II, Part V, Chap. 1, 77 pp., Leiden.

Wada, E. and Hattori, A. (1972). *Deep-Sea Res.* **19**, 123.

Wolgemuth, K., Thurber, D. and Broecker, W. (1971). Unpublished work.

Wooster, W. S., Chow, T. J. and Barrett, I. (1965). *J. Mar. Res.* **23**, 210.

Chapter 17

Marine Pollution

E. D. GOLDBERG

Scripps Institution of Oceanography, La Jolla, California, U.S.A.

SECTION 17.1. INTRODUCTION 39

17.2. TRANSPORT PROCESSES 42
17.2.1. Wind systems 42
17.2.2. River systems 44
17.2.3. Outfalls 47
17.2.4. Ships 47

17.3. RADIOACTIVITY 48
17.3.1. Nuclear detonations 49

17.4. PETROLEUM 53
17.4.1. Man-generated fluxes 55
17.4.2. Natural fluxes 56
17.4.3. Losses during World War II 57
17.4.4. Behaviour of petroleum products in the marine environment...... 57
17.4.5. Persistence of petroleum in the oceans 58
17.4.6. Effects on marine organisms 58
17.4.7. The baseline for petroleum pollution 59

17.5. METALS 60
17.5.1. Mercury 61
17.5.2. Lead 63
17.5.3. Other heavy metals 67

17.6. SYNTHETIC ORGANIC CHEMICALS 70
17.6.1. DDT and its residues 70
17.6.2. PCBs 72
17.6.3. Low molecular weight halogenated hydrocarbons 81

17.7. MARINE LITTER 82

17.8. CONTROL OF ANTHROPOGENIC INPUTS TO THE ENVIRONMENT.............. 84

REFERENCES 86

17.1. INTRODUCTION

The chemical compositions of ocean waters, of presently depositing sediments, and of marine organisms are in very small part, influenced by man's

39

activities. The use of materials in industry, agriculture, and everyday life is responsible for the mobilization of about 3×10^9 tons of material annually on the earth's surface. In addition, the combustion of fossil fuels introduces annually to the atmosphere 13.4×10^9 tons of carbon dioxide (1967 value from SCEP, 1970) and lesser amounts of other gases; much of these eventually enter the sea. As a consequence, the makeup of our surroundings is altered measurably.

The oceans receive a substantial fraction of these materials either by the deliberate or unintentional actions of man. In addition to the three main routes of transport for substances from the continents to the oceans (winds, rivers and glaciers), society has added two more: ships and outfall pipes which carry domestic and industrial wastes. The 3×10^9 ton annual dispersion by man is complemented by about 25×10^9 tons of solids carried by the rivers, winds and glaciers (Table 17.1). Thus, at the present time man appears

TABLE 17.1

Some fluxes in the major sedimentary cycle. Adapted in part from Goldberg (1972)

Material	Geosphere	Flux in 10^{14} g y^{-1}
Suspended river solids	Oceans	180
Dissolved river solids	Oceans	39
Continental rock and soil particles	Atmosphere	1–5
Sea salt from ocean surface subsequently transported over long distances	Atmosphere	3
Volcanic debris	Stratosphere	0.036
Volcanic debris	Atmosphere	<1.5
Glacial debris from Antarctica	Oceans	20*

* Garrels and Mackenzie (1971)

to be responsible for the movement of materials at a rate one tenth of that which occurs in the major sedimentary cycle. The increasing per capita consumption of materials and energy, combined with an expanding world population, will further enhance man's role as a geological agent in the future.

Two characteristics of marine systems influence their ability to accommodate the wastes produced by society. Firstly, the time scale for the retention of materials is long, and whereas rivers or wind systems retain materials introduced to them for a period of only days or months, the oceans can retain both dissolved and particulate phases for times ranging from years to hundreds of millions of years. Thus, substances injected today may still be evident many generations later. Secondly, biological and geological processes occurring in the marine environment are able to concentrate materials from sea water into living and non-living phases respectively to a remarkable extent. During their formation the ferromanganese minerals, which coat much of the sea floor, concentrate such elements as cobalt, nickel and copper from sea water by factors of over a million. The aromatic halogenated hydrocarbons such as DDT and the polychlorinated biphenyls may be enriched by factors of many thousands, in animals, particularly those which are rich in fat. There is always a serious risk if the oceans are used as a disposal site that highly toxic materials will be concentrated in shellfish and fish rather than harmlessly diluted. Thus, considerations of the potential return to man through the ingestion of sea-food have resulted in the formulation of legislation governing discharges of such toxic materials as artificial radionuclides and mercury to the oceans.

The materials injected by man fall into two general categories: (i) those that already exist in the oceans and whose concentrations are elevated (or perhaps in some cases diminished*) by the activities of man and (ii) those species, such as synthetic organic compounds and artificial radionuclides that are alien to the marine environment. Prediction of the fates of substances in both of these categories is a primary concern of those involved in the management of marine resources. The number of different compounds introduced annually to the oceans by human society is probably of the order of hundreds of thousands. Most of these are added in extremely small amounts and offer no threat to living organisms or to non-living resources. Prediction of the environmental fates of those substances which are deemed to be potentially dangerous is essential in rational management programmes. The environmental behaviour of a substance can often be foretold on the basis of its chemical, biochemical and physical properties. It can be predicted with more certainty if the behaviours of similar compounds are known. However, a surprise factor is often present in such studies. The marked concentration of ruthenium by the seaweed *Porphyra* and the environmental stability of DDT and its residues were not predicted by the scientific community.

In discussing marine pollution it is convenient to first consider the transport paths to the oceans from the continents, emphasizing both global dispersion resulting from atmospheric transport and the more localized effects produced by rivers and outfalls. Following this, the various classes of pollutants themselves will be considered: radioactivity, petroleum, metals, synthetic organic chemicals and litter. The rationale for this method of treatment arises from the patterns of recent investigations and the generalizations resulting

---

* It is conceivable for example that the injection of large amounts of solids, such as iron oxides, will scavenge dissolved species from the waters through which they settle.

therefrom. Pollutants are usually studied in groups, which are determined by the instrument with which the analyses are being performed or by the types of samples analysed. As a consequence, the most effective and concise approach to data collation and presentation involves such groupings.*

17.2. TRANSPORT PROCESSES

The amount of material transported by rivers exceeds that carried by wind by an order of magnitude (Table 17.1), whereas glacial transport accounts for an intermediate amount. Materials which enter the marine environment from rivers or outfalls or via glaciers are locally dispersed initially, whereas wind transported material may be carried across the continents and oceans. For example, the tropospheric radioactive debris, introduced by a Chinese nuclear explosion at Lop Nor (40°N; 90°E) in May 1965, circled the world in about three weeks with an average velocity of $16 \, m \, s^{-1}$ (Cooper and Kuroda, 1966). Fallout in rain was measured in Tokyo (36°N; 140°E) and Fayettesville, Arkansas (36°N; 94°W). Discharges or losses from ships can also result in a widespread dissemination of materials. Petroleum wastes contaminate the shipping lanes of both the Atlantic and Pacific Oceans.

17.2.1. WIND SYSTEMS

The main wind systems of the world in the lower atmosphere tend to carry contaminants along lines of latitude. The directions and velocities of these winds are variable, and there is a spreading of materials both to the north and to the south. A consideration of the three principal wind systems is of value in understanding the involvement of wind in global atmospheric transport.

(a) *The Trades.* Easterly winds prevail between 30°N and 30°S with intensities decreasing with altitude. Reversals can take place in the upper troposphere, except in equatorial regions in which their directions are nearly always easterly.

(b) *The Westerlies.* The so-called jet streams exist between 30°N and 70°N and between 30°S and 65°S and often have greater intensities in the upper than in the lower troposphere. Wind speeds of over $100 \, m \, s^{-1}$ have been measured, and speeds of $40 \, m \, s^{-1}$ are common in the central parts of the Westerlies.

(c) *The Polar Easterlies.* Near surface winds which decrease in intensity with height. At above ca. 3 km, they reverse their directions. They occur at 70°N–90°N and between 65°S–90°S.

* Duursma and Marchand (1974) have recently published a comprehensive review on the occurrence of organic pollutants in the sea.

In addition to these major systems there are the continental monsoons, the movements of which are determined by continent–ocean temperature differences. They can reverse their direction from summer to winter.

Meanderings of the three principal wind systems cause latitudinal spreading of the loads. In addition, cyclonic and anticyclonic eddies, which occur seasonally, can cause north to south movements. In general, there is a slow net poleward drift of the air in the lower stratosphere between the equator and 30° north or south latitudes. Particles introduced into the lower stratosphere or upper troposphere at the equator can be dispersed globally and particles introduced at mid-latitudes can extend to areas between 30° and the poles in either hemisphere.

The geographical extent of the prevailing westerlies in the northern hemisphere coincides with the important injection sites for the wastes of society.

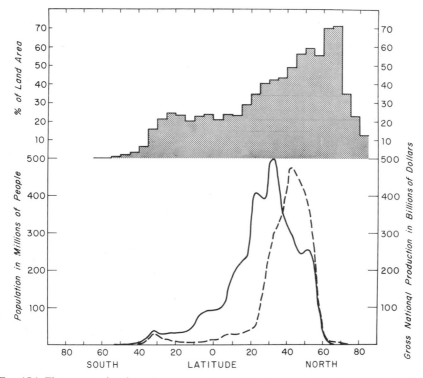

Fig. 17.1. The gross national product, population and land area as a function of latitude (data from The Book of the World, 1971, (John Montague, ed) Tom Stacey Ltd., London). The maxima of the gross national product and population in the northern hemisphere ocean coincide with the latitudes of the prevailing westerly winds. Continuous line = population curve; broken line = GNP curve.

A nation's potential for pollution may be measured by its Gross National Product (GNP) which is related to material flow and to energy utilization, the latter primarily through the combustion of fossil fuels. Those nations with the highest GNPs are situated within the mid-latitudes of the northern hemisphere (Fig. 17.1). The latitudinal zonation of man's releases of materials to the atmosphere combines with the prevailing westerly flow patterns of the jet streams to generate a pollution belt in the northern hemisphere.

17.2.2. RIVER SYSTEMS

The Amazon and the Congo – the two largest rivers of the world (on the basis of their annual discharges) drain into the equatorial Atlantic (Fig. 17.2). However, their drainage basins do not include any of the highly industrialized countries. The Atlantic Ocean receives most of the river discharge and only five of the first twenty-five major rivers drain into the Pacific Ocean directly.

Many of the world's rivers in the northern hemisphere are highly polluted by man. Rivers draining industrialized areas have significantly higher sulphate contents than those flowing through the lands of less developed countries (Berner, 1971). This can be clearly seen if the compositions of rivers passing through Europe and North America are compared with those of Asia, Africa and South America since the former are more strongly influenced by the activities of society (Table 17.2). The sulphate and chloride concentrations are essentially the same in rivers from the latter continents, whereas there is clearly far more sulphate than chloride in the rivers of Europe and North America. If it is assumed that the sulphate/chloride ratio of the rivers of Asia, Africa and South America reflect little pollution, a correction factor can be obtained and applied to European and North American rivers to calculate the sulphate introduced by man (Table 17.2). On a global basis it thus appears that 100 of the 368 megatons of sulphate carried to the oceans

TABLE 17.2

Sulphate fluxes of rivers. Adapted from Berner (1971)

	Total Cl^{-1} $(mg\,l^{-1})$	Total SO_4^{2-} $(mg\,l^{-1})$	Pollutant SO_4^{2-} $(mg\,l^{-1})$	Flux to oceans in 10^{12} g y^{-1} H_2O	SO_4^{2-} (natural)	SO_4^{2-} (pollution)
Europe	6·9	24	17	$2\cdot5 \times 10^6$	17	45
N. America	8·0	20	12	$4\cdot6 \times 10^6$	37	55
S. America	4·9	4·8	0	$8\cdot2 \times 10^6$	39	0
Africa	12·1	13·5	0	$6\cdot0 \times 10^6$	81	0
Asia	8·7	8·4	0	$11\cdot2 \times 10^6$	94	0
Total				$32\cdot5 \times 10^6$	268	100

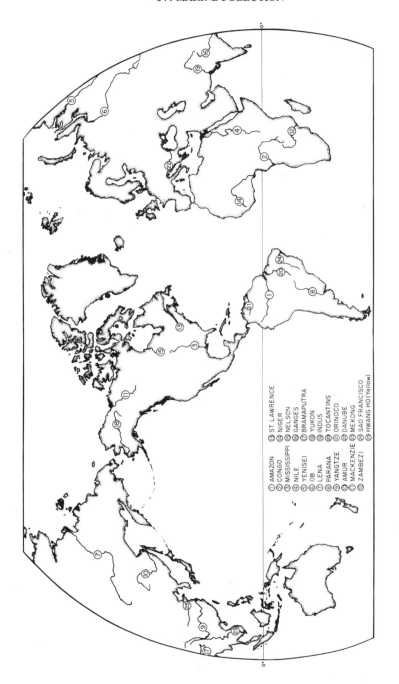

FIG. 17.2. The rivers of the world ranked in order of annual water discharge.

annually can be attributed to human activity. Since many of the analyses
incorporated into Table 17.2 were obtained before 1900, this sulphate
pollution value is most probably a lower limit. Most of the sulphate generated
by society has its source in the combustion of fossil fuels.

The radioactive pollutants (by-products of nuclear reactors) introduced
into rivers provide tags to measure the penetration of their dissolved and
particulate burdens into the marine environment. Thus, Cr-51 (as chromate)
with a radioactive half-life of 28 days produced by reactors at Hanford has
been used to study mixing of Columbia River water with that of the Pacific
Ocean for distances of 350 km (Osterberg et al., 1965). This nuclide acts as

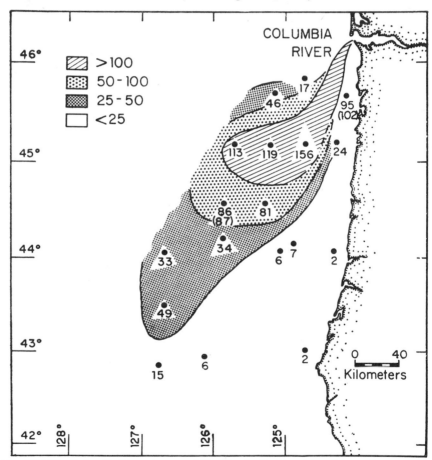

FIG. 17.3. The Columbia River plume as defined by the Cr-51 activity of the surface sea water
(counts per minute per 100 litres of water). Numbers in parentheses indicate duplicate samples
(Osterberg et al., 1965).

an effective tracer for the river water and its dissolved constituents since it remains principally in solution and is not taken up by the biota or sorbed upon particulates to any great extent. The Columbia River plume, defined by Cr-51 (Fig. 17.3) had a movement of $11\cdot4\,\text{km day}^{-1}$, a minimum value inasmuch as any losses of Cr-51 to the biosphere or to the sediments would decrease the apparent rate of flow.

The transport of sediment along the continental shelf has been followed by means of the activities of the radionuclides Zn-65 and Co-60 adsorbed nearly irreversibly on river-borne particulates after release from the Hanford reactors (Gross and Nelson, 1966). The half-period of the activity ratio which is about 280 days allows sediment transport to be measured for periods of several years. The sediments moved northward at rates of $12\text{--}30\,\text{km y}^{-1}$ along the shelf, and $2\cdot5\text{--}10\,\text{km y}^{-1}$ westward away from the coast.

The geographical delineation of river–ocean interactions, especially of their regional character over times of the order of years, has been established by studies of man's injections of easily measurable radioactive species to his surroundings. Other equally effective tracers of river waters and their dissolved and particulate loads may be found among synthetic organic chemicals foreign to the ocean system.

17.2.3. OUTFALLS

Although organic matter and micro-organisms have been used to define the impact of an outfall upon the environment into which discharge takes place, recent studies have demonstrated that heavy metals, such as mercury, can also be effectively employed for this purpose. For example, investigations of the mercury distribution about sewers in New Haven (Connecticut) Harbour (Applequist et al., 1972) and Los Angeles, California (Klein and Goldberg, 1970) indicate that abnormally high concentrations can be found up to distances of about 10 km from the injection site. The patterns of concentration distribution in the sediments reflect discharge levels and local current systems.

17.2.4. SHIPS

Dissemination of materials on both regional and global bases can result from ship discharges. The intentional dumping of materials in coastal areas continues to increase. In the U.S., for example, the tonnage discharged between 1949 and 1968 has quadrupled (CEQ, 1970). Dredge spoils account for about 80 percent of the total, and industrial and domestic sewage wastes each contribute about 10 percent to the total weight of dumped materials. There are minor entries from construction debris and explosives.

c

About 8.6 million tons of waste solids (excluding rubbish and floatable debris) were dumped each year in the coastal waters of the Atlantic near New York City and in western Long Island Sound between 1964 and 1968. This appears to be the largest single sediment source from North America entering the Atlantic Ocean (Gross, 1970). The wastes come from a population of about 9 million people at a rate of about 2 kg per person per day. They contain a vast number of substances including halogenated hydrocarbons, heavy metals, plant materials, plastic fragments and petroleum products. Many of these materials are chemically refractory. Volatile matter (measured by ignition loss), total carbon, lead and copper concentrations provide a measure of the wastes introduced to the deposits (Gross, 1972).

There is extensive dumping of drums containing chemical wastes into the North Sea (Greve, 1971). The materials have included a variety of substances including lower chlorinated alphatic compounds (the so-called EDC tar, a by-product of vinyl chloride production), vinyl esters, chlorinated aromatic amines and nitrocompounds and the insecticide Endosulfan. The drums corrode with time and sooner or later their contents will enter the waters, biota and sediments.

About 5 million wet tons of sewage arising from the London area is dumped into the outer Thames Estuary each year (Shelton, 1973). Again, as with the New York dumping sites, the sediments maintain a record of this activity. Shelton observed that the husks of tomato seeds were a common ingredient of the sediments in the dumping areas. Their distribution corresponds roughly to that of the deposits containing large amounts of organic matter. Since the seeds are probably less mobile than the finer organic components, they can only be used as an approximate index for the behaviour and distribution of sewage sludge in the dumping areas.

17.3. RADIOACTIVITY

Nuclear explosions and discharges from nuclear reactors have measurably increased the levels of radioactivity in the marine environment (see Chapter 18). The first significant releases took place in 1944 with the discharge of wastes from the plutonium production plants in Hanford, Washington to the northeast Pacific Ocean via the Columbia River (Seymour, 1971). Recognition of possible threats to the health of man through exposure to, or ingestion of, radionuclides entering the oceans has resulted in regulations controlling their release. In addition, the need to understand the behaviour of radionuclides in the ocean system has led to extensive investigations of their interactions with the sediments, organisms, waters and atmospheres.

Such studies began soon after the commencement of nuclear testing at Bikini Atoll in 1946.

The greatest contribution to the artifical radioactivity burden of the oceans is the detonation of nuclear devices (Table 17.3). Controlled discharges of

TABLE 17.3

Inventory of artificial radionuclides in the world ocean in 1970 (Preston, 1972).

Nuclear explosions	
Fission products exclusive of titrium	2.6×10^8 curies
Tritium	10^9
Reactors and reprocessing of fuel	
Fission and activation products exclusive of tritium	3×10^5
Tritium	3×10^5
Natural potassium-40	5×10^{11}

radioactive wastes from outfalls and from ships have sometimes resulted in high local concentrations of radioactive nuclides, whereas atmospheric fallout of matter introduced by bomb explosions has been relatively uniformly dispersed. As a result of improvements in waste management, the amounts of radioactivity in the oceans for the next three decades are not expected to increase substantially (Preston *et al.*, 1971), although there will be a marked growth in the production of nuclear power. This prediction assumes the continuation of the present low rate of nuclear detonations.

17.3.1. NUCLEAR DETONATIONS

Up to the end of 1968 470 nuclear explosions had been carried out by the United States, the Soviet Union, the United Kingdom, France and China. The world oceans received a substantial part of the debris from these explosions, except for those conducted underground or in outer space (Joseph *et al.*, 1971). The amount of fission material produced in this way has been estimated at 2.8×10^{28} fissioning atoms of uranium or plutonium. Two of the fission products which have entered the oceans with high activities, Cs-137 (half-life of 30 years) and Sr-90 (half-life of 28 years), have been produced at levels of 21 and 34 megacuries respectively (Joseph *et al.*). Thermonuclear explosions have introduced transuranic isotopes, including those of plutonium and neptunium. Of the explosions, 140 megatons or 72 percent of the total yield, were produced as airbursts in which the fireball did not intersect the ground (Joseph *et al.*).

Underwater detonations introduce radioactive species to a rather localized area initially, although subsequent mixing in the oceans can spread the nuclides over vast distances. Many radionuclides from surface or air detonations enter the stratosphere where they may remain for periods of several years before returning to the earth's surface. Usually, the fallout from the stratosphere takes place in the hemisphere of introduction, although stratospheric mixing processes do allow for inter-hemispheric transfer. The radionuclide W-185 was injected into the atmosphere at 12°N in the western Pacific in 1958. A year and a half later, there were secondary maxima at heights of 15 to 18 km over latitudes 50 to 90° in both hemispheres, although most of the activity was still near the input site at 0 to 5°N (Martell, 1968).

Removal of materials from the stratosphere to the troposphere usually takes place at upper and middle latitudes of both hemispheres in the late winter and early spring. After 30 days or less in the troposphere depending on the meteorological conditions in the area of input the nuclides are removed by precipitation. The worldwide production and deposition of Sr-90, the most widely studied of the fission products is shown in Table 17.4 The time

TABLE 17.4

Worldwide Production and Deposition of ^{90}Sr (in MCi)
(Joseph et al., 1971)

Year	^{90}Sr produced	^{90}Sr deposited
1945–1958	9·1	5·6
1959	0	1·1
1960	0	0·4
1961	2·5	0·4
1962	7·6	1·6
1963	0	2·6
1964	0	1·9
1965	0	1·0
1966	0	0·4

lag between production in nuclear tests and deposition appears to be several years because of residence in the stratosphere.

Sr-90 and Cs-137 are the most studied of the fission products in the oceanic system. The relative concentrations of the dissolved nuclides reflect their production ratios, when account is taken of loss by radioactive decay. It is believed that the dissolved nuclides are transported solely by water circulation processes and for this reason they are often used as tracers of water masses (Volchock *et al.*, 1971). In the Atlantic the Sr-90 concentration decreases more or less regularly with depth down to 700 m. The maximum is at the surface and the concentration rarely fails below 10 percent of the surface value (Fig. 17.4). In contrast, in the Northwest Pacific, the 700 m values

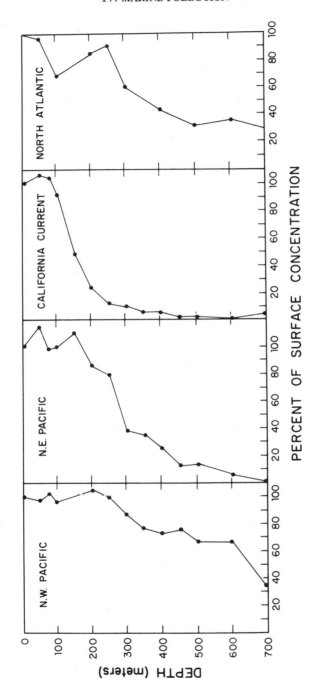

FIG. 17.4. Sr-90 concentrations in the Atlantic and in the Pacific (Volchok *et al.*, 1971).

are about 35% of the surface values. Penetrations below a depth of 700 m have not occurred in the Northeast Pacific. In these waters the surface values are maintained to about 150 m after which there is a very rapid falloff to 700 m. In Californian coastal waters, the falloff is even more rapid with practically no Sr-90 being found below 400 m.

There is considerable disagreement about the occurrence of Sr-90 and Cs-137 in waters deeper than 1000 m (Volchok et al., 1971). This stems from differences of opinion between various laboratories about how the blanks should be measured. This problem is unlikely to be resolved in the near future.

The global inventory of tritium now contains greater contributions from the detonation of megaton fusion bombs than from its natural production through the interaction of atmospheric gases with cosmic rays. There was a hundredfold increase in the H-3 level as a result of hydrogen bomb testing in 1962 and 1963. Nearly all of the H-3, independent of its mode of formation, is in the form of tritiated water (Martell, 1963). Hence, it too can be used to trace the movements of water masses. The pulses of tritium that enter the ocean from stratospheric fallout in late winter and early spring are rapidly accommodated throughout the mixed layer in the North Pacific (Dockins et al., 1967). In the Atlantic, there is a penetration of tritium below the thermocline (Roether and Munnich, 1967).

Bomb detonations have only led to the doubling of the natural, dissolved C-14 levels (Lal and Suess, 1968). Penetration of C-14 below 200 m was not observed in the Pacific (Bien and Suess, 1967). Lower activities of C-14 are found in upwelling areas since deeper waters of relatively low C-14 activity dilute the bomb-produced C-14 activity of the surface waters.

A group of fission products, Zr-95/Nb-95, Ru-103, Ru-106, Ce-141, Ce-144 and Pm-147, after entry to the marine environment, become associated with particulate phases (Volchock et al., 1971). Such nuclides show irregular depth profiles, sometimes with secondary maxima in deep waters. It is not yet known whether these particles have sunk with the debris of atmospheric tests or whether their conveyance to deeper waters involves the biosphere or inorganic particles.

There is also a group of radionuclides which enters the oceans not from atmospheric fallout but in releases from nuclear reactors or nuclear vessels. For example, the Cr-51 mentioned above is an activation product produced in the coolant system of the nuclear reactors at Hanford, Washington.

Three radionuclides (Cs-134, Cs-137 and Ru-106) produced in nuclear reactors have concerned the British Ministry of Agriculture, Fisheries and Food which is responsible for monitoring the radioactivity of British coastal waters (MAFF, 1971). The caesium isotopes are enriched in fish and the

Ru-106 in *Porphyra*, a seaweed which is utilized in the production of a food product, laverbread. Consideration of the levels of these isotopes in the food products is used to establish the permissible rates and amounts of reactor waste discharge.

The future accumulation of plutonium isotopes in the ocean system may create a serious problem of pollution. This element and its compounds are among the most toxic substances known to man (Olafson and Larson, 1963). The large scale nuclear programmes projected for the future will expand the production and utilization of the metal. Losses to the environment will inevitably occur and it is likely that the oceans will be an ultimate or a penultimate sink for it and this will pose threats to life. Its marine chemical behaviour may influence the choice of procedures used in its handling.

Injection to the atmosphere of the three isotopes Pu-238 (half-life 86 years), Pu-239 (half-life 24,400 years) and Pu-240 (half-life 6580 years) has resulted from atmospheric nuclear weapons tests. In addition, some contamination may arise from the use of Pu-238 and Pu-239 as nuclear fuels. A navigational satellite using Pu-238 as a power source (SNAP device, Systems for Nuclear Auxiliary Power) failed in 1964 to go into orbit and the Pu-238 was burned up at a height of about 50 km in the southern hemisphere. By mid-1970 95% of this Pu-238 had fallen back to the earth's surface (Hardy *et al.*, 1973). The fallout was 2·5 times greater in the southern than in the northern hemisphere. On the other hand, most of the nuclear bombs were detonated in the northern hemisphere where the heaviest fallout of Pu isotopes has taken place. A study of the accumulation of the plutonium in soils has shown that the greatest input to the earth's surface from the atmosphere occurred at mid-latitudes, and the least in the equatorial regions (Hardy *et al.*). Fallout to the oceans should be latitudinally similar to that on land.

Pu-239 activities in sea water are only a few percent of those of Sr-90 and Cs-137 and that of Pu-238 is only a few percent of that of Pu-239. The plutonium residence time in the upper layers of the ocean is about 4 years, in contrast to that of ten years for caesium (T. Folsom, personal communication). Plutonium in sea water is highly reactive geochemically and its chemistry is similar to that of polonium (Folsom, personal communication). Like polonium its conveyance out of surface waters probably results from its tendency to adhere to both organic and inorganic surfaces. It is rapidly taken up on the surfaces of large algae.

17.4. PETROLEUM

There is abundant evidence of the extent to which the marine environment has become contaminated with petroleum hydrocarbons the soiling of beaches, the staining of surface waters with films and tar-balls, and dead or moribund birds. Recently it has become possible to express this contamination at least semi-quantitatively. For example, the surface waters of the entire Baltic have a petroleum load ranging between 0·3 and 1·0 mg 1^{-1} (Simonov *et al.*, 1972) which originates from industrial activity in East and West Germany, Sweden, Finland and Russia. With increasing future demands for petroleum there will be more offshore drilling and more transoceanic shipments. As a consequence there will most probably be increased injections into sea waters. Already marine organisms carry a body burden of man-disseminated petroleum hydrocarbons; however, their effects on marine life processes have not yet been defined. It seems probable that a continued build-up of man-generated petroleum and its by-products in organisms and sediments will occur and this may have deleterious effects on marine life.

The scale of oil pollution can perhaps be assessed through the recent assays of petroleum lumps (tar balls) captured in neuston nets (Morris, 1971; Horn *et al.*, 1970). These black or brown irregularly shaped artifacts ranged in diameter from 1 or 2 mm to about 10 cm, and were evident in at least 75 percent of the neuston tows made in the North Atlantic. Their abundance there appears to be about 0·93 mg m^{-2}; in the Mediterranean it was 19·6 mg m^{-2}. The former number can be brought into perspective by noting that it represents about 0·1 percent of the weight per unit area of the *Sargassum* weed which abounds in the Sargasso Sea.

Horn *et al.*, (loc. cit.) have described many of the characteristics of the tar balls. The lumps contain the low boiling fraction of the crude oils, suggesting that the particles are relatively young, with ages of perhaps weeks. They are covered with grey films, which were assumed to be composed of bacteria. Isopods and goose barnacles were found adhering to the lumps. These animals suffered no evident disabilities from this association. Tar balls were found in the stomach contents of three out of ten sauries (epipelagic fish). These fish feed upon small crustaceans and in turn are consumed by porpoises and larger predacious fish. Through them, the tar balls can be introduced into the marine food chain.

The pelagic *Sargassum* weeds and the animals associated with them were assayed for their hydrocarbon contents by Burns *et al.* (1972) to ascertain the extent of pollution. All organisms displayed evidence of petroleum contamination. The hydrocarbon content of the animals was derived from petroleum, whereas the *Sargassum* contained substantial amounts of natural hydrocarbon. No relationship could be found between the animals' trophic level and its content of polluting hydrocarbon.

17.4.1. MAN-GENERATED FLUXES

Of the 2×10^9 tons of crude oil produced in 1969, 1.3×10^9 ton, or 65% was transported in ocean-going tankers. The probable 1980 production is predicted to be 4.4×10^9 tons – twice the 1969 value and it is likely that the amount being shipped will increase proportionately. The annual losses of petroleum to the sea as a result of man's activities have been estimated by Revelle *et al.* (1971) to be 2.2×10^6 tons (Table 17.5). Shipping operations are

TABLE 17.5

Annual entries of oil in millions of tons to the oceans through man's activities. From Revelle et al., 1971

Accidental spills	0·2
Tanker operations	0·5
Other ships	0·5
Offshore production	0·1
Refinery operations	0·3
Industrial and automotive wastes	0·6
Total	2·2

responsible for about one half of the total. These chronic discharges involve accidental spills or deliberate flushing of oil tanks at sea. The spectacular break-up of the tanker Torrey Canyon off the coast of Britain introduced 118,000 tons of oil into the sea, and other accidents of this type have released somewhat smaller amounts.

The pumping of bilges and other cleaning operations on dry cargo ships with gross tonnages greater than 100 tons have introduced an estimated 0.5×10^6 tons of oil to the sea annually, an amount similar to that derived from tankers. Offshore oil production, which accounts for about 16% of the total crude recovery, appears to discharge about 100,000 tons per year. Accidental losses from individual offshore oil-well drilling are reported to be lower than those from individual ship disasters. The Santa Barbara, California blowout injected about 11,000 tons of oil into the oceans. Clearly, a potentially high injection rate into the oceans will result as more coastal oil fields are exploited in the future if no improvements are made in operational techniques.

The sea receives about 300,000 tons of oil per year through losses from refineries and petrochemical plants (about 5% of the total oil production is involved in the manufacture of synthetic organic chemicals). Finally, industrial and automotive waste oils and greases are estimated by Revelle *et al.* to account for an entry of about 670,000 tons per year, about 450,000 tons of which comes via the rivers and 100,000 tons via municipal sewage effluents. The total direct losses of petroleum to the marine environment thus total about $2 \cdot 2 \times 10^6$ tons per year.

The atmospheric transport of petroleum hydrocarbons to the sea may be of the same order of magnitude or even greater. The emission of hydrocarbons in the U.S. from fuel combustion and manufacturing is about 30 million tons per year (SCEP, 1970, p. 296). Since the U.S. is responsible for about one-third of the energy production in the world, the total global atmospheric input will be about 90 million tons per year. If only a few percent of these hydrocarbons reach the oceans through the wind systems and subsequent washout, the atmospheric inputs will more than equal those of the processes considered above.

17.4.2. NATURAL FLUXES

Although natural seepages of oil into coastal waters have been observed in the past, few quantitative estimates of their magnitudes have been made. One set of measurements involving exudations into the Santa Barbara Channel, off the coast of California suggests that between 8000 to 11,000 l of oil per day (3–4×10^4 tons per year) enters the waters from sea floor openings (about $0 \cdot 5$ cm in diameter) in the unconsolidated sediments (Allen *et al.*, 1970). These point sources can be highly dispersed, or they can attain densities of 100 separate openings m^{-2}. Individual globules released from these holes contain between $1 \cdot 7$ and $3 \cdot 9$ g of oil (average about $2 \cdot 5$ g). They are released at intervals ranging from 15 s to 5 min and rise to the surface where they form slicks several hundred m wide and 10^{-5} cm in thickness.

The estimated $2 \cdot 2$ million ton direct annual injection of petroleum products into the oceans through human activity appears to far exceed the entries from natural seepages (Blumer, 1972; Revelle *et al.*, 1971). It is estimated that the reserves of offshore oil total 10^{11} tons. If these liquids were to seep into the ocean at a rate similar to that estimated for man's input, all these oil reserves would be lost in a mere 50,000 years. Clearly, such a natural flow rate is unlikely since the ages of crude oils vary between 2×10^6 and 6×10^8 years (i.e. Pliocene to Cambrian). The present 40 kiloton annual seepage into the Santa Barbara Channel may be atypical of past times. For if there were 25 other similar natural petroleum leakages, the total oil reserves would have been lost in one hundred thousand years. Most probably natural seepages of oil are at least several orders of magnitude less than the inputs by man.

17.4.3. LOSSES DURING WORLD WAR II

During World War II substantial amounts of oil products were introduced into the oceans through the sinking of ships. According to Revelle *et al.* (1971) 98 U.S.-controlled tankers, each containing about 10,000 tons of oil, were sunk or seriously damaged. To this injection of about one million tons should be added a further 3 million tons lost through other sinkings. Thus, altogether more than 4 million tons were lost, or about twice the normal annual input through man's activities. However, Revelle *et al.* have pointed out that there is no indication that this has damaged the ocean ecosystem.

17.4.4. BEHAVIOUR OF PETROLEUM PRODUCTS IN THE MARINE ENVIRONMENT

The composition of injected petroleum may be altered in the following ways:
Evaporation. Significant volatilization of the low molecular weight species (up to and including C_{12}) may occur with those substances that accumulate at the ocean–atmosphere interface. Since the C_4–C_{12} constituents make up more than 40% of a mature petroleum, evaporative losses can be quite substantial.
Dissolution. All hydrocarbons are slightly soluble in water. The solubility decreases with increasing molecular weight.
Surface film formation. The formation of monomolecular or multimolecular films (the so-called slicks) results in the areal dispersion of petroleum products.
Oxidation. Exposure of petroleum constituents to oxygen and light can result in their oxidation, this process sometimes being photochemically mediated. There is, as yet, no evidence for a biological oxidation in the absence of light.
Interactions with the biosphere. In aerobic zones, micro-organisms can degrade or totally decompose petroleum. Petroleum products can also be ingested by higher organisms and subsequently be degraded or decomposed.
Sedimentation. Oil components can become aggregated into tarry lumps or emulsions. When the densities exceed those of sea water, perhaps as a result of the uptake of mineral matter, they will fall to the sea floor.

The relative importance of these various processes is not yet established (IDOE, 1972). It is often thought that petroleum is entirely biodegradable and that the bacteria in the sea are primarily responsible for its decomposition. However, Floodgate (1972) has produced evidence which casts considerable doubt on this assumption. Laboratory experiments with various types of oils have produced data which are difficult to interpret. The use of such data to assess the breakdown of oil in the marine environment can lead to erroneous conclusions. At present it can only be stated that marine bacteria are capable of degrading some components of oils, but that the rates of decomposition under the natural conditions are poorly known.

17.4.5. PERSISTENCE OF PETROLEUM IN THE OCEANS

Detailed studies of a spill of fuel oil which occurred in September 1969 in Buzzards Bay, Massachusetts have shown that petroleum hydrocarbons have a considerable persistence in the marine environment (Blumer and Sass, 1972a). In the sediments, the disappearance of the oils resulted both from microbial degradation and through dissolution, the latter being quantitatively the more important. This was evident from the increase between March 1970 and April 1971 in the proportions of the more highly substituted benzenes, naphthalenes and tetrahydronaphthalenes relative to the more soluble lower homologues such as naphthalene and the C_1 to C_3 alkylnaphthalenes.

N-alkanes persisted for at least two years in the sediments. The rates of decomposition of the branched and cyclic hydrocarbons were slower than those of their normal counterparts. For periods of two years after introduction of the oil, the isoprenoids (phytane, pristane, and the C_{18} homologue) and alicyclic and aromatic hydrocarbons were easily measurable in the deposits.

Petroleum hydrocarbons are preserved in the sedimentary column for geological periods, and most probably the oil pollutants introduced by man will be evident in the geological record for similar lengths of time.

17.4.6. EFFECTS ON MARINE ORGANISMS

A recent compilation of the oil burdens of a variety of organisms (Table 17.6) clearly demonstrates the entry of petroleum hydrocarbons into the marine food web. Concentrations of petroleum hydrocarbons in the organisms in the parts per million range contrast with the parts per billion levels in sea water. The effects, if any, of such concentrations in the organisms have not yet been determined. In contrast, oil spills, which introduce large amounts of the pollutants to the environment, are responsible for mass mortalities in benthic communities.

The effects of petroleum hydrocarbons upon the metabolism of organisms may be diverse. For example, certain petroleum components interfere with the processes of chemoreception through the blocking of the detection organs. An individual animal so affected may be at a competitive disadvantage in its search for food (IDOE, 1972). Reproductive processes can be hampered by the masking of the presence of pheromones (chemicals produced by organisms to communicate information). For lobsters these exudations of the pheromones by the female are necessary to induce copulation with the males. Pheromone activity in other organisms is still poorly studied. The 1972 IDOE Report also suggests that it is possible that hydrocarbons at sub-lethal levels may inhibit reproduction either through reduction in the viability of the gametes produced or by acting lethally upon early life stages of animals.

TABLE 17.6
Petroleum hydrocarbon contamination in marine organisms (IDOE, 1972)

Location	Concentration	Boiling range[1]
Plankton[2]		
Louisiana coast	100 ppm (wet wt.)	$nC_{16}-nC_{36}$
Seston, open ocean		
2 samples North Atlantic	0·3–20 ppm	$nC_{16}-nC_{28}$
1 sample South Atlantic	(wet wt.)	
Sargassum community		
(plants and animals),	1–34 ppm	
Sargasso Sea	(wet. wt.)	
2 fish livers, Georges Bank (New England	5·19 ppm	$nC_{16}-nC_{28}$
Continental Shelf)	(wet wt.)	
Water, Louisiana coast		
(1 sample)	0·63 μg l^{-1}	$nC_{16}-nC_{34}$
Water, Gulf of Mexico		
(2 samples)	0·03, 30 μg l^{-1}	nC_1-nC_3

[1] Boiling range: boiling between the *n*-paraffins listed.
[2] Contained primarily in the plankton itself. However, petroleum hydrocarbons could be contributed by small tar balls in the sample.

17.4.7. THE BASELINE FOR PETROLEUM POLLUTION

Gas chromatographic analyses of hydrocarbons in the sediments underlying the area of the Buzzards Bay oil spill provided Blumer and Sass (1972b) with criteria which enabled them to distinguish between hydrocarbons of a biological origin (i.e. those which have a source in the present marine biota) and petroleum hydrocarbons dispersed by man. Materials extracted from the sediments at depths greater than 7·5 cm contained little, or no, fuel-derived hydrocarbons, whereas the surface layers of the deposits contained substantial quantities of such oils.

In the "biological" hydrocarbons there is a marked predominance of normal paraffins with an odd number of carbon atoms, especially those near C_{21}. Except for some very young oils, petroleums show no such trends in their normal paraffin contents. The surface sediments contained hydrocarbons boiling within the temperature range of fuel oils, whereas the hydrocarbons present at greater depths consisted predominantly of compounds boiling above this range. In addition, pristane, the principal resolved component in gas chromatographic records of partially degraded fuel oils, was found in very minor amounts in the deeper sections, although the surface layers contained it as a major component along with the adjacent C_{18} and C_{20} isoprenoids. Hydrocarbon separates from sediments containing fuel oils show an unresolved envelope in their gas chromatographic patterns for the hydrocarbons of the fuel oil boiling range, a characteristic not found for separates from sediments where there is no evidence for fuel oils. Finally, the gas chromatograms of hydrocarbons from the upper sediments showed "fingerprint peaks" corresponding to specific components of their fuel oil contaminants.

17.5. METALS

Some metals are being mobilized by man to the atmosphere and hydrosphere at rates comparable to, and sometimes exceeding, those by weathering processes. As a consequence, modifications of their concentrations in marine waters, sediments and organisms are to be expected. The existence of such changes is often difficult to establish since the normal levels and their variabilities are often poorly known, especially in waters. Anthropogenic metals, following their introduction to the oceans through atmospheric fallout or river injection, can be concentrated by marine organisms or by sedimentation processes, or dispersed throughout the oceans by physical mixing. The two most studied metals, lead and mercury, have unique behaviour patterns which were not predictable. Other metals, as they are investigated in greater detail, will most probably also show unexpected behaviours in the sea water system.

Insight into the heavy metal interactions between the constituents of river waters and those of sea water has been provided by pollution studies. For example, when river particulates enter the ocean their adsorbed metals may be desorbed (deGroot, 1973). This is especially evident for mercury entering the Rhine Estuary and Wadden Sea from the Rhine river. The average mercury content of the suspended solids in the river waters in 1969–1970 was 23·3 ppm. The mercury content of the suspended solids of the coastal waters (Cl = 16‰) varied between 1·6 and 3·0 ppm. Thus, a mobilization to the water of 89–94% of the sorbed mercury is indicated. The amount of mercury in the suspended solids in 1970 was 28% greater than the average for 1958–1960. This suggests that much of this river-borne mercury is man-generated.

The extent of desorption of metals from river particulates on entering the saline environment varies widely from one metal to another, from about 90% for mercury to no mobilization for samarium, manganese, scandium and lanthanum. For the Rhine Estuary the order of ease of desorption is mercury, copper, zinc, lead, chromium and arsenic; cobalt and iron are somewhat less easily desorbed. DeGroot (1973) has argued that the cause of these mobilization processes is the intensive decomposition of organic matter, the products of which would include species which act as ligands and form stable complexes with the metals from the sediments. Fulvic acids (soluble in both acidic and alkaline solutions) are reported to be mainly responsible for these metal mobilizations.

17.5.1. MERCURY

The world production of mercury, 8·8 kilotons per year, is small in comparison with the flux from the continents to the atmosphere through crustal degassing which is estimated to be somewhere between 25 and 150 kiltons per year (Weiss et al., 1971, Table 17.7). This flux has been calculated on the basis of

TABLE 17.7

Environmental mercury fluxes (Weiss et al., 1971).

	Flux in g y^{-1}
Natural flows	
Continents to atmosphere	
Basis of precipitation with rain	$8·4 \times 10^{10}$
Basis of atmospheric content	$1·5 \times 10^{11}$
Basis of content in Greenland Glacier	$2·5 \times 10^{10}$
River transport to oceans	$<3·8 \times 10^9$
Flows involving man	
World production (1968)	$8·8 \times 10^9$
Entry to atmosphere from fossil fuel combustion	$1·6 \times 10^9$
Entry to atmosphere during cement manufacture	$1·0 \times 10^8$
Losses in industrial and agricultural usage	$4·0 \times 10^9$

the concentration of mercury in glacial waters, the rates of accumulation of which are known, and its concentration in the atmosphere and in precipitation. It thus appears that industrial activity, fossil fuel combustion and other activities of man would not have a measurable effect on the overall concentration of mercury in the ocean. The major marine impacts have been local, being restricted to estuaries, in which localized high concentrations have led to epidemic poisonings, such as those associated with Minimata Bay and Niigata City.

The ppm (wet weight) concentration levels found in many living marine animals appear to be natural and not to have resulted from pollution. Analyses of museum specimens of tuna caught 62 to 93 years ago and a swordfish caught 25 years ago reveal mercury levels similar to those of their present day counterparts (Miller *et al.,* 1972). Barber *et al.* (1972) have studied the mercury concentrations of bottom-dwelling fish. They have found a correlation between size and mercury concentration, with the larger individuals having mercury levels up to 0·8 ppm (wet weight). The 90 year specimens closely fit the size-concentration regression curve of the nine recent individuals of the same species.

The Minimata Bay episode presents a typical example of the "surprise factor" in environmental problems. In 1953 severe neurological disorders were initially recognized among the citizens living around Minimata Bay, southwestern Kyushu, Japan. By the end of 1970, 111 cases of the disease had been recognized and 41 deaths had occurred. The principal victims were fishermen and their families. Pet cats were afflicted similarly. From the start, the disease appeared to be non-infectious. Those afflicted had one common trait – they were consumers of fish. A second occurrence of the Minimata Bay afflication occurred in 1965 in Niigata, Japan, far removed from Minimata Bay. Twenty six cases were reported with five deaths occurring among its victims.

In 1959, mercury was discovered in high concentrations in fish, and in 1963 the active agent causing the disease was identified as methyl mercuric chloride. The source of the mercury was the Minimata Factory which was involved in the manufacture of PVC (polyvinylchloride) resin and the production of octanol and dioctyl phthalate using acetaldehyde as the main starting material. Two different mercury compounds were used in the manufacturing processes: mercury(II) chloride, adsorbed on activated carbon in the synthesis of vinyl chloride; and mercury(II) sulphate, used in the production of acetaldehyde. Liquid and solid wastes discharged into Minimata Bay were shown to contain organic mercury compounds. Methyl mercuric chloride was shown to be present in the effluent from the acetaldehyde synthesis. The organic mercury was taken up by the fish and shellfish. Following their consumption by man, the disease appeared. Two factors combined to produce this tragedy. Firstly, there was the production of organic compounds containing mercury in the wastes. Secondly, there was the marked accumulation of methyl mercury in the biota.

On the heels of the Minimata Bay disaster came the observation by Westöö (1966) that the principal form of mercury in fish was methyl mercury. This was followed by studies which indicated that microorganisms are capable of producing methyl mercury and dimethyl mercury from inorganic compounds. Wood *et al.* (1968) were able to show that both of these compounds could be synthesized in both enzymatic and non-enzymatic systems in the presence of methyl cobalamin.

17.5.2. LEAD

Lead is being introduced into the major sedimentary cycle by man at a rate that rivals those of natural processes. The primary anthropogenic input is its emission from internal combustion engines in which it is used in the form of lead alkyls as an anti-knock additive in the fuel. The river fluxes of lead are estimated to be only a few times higher than the rate of combustion of lead alkyls in the northern hemisphere (Table 17.8a). About 75% of the lead issues

TABLE 17.8a
Annual lead budget (Murozumi et al., 1969).

	10^{12} g y^{-1}
World lead production (1966)	3·5
Northern hemisphere production	3·1
Lead burnt as alkyls	0·31
River-borne input of soluble lead to marine environment	0·24
River-borne input of particulate lead to marine environment	0·50

from the exhausts of the engines, the remainder being incorporated in the oil, oil filter, engine, exhaust system and silencer (Huntzicher and Friedlander, 1973). The atmospherically injected lead takes the form of an aerosol and is subsequently returned to the earth's surface either as dry fallout or as washout with rain. Over half is probably deposited in the vicinity of the injection site.

Lead concentrations in surface sea water have been significantly altered by the entry of lead aerosols in the coastal zones of the Pacific, Atlantic and Mediterranean (Chow and Patterson, 1966). In contrast to barium (an element which resembles it in geochemical behaviour and which shows monotonic increases with depth), the concentrations of lead are much higher in surface waters (Fig. 17.5). The Pacific and Mediterranean coastal upper levels have been affected to depths of about 500 m, but the impact upon the Atlantic is evident only to a shallower depth. Chow and Patterson have shown that the average lead concentration today in the coastal surface waters of the northern hemisphere is 0·07 µg kg^{-1}, which should be compared with those which have been estimated for the period before the use of lead alkyls (ca. 0·01–0·02 µg kg^{-1}). Due to the difficulty of preventing lead contamination during collection and analysis these reported values may be somewhat high (Patterson, personal communication, 1974).

Coastal sediments afford a measure of the atmospheric lead flux to the sea surface as a result of the subsequent incorporation of lead in the solid phases. Studies with Pb-210, a radioactive isotope produced in the U-238 series (Koide *et al.*, 1973), suggest that lead is removed very rapidly from the

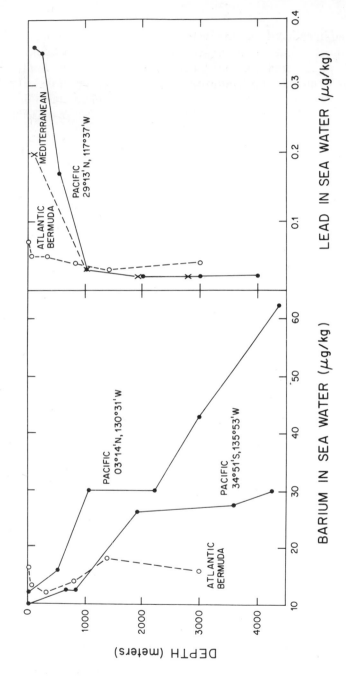

FIG. 17.5. Lead and barium profiles in sea water (Chow and Patterson, 1966).

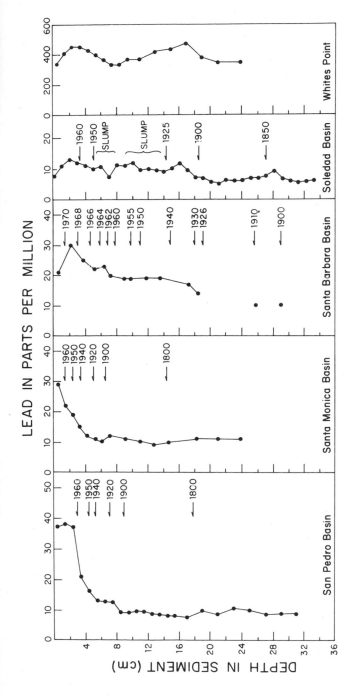

FIG. 17.6. Lead concentrations in sediments from Santa Monica, San Pedro and Santa Barbara Basins off Southern California and from the Soledad Basin off the coast of Baja California and from a site near the Whites Point Los Angeles County Outfall. The Soledad Basin may be considered a benchmark basin, as it is not adjacent to continental areas of intense industrial activity as are the Southern California Basins. (Chow et al., 1973).

water column to the sediments, probably being transferred along with biological debris. Chow *et al.* (1973) have measured by radiometric means the rates of lead accumulation in deposits off the coast of southern California (Fig. 17.6). Increased rates of lead accumulation became evident in the late 1940's, about 25 years after the introduction of lead alkyls into gasolines. Much higher values of lead were found in sediments immediately off a sewer outfall (Whites Point) than in those much further removed. This suggests that there is an additional input of lead from storm runoff and from domestic and industrial wastes introduced into the sewers. Furthermore, there are influxes to these coastal deposits from rivers. Huntzicker and Friedlander (1973) estimate that 0·4 tons of lead fall out daily from the atmosphere to a coastal area of 12,000 km^2 off Los Angeles, a value comparable to the 0·5 tons computed on the basis of the accumulation of man-generated lead in the sediments. The lead discharged through the municipal waste outfalls is estimated to be 0·55 tons per day, with the storm and river runoff of lead averaging about 0·24 tons per day (SC, 1972). All of these values have been determined independently.

The isotopic composition of lead in the marine environment can be indicative of its origin (Table 17.8b).

TABLE 17.8b
Comparative data on lead isotope ratios

Source	Ratio of Pb-206 to			Reference
	Pb-204	Pb-207	Pb-208	
California gasolines—average				
of those sold in 1964	17·92	1·145	0·4728	Chow and Johnstone, 1965
of those sold in 1968	18·08	1·555	0·4756	Chow, 1971
San Pedro sediment, 0–2 cm	18·36	1·176	0·4782	Chow *et al.*, 1973
San Pedro sediment, 30–32 cm	19·11	1·208	0·4845	
Whites Point 0–1 cm	18·09	1·166	0·4801	

The lead in gasolines is usually derived from tertiary or older lead ores and has a quite distinctive isotopic composition. The Whites Point sediments, adjacent to a sewer outfall, appear to contain lead primarily derived from gasolines. The isotopic composition of the lead of recently deposited sediments of the San Pedro Basin has been influenced by gasoline lead. The isotopic composition of lead derived from weathering processes is represented by that of the sediment at a depth of 30–32 cm deposits which accumulated before 1800.

No change in open ocean lead concentrations due to man's activities has yet been found. However, the composition of the atmosphere reflects gasoline combustion products.

Airborne particulate Pb concentrations in mid-Pacific air are about 100 times higher than those expected from crustal abundances (IDOE, 1972).

17.5.3. OTHER HEAVY METALS

The existence of anthropogenic changes in the open ocean concentrations of heavy metals has not as yet been directly established. Nonetheless, in the vicinities of river outfalls and sewers, higher concentrations are often found, not only in the water, but also in the sediments and in the organisms.

On the basis of the atmospheric burden of polluting trace metals, it is possible to estimate which ones might be expected to affect the natural levels of the upper 200 m of the ocean (IDOE, 1972 and Table 17.9). Most probable increases, as much as 30 percent, are predicted for aluminium, with cadmium, scandium, tin and manganese above 10%. However, much higher values could come about; for example, if higher atmospheric burdens are used for cadmium, a 400% increase appears possible. Since atmospheric contaminants often appear in rivers and sewage outfalls, these values give a guide to the changes which might be expected for such discharges in near-shore waters.

The history of trace metal pollution in coastal zones can be recorded in the sediments, especially for those species which are rapidly removed from the water column. Those deposits with high organic contents or with substantial contributions of fine grained sediments are especially valuable for such studies. The inner basins off the coast of California, with organic contents up to ten percent and with high contents of clays, have provided evidence of significant contributions of man-generated metals (Chow et al., 1973; Bruland et al., 1974). The deeper parts of the deposit provide baseline measurements for the chemical species under consideration. Man's influence can be seen through a comparison of concentrations in the more recently deposited materials as compared with those in the older sediments. The times at which the various strata were deposited were determined by Pb-210 and Th-228/Th-232 dating and by varve counting. Sediments from three basins off the Los Angeles area (San Pedro, Santa Monica and Santa Barbara) were assayed along with those from the Soledad Basin, off the coast of Baja California. This latter deposit, which is little influenced by industrial or agricultural activity, provides a benchmark set of data against which the influence of human activity on the southern Californian deposits can be compared. The elements Pb, Cr, Zn, Cu, Ag, V and Mo are now being introduced to the Californian Basins at a greater rate than they were several years ago. In contrast, man appears to have had little influence on the fluxes of nickel, cobalt, manganese and iron (Table 17.10).

TABLE 17.9

Possible impact of atmospheric pollutants on the marine environment (IDOE, 1972).

Element	Concentration				Estimated % increase of trace elements in upper 200 m of ocean[2]	
	Open ocean ($\mu g\,l^{-1}$)	U.S. urban air ($ng\,m^{-3}$)	Enhancement ratio in air[1]	Ratio air/ocean	High	Most probable
Pb	0·02	1000	2300	50,000	—	—
Al	1	1500	0·5	1500	200	30
Cd	0·02	20	1900	1000	400	20
Sc	0·001	1	1	1000	80	20
Sn	0·02	20	280	1000	—	20
Mn	0·3	200	6	700	60	15
Fe	5	2000	1·00	400	80	8
La	0·01	3	3	300	—	6
V	1	200	42	200	40	4
Zn	3	700	270	200	25	4
Cu	2	200	83	100	100	2
Ag	0·01	2	830	200	—	4
Cr	0·3	40	11	130	20	3
Be	0·005	1	10	200	—	4
Sb	0·2	20	2800	100	10	2
In	0·001	0·1	29	100	—	2
Ti	1	100	0·5	100	4	2
Co	0·03	2	2	70	4	1
Se	0·1	4	2500	40	5	0·8
Hg	0·1	3	1100	30	8	0·6
W	0·1	5	93	50	—	1
Ga	0·02	1	2	50	—	1
Ni	2	30	12	11	4	0·3
Cs	0·3	4	37	10	—	0·3
Ta	0·02	0·2	3	10	—	0·3
As	2	20	310	10	0·8	0·2
Mo	10	10	190	1	—	0·02
U	3	0·1	· 1	0·03	—	Negligible

[1] The ratio of the concentration in urban particulates of the specific element to iron, divided by the ratio of the same elements in average crustal material.

[2] The % increase represents the magnitude by which these trace elements could have been increased in the upper 200 m of the ocean based on the anthropogenic lead concentration in this layer. The high estimates are those which might be expected in oceanic regions adjacent to certain large urban areas.

Many of these heavy metals appear to be atmospherically transported from the continents to the oceans. Their fluxes to the sediments are similar to those found for heavy elements in atmospheric precipitation collected near these basins (on Santa Catalina Island) during a six-month period in 1966

TABLE 17.10

Fluxes of heavy metals into coastal marine basins (Bruland et al., 1974).

Element	Flux	San Pedro	Santa Monica	Santa Barbara	Soledad	Rainfall
				(Fluxes in $\mu g\,cm^{-2}\,y^{-1}$)		
Pb	Anthropogenic	1·7	0·9	1·8		
	Natural	0·26	0·24	1·2	0·23	
	rainfall					1·3
Cr	Anthropogenic	3·1	2·6	2·2		
	Natural	2·8	2·1	10·7	4·6	
Zn	Anthropogenic	1·9	2·3	1·7		
	Natural	3·1	2·8	9·7		
	rainfall					2·8
Cu	Anthropogenic	1·4	1·1	1·1		
	Natural	1·2	1·0	2·6	1·4	
	rainfall					0·5
Ag	Anthropogenic	0·09	0·09	0·10		
	Natural	0·05	0·03	0·11		
V	Anthropogenic	1·5	2·6	7·8		
	Natural	3·5	3·4	13·6	4·6	
Cd	Anthropogenic			0·07		
	Natural			0·14		
Ni	Natural	1·6	1·3	4·1	2·3	
	rainfall					0·24
Co	Natural	0·33	0·26	1·0	0·17	
Mn	Natural	13·0	8·0	24·0	7·0	
	rainfall					0·36
Fe	Natural	1260·0	1200·0	3060·0	840·0	
Mo	Anthropogenic		0·82			
	Natural		0·08			

and 1967 (Lazrus *et al.*, 1970). From a comparison of the relative amounts of trace metals in precipitation from a variety of areas with the varying industrial activity, the latter investigators concluded that lead, zinc, copper, iron and manganese in atmospheric precipitation are derived primarily from human activity. In contrast, the sole source of nickel appears to be weathering processes. The natural fluxes of iron and perhaps, manganese are apparently high enough to mask any contributions from industrial activity.

17.6 Synthetic Organic Chemicals

In the U.S. about 100×10^6 tons of synthetic organic chemicals, tar, tar crudes and crude products from petroleum and natural gas were produced in 1968 (USTC, 1971). The world figure is probably about three times this value i.e. about 300×10^6 tons per year. Of the many thousands of organic chemicals produced, only a very small percentage have escaped to the surroundings and posed threats to the well-being of living organisms. Of these, a small percentage persists in the oceans and this has led to concern about the well-being of ecosystems and about the possible return of these compounds to man through ingestion of food from the sea. Two groups of halogenated hydrocarbons have been the focus of attention: the DDTs and the PCBs (polychlorinated biphenyls). The former have been implicated in the decline in the populations of marine birds by their interference with hormonal activities; the latter have been associated with one epidemic in Japan (Yusho).

17.6.1. DDT and its residues

DDT and its metabolites have been dispersed throughout the marine environment following use of the parent substance in agriculture and public health activities over the past three decades. No figures are available for the world production of DDT, and most assessments have involved extrapolations of the U.S. data (Table 17.11). Restrictions upon the use of DDT in the U.S., Europe and Asia in the late 1960s and early 1970s, have probably resulted in a decrease in world production and utilization. The annual production rate in the late 1960s was probably about 10^{11} g y^{-1} and the integrated world production up to that time was ca. 2×10^{12} g (NAS, 1971).

The existing evidence supports the theory that the atmosphere is the primary transport path for DDT residues from the continents to the oceans. Vaporization from plants and soils, or direct entry to the atmosphere during its application in agriculture or public health activities, initiates the transfer process. It is difficult to determine the rate at which DDT is transferred to the ocean. One way in which this can be done is to couple the DDT residue concentration in rain with the annual precipitation over the oceans of $3 \cdot 0 \times 10^{20}$ ml y^{-1} (NAS, 1971). In the initial calculations the average concentration of 80 ng of DDT residues per litre in rain over Great Britain

TABLE 17.11

U.S. production of chlorinated hydrocarbons, kilotons y^{-1}.

Year	DDT[a]	Aldrin-toxaphene group[a]	PCBs
1971			18·4
1970			38·6
1969			34·6
1968	63·4	52·7	37·6
1967	47·0	54·6	34·2
1966	64·2	59·3	30·0
1965	64·0	54·0	27·4
1964	56·2	47·9	23·1
1963	81·3	48·2	20·3
1962	75·9	48·3	19·0
1961	77·9	47·2	18·4
1960	74·6	41·2	18·9
1959	71·2	39·5	
1958	66·0	44·7	
1957	56·6	34·3	
1956	62·6	39·4	
1955	59·0	35·0	
1954	44·2	20·5	
1953	38·4	—	
1952	45·4	—	
1951	48·2	—	
1950	35·5	—	—
1949	17·2	—	—
1948	9·2	—	—
1947	22·5	—	—
1946	20·7	—	—
1945	15·1	—	—
1944	4·4	—	—
Total	1,220·0	670·0	—

[a] Chemical Economics Service, Department of Business and Industrial Economics, Stanford Research Institute (1951).

in 1966 and 1967 was employed. This figure suggests that a total of $2·4 \times 10^{10}$ g of DDT residues was being transported to the oceans annually; this amounts to about one quarter of the estimated yearly production at that time. The order of magnitude of the value used in this approximate calculation was later complemented by other measurements of DDT concentrations in rain of Hawaii in 1970–71 (range $1–13$ ng 1^{-1}, average value of 5 ng 1^{-1}); of Ohio in 1965 (range $70–340$ ng 1^{-1} with an average value of 187 ng 1^{-1}); and of

Central England in 1965 (range of 2–4 ng l^{-1} with an average value of 3 ng l^{-1}) (Bevenue *et al.*, 1972). Even though there is a wide variation of DDT residue concentrations in rain, the high values in the late 1960s clearly indicate the existence of an atmospheric veil of halogenated hydrocarbon pesticides.

DDT has been found to be associated with atmospheric dusts collected from a site facing into the North East Trades at Barbados (Risebrough *et al.*, 1968; Seba and Prospero, 1971). The dusts consist principally of crustal rock debris, which derived from a different area from that at which the DDT originated. Risebrough *et al.* (op. cit.) have calculated that 6×10^5 g of aerosol-borne chlorinated hydrocarbons are annually deposited into the Equatorial Atlantic following transport in the Northeast Trades. This value may be a considerable underestimate as only DDT residues adsorbed to the dust were measured and any present as vapour or very small particles may well have passed through the collecting screen.

The amount transported by rivers appears to be several orders of magnitude less than that carried in the atmosphere. The NAS (1971) report suggests a maximum of 3.7×10^9 g y^{-1} or about 3% of the annual production in the late 1960s, could be transported to the oceans via the rivers, if the average concentration of DDT residues in river waters is 100 ng l^{-1}, the maximum concentration found in a survey of the western U.S. rivers. However, Mississippi delta water contains only about 5 ng l^{-1} (Seba and Corcoran, 1969). If this is a reasonable estimate for the concentration in the rivers of the world, the total amount transported by them would be of the order of 2×10^8 g per year. There is, in fact, no evidence that river run-off or sewer discharges compete with the atmosphere in moving DDT and its metabolites about the environment.

17.6.2. PCBs

The term PCBs (polychlorinated biphenyls) refers to complex mixtures of partially or wholly chlorinated biphenyl. In the U.S., the sole producer is the Monsanto Chemical Corporation which markets PCB mixtures under the trade name of "Aroclor". These commercial products, which generally contain between 40 and 60% of chlorine, are characterized by a four digit number; the first two digits indicate that the compounds involved are polychlorinated biphenyls (12) and the last two digits give the approximate percentage of chlorine in the mixture. Thus, Aroclor 1254 is a polychlorinated biphenyl containing 54% (wt) chlorine. These products were introduced in 1930 and peak U.S. production of 38,000 metric tons occurred in 1970 (Table 17.11). In 1970, Monsanto restricted sales of these compounds to "closed system" uses as a consequence of widespread concern about the build-up of PCBs in the environment.

The important industrial uses of PCBs depend upon their unique character-istics of high stability, noninflammability, low water solubility, low volatility, high dielectric constant and plasticizing ability. As a result, they have been used as dielectric fluids in capacitors and transformers, in hydraulic fluids, especially in systems involving high temperatures, and in aircraft as heat transfer fluids. In addition, they have been utilized as plasticizers and resin extenders in adhesives, sealants, paints and printing inks. Before the restric-tions upon sales, about 60% of the PCBs were used for closed-system electric and heat transfer equipment, 25% for plasticizers, 10% for hydraulic fluids and lubricants, and less than 5% for such applications as surface coatings, adhesives, printing inks and pesticide extenders.

It is difficult to compute the past or present world production rate. PCBs have been, or are being, manufactured in Japan, Spain, Italy, France, England, West Germany, East Germany and the Soviet Union. To a first approxima-tion the world rate is probably 2–3 times the U.S. rate and was about 80–120 thousand tons per year for the years immediately before 1970.

The routes of the PCBs to the oceans have been identified and semi-quantitatively assessed by Nisbet and Sarofim (1972). The entries through the atmosphere, rivers and domestic and industrial sewage outfalls are the dominant ones. Incineration of wastes appears to be an important mechanism for mobilizing substantial amounts of these substances to the atmosphere. Nisbet and Sarofim (op. cit.) have estimated that an amount equal to 80% of the 1970 sales in North America, or about 28×10^3 tons per year was eventually discharged to the environment in the following ways:

$1–2 \times 10^3$ tons	vaporization of plasticizers
$4–5 \times 10^3$ tons	leakage and disposal of hydraulic fluids and lubricants and of smaller amounts of heat transfer and transformer oils
22×10^3 tons	disposal via incineration, dumping and sanitary landfilling. Perhaps 10–20% (3×10^3 tons) were destroyed by burning

These figures can be transposed into estimates of environmental fluxes as follows:

$1–2 \times 10^3$ tons year^{-1}	into the atmosphere
$4–5 \times 10^3$ tons year^{-1}	into fresh and coastal waters
18×10^3 tons year^{-1}	into dumps and landfills.

It is difficult to establish incontrovertible models for describing the distribu-tion and dissemination of a pollutant throughout the environment because of a shortage of information about the amounts of it entering and exchanging between the various reservoirs. In addition, information is often lacking for the persistence of the compound in the reservoirs. Nevertheless such models are useful because they draw attention to information gaps and because they assist understanding of pollutant–environment interaction. The model developed by Nisbet and Sarofin (1972) for the environmental distribution of PCBs has not only led to some reasonable, and in one case verifiable, conclusions, but also to the recognition of a need for future research.

The possible routes of transport of PCBs in the environment are indicated in Fig. 17.7, and Table 17.12 gives estimates of the rates of PCB input and the total accumulation by 1970 in various reservoirs. Most of the PCBs entering the atmosphere were Aroclors 1248 to 1260 from plastic resin vaporization together with some Aroclor 1242 volatilized in dump-burning. The losses

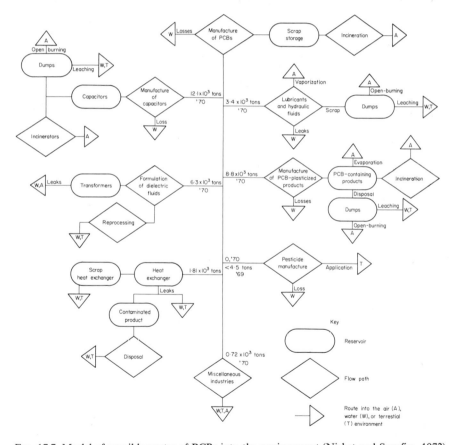

FIG. 17.7. Model of possible routes of PCBs into the environment (Nisbet and Sarofim, 1972).

TABLE 17.12

Gross estimates of rates of input and accumulation of PCBs in North America in 1970 (Nisbet and Sarofim, 1972).

Input	Rates (tons y^{-1})	PCB grade
Vaporization of plasticizers	$1-2 \times 10^3$	Mainly 1248 to 1260
Vaporization during open-burning	4×10^2	Mainly 1242
Leaks and disposal of industrial fluids	$4-5 \times 10^3$	1242 to 1260
Destruction by incineration and open burning	3×10^3	Mainly 1242
Disposal in dumps and landfills	$1\cdot8 \times 10^4$	1242–1260
Accumulation in service	7×10^3	1242–1254

Reservoir	Total accumulation (tons)	
	PCBs	ΣDDT
Soil (excluding dumps)	$1\cdot5 \times 10^4$	3×10^5
Oceans (adjacent to North America)	$1\cdot5 \times 10^4$	10^5
Fresh water (dissolved or in suspension)	10^2	$\sim 10^2$
Fresh water sediment	2×10^4	?
Biota	$< 10^3$	$< 10^3$

into waters involve Aroclors (1242–1260) which are used as hydraulic fluids and lubricants.

Of the yearly 1·5 to 2·5 kilotons entering the atmosphere from the U.S., most will be deposited within 2 or 3 days onto the land mass and coastal areas adjacent to the urban sites of entry. The terrestrial fallout is estimated to be 1 to 2 kilotons per year with about 0·5 kilotons entering the marine environment. The amount of PCBs transported to the ocean dissolved or in suspension in river waters is estimated to be 0·2 kilotons per year.

Nisbet and Sarofim have suggested three processes by which PCB mixtures may change after release into the environment:

(a) Fractionation. The water solubilities and vapour pressures decrease with increasing chlorine content. For these reasons the less chlorinated PCBs will be more widely disseminated.

(b) Photolysis and chemical decomposition. The more highly chlorinated isomers are more readily broken down than the less chlorinated ones; this would be expected to lead to a decrease in the amounts of the former relative to the latter in the environment.

(c) Metabolism and excretion. The more highly chlorinated PCBs are, in general, taken up and retained more strongly than the less chlorinated ones. It is not clear whether this deficiency in less chlorinated isomers is due to preferential metabolism or to excretion.

A survey of the concentrations of the various PCB isomers in water and organisms by Nisbet and Sarofim has indicated that the isomers containing few chlorine atoms are disappearing; probably as a result of differential metabolism, perhaps microbially mediated.

Since the initial production of these compounds in 1930, 5×10^5 tons of PCBs have been sold in North America. The cumulative losses to the environment are estimated to be about 3×10^4 tons to the air, 6×10^4 tons to fresh and coastal waters and 3×10^5 tons to dumps and landfills. On the basis of their degradation in the environment, it is estimated that about one-third of the PCBs entering the air and about one-half of those entering natural waters have disappeared. By analogy with DDT, Nisbet and Sarofim have estimated that of the total PCBs introduced to the air (5×10^3 tons), one quarter will have been transferred to the sea. Because of the prevailing westerlies over the U.S., most of this would have been deposited in the Atlantic Ocean. Somewhat later, Harvey *et al.* (1973) estimated, on the basis of their own measurements, that there are 2×10^4 tons of PCBs in the upper 200 m of Altantic waters. This is, perhaps, a substantial underestimate for the whole ocean, because if PCBs have a residence time in surface waters of only a few years and a persistence somewhat longer, the bulk of the PCBs would be in deeper waters.

Nisbet and Sarofim (op. cit.) also compared the PCB and DDT residue (ΣDDT) concentrations in the various natural reservoirs to ascertain the internal consistency of their model and to define more clearly the impact that halogenated hydrocarbons of this type are making on the environment. By the late 1960's the input of PCBs to the seas around North America, primarily the Atlantic Ocean, through localized discharge and aerial fallout, totalled about $1 \cdot 5 \times 10^4$ tons; the total DDT influx also principally into the North Atlantic was probably close to 5×10^5 tons. Thus, the DDT/PCB ratio should be about 30. However, the ratio observed for marine invertebrates (Table 17.13) appears to be at least an order of magnitude less. Explanations for this apparent discrepancy involve differences in behaviours of the collective of DDT species and of the PCBs either in the atmosphere and waters or in biological processes. In any event, the DDT residues appear to be more easily degraded than the PCB's.

TABLE 17.13

PCB levels and ΣDDT/PCB ratios in marine vertebrates (Nisbet and Sarofim, 1973).

Area	Mean PCB level in fish (ppm)	ΣDDT/PCB ratio in fish, fish-eating birds and/or mammals
Long Island Sound	1·2	0·08–0·18
Bay of Fundy	0·5	0·4
Atlantic Ocean	0·1	0·2–0·5
Puget Sound	0·16	1·1
San Francisco Bay	0·1–1·2	1–3
Californian Coast	0·02–1	5
Gulf of California	—	10
Gulf of Panama	—	1–2
Pacific Ocean (Galapagos and Hawaii)	0·03	> 10

Atmospheric transport provides a mechanism which accounts for the high PCB levels in marine organisms from both hemispheres, even though the primary source is in the northern hemisphere. The vapour pressure of Aroclor 1254 is similar to that of DDT (Nisbet and Sarofim, 1972), and it is reasonable to assume that the paths of the PCBs from the continents to the oceans are similar to those of DDT and its residues. All these halogenated hydrocarbons can associate with particulates in the atmosphere or form aerosols and subsequently be removed during precipitation. Once these compounds reach the sea surface, they may be incorporated in surface films (slicks). As with DDT and its metabolites, the entry of PCBs into the food chain may occur through sorption or ingestion of these slick materials by the myriads of micro-organisms that inhabit the upper few millimetres of the sea.

Analyses of zooplankton indicate PCB concentrations spanning three orders of magnitude (Table 17.14) from 0·7–1300 ppb (wet weight). The DDT residue levels in the samples range from 0·2–20·6 ppb (wet weight). These measurements give a general idea of the concentrations in plankton. However, there is considerable uncertainty about the data. Different net mesh sizes, mesh materials, depths of tow, etc. were employed in the sample collections. In addition, the recovered materials contained paint flakes and tar balls, either of which could have introduced the halogenated hydrocarbons to the analysed samples. In general, the concentrations of PCBs and DDT residues in the seston are similar; those of the PCBs being slightly higher.

TABLE 17.14

DDT and PCB in Plankton (IDOE, 1972).

Area	No. of samples	Totals ($\mu g\, kg^{-1}$, wet weight)	
		DDT	PCB
Sargasso Sea	4	0·7	7–450
S. Atlantic	4	0·2–2·6	19–638
N.E. Atlantic	22	2–26	10–110
Clyde, Scotland	15	6–130	40–230
California, U.S.A.	15	0·2–206	0·7–30
California, U.S.A.	250	–	100–1300
Iceland (phytoplankton)	1	–	1500

PCBs and DDT residues have been found in nearly every individual species of fish examined from the North and South Atlantic, Denmark Straits, Gulf of Mexico, Caribbean Sea, Northeast Pacific, Scottish west coast and the Baltic Sea (IDOE, 1972). Unexpectedly, samples from the open South Atlantic had concentrations of PCBs and DDT residues nearly as high as those from the North Atlantic (Table 17.15). In contrast, seabirds which feed upon fish in the southern hemisphere have lower levels of PCBs than do their North Atlantic counterparts. The explanation probably lies in the rates of meridional mixing of atmospheric masses and the consequent transfer of materials between hemispheres (IDOE, 1972).

Fish from the open ocean are richer in PCBs than in DDT residues; in coastal areas the body burdens of the two types of compounds are more nearly equal. The reason for this is at present unknown. The absolute concentrations of both groups of compounds are usually higher in organisms from coastal areas, near the sites of continental release.

It seems probable that the concentrations of PCBs and DDT residues in the lipids do not increase as the food chain is ascended (IDOE, 1972). Possibly the organisms can take up and lose these halogenated hydrocarbons by partitioning across a permeable body surface, a process that could occur at rates and concentrations unrelated to the trophic position of the organism.

The fact that PCBs at very low levels can have an impact on marine organisms has been demonstrated by laboratory experiments. Thus, they inhibited the growth of the diatom *Thalassiosira pseudonana*, in unialgal culture in sea water at concentrations of 25 ppb, but not at 10 ppb. In contrast, growth was inhibited by DDT at 100 ppb, diminished at 50 ppb, and not affected at 25 ppb in similar experiments (Mosser *et al.*, 1972). The green alga

TABLE 17.15

DDT and PCB concentrations in fish, shellfish and zooplankton (IDOE, 1972).

Region	Material	Concentration	
		PCBs	DDT residues
		$\mu g\, kg^{-1}$ (wet weight)	
Open North	Pelagic fish		
Atlantic	muscle	1–10	0·6–3
	liver	1000–6000	95–4800
	Midwater fish	8–59	3–12
	and crustacea		
Open South	Midwater fish	2–14	1–8
Atlantic	and crustacea		
Denmark	Groundfish		
Strait	muscle	2–360	3–30
	liver	300–1000	9–260
Northwest	Groundfish		
Atlantic	muscle	37–187	3–74
shelf	liver	1870–21,000	390–2680
Gulf of	Whole fish	<1–530	1–150
Mexico	or muscle		
Northeast	Euphausiids	9·2 (mean)	2·7 (mean)
Pacific	Pink shrimp	23 (mean)	2·5 (mean)
	Flatfish	23 (mean)	10·8 (mean)
Scottish	Fish muscle	<100–1500	<30–480
west coast	Fish liver	200–42600	70–5800
Baltic Sea	Herring	150–1500	100–1500
	Cod	16–180	9–340

Dunaliella tertiolecta in unialgal culture was not affected by either of the above chemicals at the concentrations cited. In mixed cultures, exposure to PCBs or DDT significantly diminished the competitive success of *T. pseudonana* and increased that of *D. tertiolecta*. If halogenated hydrocarbons were present in the sea at these concentrations they could influence the makeup of the algal community by suppressing sensitive species and permitting resistant forms to become dominant. Since many species of zooplankton graze preferentially upon specific species, the health and distribution of organisms higher in the food web could be affected.

Juvenile shrimps, when exposed to 5 ppb of Aroclor 1254 in flowing sea water, suffered a 72% mortality within 20 days. A composite sample of these shrimps contained 16 ppm (wet weight) of the PCB (Duke *et al.*, 1970). These experiments were carried out following an industrial leakage of PCBs into Escambia Bay, Florida, where waters contained less than 1 ppb and the shrimps contained 2·5 ppm (wet weight), values below lethal ones. Juvenile blue crabs were not as sensitive to the shrimp to the PCB exposure, only one of the 20 crabs dying during the 20-day experiment. The average whole body residue in 5 of the crabs, was 23 ppm (wet weight) with a range of 18 to 27 ppm. The persistence of the PCBs in animal tissues is indicated by an experiment in which 6 crabs were held in clean water for four weeks following the PCB exposure and the average body burdens only dropped from 23 to 11 ppm.

Human deaths and illness have occurred as a result of the ingestion of PCBs (Kuratsune et al., 1972). A rice oil in Japan was contaminated through leakage of PCBs from heat-exchangers used in the manufacturing process during 1968. In some instances 0.5–2 g of PCBs were eaten. More than 5000 people allegedly suffered from a chloracne-like skin eruption. In addition, some of those who consumed the rice oil showed such symptoms as palsy, fatigue and vomiting spells. The deaths of two adults and of two babies were attributed to the uptake of these chemicals. Several babies were born with abnormal pigmentation of the skin. There have been no other reports of effects on human health and there is no evidence that present body burdens give cause for concern.

PCBs and DDT residues have been measured during 1972 in North Atlantic Ocean waters collected from the surface and from deep water at latitudes between 26° and 63° N (Harvey, et al., 1973, see also Section 12.3.2.2). For the PCBs the closest matching commercial mixture, Aroclor 1254 (containing 54% chlorine) was used as a standard. An average PCB concentration of 20 ngl^{-1} was found for the upper 20 m with 35 ngl^{-1} in the surface and 10 ngl^{-1} at 200 m. PCBs were observed in samples taken as deep as 3000 m. Their concentrations in the surface waters of the Sargasso Sea were slightly lower (27 ngl^{-1}) than those in other parts of the North Atlantic. If the volume of the North Atlantic is assumed to be 10^{18} l, then the burden of PCBs in these waters is 2×10^4 tons, a value similar to the U.S. annual production in 1971 (Table 17.11), and perhaps a third of the world production during that year. If a similar amount occurs in the North Pacific Ocean, it is evident that a substantial proportion of the production of PCBs enters the marine environment.

No gradients between land and the ocean were observed by Harvey et al. for the PCBs. This, coupled with the wide distribution of these compounds supports the hypothesis that the atmosphere is the predominant transport path.

The total DDT/PCB ratio was found to be less than 0·05. The DDT family, if present, was at concentrations less than 1 ngl^{-1}. U.S. production ratios (DDT/PCB) in the 1960s appeared to be around 2 to 4. Since production of

DDT was reduced in 1969, a much lower value for the ratio might be expected for the 1970s. If the primary sources of the DDT and the PCBs were in the U.S. and subsequent fallout occurred into the North Atlantic, then the observed ratio is reasonable.

The residence times of these halogenated hydrocarbons in surface waters must be short, most probably of the order of years. Association with living organisms or sorption on particulates may lead to their transport to deeper waters

17.6.3. LOW MOLECULAR WEIGHT HALOGENATED HYDROCARBONS

Low molecular weight aliphatic halogenated hydrocarbons are introduced to the atmosphere through volatilization after being used as intermediates in chemical synthesis, as solvents, as aerosol propellants, as refrigerants and as cleaners. A substantial fraction of these chemicals, other than those used in the chemical industry, will eventually enter the atmosphere, and part will subsequently dissolve in the surface waters of the ocean. Many of these compounds are refractory to biological and photo-chemical degradation and may have long lifetimes in both the atmosphere and oceans.

Murray and Riley (1973) have examined the concentrations of these compounds in the atmosphere and over the sea close to Great Britain. Compounds detected included chloroform, carbon tetrachloride, trichloroethylene and perchloroethylene. The samples also showed evidence of other, as yet unidentified, halogenated hydrocarbons. The average concentrations in marine air and waters were:

	$CHCl_3$	CCl_4	$CHCl = CCl_2$	$CCl_2 = CCl_2$
Airs (ng m^{-3})*.	–	410	200	600
Waters (ng l^{-1})	8	0·14	7	0·5

The fluorocarbons (CCl_2F_2 (Freon-12) and CCl_3F (Freon-11) have been used extensively during the past decade as propellant solvents in aerosol dispensers, and have been dispersed to the atmosphere and surface marine waters (Lovelock et al., 1973). U.S. production in 1969 was 0·17 megaton for Freon–12 and 0·11 megaton for Freon–11 (USTC, 1971) and the world production was perhaps 0·5 and 0·3 megatons respectively. The levels of the Freon-11, methyl iodide and carbon tetrachloride were measured between Great Britain and Antarctica in 1971 and 1972 by Lovelock et al. as follows:

Substance	CCl_3F	CH_3I	CCl_4
Concentration in air in ppb	49·6	1·2	71·2
Surface water concentration (ppb)	0·0076	0·135	0·060

* Average values for 3 samples over Irish Sea (Murray, private communication).

There were no substantial changes with latitude, although there were some local variations. The source of the methyl iodide has been found to be marine algae. There is no obvious marine source for the carbon tetrachloride. Its concentrations in the southern tropical regions were as high as those in northern temperate regions. Its origin is yet to be determined.

Mixtures of short-chain chlorinated aliphatic hydrocarbons, waste by-products from the manufacture of vinyl chloride, are being dumped into the sea and have been implicated in the deaths of plankton and fish (Jernelöv *et al.*, 1972). About 75,000 tons of these wastes (EDC-tars) are produced annually in western Europe and the world production rate is probably about 300,000 tons per year. The 25 or so compounds which comprise EDC tars originate from impurities present in the ethylene which is chlorinated to vinyl-chloride for subsequent polymerization to polyvinyl chloride. After the chlorination, the waste products are separated from the vinyl chloride. When they are dumped in the sea, the EDC-tars which have a density of $1·25–1·40$ g ml^{-1} begin to sink and simultaneously to degrade. They tend to adhere to particles in the sea water and probably enter the food chains with them. Most fish specimens caught in the North Sea contain these chlorinated hydrocarbons.

17.7. MARINE LITTER

A variety of plastic, rubber and metal wastes are accumulating in all parts of the marine environment and are creating not only an aesthetic nuisance but also perhaps endangering the well-being of living organisms.* The marine litter enters the ocean system from direct discharge, dumping of wastes, sewer outflows from cities, ship discharges and atmospheric fallout. Much of this material is relatively indestructible and will persist for long periods in the sediments and the water column. A non-biodegradable plastic sheet which has settled on the sea floor can make the area uninhabitable for both plants and animals of the benthos.

Polystyrene spherules, averaging $0·5$ mm in diameter (with a range between $0·1$ to $2·0$ mm) were observed in the coastal waters off southern New England (Carpenter *et al.*, 1972). The spherules are used for the fabrication of plastic ware and their source is probably the polystyrene producers of the eastern U.S. The highest concentrations were found in the waters of Niantic Bay which contained an average of one spherule m^{-3}. The highest concentration

---

* This litter also provides habitats for a variety of organisms, from algae to barnacles. It is possible that these new ecological niches may enhance biological productivity, although the value of the litter in this respect may be more than offset by increased erosion of beaches caused by enhanced scouring. These aspects merit further study.

was 14 per m^3. PCBs were detected on the spheres at a concentration of 5 ppm and they probably were sorbed from sea water. The spherules were also found in the gut of fish from the area.

Plastic particles, many of which are cylindrical pellets 0·25 to 0·5 cm in diameter have been found at concentrations averaging 3500 pieces or 290 g km^{-2} in the western Sargasso Sea (Carpenter and Smith, 1972). Many of the plastic fragments were brittle, indicating a loss of plasticizer by weathering. Hydroids and diatoms were often attached to their surfaces.

An assortment of larger objects was sighted by Venrick *et al.* (1973) during a sweep of 12·5 km^2 over the central North Pacific. Fifty three man-made objects were recorded; these included 6 plastic bottles, 22 plastic fragments, 12 glass fishing floats, 4 glass bottles, rope, an old balloon, finished wood, a shoebrush, a rubber sandal, a coffee can and 3 paper items.

Elastic bands and threads are affecting the health of fish and birds. Elastic bands have ringed sterlets in the Danube delta (Anon. 1971) and worked their way into their flesh. The bands produce mucus-covered ulcers and appear to cause damage to the gills. Rubber thread cuttings are apparently mistaken for fish by puffins which swallow them (Parslow and Jefferies, 1972). Four of six puffins found dead on beaches were found on examination for their pollutant concentration to contain strands of elastic thread in their alimentary tracts. It is unlikely that these were the cause of death. This probably resulted from oil pollution and in one instance from the striking of wires. The threads which are often stranded have a diameter of 0·8 mm and are 50 to 100 m in length They are used in the garment industry and those found in the sea are most probably wastes from the manufacturing processes. It is uncertain to what extent they affect the health of the puffins or increase the bird's mortality rate.

Drums containing a wide variety of synthetic organic chemicals are dumped into the North Sea. These compounds (Greve, 1971), will be released to the waters and sediments when their containers rupture. These drums are not only disposed of in deep water, but also in the shallow areas where they are picked up in the nets of fishermen. The drums contain such wastes as lower chlorinated aliphatic compounds, vinyl esters, chlorinated aromatic amines and nitrocompounds and the pesticide Endosulfan.

Mustard gas $[(ClCH_2CH_2)_2S]$, (bombs from German World War II ammunition depots) disposed of into deep abysses of the Baltic has been implicated in a number of cases of acute food poisoning associated with the eating of contamined fish eggs (Garner, 1973). In another type of incident a fishing vessel picked up a bomb and leakage from it contaminated the fish while the crew were cleaning them. As a result some crew members suffered skin burns. Other trawlers have hauled in such bombs but have immediately returned them to the sea. The total weight of the mustard gas bombs which have been dumped in the Baltic appears to be about 20,000 tons.

17.8. CONTROL OF ANTHROPOGENIC INPUTS TO THE ENVIRONMENT

Prime considerations in the disposal of the waste products of society must be the maintenance of the well-being of man and other living organisms, and the protection of natural resources. To minimize risk to public health, discharges of mercury and radioactive nuclides are subject to strict legal control in some countries. The dissemination of DDT in the U.S. and other countries has been restricted in order to prevent it and its residues accumulating in organisms and to maintain their viability. Oil drilling off recreational areas has been curtailed for aesthetic reasons. In the development of rational control programmes which ensure minimal risk to society from the accumulation of toxic materials in the oceans the following steps are involved.

1. Identification of the substance or substances yielding harmful or unwanted effects by ingestion in food or by exposure.

2. Determination of acceptable environmental levels.

3. Determination of major discharge sites, proper regard being paid to the ways in which the substance(s) in question are distributed to the various environmental domains.

4. Investigating the ways in which the substance(s) are concentrated within the food web.

5. Imposition of regulations controlling the discharge so that acceptable levels are not exceeded.

6. Surveillance programs.

7. Enforcement of the legal limits on the discharge.

Acceptable environmental levels of toxic materials are difficult to determine. Permissible levels are often obtained by extrapolation from the results of experiments on laboratory animals which were exposed for short periods of time to abnormally high amounts of the given substances. Investigations involving long term exposures to low levels of pollutants are often extremely costly and their results are frequently difficult to interpret. An understanding of the distribution of pollutants throughout natural reservoirs (including organisms) is an important prerequisite to a management programme. Since there are limitations on the numbers of samples which can be assayed, the "critical pathways" approach is frequently used to study the more important routes which can lead to harmful ecological effects. For example, the high level of enrichment of zinc in oysters and clams offers a means of ascertaining whether potential hazards are arising from the discharge of radioactive zinc-65 from coastally-sited nuclear reactors. The consumption of these bivalves will introduce Zn-65 to the human body. Permissible environmental levels can thus be based on considerations of the consumption of these shellfish and their Zn-65 burdens. Similarly, the permissible levels of the extremely toxic element plutonium can be assessed from its concentration in edible algae.

The currently accepted permissible environmental level of mercury is set by the danger to human health arising from the consumption of fish containing methyl mercury compounds. A detailed consideration of this general approach is worth while as it has been used in establishing acceptable levels for the rate of release of radioactive nuclides to the marine environment, and it will most probably be useful in the future for the regulation of other environmental releases.

This approach is based on a model in which man is considered as a methyl mercury reservoir (Stock, 1971). When the body burden reaches a certain value, there is a high probability that the neurological symptoms of methyl mercury poisoning (the Minimata Bay Disease) will appear. Either the highest values found in clinically asymptomatic persons or in the lowest values in persons with detectable symptoms may be used as guides to permissible levels. Analyses of the methyl mercury contents of blood cells and, less effectively, of the whole blood or the hair can serve as indices of accumulation. Asymptomatic persons in Japan, Sweden and Finland carry up to 0·6 to 0·7 parts of Hg per million parts of whole blood and up to 200 parts of Hg for each million parts of hair. In cases in which poisoning occurred in Japan, the whole blood contained more than 0·2 ppm of Hg and the hair more than 200 ppm. These data are derived from a rather small number of cases, and the overlap probably emphasizes the differences in response of any given individual. Such epidemological investigations have indicated that the lowest mercury levels for the onset of symptoms are of the order of 0·2 ppm in the whole blood, which is equivalent to about 0·4 ppm in the blood cells. Studies with fish eating populations indicate that this corresponds to prolonged consumption of 0·3 mg of methyl mercury day^{-1}.

It is assumed that, after an initial buildup period, the continued consumption of methyl mercury at a constant rate results in a "steady state" level in the body at which the intake of methyl mercury is balanced by the amount excreted. Tracer experiments conducted on humans using extremely small amounts of radioactive methyl mercury indicate that its biological half life is between 70 and 90 days. This is equivalent to the elimination of about 1 % of the body burden per day. It has been estimated that at the onset of Minimata disease the level in the brain tissue of adults who succumbed was about 5 ppm. For a brain weighing 1·5 kg this corresponds to a total of about 80 mg of Hg in the entire body for a man having a body weight of 70 kg, and this combined with an excretion or intake rate of 1 % per day leads to a dosage rate of 0·8 mg of Hg as methyl mercury chloride day^{-1}.

The higher mercury level derived from brain studies on patients who *died* is concordant with the lower one based upon patients with mild symptoms. The agreement is remarkable in view of the uncertainties surrounding both the data and the assumptions made in using them. Emphasis should be placed upon the fact that a large number of persons who did not experience symptoms had mercury blood contents greater than those of some of the diseased individuals. Such data can be used, with the application of appropriate safety factors, to take into account the variability in the response of individuals to a given body burden of methyl mercury, in obtaining estimates for an acceptable daily intake of mercury.

A safety factor of ten has been proposed by a Swedish group of experts who had access to all available epidemological and physiological information from both Sweden and Japan (Stock, 1971). It has been offered with the proviso that it is subject to re-evaluation in the light of new data. Thus, the lowest whole blood value of 0·2 ppm of Hg and the lowest exposure of 0·3 mg of Hg per day should be reduced to 0·02 ppm of Hg and 0·03 mg of Hg per day respectively to provide adequate protection.

What are the sources of mercury in the diet of man? Clearly, this varies not only from country to country, but also from one inhabitant to another in a particular country. Swedish investigators suggest that the average intake of Hg through food, excluding fish, is on the order of 0·005 mg per day. Of this probably only a small part is in the form of methyl mercury. Since methyl mercury levels are much higher in fish than in other foodstuffs, attention will be directed to the consumption of methyl mercury from fish. To obtain a conservative set of figures, it will be assumed that all of the Hg in fish is in the form of methyl mercury. The highest consumption of fish for either Japanese or Swedish individuals appears to be around 500 grams per day. On the basis of this intake, the maximum allowable concentration of methyl mercury in edible fish would be 0·06 parts of mercury per million parts of fish. In Sweden an average consumption of fish is about 30 grams per day, corresponding to slightly more than one meal per week. On such a diet, an allowable one part per million of mercury in the fish results. Marine fish from uncontaminated waters show mercury levels in the region of 0·01 to 0·1 ppm. Fish with the higher Hg content in principle would be unacceptable to those consuming 500 grams per day.

REFERENCES

Allen, A. A., Schlueter, R. S. and Mikolaj, P. G. (1970). *Science, N.Y.* **170**, 974.
Anon (1971). *Mar. Pollut. Bull.* **2**, 165.
Applequist, M. D., Katz, A. and Turekian, K. K. (1972). *Environ. Sci. Tech.* **6**, 1123.

Barber, R. T., Vijayakumar, A. and Cross, F. A. (1972). *Science, N.Y.* **178**, 636.
Berner, R. A. (1971) *J. Geophys. Res.* **76**, 6597.
Bevenue, A., Hylin, J. W., Kawano, Y. and Kelley, T. W. (1972). *Pestic. Monit. Bull.* **6**, 60.
Bien, G. S. and Suess, H. E. (1967). *Proc. Symp. Radioactiv. Dating Methods Low Level Counting.* Int. Atomic Energy Agency, Vienna.
Blumer, M. (1972). *Science, N.Y.* **176**, 1257.
Blumer, M. and Sass, J. (1972a)·*Science, N.Y.* **176**, 1120.
Blumer, M. and Sass, J. (1972b). *Mar. Pollut. Bull.* **3**, 92.
Bruland, K., Bertine, K., Koide, M. and Goldberg, E. D. (1974). *Environ. Sci. Tech.* **8**, 425
Burns, K. A. and Teal, J. M. (1972). Unpublished manuscript.
Carpenter, E. J., Anderson, S. J., Harvey, G. R., Miklas, H. P. and Peck, B. B. (1972). *Science, N.Y.* **178**, 749.
Carpenter, E. J. and Smith, K. L. (1972). *Science, N.Y.* **175**, 1240.
CEQ. (1970). Ocean dumping: a national policy. U.S. Council on Environmental Quality, October 1970.
Chow, T. J. (1971). *Proc. 2nd Int. Clear Air Congr.* pp. 348–352. Academic Press, London and New York.
Chow, T. J., Bruland, K., Bertine, K., Soutar, A., Koide, M. and Goldberg, E. D. (1973). *Science, N.Y.* **181**, 551.
Chow, T. J. and Johnston, M. S. (1965). *Science, N.Y.* **147**, 502.
Chow, T. J. and Patterson, C. C. (1966). *Earth Planet. Sci. Lett.* **1**, 397.
Cooper, W. W. and Kuroda, P. K. (1966). *J. Geophys. Res.* **71**, 5471.
Dockins, K. O., Bainbridge, A. E., Houtermans, J. C. and Suess, H. E. (1967). *Proc. Symp. Radioactiv. Dating Methods Low Level counting.* Int. Atomic Energy Agency, Vienna.
Duke, T. W., Lowe, J. I. and Wilson, A. J. (1970). *Bull. Environ. Contam. Toxicol.* **5**, 171.
Duursma, E. K. and Marchand, M. (1974). *Oceanogr. Mar. Biol. Ann. Rev.* **12**, 315.
Floodgate, G. D. (1972). *Mar. Pollut. Bull.* **3**, 41.
Garner, F. (1973). *Environment* **15**, 4.
Garrels, R. M. and Mackenzie, F. T. (1971) "Evolution of Sedimentary Rocks". W. W. Norton and Co., New York, 397 pp.
Goldberg, E. D. (1972) *In*: "The Changing Chemistry of the Oceans" (D. Dyrssen and D. Jagner, eds.). Wiley Interscience Division, New York.
Greve, P. A. (1971) *Science, N.Y.* **173**, 1021.
deGroot, A. J. (1973). In "North Sea Science" (E. D. Goldberg, ed.). MIT Press, Cambridge, Mass.
Gross, M. G. (1972). *Mar. Pollut. Bull.* **3**, 61.
Gross, M. G. (1970). *Mar. Sci. Res. Cen., Techn. Rep.* No. 2 State *Univ., New York.*
Gross, M. G. and Nelson, J. L. (1966) *Science, N.Y.* **154**, 879.
Hardy, E. P., Krey, P. W. and Volchock, H. L. (1973). *Nature, Lond.* **241**, 444.
Harvey, G. R., Steinhaver, W. G. and Teal, J. M. (1973). *Science, N.Y.,* **180**, 643.
Horn, M. H., Teal, J. M. and Backus, R. H. (1970). *Science, N.Y.* **168**, 245.
Huntzicker, J. J. and Friedlander, S. (1973). Personal communication.
IDOE (1972), Baseline studies of pollutants in the marine environment and research recommendations. Deliberations of the International Decade of Ocean Exploration (IDOE) Baseline Conference, May 24–26, 1972. New York.
Jernelöv, A., Rosenberg, R. and Jensen, S. (1972). *Water Res.* **6**, 1181.

Joseph, A. B., Gustafson, P. F., Russel, I. R., Schuert, E. A., Volchok, H. L. and Tamplin, A. (1971), *In* "Radioactivity in the Marine Environment". National Academy of Sciences, N.R.C. Washington, D.C.

Klein, D. H. and Goldberg, E. D. (1970). *Environ. Sci. Technol.* **4**, 765.

Koide, M., Bruland, K. W. and Goldberg, E. D. (1973). *Geochim. Cosmochim, Acta,* **37**, 1171.

Kuratsune, M. Yoshimura, T., Matsuzaka, J. and Yamaguchi, A. (1972). *Environ. Health Persp.* **1**, 119.

Lal, D. and Suess, H. E. (1968). *Ann. Rev. Nucl. Sci.* **18**, 407.

Lazrus, A. L., Lorange, E. and Lodge, J. P. (1970). *Environ. Sci. Tech.* **4**, 55.

Lovelock, J. E., Maggs, R. J. and Wade, R. J. (1973). *Nature, Lond.* **241**, 194.

MAFF (1971). Technical Report FRL 8, Fisheries Biological Laboratory, Lowestoft, Great Britain.

Martell, E. A. (1968). *J. Atmos. Sci.* **25**, 113.

Martell, E. A. (1963) *J. Geophys. Res.* **68**, 3759.

Miller, G. E., Grant, P. M., Kishore, R., Steinkruger, F. J., Rowland, F. S. and Guinn, V. P. (1972). *Science, N.Y.* **175**, 1121.

Morris, B. F. (1971). *Science, N.Y.* **173**, 430.

Mosser, J. L., Fisher, N. S. and Wurster, C. F. (1972). *Science, N.Y.* **176**, 533.

Murozumi, J., Chow, T. J. and Patterson, C. (1969). *Geochim. cosmochim. Acta* **33**, 1247.

Murray, A. J. and Riley, J. P. (1973) *Nature, Lond.* **242**, 37.

NAS (1971). "Chlorinated hydrocarbons in the marine environment". U.S. National Academy of Sciences, 1971.

Nisbet, I. C. T. and Sarofim, A. F. (1972). *Environ. Health Persp.* **1**, 21.

Olafson, J. H. and Larson, K. H. (1963). *In* "Radioecology" (V. Schultz and A. W. Klement, eds.). Reinhold Publishing Co., New York.

Osterberg, C., Cutshall, N. and Cronin, J. (1965). *Science, N.Y.* **150**, 1585.

Parslow, J. L. F. and Jefferies, D. J. (1972) *Mar. Pollut. Bull.* **3**, 43.

Preston, A. (1972) *Underwater J.* 49.

Preston, A., Fukai, R., Volchok, H. L. and Yamagata, N. (1971). *FAO Fisheries Reports No. 99, Suppl. 1,* 87, FAO, Rome.

Revelle, R., Wenk, E., Ketchum, B. H. and Corino, E. R. (1971). *In* "Man's Impact on Terrestrial and Oceanic Ecosystems" (W. H. Matthews, F. E. Smith and E. D. Goldberg, eds.). MIT Press, Cambridge, Mass., pp. 297–318.

Risebrough, R. W., Huggett, R. J., Griffin, J. J. and Goldberg, E. D. (1968). *Science, N.Y.* **159**, 1233.

Roether, W. and Munnich, K. O. (1967). Cited in Volchock *et al.* (1971).

SC (1972). Southern California Coastal Water Research Project Report "The Ecology of the Southern California Bight: Implications for Water Quality Management", Vol. 1, October (1972).

SCEP (1970). Man's Impact on the Global Environment. Report of the Study of Critical Environmental Problems. MIT Press, Cambridge, Mass. 319 pp. (1971).

Seba, D. B. and Corcoran, E. F. (1969). *Pest. Monit. J.* **3**, 190.

Seba, D. B. and Prospero, J. M. (1971) *Atmos. Environ.* **5**, 1043.

Seymour, A. H. (1971). *In* "Radioactivity in the Marine Environment". National Academy of Sciences, Washington, D.C.

Shelton, R. G. J. (1973). *In* "North Sea Science" (E. D. Goldberg, ed.). MIT Press, Cambridge, Mass.

Simonov, A. J. and Justchak, A. A. (1972). *Ambio Special Report No. 1.*

Stock (1971). *Nord. Hyg. Tidskr.*, Suppl. 4, 364 pp. Stockholm.

USTC (1971). *United States Tariff Commision Publication* 412.

Venrick, E. L., Backman, T. W., Bartram, W. C., Platt, C. J., Thornhill, M. S. and Yates, R. E. (1973). *Nature, Lond.* **241.**

Volchok, H. L., Bowen, V. T., Folsom, T. R., Broecker, W. S., Schuert, E. A. and Bien G. S. (1971). *In* "Radioactivity in the Marine Environment". National Academy of Sciences, Washington, D.C.

Weiss, H. V., Koide, M. and Goldberg, E. D. (1971) *Science, N.Y.* **174,** 692.

Westöö, G. (1966). *Acta Chem. Scand.* **20,** 2131.

Wood, J. M., Kennedy, F. S. and Rosen, C. G. (1968). *Nature, Lond.* **22,** 173.

Chapter 18

Radioactive Nuclides in the Marine Environment

J. D. BURTON

Department of Oceanography, The University, Southampton, England

SECTION 18.1. INTRODUCTION 91
18.2. LONG-LIVED NUCLIDES OF ELEMENTS WITH STABLE ISOTOPES 93
    18.2.1. Potassium-40 93
    18.2.2. Rubidium-87 95
    18.2.3. Other nuclides 96

18.3. URANIUM, THORIUM AND ACTINO-URANIUM SERIES 97
    18.3.1. Uranium 97
    18.3.2. Thorium and protactinium 113
    18.3.3. Radium 120
    18.3.4. Radon-222 125
    18.3.5. Lead-210 and polonium-210 128
    18.3.6. Applications to the geochronology of marine sediments 131

18.4. NUCLIDES PRODUCED BY COSMIC RADIATION 139
    18.4.1. Introduction 139
    18.4.2. Carbon-14 141
    18.4.3. Tritium 150
    18.4.4. Beryllium-7, beryllium-10 and aluminium-26 153
    18.4.5. Silicon-32 155

18.5. ARTIFICIALLY PRODUCED NUCLIDES 157
    18.5.1. Introduction 157
    18.5.2. Principal artificially produced nuclides and their physico-chemical
        forms in sea water 158
    18.5.3. Some radiological and radioecological aspects of artificially pro-
        duced nuclides 162
    18.5.4. Occurrence and distribution of some artificially produced nuclides
        in ocean waters 165
REFERENCES 177

18.1. INTRODUCTION

Radioactive nuclides,* in their passage through marine environments, are

---

* The term nuclide refers to a nuclear species characterized by its atomic number and atomic mass. The term isotope strictly refers only to nuclides with the same atomic number, i.e. nuclides of a single element, and is used here in this way.

acted upon by the same processes which affect stable elements, and the behaviour of a primary, long-lived nuclide such as uranium-238 can be viewed in the same terms as that of any other trace metal. With most radio-active nuclides, however, their half-lives, genetic relations to other nuclides and consequent specialized modes of production, introduce additional aspects to their behaviour which are not paralleled among the stable elements. These differences can afford useful insights into geochemical processes. Furthermore, because of their radioactive decay properties, many nuclides have important applications as indicators of the time-scales of various oceanic processes, such as water mixing and sediment accumulation. Additional possibilities for such applications have arisen through anthropogenic inputs of radioactive nuclides over recent decades; some of the opportunities afforded may be unique to the present time. The man-introduced nuclides also form a group of environmental pollutants which raise special considerations.

For all these reasons it is convenient to discuss the radioactive nuclides in this separate chapter. The present account, although intended to be reasonably self-contained, emphasizes the work of the last decade. More information on earlier investigations can be found in the previous edition of this work (Burton, 1965) and in other sources which include Suess (1958), Koczy and Rosholt (1962) and the International Union of Geodesy and Geophysics (1963).

The radioactive nuclides in the marine environment are considered here in four categories.

(i) Primordial nuclides of elements which also have stable isotopes. Those of importance are potassium-40 and rubidium-87.

(ii) The primordial parent nuclides of the three natural radioactive decay series, uranium-238, thorium-232 and uranium-235, and the families of shorter lived daughter nuclides which are continually renewed by their decay.

(iii) Naturally occurring nuclides, other than the daughter products of the primordial nuclides, which have short half-lives on the geological time-scale, but which are continually produced by nuclear processes. Those important from the present viewpoint are produced by the interaction of cosmic radiation with either atmospheric constituents or the extra-terrestrial material accreted by the earth.

(iv) Artificially produced nuclides released through the explosion of nuclear weapons, or through planned or accidental discharges of material from nuclear installations. Of these nuclides, carbon-14 and tritium add to an appreciable inventory already present in nature. Natural steady-state concentrations of other nuclides in this group, arising from nuclear processes in nature, such as spontaneous fission and neutron capture, are minute compared with the amounts produced artificially.

In this account the second and third categories are treated in more detail than are the others. The distributions and behaviours of nuclides in the first category follow those of the stable forms which have been discussed in other chapters; marine applications of the radioactive nuclides can be summarised quite briefly. Regarding the artificial nuclides there is now a vast literature, but much of it is concerned with radiological and radiobiological aspects which lie largely outside the scope of this work; discussion of these nuclides here is mainly concentrated on aspects which are of general marine geo-chemical interest.

Unless otherwise stated, nuclear data have been derived from values listed by Lederer *et al.* (1967). Where values for concentrations of nuclides are given in terms of radioactivity, rather than mass, two different units are used, namely the picocurie (pCi) and the number of disintegrations per minute (dpm). One picocurie corresponds to 2·22 dpm.

18.2. Long-lived Nuclides of Elements with Stable Isotopes

18.2.1. Potassium-40

Natural potassium contains 0·0118 % of potassium-40 (half-life 1·26 × 10^9 years). For sea water of salinity 35‰, which contains 0·40 g K l^{-1}, the radio-activity due to ^{40}K is 331 pCi l^{-1}. This accounts for over 90 % of the total radioactivity of sea water. At depths greater than about 100 m, above which cosmic rays and photolysis of organic compounds are of increasing import-ance, the decay of ^{40}K is the dominant source of hydrated electrons, sus-taining a steady-state concentration of about 6 × 10^{-26} M in these waters, according to Swallow (1969).

The decay of ^{40}K to ^{40}Ar provides the basis for one of the most important geochronological methods, which is applicable to a wide variety of geo-logical materials (see, for example, Schaeffer and Zähringer, 1966). If at the time of formation from magma a given mass of rock contained some ^{40}K and was free of daughter nuclides, then the number (D) of stable radiogenic daughter atoms present after a time t is related to the number of ^{40}K atoms (N) present at this time by the expression

$$\frac{N}{N + D} = e^{-\lambda t} \tag{18.1}$$

where λ is the radioactive decay constant for ^{40}K. Provided that the system is closed (i.e. ^{40}Ar is neither lost nor gained from other sources, and there is no significant change in potassium content) measurements of potassium and

^{40}Ar can be used to calculate the age of igneous material from the relationship

$$t = \frac{1}{\lambda} \ln \left\{ 1 + \frac{D_{Ar}}{N} \left(1 + \frac{1}{R} \right) \right\} \qquad (18.2)$$

where D_{Ar} is the number of atoms of ^{40}Ar and R is the ratio of the decay constant for electron-capture decay of ^{40}K to that for beta decay. This branching ratio is needed to allow for the fact that only a proportion of the disintegrations of ^{40}K atoms gives rise to ^{40}Ar. If at the magma consolidation stage a significant amount of ^{40}Ar was retained, the derived age will represent only a maximum value unless allowance can be made for this effect.

The K/Ar method has found applications to marine igneous material and has assisted studies of sea floor spreading. Dymond and Windom (1968) found that for separated minerals and whole rocks from seamounts the least altered phases gave the oldest and most concordant ages; this suggested that losses of argon on alteration might affect the validity of derived ages more than does initial ^{40}Ar retention, although the latter factor had to be considered. The importance of retention of ^{40}Ar already present in magma has been emphasised in other studies (Dalrymple and Moore, 1968; Noble and Naughton, 1968; Funkhouser et al., 1968; Dymond, 1970). The rapidity of cooling in sea water favours retention to an extent which increases with depth of occurrence as a result of hydrostatic pressure effects. The contrasting distributions of isotopes of argon between glassy crusts and interior zones of basalts have been discussed by Fisher (1971). For the whole rock system it is possible that both initial retention of ^{40}Ar and post-crystallization losses may occur. The interpretation of apparent age values can thus be particularly difficult for materials formed under submarine conditions. Fisher (1972) has also discussed various criteria of validity for derived ages, and proposed the use in this connection of U/He ratios; age values by this latter method are insufficiently accurate to be generally of independent value, but concordance of K/Ar and U/He ages may be taken to indicate that the K/Ar values are not invalidated by the presence of excess ^{40}Ar. Accounts of the applications of K/Ar dating in various sea-floor regions include those of Aumento et al. (1968) for the mid-Atlantic Ridge, Ozima et al. (1968, 1970) for the East Pacific Rise and western Pacific seamounts, Fisher et al. (1968) for the East Pacific Rise, and Turner et al. (1973) for the Kodiak Seamount and Giacomini Guyot in the Gulf of Alaska.

The assumption of a closed system is less tenable for many sedimentary materials, but the method may still be used to give minimum ages and, possibly, information on the provenance of lithogenous material in marine sediments. Krylov et al. (1961) used this approach, which has also been applied by Hurley et al. (1963) to study potassium-bearing minerals, mainly illites, in pelagic sediments; the results confirmed the essentially lithogenous character of these phases in North Atlantic sediments.

Dating by the K/Ar method has been applied to glasses and feldspars separated from volcanic ash layers in pelagic sediments (Dymond, 1966, 1969). Sedimentation rates can be derived from the ages of these layers by making the additional assumptions that the volcanic material reached the sediment surface soon after its formation from magma and that the analysed material is free from significant amounts of non-volcanic material. The method has particular value in that its range covers the whole of the Tertiary with a possible lower limit of about 10^5 years. The general validity of the approach is supported by the concordant values obtained for coexisting igneous minerals in certain layers and by comparisons with core dating by other techniques. This approach has also been used to date volcanic material forming the nuclei of ferro-manganese concretions (Barnes and Dymond, 1967). Minimum accretion rates, estimated on the basis that deposition began soon after the nucleus was formed from magma, were $0.5-3.5 \times 10^{-6}$ mm year^{-1}, values which are compatible with rates estimated by several other radiometric methods. Very similar values were found, using the same approach, for ferro-manganese coatings on rocks from the mid-Atlantic Ridge (Aumento, 1969).

An examination of K/Ar ratios in phillipsites from sediment cores has been reported by Bernat et al. (1970). For surface layers the mineral gives high apparent ages, attributable to the presence of detrital material. As phillipsite grows within the deposit the new material incorporates potassium, and below the surface layers there is thus, initially, a decrease in apparent age of the phillipsite with depth, the values at a given depth varying also with particle size. Subsequently, the production of ^{40}Ar predominates over dilution by crystal growth and the apparent age increases. These variations are described by a model for the growth of phillipsite developed by Bernat et al. (1970), but any use of this open-system mineral in dating by the K/Ar method would require a very detailed appreciation of the complex factors influencing the ratio.

18.2.2. RUBIDIUM-87

Natural rubidium contains 27.85% of the radioactive isotope of mass number 87 with a half-life of 4.7×10^{10} years. On the basis of a rubidium concentration in sea water of $120 \, \mu g \, l^{-1}$, the radioactivity from this source is $2.9 \, pCi \, l^{-1}$, similar to that contributed by uranium and less than 1% of that due to ^{40}K.

The geochronological method based on the accumulation of radiogenic

^{87}Sr is analogous to the K/Ar method already described; allowance must be made for non-radiogenic ^{87}Sr. A full account is given by Faure and Powell (1972). The technique finds its principal applications in the dating of igneous material, but has been used also for limestones and other sedimentary deposits. The existence of significant differences in the ratios of ^{87}Sr to ^{86}Sr in various materials has led to a number of isotopic geological applications which can be only briefly indicated here by reference to some marine studies.

The ratio of ^{87}Sr to ^{86}Sr for dissolved strontium in present-day sea water from the major oceans is well established at a uniform value of about 0·709 (see, for example, Murthy and Beiser, 1968). Faure et al. (1967) showed that the ratios in *Mytilus* shells and in sea water from Hudson Bay were identical (within experimental limits) with the value for North Atlantic Ocean water, reflecting the small contribution of fresh water sources to the total strontium present even at considerably reduced salinities. The ratio reflects the origin of the strontium now present in the oceanic reservoir, and is consistent with the view that the element has been derived mainly from marine carbonate rocks (Faure et al., 1965). Changes in the ratio over a period of about 4·5 × 10^8 years have been investigated by Peterman et al. (1970) by measurements on fossil carbonates. Variations were attributed to differences in the isotopic composition of the dominant material undergoing erosion on land at different periods (see also Armstrong, 1971), but Hart (1973) has suggested that they could be influenced by differences in the amounts of strontium introduced by submarine weathering of basalts and thus by differences in rates of sea-floor spreading.

The close similarity between the ratio of ^{87}Sr to ^{86}Sr in ocean water and that in phillipsite is one indication of the major interaction with sea water involved in the formation of this mineral (Pushkar and Peterson, 1967). In detrital aluminosilicates in sediments the ratios of strontium isotopes reflect, in a complex way, the age and type of source rocks and the weathering processes. The isotopic ratios in these deposits, and their apparent ages as derived from isochron diagrams, can be used to derive information on the geological provenance of the sediments (Dasch, 1969; Biscaye and Dasch, 1971).

18.2.3. OTHER NUCLIDES

Other nuclides in this group, with half-lives ranging between about 10^{10} and about 10^{15} years, are ^{50}V, ^{115}In, ^{138}La, ^{144}Nd, ^{147}Sm, ^{152}Gd, ^{176}Lu, ^{174}Hf, ^{187}Re, ^{190}Pt and ^{192}Pt. The list may be modified by further investigations. For many of these weakly radioactive elements, information on their occurrence in the hydrosphere is very limited, and the nuclides listed are of no immediate oceanographic interest.

18.3. URANIUM, THORIUM AND ACTINO-URANIUM SERIES

The uranium (4n + 2), thorium (4n) and actino-uranium (4n + 3) series have as their parent nuclides uranium-238, thorium-232 and uranium-235, respectively. Each parent gives rise to a complex chain of decay products. The parts of each series which are relevant from the present standpoint are shown in Table 18.1; where nuclides have more than one mode of decay only the principal route is shown.

In a closed system of sufficient age each daughter product in a given series would be in secular equilibrium (i.e. at a steady state) so that the radioactive decay rates of each would equal that of the primary parent and the total activity would decrease only with the half-life of the primary parent. Under these circumstances each daughter product is produced at the same rate at which it decays. On separation of a shorter-lived nuclide from its immediate parent, the unsupported daughter nuclide decays with its own half-life. The parent nuclide regenerates the daughter at a rate which is governed by the half-life of the daughter and secular equilibrium will be reattained if the system remains closed. A period of about six daughter half-lives gives practically complete restoration of secular equilibrium and this period allows decay of an unsupported nuclide to only a few per cent of its original concentration. Other circumstances may arise in which there are different relationships between parent and daughter half-lives but the case of secular equilibrium is the most important from the present standpoint.

In the marine environment the equilibria in the radioactive families shown in Table 18.1 are profoundly disrupted by geochemical processes. The decay of unsupported nuclides and regrowth of daughter products then provide many approaches to the measurement of time-scales of environmental processes. Because two isotopes may have different genetic histories their geochemical pathways can be totally different, and considerable complications of interpretation can arise on this account. For example, two isotopes may show major differences in their degree of association with the authigenic and lithogenous fractions in sediments.

18.3.1. URANIUM

18.3.1.1. *Sea and river waters*
The results of recent determinations of uranium in sea water are shown in

TABLE 18.1

Important relationships in the uranium, thorium and actino-uranium series

---

Uranium series

$$^{238}\text{U} \xrightarrow[\alpha]{4\cdot51 \times 10^9 \text{ year}} {}^{234}\text{Th} \xrightarrow[\beta^-]{24\cdot1 \text{ day}} {}^{234m}\text{Pa} \xrightarrow[\beta^-]{1\cdot17 \text{ min}} {}^{234}\text{U}$$

$$(99\cdot274\% \qquad\qquad\qquad\qquad\qquad (0\cdot0056\%$$
$$\text{of U}) \qquad\qquad\qquad\qquad\qquad\qquad \text{of U})$$

$$\xrightarrow[\alpha]{2\cdot47 \times 10^5 \text{ year}} {}^{230}\text{Th}^a \xrightarrow[\alpha]{7\cdot52 \times 10^4 \text{ year}} {}^{226}\text{Ra} \xrightarrow[\alpha]{1602 \text{ year}} {}^{222}\text{Rn} \xrightarrow[\alpha]{3\cdot82 \text{ day}} {}^{218}\text{Po}$$

$$\xrightarrow[\text{intermediates}]{\text{through further short-lived}} {}^{210}\text{Pb} \xrightarrow[\beta^-]{21 \text{ year}} {}^{210}\text{Bi} \xrightarrow[\beta^-]{5\cdot01 \text{ day}} {}^{210}\text{Po} \xrightarrow[\alpha]{138 \text{ day}} {}^{206}\text{Pb(stable)}$$

Thorium series

$$^{232}\text{Th} \xrightarrow[\alpha]{1\cdot41 \times 10^{10} \text{ year}} {}^{228}\text{Ra} \xrightarrow[\beta^-]{6\cdot7 \text{ year}} {}^{228}\text{Ac} \xrightarrow[\beta^-]{6\cdot13 \text{ hour}} {}^{228}\text{Th} \xrightarrow[\alpha]{1\cdot91 \text{ year}} {}^{224}\text{Ra}$$

$$(100\%$$
$$\text{of Th})$$

$$\xrightarrow[\alpha]{3\cdot64 \text{ day}} {}^{220}\text{Rn} \xrightarrow[\text{intermediates}]{\text{through further short-lived}} {}^{208}\text{Pb(stable)}$$

Actino-uranium series

$$^{235}\text{U} \xrightarrow[\alpha]{7\cdot1 \times 10^8 \text{ year}} {}^{231}\text{Th} \xrightarrow[\beta^-]{25\cdot5 \text{ hour}} {}^{231}\text{Pa} \xrightarrow[\alpha]{3\cdot25 \times 10^4 \text{ year}} {}^{227}\text{Ac} \xrightarrow[\beta]{21\cdot6 \text{ year}} {}^{227}\text{Th}$$

$$(0\cdot720\%$$
$$\text{of U})$$

$$\xrightarrow[\alpha]{18\cdot2 \text{ day}} {}^{223}\text{Ra} \xrightarrow[\text{intermediates}]{\text{through further short-lived}} {}^{207}\text{Pb(stable)}$$

---

[a] The half-life shown for ${}^{230}\text{Th}$ is that reported by Attree *et al.* (1962).

Table 18.2, together with some earlier measurements. There is considerable agreement between the average values for oceanic waters around an overall mean of $3\cdot3 \ \mu\text{g l}^{-1}$, which corresponds to a radioactivity of $2\cdot2 \ \text{pCi l}^{-1}$. The dispersion of values, however, varies significantly between investigations. Values obtained by neutron activation techniques, alpha spectrometry, and isotopic dilution using mass spectrometry, generally show low dispersions not greatly outside the analytical variability, but the extensive spectro-photometric determinations on western North Pacific waters, show wider variations, some of which have been considered to show significant trends with depth (Miyake and Sugimura, 1964; Miyake *et al.*, 1966).

Although the extent of such variations remains to be defined by further work, it is clear that uranium has a more uniform distribution in open ocean waters than do the majority of trace elements, and this is probably attributable to reduction in its reactivity through formation of the stable $[UO_2(CO_3)_3]^{4-}$ complex (Starik and Kolyadin, 1957). Miyake *et al.* (1972) measured the amounts of uranium in the particulate fraction from several samples from the North Pacific Ocean and found a concentration of about $0.01\,\mu g\,l^{-1}$ which is negligible compared with the dissolved fraction. The generally uniform distribution of uranium in the ocean is a very significant phenomenon since it leads to the uniform generation *in situ* of a number of important daughter products.

Knowledge of the input of uranium to the ocean in river waters is needed for the interpretation of events in marginal waters affected by run-off and for an understanding of the geochemical balance of the element. There is, however, some difficulty in estimating a representative value for the concentration of uranium in fresh-water sources. From earlier data summarized by Livingstone (1963) no meaningful average could be derived, although the most probable range for such a value appeared to be $0.1-1\,\mu g\,l^{-1}$. Miyake *et al.* (1964b) found a range of $0.34-1.2\,\mu g\,l^{-1}$ for ten Japanese rivers with a weighted mean of $0.57\,\mu g\,l^{-1}$. A very wide range of concentrations ($0.01-7.0\,\mu g\,l^{-1}$) has been reported for Indian rivers (Bhat and Krishnaswamy, 1969). Extensive analyses by Bertine *et al.* (1970) showed a range of $<0.01-1.2\,\mu g\,l^{-1}$, with a mean of $0.27\,\mu g\,l^{-1}$. Measurements on the Mackenzie River system suggest an average value of $0.6\,\mu g\,l^{-1}$ (Turekian and Chan, 1971).

There is thus wide variability between river sources, and the estimation of a global average value is rendered more difficult by uncertainty as to recent human influences on the amounts of uranium in certain river systems, through mineral workings and, more widely, the leaching of the element from phosphatic fertilizers applied in the drainage region. The latter influence appears to have affected concentrations in some North American rivers (Adams, 1954; Sackett and Cook, 1969; Spalding and Sackett, 1972), but uncertainty surrounds the interpretation of values obtained for the Mississippi River. Moore (1967) found a concentration of $1.0\,\mu g\,l^{-1}$ for the river in 1965 and, in comparing this with the value of $0.04\,\mu g\,l^{-1}$ found earlier by Rona and Urry (1952), suggested that there had been an anthropogenic modification. Such an interpretation is far from unequivocal since a value of $1.8\,\mu g\,l^{-1}$ had been reported by Rona *et al.* (1956), and Adams (1954) had noted unsystematic variations from 0.08 to $0.82\,\mu g\,l^{-1}$ at a single station on this river. Anomalous ratios of ^{230}Th to ^{232}Th, explicable in terms of a man-influenced source of uranium, have been found in the river

TABLE 18.2

Uranium in sea water

Origin of samples[a]	Number of samples	U (μg l^{-1}) range	mean
N. Pacific, N. Atlantic, Gulf of Mexico and Straits of Florida	9	3·21–3·60	3·39
Indian Ocean	28	1·4–3·4	2·2
Bay of Biscay and English Channel	10	3·2–3·6	3·3
Black Sea	53	1·5–2·8	2·0 ± 0
N.W. Pacific—Oceanic waters 0–110 m	5	1·9–2·6	2·3
370–7000 m	20	2·0–4·7	3·3
Coastal waters 0–200 m	5	1·6–2·0	1·9
400–1000 m	5	2·0–5·0	3·3
Indian and Southern Oceans, S. China Sea	100	2·5–3·5	3·0
Southern Ocean sub-surface waters	25	2·8–3·5	3·1
S. Pacific, Atlantic, Caribbean and Arctic	—	1·5–3·3	—
N. Atlantic	2	both 3·7	3·7
U.S.A. coastal waters, some considerably reduced in salinity	7	0·98–4·5	1·8
	15	0·98–6·4	2·2
N.W. Pacific coastal waters	2	2·9–3·2	3·0
N.W. Pacific—Surface waters	6	3·52–3·78	3·63
630–6780 m	12	1·96–3·79	3·15
Black Sea—Near bottom waters	16	0·7–7·0	2·0
Other depths	36	1·2–5·0	2·9
Gulf of Mexico—Oceanic waters	—	3–4	—
Coastal and shelf waters	—	2–3	—
Arabian Sea coastal waters	4	2·7–3·0	2·8
Gulf of Mexico—Oceanic waters	6	3·4–3·6	3·5
Coastal waters[c]	9	2·1–17·3	—
Indian Ocean open waters[d]	—	—	3·5 ± 0
N.W. Pacific—Surface oceanic waters	22	2·6–3·7	3·4
Oceanic waters below 500 m	9	2·9–3·6	3·3
Coastal waters	3	2·4–3·2	2·8
Unstated	1		3·31
N. Pacific, GEOSECS station, 10–4000 m	16	2·92–3·52[e]	3·16
Long Island Sound, reduced salinity	1		2·87
Japan coastal water	4	3·28–3·48	3·40 ± 0
Unstated	1		3·1
N.W. Pacific—Surface waters, including Japan Sea	24	2·87–3·78	3·41
Waters below 500 m	20	2·40–3·42	3·01
Coastal waters	9	2·39–3·62	3·23

[a] Where depth is not stated, samples were generally of surface water, but in some cases values for surface and sub-surface waters have been combined.

[b] Where sufficient details are available, the method of preliminary concentration is indicated first, followed by the technique used for determination.

Analytical procedure[b]	Reference
~ent extraction, using di-(2-ethyl hexyl) monohydrogen ⬤hosphate; isotopic dilution, using mass spectrometry	} Rona *et al.* (1956)
~rystallization, using methyl violet; fluorimetry	Baranov and Khristianova (1959)
~rcomparison of methods. Solvent extraction, using ~hydroxyquinoline; coprecipitation with $AlPO_4$. Fluorimetry ⬤sing NaF, and pulse polarography, with isotopic dilution, ⬤sing ^{237}U; isotopic dilution, using mass spectrometry	} Wilson *et al.* (1960)
~orimetry	Nikolaev *et al.* (1960)
⬤recipitation with $Fe(OH)_3$. Spectrophotometry using ⬤ibenzoylmethane	} Miyake and Sugimura (1964) Miyake *et al.* (1964a)
⬤recipitation with $AlPO_4$; fluorimetry, using NaF	} Torii and Murata (1964)
—	Thurber (1964)
—	Moore and Sackett (1964)
⬤recipitation with $AlPO_4$; fluorimetry, using NaF	Blanchard (1965)
	Blanchard and Oakes (1965)
⬤recipitation with $AlPO_4$; alpha counting	Umemoto (1965)
⬤lating ion exchange; spectrophotometry, using Arsenazo-III	} Miyake *et al.* (1966)
~orimetry	} Baturin *et al.* (1966)
~ha spectrometry	Noakes *et al.* (1967a)
⬤recipitation with $Fe(OH)_3$; fluorimetry	Sarma and Krishnamoorthy (1968)
⬤recipitation with $Fe(OH)_3$; alpha spectrometry	Sackett and Cook (1969)
⬤recipitation with $Fe(OH)_3$; alpha counting; isotopic ~ilution, using ^{232}U	Bhat *et al.* (1969)
~on exchange; spectrophotometry, using Arsenazo-III	Miyake *et al.* (1970a)
~utron activation, fission track analysis, using freeze-dried ~ample	Bertine *et al.* (1970)
Bertine *et al.* (1970), calibrated against alpha spectrometric ~etermination on inshore sample	Turekian and Chan (1971)
⬤recipitation with $AlPO_4$; neutron activation, fission track ~nalysis	Hashimoto (1971)
~loid flotation; spectrofluorimetry, using Rhodamine-B	Leung *et al.* (1972)
~lating ion exchange; spectrophotometry, using ~rsenazo-III	} Miyake *et al.* (1972)

[c] Includes estuarine waters, some artificially influenced.

[d] Values for coastal waters ranged down to 2 $\mu g\, l^{-1}$.

[e] Range is for individual samples which received different treatments regarding storage method and filtration. A mean value of 3·4 $\mu g\, l^{-1}$ was obtained independently by neutron activation of evaporated samples and gamma spectrometry (Spencer *et al.*, 1970).

sediments (Noakes *et al.*, 1967b) but were probably associated with local ore residues.

In view of the natural variability and uncertain extent of human influences, and the lack of data on some important river systems, any estimate of a representative concentration of uranium in river waters must be regarded as tentative. Turekian and Chan (1971) take a value of $0.3 \, \mu g \, l^{-1}$ for the concentration in the global river supply. This is similar to the estimate of the global mean of 0.5–$0.55 \, \mu g \, l^{-1}$ made by Baturin and Kochenov (1969) on the basis of analyses on Russian rivers, and seems a reasonable estimate on present evidence. It is an order of magnitude greater than the value taken by Moore (1967) as representative of the fresh water input to the Atlantic Ocean.

Relationships between the concentration of uranium and the salinity in coastal and estuarine waters have not been investigated in detail. Such waters typically show somewhat lower uranium concentrations, as is to be expected from the difference in concentration between ocean water and most river waters. Occasionally, high values have been observed (see Table 18.2) some of which, at least, are probably attributable to man's influence on the river inputs. There is some indication that uranium may behave non-conservatively in estuarine mixing (Sackett and Cook, 1969).

It has been suggested that the behaviour of uranium in some coastal waters, such as those of the Baltic Sea, may be influenced by alterations in speciation under reducing conditions (Koczy *et al.*, 1957). Although the enhancement of uranium in sediments deposited in anoxic regimes is well established there is, as yet, no unequivocal evidence that reduction of uranium in the overlying waters is a contributing process. Thus, Kolyadin *et al.* (1960) have stated that uranium is not reduced in anoxic Black Sea waters, but according to Agamirov (1963), such reduction is possible.

For a system in secular equilibrium the activity ratio of uranium-234 to uranium-238 is unity. In natural systems, departures from the equilibrium condition occur as was first reported by Cherdyntsev and Chalov (see Cherdyntsev, 1955). The extensive literature which has arisen from investigations seeking to exploit this valuable indicator of geochemical processes has been reviewed by Cherdyntsev (1969). The earliest observations on marine samples were those of Thurber (1962) who found activity ratios in recent marine carbonates of about 1·15. Direct determinations on sea water, summarized in Table 18.3, have confirmed the presence of an excess of ^{234}U. Most workers have found no significant variations in the ratio between various oceanic regions and, within the limits of error, the average values usually agree with an overall mean value of 1·15. This also appears to apply generally to coastal waters, although Koide and Goldberg (1965) found a

TABLE 18.3

Activity ratios of ^{234}U to ^{238}U in sea water

Origin of samples[a]	Number of samples	^{234}U/^{238}U Range	Mean	Reference
Pacific and Atlantic	4	—	$1·15 \pm 0·02$	Thurber (1963)
S. Pacific, Atlantic, Caribbean and Arctic	—	—	1·15	Thurber (1964)
N. Atlantic	2	—	1·14	Moore and Sackett (1964)
Black Sea	2	1·05–1·15	—	Kazachevskii *et al.*
Sea of Azov	1	—	1·20	(1964)[b]
Sea of Aral	1	—	1·19	
Pacific, Atlantic, Indian Ocean and Mediterranean	19	1·13–1·17	$1·14 \pm 0·014$	Koide and
Red Sea	1	—	$1·18 \pm 0·01$	Goldberg (1965)
N.W. Pacific	2	1·17–1·18	1·18	Umemoto (1965)
U.S.A. coastal waters	7	1·13–1·16	1·15	Blanchard (1965)
Pacific	4	1·14–1·16	1·15	Somayajulu and Goldberg (1966)
N.W. Pacific	18	1·02–1·20	$1·09 \pm 0·05$	Miyake *et al.* (1966)
Arabian Sea coastal waters	4	1·15–1·18	1·16	Sarma and Krishnamoorthy (1968)
E. Pacific—				
N. of Antarctic convergence	[c]	—	$1·13 \pm 0·02$	Veeh (1968)
S. of Antarctic convergence	[c]	—	$1·15 \pm 0·02$	
Gulf of Mexico oceanic waters	7	1·14–1·18	1·16	Sackett and Cook (1969)
N.W. Pacific	—	—	$1·17 \pm 0·007$	Cherdyntsev
Black Sea	—	—	$1·17 \pm 0·01$	(1969)
N.W. Pacific	34	1·07–1·20	$1·13 \pm 0·04$	Miyake *et al.* (1970a)
S. Pacific (50–3500 m)	14	1·11–1·17	$1·14 \pm 0·02$	Krishnaswamy *et al.* (1970)
N.W. Pacific	53	0·99–1·20	$1·12 \pm 0·04$	Miyake *et al.* (1972)

[a] In some cases surface and sub-surface waters were analysed.
[b] Higher values for the Black Sea and the Sea of Azov have been reported by Nikolaev *et al.* (1965).
[c] Composite sample.

significantly higher ratio of 1·18 for the Red Sea and more marked divergences may occur in waters influenced by local land drainage. The results obtained by Miyake *et al.* (1966, 1970a, 1972) show much wider variations for oceanic samples than have been found by other workers and they also give a lower average ratio of 1·12. These workers consider that these variations are to some extent systematic and may be correlated with features such as the oxygen minimum zone (Miyake et al., 1966).

The excess ^{234}U in the dissolved uranium in the ocean reflects preferential mobilization of this isotope in solutions arising from continental weathering processes and in interstitial waters of sediments, as a result of recoil effects and changes in the physico-chemical state of the decay product. According to Cherdyntsev (1969), a ratio of 1·25 is typical for continental surface waters, but the values for individual streams and rivers may vary greatly. Chalov et al. (1964) gave a range of 1·005–1·58 for Russian rivers. Waters draining sedimentary terrains may usually be expected to show values in this range; much higher values have been recorded particularly for waters issuing from igneous regions. Regional variations associated with differences in the geological characteristics of the drainage basins have been noted also for Indian rivers by Bhat and Krishnaswamy (1969); activity ratios ranged from 1·03 to 1·58 with a mean of 1·2. These workers concluded that a value of 1·3 represents the maximum activity ratio for the global river input and they showed that the observed value of 1·15 for ocean water is incompatible with supply of uranium solely by rivers. It was suggested that additional possible sources of uranium with a high activity ratio could be the *in situ* weathering of wind- or river-borne solids, and supply from interstitial waters of sediments. A detailed model for input from the latter source, based on observations of disequilibrium in pelagic sediment cores, has been given by Ku (1965) (see p. 111).

18.3.1.2. *Sediments*

Measurements made up to 1963 of the concentrations of uranium in deep-sea sediments were discussed by Burton (1965). They indicated that a wide variety of deposits contained from 0·6–6 $\mu g\, g^{-1}$, with an average concentration of 2–3 $\mu g\, g^{-1}$; higher values of up to 31 $\mu g\, g^{-1}$ were recorded in some sections of cores. Subsequent measurements have not greatly altered this picture but have provided more detailed information especially since accurate determinations on a routine basis have become available with the development of neutron activation techniques using delayed neutron counting or fission track analysis. Bertine et al. (1970) found values of 0·25–7·09 $\mu g\, g^{-1}$ (0·61–40 $\mu g\, g^{-1}$ on a $CaCO_3$-free basis) for non-reducing pelagic sediments. The results support an average value of 2 $\mu g\, g^{-1}$ for deep-sea sediments. This value is supported also by the determinations by Mo et al. (1973) on numerous samples from the main ocean basins; these show a range of 0·57–4·3 $\mu g\, g^{-1}$, with an average of 1·6 $\mu g\, g^{-1}$. Ku (1965) had previously reported values lying mostly within the range of $0.7–3$ μg g^{-1} for sections of a number of cores. In the northwestern Indian Ocean calcareous muds and muds low in calcium carbonate showed mean concentrations of 1.6 and 3.4 μg g^{-1}, respectively (Baturin, 1969).

In pelagic sediments the carbonate fraction can usually be regarded as diluting material of low uranium content. As discussed below, foraminiferal and coccolithal material appears to contain only about 0.1 μg g^{-1}. Sediments from the Gulf of Mexico, analysed by Mo *et al.* (1973) showed a higher range ($1.2–6$ μg g^{-1}) and mean (3 μg g^{-1}) than they found for the main ocean basins (see above), and in these sediments it appears that the carbonate fraction contains 1 μg g^{-1} or more of uranium. This confirmed the findings of Sackett and Cook (1969) and may reflect in part the presence of pteropod, molluscan and coralline skeletal material in the marginal deposits.

Significant variations in the concentration of uranium in sediments may thus occur which are attributable simply to differing proportions of clay and carbonate fractions and to differences in the origin of the calcareous material. There are, however, other factors which may influence the concentration, and there has been particular interest in the increased abundance of uranium in deposits associated with oceanic ridge and other geothermal areas such as the hot brine region of the Red Sea. Fisher and Boström (1969) found that sediments, many of which were highly calcareous, from a section including the East Pacific Rise contained from 0.24 to 10.6 μg g^{-1}, with the greatest concentrations occurring on the Rise. Associations of enhanced concentrations of uranium with ridge sediments were also indicated by the work of Baturin (1969), Bertine *et al.* (1970) and Turekian and Bertine (1971). In the Indian Ocean, Boström and Fisher (1971) found an average concentration, on a minerogen basis, of 4.6 μg g^{-1} for sediments from active ridge regions, compared with mean values of 2.3 μg g^{-1} for coastal sediments and 1.7 μg g^{-1} for basin and inactive ridge deposits. There is further evidence of enhancement in sediments of active ridge areas from the results of Mo *et al.* (1973). Elevated concentrations of $5.2–24$ μg g^{-1} occur in sediments from the areas of hot brine accumulation in the Red Sea (Ku, 1969; Bertine *et al.*, 1970). Additional data for materials influenced by volcanic or hydrothermal sources are summarized by Rydell and Bonatti (1973).

There has been much discussion of the origin of uranium in these deposits. The ratios of ^{234}U to ^{238}U are consistent with a sea water origin for the uranium in those deposits which show significant enrichment with the element (Ku, 1969; Bender *et al.*, 1971; Veeh and Boström, 1971). Some samples of volcanic or hydrothermal origin with lower concentrations of uranium do, however, show ratios of ^{234}U to ^{238}U which are higher than that in sea water (Veeh and Boström, 1971; Bonatti *et al.*, 1972; Scott *et al.*,

1972; see also Rydell and Bonatti, 1973). Several authors (Ku, 1969; Bender *et al.*, 1971) have suggested that increases in the concentration of uranium in active ridge deposits reflect hydrothermal activity which leads to enrichment with iron and manganese; incorporation of uranium derived from sea water into phases such as hydrous ferric oxide could then explain both the enhanced concentration and the isotopic composition of the uranium. Rydell and Bonatti (1973) give data showing that uranium with an isotopic composition corresponding to that in sea water is present in certain deposits which are indubitably of hydrothermal origin. However, consistency of isotopic ratios between sediment and sea water can obviously arise through other incorporation mechanisms, and Bertine *et al.* (1970) have pointed out that variations in the concentration of uranium with depth in cores from ridge areas could just as well have arisen through intermittent periods of anoxic conditions in ridge basins as by intermittent hydrothermal activity.

The formation of deposits which contain low or average amounts of uranium with ratios of ^{234}U to ^{238}U higher than that in sea water clearly requires the operation of a mechanism which differs from those just discussed. Rydell and Bonatti (1973) have suggested that, depending upon the redox conditions pertaining, the interaction of thermal waters, consisting essentially of sea water, with basaltic material can produce either a hydrothermal solution in which uranium has not been leached from the basalt, or one in which there has been a contribution by leaching. In the first case the uranium in solution will have an isotopic composition similar to that in sea water whereas a higher ratio of ^{234}U to ^{238}U could be produced in the second instance and may be reflected in deposits which subsequently derive uranium primarily from such a solution.

Another specialized mode of deposition of uranium is by its incorporation into organic rich anoxic muds. It has been observed that when reducing conditions develop below the surface within a column of sediment, enhancement of uranium in the reduced zone can occur. Thus, Bonatti *et al.* (1971) found a steady increase in concentration with depth in the reduced zone of a hemipelagic core, with values of 0.45–0.63 µg g^{-1} in the overlying oxidized and transitional sections, and of 0.61–1.37 µg g^{-1} in the reduced layers. This was attributed to a gradual reduction of U(VI) and fixation of the element. Increases in uranium concentration are more marked in sediments accumulating under anoxic conditions. Concentrations of up to 60 and 32 µg g^{-1}, respectively, have been reported for black muds in land-locked Norwegian fjords (Strøm, 1948) and anoxic Baltic Sea sediments (Manheim, 1961), and concentrations up to 35 µg g^{-1} have been measured in deposits from the Black Sea (Baturin *et al.*, 1967). Anoxic sediments from various areas examined by Veeh (1967) contained 5–39 µg g^{-1}. Baturin (1968) has reported values of $1-16\,\mu g\,g^{-1}$ for Baltic Sea sediments, the lower values occurring in sands and sandy silts.

The contrasts between reduced and oxidized sediments are strikingly displayed in deposits from the Cariaco Trench (Dorta and Rona, 1971). The reduced sediments show concentrations about an order of magnitude greater than those in cores from the Caribbean Sea, with values on a $CaCO_3$-free basis of up to $25\,\mu g\,g^{-1}$. In an oxidized zone formed during the last glaciation, below the reduced layers, the concentration fell abruptly to $4\cdot7\,\mu g\,g^{-1}$ as compared with $17\,\mu g\,g^{-1}$ in the reduced zone above.

Relationships between the concentrations of uranium and organic carbon in sediments have been examined in several locations, particularly in anoxic environments, by Baturin and his colleagues (Kochenov et al., 1965; Kochenov and Baturin, 1967; Baturin et al., 1967; Baturin, 1968; Baturin, 1969). Positive correlations were found not only in regions of pronounced enhancement of uranium and organic matter such as the Black and Baltic Seas, but also in the northwestern Indian Ocean and to a less marked degree in the Mediterranean Sea. Relationships of this kind were recognized earlier in black shales (Swanson, 1961). The processes underlying these correlations remain far from clear, and it is uncertain whether there is any direct involvement of organic matter in the uptake of uranium although such a process has been suggested by a number of workers. In short-term laboratory experiments, Baturin et al. (1966) showed that uranium could be taken up from anoxic water by reduced sediments and released to solution on reoxygenation. These workers also noted a significant reduction in the concentration of uranium in waters sampled 2–5 m off the bottom of the Black Sea. They consider that organic material is involved in the process of uptake, but do not postulate a detailed mechanism. Chitin separated from anoxic sediments of the Black Sea has been found to be greatly enriched in uranium in comparison to similar material from other deposits (Kochenov et al., 1965). Kolodny and Kaplan (1973) suggest that organic uranyl complexes are involved in the removal of uranium into organic rich sediments of the Saanich Inlet, the process being accompanied by partial reduction of U(VI) during early diagenesis. They concluded that about 50% of the authigenic uranium in these deposits is associated with organic material, and that of this fraction some 50% is probably complexed with humic acids.

A better appreciation of the processes involved in the incorporation of uranium in anoxic sediments requires more knowledge of the redox relations of the element in anoxic waters (see p. 102) as well as of the organic geochemistry of the element. Association of uranium with phosphatic material must also be considered as an enhancement process in some reduced sediments as shown by the work of Baturin et al. (1971) on organic rich diatomaceous oozes in the upwelling regions off the western coast of Africa. Concentrations in phosphate minerals are discussed below.

In marine ferro-manganese concretions the concentrations of uranium are generally within the range of 2–18 $\mu g\,g^{-1}$ (Tatsumoto and Goldberg, 1959; Nikolayev and Yefimova, 1963; Ku and Broecker, 1969; Krishnaswamy and Lal, 1972; Mo et al., 1973). A correlation between concentrations of uranium and calcium in the concretions was noted by Mo et al. (1973) with both decreasing as the depth of overlying water increased; they considered that there may be a significant association of uranium with apatite contained in the ferro-manganese nodules. Sections of two epicontinental nodules analysed by Ku and Glasby (1972) showed concentrations of 10–22 $\mu g\,g^{-1}$; this feature contrasts with the behaviour of most trace metals, including thorium, which are lower in such nodules than in oceanic concretions.

Analyses of uranium in barite have been made by Goldberg et al. (1969) and by Church and Bernat (1972). Barites from marine environments contained 0·2–5·0 $\mu g\,g^{-1}$ in contrast with those from continental sources which contained less than 0·05 $\mu g\,g^{-1}$. However, uranium appears less useful than thorium as an indicator of the origin of barite concretions occurring in coastal water.

Phillipsites from two sediment cores from the Pacific Ocean were examined by Bernat et al. (1970) who found values for uranium in the mineral from surface layers to be 0·5–0·7 $\mu g\,g^{-1}$, whereas the bulk sediment contained 2·6–3·1 $\mu g\,g^{-1}$. The concentration of the element in the phillipsite decreased with depth below the surface, whereas the concentration of phillipsite in the sediment increased; the concentration of uranium in the bulk sediment remained uniform. This effect can be satisfactorily explained only by the continuing growth of phillipsite within the column (see also Section 18.2.1).

Concentrations of uranium in authigenic phosphate minerals and skeletal apatite are markedly higher in reducing environments than in those subject to oxidizing conditions (Arrhenius, 1963). Kolodny and Kaplan (1970) reported values for over 30 sea floor phosphorites ranging from 6–524 $\mu g\,g^{-1}$ and Veeh et al. (1973) found a range of 7–168 $\mu g\,g^{-1}$ for Recent phosphorite nodules from Peruvian coastal waters. In the former work a correlation was found between high uranium concentrations and high proportions of U(IV), the latter usually accounting for between 38–86 % of the total. This accords with the ready substitution of U(IV) for calcium ions in the apatite crystal lattice (Clarke and Altschuler, 1958). Concentrations of uranium in fish teeth from different depths in two sediment cores studied by Bernat et al. (1970) were very uniform in each case, the overall range being 21–49 $\mu g\,g^{-1}$; these values are much higher than those for the living material. These features signify that the uptake of uranium occurred mainly while the skeletal debris was exposed at the sediment surface. Enrichment under reducing conditions is apparent in the values obtained by Baturin *et al.* (1971) for fish bone material from oozes rich in diatoms and organic matter in an upwelling region; concentrations of up to 700 µg g^{-1} occurred in phosphatic skeletal remains. Fossil skeletal apatite can show considerable secondary enrichment with uranium as found, for example, by Bowie and Atkin (1956).

Recently deposited biogenous calcium carbonate, other than that formed by corals, usually contains less than 0·3 µg g^{-1} of uranium, and concentrations below 0·01 µg g^{-1} are not uncommon. Ku (1965) found less than 0·1 µg g^{-1} in foraminiferal and coccolithal material and a value of 0·12 µg g^{-1} for foraminifera is reported by Holmes *et al.* (1968). Of numerous recent molluscan shells analysed by Broecker (1963a) and Blanchard *et al.* (1967) some 85–90% contained less than 0·3 µg g^{-1}. Older materials typically show considerable secondary enrichment with uranium. Kaufman *et al.* (1971) have summarised the results for over 700 marine molluscan samples and concluded that the addition mainly occurs within 10^4 years with some less marked subsequent enrichment.

In contrast, living and recent corals show higher uranium concentrations with low dispersion and a lesser tendency to secondary alteration of concentration. Kaufman *et al.* (1971) have summarized the results for over 140 samples of different ages; specimens containing more than 20% calcite were excluded as uranium is lost during the inversion of aragonite. Of the living corals, 90% show concentrations of 2–4 µg g^{-1} and all lie between 1 and 5 µg g^{-1}. There is no evidence for significant change in the uranium concentration for corals of Pleistocene age. Tertiary corals mainly have concentrations in the range 1–2 µg g^{-1}, but some show increased concentrations of up to 15 µg g^{-1}. Concentrations reported for oolites are 2·4–3·8 µg g^{-1}, similar to those in corals (Broecker, 1963a; Osmond *et al.*, 1965).

In the deposition of most biogenous carbonates there is, as can be calculated from the data already summarised, discrimination against the incorporation of uranium into the calcareous structure. For skeletal materials other than those laid down by corals the discrimination factor* (ratio of uranium to calcium in skeletal material/ratio in sea water) is below 0·5 and usually below 0·1 (Broecker, 1963a; Blanchard and Oakes, 1965). Calcitic molluscan shells show more discrimination against uranium than do

---

* The term discrimination factor is used here with reference to the overall discrimination observed in the movement of the elements concerned from a source material to an organism or tissue. Such overall discrimination is usually the result of several physiological stages and the factors should be regarded as observed ratios rather than discrimination factors in a strict physiological sense.

aragonitic types (by a factor of about 7 on average; see Blanchard and Oakes, 1965). Discrimination factors reported for corals (Sackett and Potratz, 1963; Broecker, 1963a; Veeh and Turekian, 1968; Thompson and Livingston, 1970) are in a higher range of 0·4–1·4. Aragonite precipitated from sea water in the laboratory showed a discrimination factor of 1·1 (Tatsumoto and Goldberg, 1959) which is in general agreement with the observed range of concentrations of uranium in oolites.

It appears that within individual skeletal structures of both molluscs and corals, significant inhomogeneities in the distribution of uranium can occur (Lahoud et al., 1966; Schroeder et al., 1970). Environmental factors and variations in skeletal formation rates may be causes of these variations, and Thompson and Livingston (1970) invoked the latter factor to explain their findings of greater overall discrimination against uranium by shallow water hermatypic corals than by other groups. Amiel et al. (1973) showed that in corals containing 3 μg g^{-1} in the aragonite lattice a much higher concentration of 40–70 μg g^{-1} occurred in the chitinous organic filaments which constitute a small fraction of the total material.

In sediments which contain either entirely or mainly authigenic uranium, the ratio of ^{234}U to ^{238}U in recently deposited material is almost invariably close to the average ratio in sea water, although atypically high values may occur in marginal environments (Blanchard et al., 1967). The former situation has been described for calcareous materials (Thurber, 1962; Koide and Goldberg, 1965; Veeh, 1966, 1968; Broecker and Thurber, 1965; Broecker et al., 1968b; Ku, 1968), phosphorite nodules (Baturin et al., 1972; Veeh et al., 1973), ferro-manganese phases (Barnes and Dymond, 1967; Ku and Broecker, 1967a), anoxic muds with high uranium contents (Veeh, 1967) and organic-rich diatom oozes (Baturin et al., 1971, 1972). The value of the ratio as an indicator of the origin of uranium in geothermal regions has been discussed previously (p. 105). Very marked isotopic fractionation between the uranium present in different oxidation states in sea-floor phosphorites was discovered by Kolodny and Kaplan (1970), the ratio of ^{234}U to ^{238}U being, on average, 0·71 for the U(IV) and 1·57 for the U(VI). The presence of excess ^{234}U potentially provides a method of dating deposits, but in practice this approach can be complicated through secondary changes and migration (see Section 18.3.6.2).

In lithogenous material entering the ocean, ratios of ^{234}U to ^{238}U of less than unity, typically about 0·94, are common (Blanchard, 1965; Moore, 1967; Scott, 1968). In sediments containing much lithogenous material accumulated more than about 10^6 years ago the deposited material would, in the absence of migratory processes, have grown ^{234}U to give a ratio of unity, but Ku (1965) has shown that ratios below unity occur at such levels in pelagic cores. The distribution of the ratio, showing a minimum value at an intermediate depth below the sediment surface and then increasing toward equilibrium at greater depth, is explicable in terms of upward migration of ^{234}U in the sediments. Ku (1965) has developed a model for this process from which a diffusion coefficient for ^{234}U of 10^{-8} to 10^{-9} cm^2 sec^{-1} is obtained. The low magnitude of this coefficient compared with values for ionic diffusion implies that there is substantial hold-up of the migration of ^{234}U by interaction with solid phases, as is also the case with radium (see p. 124). Kigoshi (1971) has shown that the fraction of ^{234}U required to be mobile in Ku's model can be explained on the basis of the alpha recoil behaviour of the ^{234}Th atoms initially produced by radioactive decay of ^{238}U.

18.3.1.3. *Geochemical balance.*

The geochemical balance and residence time of dissolved uranium in the ocean are matters on which there is much uncertainty, stemming in part from the difficulty, discussed on p. 99, of estimating accurately the input of the element by the global river supply. Moore (1967) calculated the oceanic residence time of uranium as 4×10^6 years, taking an average concentration in river water of 0·03 μg l^{-1}; use of the average estimated by Bertine *et al.* (1970) gives a value of 4×10^5 years. Using the alternative approach through estimation of the rate of authigenic removal, Krishnaswamy and Lal (1972) estimated the residence time as 2×10^7 years, on the basis that the average concentration of authigenic uranium in sediments is 0·44 μg g^{-1}. In calculating authigenic removal, however, account must be taken of the possible importance of more specialized routes of removal.

The significance in this respect of the high amounts of uranium in anoxic sediments has been examined by Veeh (1967). Any conclusion is again dependent upon the estimate for the river water input, and he considered an indirect means of calculating this from data on the supply of ^{234}U to the ocean. Assuming a mean activity ratio of ^{234}U to ^{238}U of 1·20 in the river supply and, following Ku (1966), a value of 0·3 × 10^{-3} dpm cm^{-2} yr^{-1} for the rate of input of ^{234}U by diffusion from pelagic sediments, the maintenance of the observed oceanic activity ratio of 1·15 requires an input of uranium corresponding to a mean concentration in river water of 0·2 μg l^{-1}. An increase in the estimate of diffusion rate by 33%, however, decreases the corresponding value to 0·04 μg l^{-1}, so that, without a rather accurate knowledge of the diffusive flux of ^{234}U, this approach places rather wide limits on the possible concentration of uranium in river water.

From a comparison of the lower value for river input with a calculated removal rate in near-shore anoxic sediments of 0·1 μg cm^{-2} year^{-1}, Veeh (1967) concluded that this removal process could balance the total input if

E

0.4% of the oceanic area were covered by such deposits. Turekian and Chan (1971) have estimated, on the basis of the more probable average concentration in river water of 0.3 $\mu g\,l^{-1}$, that to account for removal of the input in this way several per cent of the sea-floor area would have to consist of anoxic or other equally enriched sediments, even when account is taken of authigenic removal in ridge areas. Another specialised route of removal, which has received little attention, is via the carbonate sediments of coastal areas (Sackett and Cook, 1969; Mo et al., 1973).

Several matters must be clarified before the geochemical balance of uranium can begin to be properly assessed. They include the resolution of uncertainties about the input by rivers, investigation of possible removal in estuarine mixing and assessment of inputs other than in terrestrial drainage. Furthermore, the possible scale of removal in submarine weathering reactions of basalts remains to be fully assessed. According to Aumento (1971), such processes increase the concentration of uranium in tholeiitic ridge basalts by at least 1 $\mu g\,g^{-1}$ in 10^7 years.

Some inferences may be drawn about changes in the oceanic concentration of uranium on geological time scales. The data on corals discussed above (p. 109) and in Section 18.3.6.3, suggest that the ratio of uranium to calcium and the ratio of ^{234}U to ^{238}U in ocean waters have remained essentially constant since at least the beginning of the Pleistocene (see also Ku and Broecker, 1966). From studies on the composition of ferro-manganese concretions, Krishnaswamy and Lal (1972) suggested that there had been negligible changes in the oceanic concentration of uranium over the last 1.5×10^7 years but that 10^8 years ago the concentration was about one half of the present value.

18.3.1.4. *Soft tissues of organisms.*

There is relatively little information on the amounts of uranium in the marine biosphere, other than that relating to skeletal material. Aten et al. (1961) give an average concentration for fish muscle of 0.02 $\mu g\,g^{-1}$ wet weight. For samples of plankton, including phyto- and zooplankton, Miyake et al. (1970a) found 0.17–0.78 $\mu g\,g^{-1}$ dry weight, and fixed algae contained 0.04–2.35 $\mu g\,g^{-1}$; the ratios of ^{234}U to ^{238}U in these tissues reflected, as expected, the excess ^{234}U in sea water. Various marine algae, representing widely different degrees of calcification, were analysed by Edgington et al. (1970). Concentrations ranged from 0.07–1.64 $\mu g\,g^{-1}$ of the dry acid-soluble material. Although those algae containing 10% or more of $CaCO_3$ mostly discriminated against uranium relative to calcium, preferential uptake of uranium occurred in the algae which were less calcified. This was attributed to the greater importance of organic complexation in the uptake processes of the latter species, as compared with the dominant association with the calcareous matrix in the more calcified forms.

8.3.2. THORIUM AND PROTACTINIUM

Investigations of isotopes of thorium in the marine environment have been mostly concerned with ^{232}Th and ^{230}Th, but some information on the distributions of ^{228}Th and ^{234}Th is also available. Thorium-232 is the parent of the thorium (4n) series. Its daughter, ^{228}Th, is formed through two intermediate nuclides, one of which is ^{228}Ra. Radium is a far more mobile element than is thorium, and the half-life of ^{228}Ra (6·7 years) is sufficiently long to allow significant separation of ^{228}Th from the parent ^{232}Th. Although ^{228}Th is acted on by geochemical processes in the same way as is ^{232}Th, its distribution is thus partially independent of the parent isotope and is more closely governed by the behaviour of ^{228}Ra.

Thorium-230 and thorium-234 are each daughter products of uranium isotopes, although with widely differing half-lives of $7·5 \times 10^4$ years and 24 days, respectively. They thus have a quite uniform source in sea water, and the most important facet of their behaviour lies in their separation from the parent nuclides by geochemical processes. In this they closely resemble protactinium-231.

18.3.2.1. Thorium.

The measurement of the very low concentrations of the isotopes of thorium in sea water has proved a difficult task. A summary of the most important results is given in Table 18.4. The values show considerable variations, some of which may be associated with different degrees of separation of particulate and dissolved fractions.

Kaufman (1969b) has discussed critically the interpretation of alpha spectra in relation to the determination of ^{232}Th, measurements of the element having usually been made by this technique after concentration and purification. His value for ^{232}Th of $0·7 \times 10^{-10}$ g l^{-1}, based on a composite sample representing several principal regions, agrees with the upper limit of $0·8 \times 10^{-10}$ g l^{-1} determined by Krishnaswamy et al. (1972) for North Pacific Ocean waters by in situ sampling. These values are somewhat lower than those found by Somayajulu and Goldberg (1966), but their results, together with the lower of the values obtained by Moore and Sackett (1964), are of the same order of magnitude (about 10^{-10} g l^{-1}). Concentrations in inshore waters may be about two orders of magnitude greater than those in oceanic water. High values found near the coast of the Azov Sea were attributed to high concentrations of suspended material (Nikolayev et al.,

TABLE 18.4

Isotopes of thorium in sea water[a]

Origin of samples	^{232}Th $(10^{-10}\,g\,l^{-1})$
Inland seas and inshore waters	
Black Sea (9 samples)	24 (14–39)
Black Sea	(14–73)
Sea of Azov (7 stations)	(40–2190
Californian coastal water (1 sample)	130
Japanese coastal waters (5 samples)	150 (50–220)[c]
N. Adriatic (S = 21–24‰) unfiltered (5 samples)	14 (0·2–80)
filtered (0·45 µm) (5 samples)	0·003 (<0·000:
(S ca. 36‰) unfiltered (8 samples)	25 (1–60)
filtered (0·45 µm) (8 samples)	0·004 (0·002–0·
Oceanic waters	
N. Atlantic (surface), Caribbean (surface and 800 m) (3 samples)	5·5 (3·6–6·4)
N. Atlantic, 4500 m (1 sample)	45
Pacific, 0–2500 m (3 samples)	3·9 (2·0–6·5)
Equatorial Pacific, 0–4800 m (8 composite samples)	39 (10–79)
Pacific, N.W. Atlantic and adjacent regions (composite from 24 samples)	0·69
N.W. Pacific, E. China and Japan Seas (27 surface and sub-surface samples)	42 (1–280)
N. Pacific (*in situ* sampling)	<0·8
N. Atlantic, 3–5400 m (16 samples)	
S. Atlantic (Walvis ridge area) (8 samples)	
N. Atlantic, Caribbean, Gulf of Mexico (37 samples)[g]	
S. Atlantic (4 samples)	
Indian Ocean (15 samples)	
N. Pacific (17 samples)	
S. Pacific (10 samples)	
Arctic (1 sample)	
Particulate concentrations	
Japanese coastal water (1 sample)	200
N.W. Pacific (2 samples)	7 (6·6–7·4)
Indian and Antarctic Oceans (37 samples)[h]	3·2 (0·2–23)

[a] Ranges of observed concentrations are given in parentheses.
[b] Determined on a single large sample.
[c] Higher values for bay waters were reported by Miyake *et al.* (1964a).
[d] Separate measurements were made on 7 samples for this isotope.
[e] Activity ratio $^{228}Th/^{230}Th$ was 1·5.
[f] For other samples from the Atlantic, off S. Africa, values of $< 340 \times 10^{-18}\,g\,l^{-1}$ were found, upper limit reflecting the small volumes analysed.
[g] These samples, and the others analysed by Broecker *et al.* (1973), were of surface waters.
[h] Of these samples, 18 were analysed for ^{230}Th.

$_3 1^{-1}$)	^{228}Th $(10^{-18} \text{ g l}^{-1})$	Activity ratio ^{230}Th/^{232}Th	^{228}Th/^{232}Th	Reference
5[b]		ca. 30[b]		Starik et al. (1959)
10				Lazarev et al. (1961)
	9·8[b]	2·7[b]	1·4[b]	Nikolayev et al. (1961)
	1·8	0·94	1·0	Somayajulu and Goldberg (1966)
18)[c]	16 (5–25)	1·4 (0·8–2·4)	7·7 (4·3–9·1)	Miyake et al. (1970b)
				}Strohal and Pinter (1973)
7–1·0)	1·1 (0·6–1·4)⎱ 7·7	2·6 (1·5–3·5)}	14 (12–16) }	Moore and Sackett (1964)
2–3·7)	0·9 (0·6–1·1)	9·1 (6·8–10·5)	19 (10–25)	Somayajulu and Goldberg (1966)
–204)		61 (13–163)		Kuznetsov et al. (1966)
	2·0 (0·3–4·8)[d]		210	Kaufman (1969b)
2–9·7)	3·7 (0·3–12)	1·9 (0·5–5·4) [e]	10 (1·7–36) [e]	Miyake et al. (1970b)
				Krishnaswamy et al. (1972)
	3·0 (1·2–7·7)			Moore (1969a)
	2700 (800–4800)[f]			Cherry et al. (1969)
	3·5 (1·1–6·2)			
	1·4 (1·2–1·7)			
	3·2 (0·15–7)			} Broecker et al. (1973)
	0·7 (0·2–1·6)			
	<0·4 (<0·2–0·6)			
	15			
	3·2	1·4	1·2	} Miyake et. al. (1970b)
53–0·56)	0·28 (0·25–0·32)	1·4 (1·4–1·5)	3·0 (2·8–3·2)	
27–3·7)		11 (1·9–27)		Kuznetsov et al. (1964)

1961). Miyake *et al.* (1970b) observed that the concentrations of ^{232}Th were similar in both the particulate and the dissolved fractions of inshore waters. Suspended material was found by these workers to contain about 0·1–0·2 μg g^{-1} of ^{232}Th, considerably less than is found in bottom deposits.

Concentrations of ^{232}Th in river waters of about 10^{-7} to 10^{-8} g l^{-1} have been reported (Miyake *et al.*, 1964b; Moore, 1967). The oceanic residence time has been estimated as about 60 years (Somayajulu and Goldberg, 1966). Although such a low value can have little real meaning in terms of the model from which it is derived, it indicates very rapid removal of ^{232}Th from sea water. This behaviour of thorium is strikingly shown also by the oceanic concentrations of ^{230}Th (see Table 18.4). These amount to only a small fraction, probably less than 0·1 % on average, of the concentration corresponding to secular equilibrium with the ^{234}U present, this marked disequilibrium reflecting efficient scavenging by clays and ferro-manganese phases. Estimates of the residence time of ^{230}Th are of similar magnitude to that for ^{232}Th (Moore and Sackett, 1964; Somayajulu and Goldberg, 1966). *In situ* formation is the main source of ^{230}Th in the ocean, the river input accounting probably for less than 4 % of the total supply (Moore, 1967).

The processes removing ^{230}Th act also on ^{234}Th, the daughter of ^{238}U, but because of its much shorter half-life of 24 days and consequent rapid regeneration, the consequences of interactions with particulate material are different. This is shown by the work of Bhat *et al.* (1969) on Indian Ocean waters. Surface waters show variable activity ratios of ^{234}Th to ^{238}U, the ratios being very low in coastal waters, as a result of scavenging of ^{234}Th by suspended material, whereas in more open surface waters they tend to increase and values above and below unity occur. Depth profiles in open waters show activity ratios decreasing from around the equilibrium value at the surface to 0·5–0·7 at 25–75 m; some waters between 100–150 m show excess ^{234}Th. This distribution is consistent with a particulate transference mechanism with some hold-up in the surface layer, possibly by association with plankton.

The amounts of ^{228}Th (half-life 1·9 years) in ocean waters are substantially in excess of those which could be supported by the ^{232}Th present (see Table 18.4). The excess of ^{228}Th in the few river waters which have been examined amounts to about 20–40 % of the equilibrium value (Moore, 1967), and this source cannot sustain the observed excess in the ocean. The oceanic excess of ^{228}Th originates from *in situ* production from ^{228}Ra which appears to be supplied dominantly by release from sediments into marginal and deep-sea waters. This question is fully discussed in Section 18.3.2. Moore (1969a) showed that the distribution of ^{228}Th is consistent with that of ^{228}Ra modified by the geochemical fractionation of thorium. Thus, activity ratios of ^{228}Th to ^{228}Ra are below unity in surface waters but at intermediate depths, where concentrations of ^{228}Ra are minimal, there is an excess of ^{228}Th which is attributed to preferential transference from surface waters by particulate material. Measurements by Broecker *et al.* (1973) of the activity ratio of ^{228}Th to ^{228}Ra in open oceanic surface waters show a relatively low dispersion about a mean value of 0·2; this value indicates a residence time of thorium in surface water of about 0·7 year which these authors considered might imply the involvement of phytoplankton in the transference processes. In bottom water where the concentration of ^{228}Ra is increased by release from sediments the activity ratio of ^{228}Th to ^{228}Ra is close to, or less, than unity.

Concentrations of ^{232}Th in pelagic sediments are usually in the range of 1–15 µg g^{-1} (see, for example, data summarised by Burton (1965) and Ku *et al.* (1968)). Thorium-232 is present almost entirely in the non-carbonate material (Heye, 1969), and a substantial fraction of the total concentration occurs in lithogenous phases. Low concentrations (0·05–0·8 µg g^{-1}) have been noted in sediments from the East Pacific Rise, where lithogenous accumulation rates are particularly low (Bender *et al.*, 1971), and in sediments of the Red Sea geothermal area (Ku, 1969).

In ferro-manganese concretions, concentrations are variable but often higher than in associated sediments. The most extensive analyses are those by Ku and Broecker (1969) who found 3–154 µg g^{-1} in the acid-soluble fraction. There is a general trend, within a given oceanic region, for the concentration to increase with the depth of occurrence, but there are also regional variations so that, for a given depth, concretions from the Atlantic Ocean contain considerably more ^{232}Th than do those from the Pacific Ocean. This effect parallels the difference in ratios of ^{230}Th to ^{232}Th found for sediments from the main ocean basins as discussed below. Two epicontinental nodules showed concentrations in different layers of 2·5–5 µg g^{-1} (Ku and Glasby, 1972).

Marine barites contain from 50–114 µg g^{-1} of ^{232}Th (Somayajulu and Goldberg, 1966; Goldberg *et al.*, 1969; Church and Bernat, 1972), considerably more than in associated bulk sediments. Thorium appears to be a good indicator element for diagnosing the origin of barites. In phillipsites, ^{232}Th shows the same trend as does uranium, decreasing in concentration as the mineral grows with depth in the core. Near-surface phillipsites examined by Bernat *et al.* (1970) contained 3–8 µg g^{-1}. These workers also found 63–230 µg g^{-1} in fish teeth from the same cores, and values up to 92 µg g^{-1} were found for fish debris from sediments by Arrhenius *et al.* (1964). Recent phosphorite nodules have been observed to contain 2–7 µg g^{-1} (Veeh *et al.*,

1973) and a higher range of 5–43 µg g^{-1} has been reported for older sea-bed phosphorites (Kolodny and Kaplan, 1970).

For corals, concentrations as high as several µg g^{-1} have occasionally been reported (Osmond et al., 1965), but most values are below 0·1 µg g^{-1} (Thurber et al., 1965; Veeh, 1966; Ku, 1968). The tendency for thorium to be excluded from the crystal structure in coral deposition, in contrast to uranium, leads to the possibility of dating these materials by the post-depositional growth of ^{230}Th (see Section 18.3.6.3).

The ^{230}Th in the upper layers of most deep-sea sediments is primarily unsupported and, unlike ^{232}Th, mainly authigenic. Ratios of ^{230}Th to ^{232}Th in surface sediments show the considerable regional variations expected from their different origins, with ^{230}Th uniformly produced in the ocean and ^{232}Th entering largely via continental drainage. These variations, first described by Goldberg and Koide (1962), are broadly correlated with the areas of land draining into the oceans relative to their size. A very high activity ratio of 250 in the acid-extractable fraction was reported by Goldberg (1965) for the East Pacific Rise and ratios exceed 100 over much of the South Pacific Ocean, in contrast with values for the North Atlantic and North West Pacific Oceans which are often below 10. There is evidence of similarity between the ratio of ^{230}Th to ^{232}Th in recently deposited sediment from a given area and that in the overlying water (Somayajulu and Goldberg, 1966); this similarity applies also to suspended material (Moore and Sackett, 1964). An unusual feature established by Goldberg and his co-workers (Goldberg and Koide, 1963; Goldberg and Griffin, 1964; Goldberg et al., 1964a) is a positive correlation between the ratio of ^{230}Th to ^{232}Th in surface sediments and the content of calcium carbonate. Goldberg et al. (1964a) developed a model to explain this relationship on the basis that there is a difference in the ratio of thorium isotopes in surface and deep water. There is evidence for such a difference as barites from pelagic sediments, which derive their material from water near the bottom, show lower ratios than do the bulk sediments (Somayajulu and Goldberg, 1966). A similar effect has been observed for phillipsite from one core top, but two others show similar ratios in this mineral and the bulk sediment (Bernat et al., 1970). The use of unsupported ^{230}Th in dating marine sediments is discussed in Section 18.3.6.1.

Apart from data relating to the skeletal tissues there is little information on thorium in the marine biosphere. Measurements on algae showing appreciable but variable degrees of calcification have shown that they concentrate ^{232}Th to a greater extent than they do calcium in uptake from sea water (Edington et al., 1970); concentrations of thorium ranged from 0·02 to 0·62 µg g^{-1} dry weight (acid-soluble). Very similar concentrations of 0·01–0·66 µg g^{-1} dry weight are reported by Strohal and Pinter (1973) for green and brown algae from the Northern Adriatic; red algae from the same region showed higher concentrations of 1–2·3 µg g^{-1} dry weight. Cherry *et al.* (1969) give values for ^{228}Th in plankton collected off South Africa. Samples from different water masses showed significant differences in concentration with a higher range of mean values for phytoplankton (9–65 × 10^{-18} g g^{-1} wet weight) than in zooplankton (2–27 × 10^{-18} g g^{-1} wet weight). Average concentrations for thorium isotopes in plankton of this region are given by Shannon (1972) as follows (values as g g^{-1} dry weight): ^{228}Th, 4·7 × 10^{-16}; ^{230}Th, 1·7 × 10^{-12}; ^{232}Th, 2·7 × 10^{-7}. The average activity ratio of ^{228}Th to ^{232}Th of about 15 in plankton is similar to that in sea water.

18.3.2.2. Protactinium.

The behaviour of protactinium-231 in sea water is closely similar to that of thorium-230. Moore and Sackett (1964) found from 1–2 × 10^{15} g l^{-1} in three samples from the North Atlantic Ocean and Caribbean Sea. This concentration is about 0·2% of that corresponding to equilibrium with the ^{235}U present. The residence time for ^{231}Pa was estimated as less than 100 years. Measurements by Kuznetsov *et al.* (1966) for water from the equatorial Pacific gave higher concentrations of from less than 3–7 × 10^{-14} g l^{-1}, which are also slightly above the upper limit reported by Sackett (1960) for coastal Pacific Ocean water. The results of Moore and Sackett (1964) indicate an activity ratio of ^{230}Th to ^{231}Pa of about 2 for sea water. As with ^{230}Th, the ^{231}Pa in upper layers of pelagic sediments is mostly unsupported.

18.3.2.3. Comparative behaviour of ^{231}Pa and ^{230}Th in sedimentation.

Interest in the comparative behaviour of ^{231}Pa and ^{230}Th centres mainly on the use of their ratio in the geochronology of sediments, as discussed in Section 18.3.6.1. This application assumes an essentially constant ratio in the freshly deposited material at any one site over the last 3 × 10^5 years. Both nuclides are rapidly and efficiently transferred to sediments and, provided that they are geochemically coherent, surface layers of sediments would be expected to contain the unsupported nuclides in the ratio in which they are produced from uranium. For activity ratios of ^{234}U to ^{238}U and ^{235}U to ^{238}U of 1·15 and 0·046 respectively, and using the half-lives shown in Table 18.1, the production ratio of ^{230}Th to ^{231}Pa is 10·8. Extrapolated surface ratios of the unsupported nuclides in ferro-manganese concretions are commonly significantly less than this value (Ku and Broecker, 1969), whereas considerably higher ratios occur in some pelagic clays, as shown for example by the data of Sackett (1964) and Ku (1965). These results indicate preferential incorporation of ^{231}Pa into ferro-manganese phases. Comparisons on the basis of the integrated amounts of unsupported nuclides in deposits confirm the fractionation effect. If no fractionation occurred the ratio of the integrated amounts would equal the oceanic activity ratio of ^{234}U to ^{235}U, which is 25. A difference is again apparent between sediment cores with an average ratio of integrated activities of 41 and ferro-manganese nodules with a corresponding ratio of 13 (Ku and Broecker, 1969; see also Sackett, 1966). Turekian and Chan (1971) have suggested that the comparative lack of discrete ferro-manganese concretions in the Atlantic Ocean allows the attainment of the theoretical ratios of ^{230}Th to ^{231}Pa in sediment cores from that region. Ferro-manganese nodules appear to account for only a small fraction (less than 5–10%) of the oceanic inventory of ^{230}Th and it is a controversial question as to whether they contain a really significant fraction of the inventory of ^{231}Pa (Ku and Broecker, 1969; Turekian and Chan, 1971).

18.3.3. RADIUM

Most of the investigations of radium in sea water have been concentrated on the longest lived isotope, ^{226}Ra, but the distribution of ^{228}Ra has recently been studied in some detail. A few measurements of ^{224}Ra have been reported (Lazarev et al., 1965) for Black Sea coastal waters in which the concentrations of this nuclide were close to equilibrium with the ^{228}Ra present.

18.3.3.1. Radium-226.

The investigations by Pettersson (1955) and Koczy (1958) into the distribution of ^{226}Ra in the ocean were the first to give a clear idea of the behaviour of this nuclide, although some features had been observed in earlier decades (see Burton, 1965). These workers showed that in most ocean regions there is a marked vertical gradient in concentration with higher values in deep water, and that the average concentration is of the order of magnitude of 10^{-13} g l^{-1}. This amounts to only about 10% or less of that corresponding to equilibrium with the uranium present, but is one or two orders of magnitude greater than could be supported by the concentration of the immediate parent, ^{230}Th, in ocean water. The concentration of ^{226}Ra in river water has been reported to be in the range $0.1–14 \times 10^{-14}$ g l^{-1} (Miyake et al., 1964b; Moore, 1967; Bhat and Krishnaswamy, 1969). This range leaves considerable uncertainty in any estimate of the river input, but confirms earlier evidence that river sources are inadequate to sustain the oceanic concentration. This was interpreted by Koczy (1958) as indicating the release of ^{226}Ra from interstitial waters of sediments to the overlying water mass, an interpretation which accords with the considerable mobility of the nuclide within the sediment column (see p. 124).

The phenomenon of addition at the sea floor, in contrast with the entry of cosmic-ray produced nuclides from the atmosphere, together with the half-life of 1620 years which is comparable with the time-scale of some deep water mixing processes, suggested that the nuclide could be useful as a tracer for water mixing; these possibilities were initially explored by Koczy (1958) and Koczy and Szabo (1962).

Subsequent investigations have confirmed many of the above features. Values close to 0.4×10^{-13} g l^{-1} have been reported for surface waters of widely separated oceanic regions, in the Indian Ocean (Koczy and Szabo, 1962), the northwestern Atlantic Ocean (Szabo, 1967; Broecker et al., 1967), the north and equatorial Pacific (Broecker et al., 1967, 1970) and north of the Antarctic Convergence at 128°E (Ku et al., 1970). The vertical gradients show considerable differences. As shown in Fig. 18.1, the concentrations for the Pacific stations increase to 1.6×10^{-13} g l^{-1}, whereas for the Atlantic stations the gradient is less marked and the deep water contains about 0.8×10^{-13} g l^{-1}. The data of Miyake and Sugimura (1964) for the western North Pacific also indicate increases to concentrations as high as 1.3×10^{-13} g l^{-1} in deep water, but show a rather wider scatter for surface waters, with an average value of about 0.5×10^{-13} g l^{-1}. Chung and Craig (1973) have found concentrations around 1.4–1.6×10^{-13} g l^{-1} for deep water in the vicinity of the East Pacific Rise, with surface concentrations of 0.26–0.37×10^{-13} g l^{-1}.

The vertical profiles have been interpreted in two ways. The first, following Koczy (1958), treats ^{226}Ra as essentially conservative, and the decrease in concentration in surface water relative to that in deep water is taken to reflect radioactive decay during the residence time of water in the surface layer. Residence times derived on this basis are, however, inconsistent with those derived from measurements of ^{14}C. From the residence times on the latter basis it can be estimated that radioactive decay of ^{226}Ra would produce only a 12 % difference in concentration between surface and deep waters in the Pacific Ocean (Broecker et al., 1967). From this, Broecker et al. (1967) and Wolgemuth and Broecker (1970) have argued that particulate transference mechanisms are important for ^{226}Ra. On this view, water mixing processes would tend to produce a more uniform distribution of ^{226}Ra, but considerable downward transport occurs as a result of association with particles in surface layers and subsequent release in deep waters. For waters north of the Antarctic Convergence, Ku et al. (1970) suggested that siliceous organisms were involved in this transference, a view supported in the discussion of their results by Edmond (1970).

Broecker et al. (1967) have shown that the results in Fig. 18.1 are compatible with a similar transfer rate of ^{226}Ra with particulate material in the different

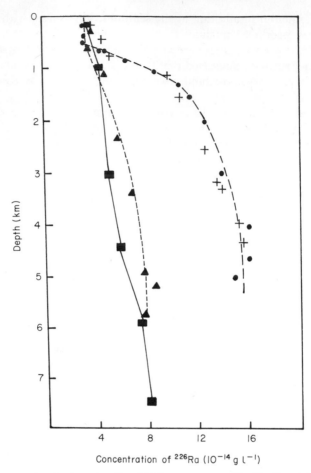

Fig. 18.1. Vertical distribution of radium-226 in ocean waters. ●—Northwest Pacific Ocean; +—East Equatorial Pacific Ocean; ▲—Northwest Atlantic Ocean; ■—Atlantic Ocean, North of the Antilles Islands. After Broecker *et al.* (1967) and Szabo *et al.* (1967).

oceanic regions, the difference in concentrations between the deep waters reflecting the longer residence time of water in the deep Pacific Ocean. The comparative behaviour of radium-226 and barium has been considered by Wolgemuth and Broecker (1970) using a two box (surface and deep reservoirs) model. This first order description shows a considerable degree of geochemical coherence between the elements, supporting the importance of a particulate transference mechanism for ^{226}Ra. The marked similarity in the profiles of Ba and ^{226}Ra at a single station is clearly shown by the results in Fig. 18.2.

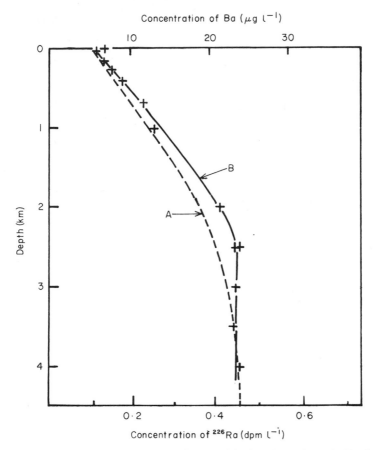

FIG. 18.2. Vertical distributions of radium-226 (A) and barium (B) at the 1969 North Pacific Ocean GEOSECS station (28° N, 122° W). After Wolgemuth (1970).

Measurements south of the Antarctic Convergence show a much more uniform vertical distribution of ^{226}Ra, as a result of upwelling, than occurs to the north, with similar deep water concentrations in both regions (Ku et al., 1970). Rather uniform profiles have also been found in areas of relatively rapid vertical mixing in the Bahamas and Caribbean regions; for example, 40 samples from various depths in the Tongue of the Ocean contained 0.35–0.53×10^{-13} g l^{-1} (Szabo, 1971). Values in the Venezuelan and Colombian Basins were also relatively uniform with average concentrations of 0.58 and 0.53×10^{-13} g l^{-1}, respectively (Szabo et al., 1967) and profiles in the Mediterranean Sea have shown quite uniform values with a mean of about 0.7×10^{-13} g l^{-1} (Koczy, 1958). The average concentration in Black

Sea waters has been reported as 1.0×10^{-13} g l^{-1} (Grashchenko et al., 1960).

Highly variable concentrations of ^{226}Ra occur in coastal waters. A range of 0.34–5.9×10^{-13} g l^{-1} is reported by Blanchard and Oakes (1965) and Koczy et al. (1957) found 0.12–0.60×10^{-13} g l^{-1} for Baltic waters.

Concentrations of ^{226}Ra in sediments were discussed in some detail by Burton (1965); only a brief summary is given here. In near-shore sediments concentrations are usually of the order of magnitude of 1×10^{-6} μg g^{-1}. In pelagic deposits surface layers may contain up to about 4×10^{-5} μg g^{-1}, the highest values being in slowly accumulating clays. These high concentrations of ^{226}Ra are attributed to its production in situ from ^{230}Th. Variations with depth in pelagic sediment cores are often erratic because of the migration of ^{226}Ra which makes it unreliable for most geochronological applications. The diffusion coefficient is similar to that for ^{234}U (see p. 111), and movement of ^{226}Ra released after formation by decay is likewise held up by exchange with solid phases (Goldberg and Koide, 1963). Phillipsite appears to be an important host mineral for migratory radium isotopes (Arrhenius et al., 1957; Bernat et al., 1970).

Recently deposited mollusc shells contain 0.03–2.5×10^{-7} μg g^{-1} of ^{226}Ra (Koczy and Titze, 1958; Broecker, 1963a; Blanchard and Oakes, 1965). The average discrimination factor (see p. 109) is higher for aragonitic (0.49) than for calcitic (0.13) shells (Blanchard and Oakes, 1965). Results summarized by Vinogradov (1953) for various marine organisms, including algae, suggest an average concentration of ^{226}Ra in soft tissue lying between 10^{-7} and 10^{-8} μg g^{-1} dry weight. Edgington et al. (1970) found that ^{226}Ra is concentrated in algae more efficiently than is calcium unless there is a high degree of calcification.

18.3.3.2. Radium-228.

Radium-228 is supplied to the oceans primarily by release from bottom sediments where it is produced by decay of thorium-232. This behaviour parallels that of ^{226}Ra, but the shorter half-life of ^{228}Ra leads to a greatly modified subsequent distribution. Because of the very low concentrations, leading to difficulties in obtaining quantitative recoveries from very large samples, in some cases exceeding 10^3 l, the general approach to measuring ^{228}Ra has been through measurements of ^{226}Ra on smaller samples and determination of ^{228}Ra/^{226}Ra ratios on radium separated from larger samples or obtained by in situ sampling.

In surface waters, concentrations varying from less than 0.3 to 190×10^{-3} dpm l^{-1} (1 dpm $l^{-1} = 1.9 \times 10^{-15}$ g l^{-1}) have been found (Kaufman, 1969a; Moore, 1969a, b; Krishnaswamy et al., 1972; Kaufman et al., 1973). The highest values occur near the continental margins and estimated concentrations derived from measurements of ^{228}Ra/^{226}Ra ratios on shells are also near or above the upper end of the above range, with especially high ^{228}Ra/^{226}Ra ratios in waters with restricted circulation (Blanchard and Oakes, 1965; Moore, 1969a, b). The lowest surface concentrations occur in South Pacific and, most pronouncedly, Antarctic waters. For open Atlantic and Indian Ocean waters the typical range is 10–35 × 10^{-3} dpm l^{-1}.

These features indicate a major input into surface ocean waters by release from inshore, shelf and slope sediments. River water inputs are insufficient to sustain the observed concentrations (Moore, 1969a), and the *in situ* supply from suspended material would require a concentration of ^{232}Th in particulate matter which is greater than any reasonable upper limit. The examination by Kaufman *et al.* (1973) of the distribution of ^{228}Ra in relation to surface ocean circulation patterns shows it to be wholly consistent with supply from marginal sediments. In the Black Sea the distribution of ^{228}Ra can be accounted for only on the basis that there is some input from sediments (Lazarev *et al.*, 1965).

For deep ocean waters collected within 200 m of the sea-bed, Moore (1969a, b) reported concentrations of 2–28 × 10^{-3} dpm l^{-1}; near-bottom Atlantic Ocean waters contain more ^{228}Ra than do those of the Pacific Ocean. The amounts in near-bottom waters must be supplied from the interstitial waters of underlying sediments and ^{228}Ra should be of use in estimating coefficients of vertical eddy diffusion in these waters on a longer time-scale than is provided by measurements of ^{222}Rn (see Section 18.3.4). Waters at intermediate depths contained only 1–4 × 10^{-3} dpm l^{-1}.

These features are confirmed in a profile at the North Atlantic 1970 GEOSECS station which showed a marked decrease within the main thermocline and an increase in near-bottom water (Trier *et al.*, 1972). At a station in the equatorial Atlantic Ocean, Moore (1972) found a rapid decrease from a surface value of 18 × 10^{-3} dpm l^{-1} to values of less than 0·3 × 10^{-3} dpm l^{-1} below the thermocline and has discussed the use of such data in studies of mixing. Subsequent investigations of oceanic distributions by Kaufman *et al* (1973) confirm the potential value of ^{228}Ra as a tracer not only for calculating vertical eddy diffusion coefficients in suitable hydrographic situations, but also in the study of horizontal transport.

18.3.4. RADON-222

Pronounced concentration gradients usually exist for radon-222 (the short-lived daughter of radium-226) at the boundaries of the ocean with both the sea floor and the atmosphere. The principal features of its distribution were first established in detail by Broecker (1965). In water immediately above deep-sea sediments an excess of ^{222}Rn over the amount in equilibrium with the dissolved ^{226}Ra is maintained by its release from the interstitial waters of sediments, in which its concentration is some orders of magnitude greater than that in normal ocean waters. Above the zone influenced by these inputs, the concentrations of ^{222}Rn correspond to equilibrium with the ^{226}Ra present. However, in near-surface waters, because the concentrations of ^{222}Rn are very much greater than those in the atmosphere, there is a continuous loss of ^{222}Rn at the interface, so that usually the concentration of the daughter is significantly below the equilibrium value. These zones are typical of waters which are deep enough to permit their development. In the shallow waters of the Bahama Banks, the input at the sea-bed maintains an excess of ^{222}Rn throughout the water column.

Broecker (1965) showed that the standing crop, M, of ^{222}Rn in water overlying sediments, as defined by the relation

$$M = C_s \sqrt{\frac{D_m}{\lambda}} \tag{18.3}$$

(where C_s is the concentration of ^{222}Rn in the sediment below the zone depleted by diffusion, D_m is the coefficient of molecular diffusion of ^{222}Rn in the sediment, and λ is the decay constant for ^{222}Rn) has a predicted typical value of 9×10^{-13} g cm^{-2}, expressed in equivalents of ^{226}Ra, whereas measured values are in the range $1\cdot5$–160×10^{-13} g cm^{-2}. On the basis that the distribution of ^{222}Rn in the overlying water is determined by its input and the coefficient of vertical eddy diffusion (D_e), i.e. neglecting advective processes, the concentration of excess ^{222}Rn (C) at a height x cm above the bottom is given by the relationship (Broecker, 1965):

$$C = M \left(\frac{\lambda}{D_e}\right)^{1/2} \exp\left[-\left(\frac{\lambda}{D_e}\right)^{1/2} x\right] \tag{18.4}$$

Measurable excesses of ^{222}Rn were found by Broecker et al. (1967) in bottom water of several regions, and this feature has been shown to be of widespread occurrence (Broecker et al., 1968a; Broecker and Kaufman, 1970; Chung and Craig, 1972; Chung, 1973). In some cases the concentrations of excess ^{222}Rn show a close approximation to an exponential decrease above the sediment, as predicted by equation (18.4), enabling values of D_e to be estimated from the observations. Examples of two such profiles are shown in Fig. 18.3. Many near-bottom profiles which show an excess of ^{222}Rn, however, are more complex; some examples are given in Fig. 18.4. Chung and Craig (1972) classified these profiles into several types and demonstrated the value of temperature profiles in their interpretation. Factors which cause departures

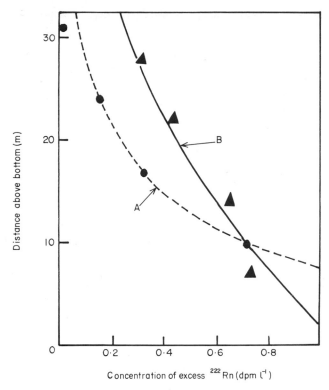

FIG. 18.3. Vertical distribution of excess radon-222 immediately above the sea bed at two stations (A—23° S, 33° W; B—46° N, 175° E) where the concentrations fit the relationship given in equation (18.4). Coefficients of eddy diffusion calculated from the results are: A, 2 cm^2 s^{-1}; B, 8 cm^2 s^{-1}. From Broecker et al. (1968a).

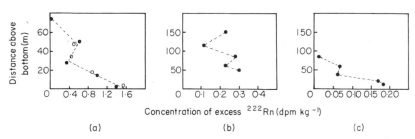

FIG. 18.4. Vertical distribution of excess radon-222 immediately above the sea bed at stations where departures occur from the behaviour predicted by equation (18.4). (a) 28° N, 122° W (1969 GEOSECS station) (b) 6° S, 107° W (Crest of East Pacific Rise) (c) 5° S, 114° W. The open and closed circles in (a) refer to different sampling occasions. After Broecker and Kaufman (1970) and Chung and Craig (1972).

from exponential distributions include variations in vertical mixing rates in different layers, horizontal advective processes, and possibly the presence of high concentrations of suspended matter in nepheloid layers, which could act as an additional source of excess ^{222}Rn. Chung (1973) has shown that in the Santa Barbara Basin the near-bottom profile of excess ^{222}Rn altered from an essentially exponential type in July to one with a maximum concentration somewhat above the sea bed in November. These studies have demonstrated the considerably potential value of measurements of ^{222}Rn near the sea floor, particularly in conjunction with temperature data, in the study of mixing processes in this boundary region.

The use of deficiency anomalies in the concentration of ^{222}Rn in surface waters to estimate coefficients of vertical eddy diffusion in these layers and the rate of exchange of gas across the ocean-atmosphere boundary was also discussed by Broecker (1965). Thirteen samples collected at 1 m in the northwestern Pacific Ocean were found by Broecker et al. (1967) to show activity ratios of ^{222}Rn to ^{226}Ra of 0·41–0·80. The vertical distribution in the upper 300 m is shown in Fig. 18.5. Deficiencies within the above range were found by Broecker and Kaufman (1970) for the 1969 North Pacific GEOSECS station, and by Broecker and Peng (1971) for the BOMEX area (15°N, 46°W). At the latter station the maintenance of a deficiency of about 25% almost uniformly from the surface to a depth of about 20 m was interpreted in terms of a coefficient of vertical eddy diffusion of more than $160 \, \text{cm}^2 \, \text{s}^{-1}$ in that layer, whereas in the zone between 30–40 m, the results were consistent with a value of less than $4 \, \text{cm}^2 \, \text{s}^{-1}$.

18.3.5. LEAD-210 AND POLONIUM-210

Concentrations of ^{210}Pb in ocean water were first reported by Rama et al. (1961a) and Goldberg (1963). The amounts in surface water off the Californian coast were about 0·1 dpm kg^{-1} (1 dpm = $5·1 \times 10^{-15}$ g), representing an excess over the concentration corresponding to equilibrium with the probable concentration of ^{226}Ra. This is attributable to an input of ^{210}Pb from the atmosphere from which it is washed out after formation from ^{222}Rn. Contributions to the inventory of ^{210}Pb from the testing of nuclear weapons appear to have been negligible (Beasley, 1969). Concentrations in waters below 1000 m ranged from 0·09 to 0·27 dpm kg^{-1}.

Latitudinal variations in the concentrations of ^{210}Pb in surface waters of the Pacific have been examined by Tsunogai and Nozaki (1971) and Nozaki and Tsunogai (1973). The highest mean concentrations of about 0·18 dpm kg^{-1} occur in northern mid-latitudes, about 30°N, with moderately high values of about 0·10 dpm kg^{-1} in southern mid-latitudes and low mean

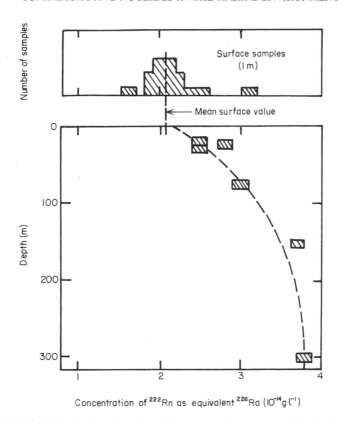

FIG. 18.5. Vertical distribution of radon-222 in near-surface waters of the Northwest Pacific Ocean between 20–30° N, 158° W–144° E. After Broecker *et al.* (1967).

concentrations (about $0.05\,\mathrm{dpm\,kg^{-1}}$) in the equatorial zone and southern South Pacific; at higher northern latitudes the average concentrations decrease to about $0.10\,\mathrm{dpm\,kg^{-1}}$. These differences broadly reflect latitudinal variations in input rate associated with the greater emanation of $^{222}\mathrm{Rn}$ from arid land areas than from other regions.

Craig *et al.* (1973) have compared the profiles of $^{210}\mathrm{Pb}$ and $^{226}\mathrm{Ra}$ at two stations. In the North Pacific Ocean the concentration of $^{210}\mathrm{Pb}$ was nearly constant for most samples below 1800 m at about $0.2\,\mathrm{dpm\,kg^{-1}}$. This represented a decrease from about 67% of the amount corresponding to equilibrium with $^{226}\mathrm{Ra}$ at 1800 m to about 43% in the near bottom water. The results for a station in the North Atlantic Ocean, as shown in Fig. 18.6, although complicated by the presence of various water masses, illustrate the features discussed above. The deficiency of $^{210}\mathrm{Pb}$ in deeper water

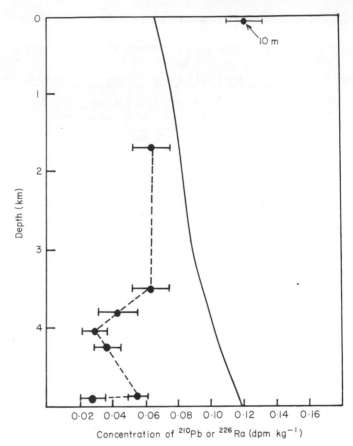

FIG. 18.6. Vertical distributions of lead-210 (--●--) and radium-226 (——) at the North Atlantic Ocean GEOSECS station (36° N, 68° W). Samples at 1700 and 4000 m are in the cores of the Labrador Sea water and Denmark Straits overflow, respectively. The deepest sample analysed for lead-210 was taken 38 m above the sea bed. From Craig *et al.* (1973).

indicates effective removal of the nuclide throughout the whole water column, probably by uptake on sinking particulate material. From the results for the North Pacific Ocean, Craig *et al.* (1973) calculated a residence time of ^{210}Pb in the deep water of 54 years, and Nozaki and Tsunogai (1973) have estimated the residence time in surface water above 100 m as 0·63 years.

Measurements of the rates of deposition of ^{210}Pb in a coastal marine environment (Koide *et al.*, 1972) gave values which are greater than can be balanced by input in rainfall and production from ^{226}Ra in the water column. To interpret this apparent anomaly there is a need for detailed studies of fluxes of ^{210}Pb through such waters and of the role of biological uptake in the cycling of the nuclide.

The concentrations of ^{210}Po in surface waters showed similar latitudinal variations to those reported for ^{210}Pb (Tsunogai and Nozaki, 1971). The mean activity ratio of ^{210}Po to ^{210}Pb, although below unity, was considerably higher than the value of about 0·1 typically found in rainwater. Measurements around the Cape of Good Hope showed an average activity ratio of ^{210}Po to ^{210}Pb of about 0·5; the average concentration of ^{210}Pb in these waters was 0·08 dpm l^{-1} (Shannon et al., 1970). Results for a single profile (Shannon, 1973) showed an increase in the concentration of ^{210}Po below the mixed layer with values of 0·033 dpm l^{-1} at the surface and 0·07 dpm l^{-1} at 300–630 m.

Measurements of ^{210}Po and ^{210}Pb have been made on various marine organisms (Shannon and Cherry, 1967; Holtzmann, 1967; Beasley et al., 1969, 1971; Shannon et al., 1970; Shannon, 1973). They showed that ^{210}Po is concentrated from sea water by organisms to a greater extent than is ^{210}Pb. There are, however, some considerable differences in the concentrations reported for various groups of organisms, and conclusions on bioaccumulation and food chain transference can be drawn only tentatively at this stage. The data of Shannon (1973) indicate that there is a general increase in the concentration of both total and unsupported ^{210}Po along the food chain from phytoplankton to zooplankton to pelagic fish, whereas the concentrations of ^{210}Pb are somewhat similar at around 60–70 dpm kg^{-1} wet weight in phytoplankton and zooplankton, but decrease in pelagic fish to an average value of about 20 dpm kg^{-1}. Although there is considerable variation between individual samples, the overall indication is that the average ratio of ^{210}Po to ^{210}Pb increases from less than 5 in phytoplankton to about 100 in pelagic fish, with a value of about 10 in zooplankton. The data of Beasley et al. (1971) suggest an order of magnitude concentration of ^{210}Po of 10^3 dpm kg^{-1} dry weight in the muscle of pelagic fish and show that there is a greater concentration in liver and viscera; except in bone the nuclide is almost wholly unsupported. These data are broadly consistent with the findings of Shannon (1973) for whole fish, but much lower concentrations are reported by Holtzmann (1967). Particularly high concentrations of unsupported ^{210}Po, averaging $1·8 \times 10^4$ dpm kg^{-1} wet weight, occur in the digestive gland of the rock lobster (Shannon, 1973).

18.3.6. APPLICATIONS TO THE GEOCHRONOLOGY OF MARINE SEDIMENTS

Disequilibria in sediments between members of the natural decay series provide the bases of several geochronological methods, which are essentially of two kinds. In the first, a time scale is provided by the extent of decay of an unsupported nuclide. If the number of atoms of an unsupported nuclide incorporated in unit mass of sediment at the time of formation is N_0, then the number (N_t) remaining at a time t after deposition is given by the general equation describing the first order radioactive decay process

$$N_t = N_0 e^{-\lambda t} \tag{18.5}$$

where λ is the decay constant for the nuclide concerned. In practice, the corresponding radioactivities, A_0 and A_t, are measured. Provided that the initial value (A_0) was the same for different sediment layers then a comparison of the present radioactivities of these layers enables the interval between their deposition to be estimated. This assumption requires, in practice, the use of cores representing uniform conditions of sedimentation over long periods. The use of ratios of nuclides which have common geochemical pathways to the sediment can theoretically overcome the need to assume a constant rate of incorporation. In this approach the concentrations of the nuclide are, in effect, normalized on the basis of the concentration of another radioactive or stable nuclide. The problem here is to identify a suitably coherent pair of nuclides. A pair of isotopes would be expected to meet the requirements on chemical grounds, but complications are often introduced through the differences in oceanic geochemical pathways for isotopes with different genetic histories and modes of input. From the geochronological standpoint, the geochemical coherence of the pair of nuclides is the primary criterion of their usefulness.

The second kind of dating procedure utilizes the growth of a daughter nuclide in material which initially contains the parent nuclide segregated from any significant amount of the daughter product. If the number of atoms of the parent and daughter nuclides in unit mass of material after time t are N and D, respectively, and the number of daughter atoms originally present was D_0, then

$$\frac{D - D_0 e^{-\lambda_d t}}{N} = \frac{\lambda_p}{\lambda_p - \lambda_d}(e^{(\lambda_p - \lambda_d)t} - 1) \tag{18.6}$$

where λ_p and λ_d are the decay constants for the parent and daughter, respectively.

If there is initially a negligible amount of the daughter present, then the elapsed time since the segregation of parent and daughter is given by

$$t = \frac{1}{\lambda_p - \lambda_d} \log_e\left(1 + \frac{D}{N}\frac{(\lambda_p - \lambda_d)}{\lambda_p}\right) \tag{18.7}$$

The marine applications of growth methods in the radioactive decay series centre largely on the use of radioactive daughter products of uranium, namely ^{230}Th and ^{231}Pa. The accumulation of helium and of lead isotopes in uranium and thorium bearing materials can also be used in dating, and the helium growth method has been applied by Fanale and Schaeffer (1965) to shells from raised beaches and to corals.

The brief discussion of geochronological methods given here emphasises the geochemical factors which affect the validity of the procedures, and a detailed examination of derived sedimentation rates is not made.

18.3.6.1. Geochronology using the decay of thorium-230 and protactinium-231
The presence of considerable amounts of unsupported ^{230}Th and ^{231}Pa in surface layers of deep-sea sediments has been discussed in Section 18.3.2. The use of unsupported ^{230}Th to date sediments was first attempted through measurements of its daughter, ^{226}Ra, but these attempts were greatly complicated by the diagenetic migration of ^{226}Ra. Present developments stem from the proposal by Picciotto and Wilgain (1954) for the use of the ratio of unsupported ^{230}Th to ^{232}Th as a geochronological index, with a range of about 4×10^5 years. The ratio required for this purpose is that in the authigenic fraction. The method assumes that the ratio in the sea water from which this fraction is derived was constant over the interval dated and that there is no significant post-depositional migration of the thorium isotopes. The second assumption appears well founded (Bernat and Goldberg, 1969; Bonatti *et al.*, 1971). Although it appears fundamentally disadvantageous that authigenic ^{230}Th and ^{232}Th have different routes of entry to the ocean and do not become well mixed before removal, the consistency of results for most cores suggests that the first assumption is also usually justified. Corrections for uranium-supported ^{230}Th are necessary for the older parts of the core. The amounts of unsupported ^{230}Th in lithogenous phases can be neglected, but a complication arises through the substantial lithogenous fraction of ^{232}Th in deep-sea sediments.

Applications of the ^{230}Th/^{232}Th method have largely followed the approach of Goldberg and Koide (1958), who used leaching of the sediment with hydrochloric acid as a means of removing primarily the authigenic thorium isotopes. This technique has been extensively applied (Goldberg and Koide, 1962, 1963; Goldberg and Griffin, 1964; Goldberg *et al.*, 1964a, b; Goldberg, 1965; Miyake and Sugimura, 1961, 1965; Blackman and Somayajulu, 1966; Sugimura and Miyake, 1968; Miyake *et al.*, 1968; Heath *et al.*, 1970; Griffin *et al.*, 1972) and has usually given consistent results, assisted undoubtedly by the general tendency towards uniformity of occurrence of ^{232}Th in many cores. The empirical nature of any extraction technique as a means of distinguishing authigenic material is generally recognised, but the approach has been shown to give more interpretable results than total dissolution (Goldberg *et al.*, 1964b), and at least takes some account of the variations which do occur in thorium content. The $^{230}Th/^{232}Th$ method has been applied also to ferro-manganese concretions (Goldberg, 1961; Barnes and Dymond, 1967; Somayajulu *et al.*, 1971; Krishnaswamy and Lal, 1972).

Koczy (1961, 1965) has argued that deposition rates obtained by this method will generally be too high as a result of the partial extraction of lithogenous ^{232}Th. It has been shown by Antal (1966) that the use of less drastic conditions for leaching sediments may fail to remove authigenic ^{230}Th to an extent which changes with the age of the deposit as a result of diagenetic processes, a condition which clearly precludes successful use in dating. One approach which has been developed to avoid the problems associated with normalization on the basis of the concentration of ^{232}Th is the use of the ratio of ^{231}Pa to ^{230}Th as a geochronological index (Rosholt *et al.*, 1961).

The choice of this pair of nuclides is based upon the similarity of their geochemical pathways to the sediment. Both nuclides are generated by decay of uranium isotopes (^{235}U and ^{234}U, respectively) which are uniformly distributed in the ocean; after formation they are rapidly removed to sediments. The unsupported concentrations of both nuclides are almost wholly authigenic. Although the geochemical coherence of the nuclides in marine sedimentary processes is not so ideally close as to allow the assumption that the theoretical production ratio of the nuclides applies to the surface layers of all sediments (see Section 18.3.2.3), the general requirement that the initial incorporation occurs in a constant ratio over the dating period appears to be fulfilled for many deposits. The ratio decreases with a half-life of $5 \cdot 72 \times 10^4$ years, giving an effective range for the method of about 3×10^5 years. The method has usually been applied on the basis of total dissolution of the sediment, with correction for the amounts of the two nuclides which are uranium supported. Applications include the investigations by Rosholt *et al.* (1961, 1962), Rona *et al.* (1965), Ku and Broecker (1967b) and Rona and Emiliani (1969).

An alternative to the methods using either $^{230}Th/^{232}Th$ or $^{231}Pa/^{230}Th$ ratios is to use simply the change in concentration of either unsupported ^{230}Th or unsupported ^{231}Pa. This requires the assumption of the incorporation of a constant concentration of the nuclide at the time of sedimentation over the period concerned. This may appear a considerable constraint on the utility of the method, but as with the other procedures the conformity of the values with a consistent decay pattern provides a good internal check on the validity of the assumption for each core examined. The approach has, in fact, proved satisfactory for estimating the long-term average deposition rates of a range of sediment cores analysed after total dissolution (Sackett, 1965; Ku and Broecker, 1966; Ku et al., 1968, 1972; Kharkar et al., 1969; Bender et al., 1971) and for ferro-manganese concretions (Ku and Broecker, 1967a, 1969; Krishnaswamy and Lal, 1972). The difference between the methods using unsupported ^{230}Th and the ^{230}Th/^{232}Th ratio is often smaller in practice than in principle, as the concentration of ^{232}Th often remains fairly uniform with depth in the core (Goldberg and Griffin, 1964). Krishnaswamy and Lal (1972) report results by the two methods for ferro-manganese concretions; they were mostly in very good agreement.

Several workers (Osmond and Pollard, 1967; Yokoyama et al., 1968; Popov and Grekov, 1971) have explored the possibilities of direct measurements of ^{230}Th or of the ratios of ^{230}Th to total ^{232}Th in sediments through gamma spectrometric measurements of daughter nuclides. In view of the problem of mobility of ^{226}Ra, interpretation through daughter activities clearly requires caution, but the method has advantages in its directness and rapidity.

The use of these different approaches and the results of comparisons of methods included in some of the above studies have led to some controversies. There has, for example, been disagreement concerning the chronology of Caribbean cores (Rona and Emiliani, 1969; Broecker and Ku, 1969; Emiliani and Rona, 1969) which stemmed primarily from discrepant uranium analyses (see also Mo et al., 1971; Rydell and Fisher, 1971). Interpretation of results by the methods using unsupported ^{230}Th and the ^{230}Th/^{232}Th ratio has also led to controversy (Ku et al., 1968; Goldberg, 1968) centred mainly upon the derivation of the best fit to the observed decrease in the concentration of ^{230}Th with depth below the sediment surface. For some cores a considerable amount of information is available, not only from the methods just discussed, but also from other methods such as ^{14}C dating, K/Ar dating, and palaeomagnetic measurements. The existence of such information is of great value in enabling anomalous values given by a particular method to be identified and in verifying the suitability of cores for dating by particular methods.

18.3.6.2. Geochronology using the decay of excess uranium-234

The presence of excess ^{234}U in authigenic marine deposits (see Section 18.3.1) potentially gives a method for dating such materials over a range of about 10^6 years, using the ratio of ^{234}U to ^{238}U. The primary assumptions are that the activity ratio of ^{234}U to ^{238}U in the ocean was essentially constant over this period, which seems reasonably well established, and that there is no diagenetic exchange of uranium or migration of the isotopes within the system to be dated. Ku (1965) showed that mobility of ^{234}U, together with the difficulty of analytically distinguishing authigenic and lithogenous material, renders the method valueless for general application to pelagic sediments, although horizons from a single core have been dated by Krishnamoorthy *et al.* (1971) using an extraction technique which removed only the carbonate fraction.

Although some values for accretion rates in surface layers of ferromanganese concretions have been obtained from $^{234}U/^{238}U$ ratios (Ku and Broecker, 1967a; Barnes and Dymond, 1967), later work (Ku and Broecker, 1969) has shown that mobility of ^{234}U in these materials limits the general applicability of the method. With carbonates the method is subject to the limitations associated with open systems, which are discussed in Section 18.3.6.3. An application of general validity has been to corals for which the method provides a useful complementary approach to the ^{230}Th growth method (Veeh, 1966; Broecker *et al.*, 1968b). Ratios in phosphorites have also been used to establish ages ranging from the late Pleistocene to very recent (Baturin *et al.*, 1972; Veeh *et al.*, 1973) and the method could also find application to other deposits enriched in authigenic uranium.

18.3.6.3. *Geochronology using the growth of thorium-230 and protactinium-231*
As discussed in Section 18.3.1, uranium is sometimes incorporated to an appreciable extent into calcareous material during its formation. This is particularly the case with corals which contain, on average, about 3 $\mu g\ g^{-1}$ of uranium. For deposits which are substantially free from detrital material the incorporation of uranium occurs with a high degree of segregation from the daughter products, ^{230}Th and ^{231}Pa. The growth of these daughters then provides a possible method of dating which was first explored by Sackett (1958), Potratz and Sackett (1959) and Tatsumoto and Goldberg (1959). Allowance must be made for the extent of disequilibrium between ^{234}U and ^{238}U for which purpose an initial activity ratio of ^{234}U to ^{238}U of 1·15 may be assumed.

The equation for the growth of ^{230}Th is (Broecker, 1963a)

$$\frac{A_{230Th}}{A_{238U}} = (1 - e^{-\lambda_{230Th}t}) + \left(\frac{A_{234U} - A_{238U}}{A_{238U}}\frac{\lambda_{230Th}}{\lambda_{230Th} - \lambda_{234U}}\left\{e^{-\lambda_{234U}t} - e^{-\lambda_{230Th}t}\right\}\right)$$

(18.8)

where A and λ are respectively the radioactivities and the decay constants of the nuclides indicated (A_{234U} is the *initial* radioactivity) and t is the time since segregation of the uranium. The decay of ^{234}U can be neglected with a maximum error in dating of only 2% for samples younger than $1·3 \times 10^5$

years (Osmond et al., 1965). The relationship can then be simplified to give for the age (t) of the deposit

$$t = \frac{1}{\lambda_{230\text{Th}}} \log_e \left(1 - \frac{A_{230\text{Th}}}{A_{234\text{U}}}\right) \tag{18.9}$$

The basic assumptions of the method are (i) that the system has remained closed for uranium, which had a constant initial ratio of ^{234}U to ^{238}U, and that significant internal migration of ^{234}U has not occurred, (ii) that the system is essentially free from ^{230}Th to begin with, remains closed with respect to ^{230}Th, and that significant migration of ^{230}Th has not occurred. Various criteria have been proposed by which to assess the suitability of material for dating, including absence of mineralogical changes, freedom from unusual amounts of ^{232}Th, absence of excess ^{228}Th and consistency of $^{234}\text{U}/^{238}\text{U}$ ratios within the material and normality of the zero-time ratio. The distribution of ^{226}Ra has also been examined to give ancillary information (Broecker, 1963a).

In practice, the only materials which appear generally suitable for dating by this method are corals. As discussed in Section 18.3.1, other calcareous materials such as mollusc shells often incorporate much less uranium when the organism is alive and, more importantly, they commonly show major secondary additions of the element. Kaufman et al. (1971) have critically examined the data on 400 mollusc shell samples and have shown that, for materials of independently known ages, the dates derived from $^{230}\text{Th}/^{234}\text{U}$ ratios are correct for only about 50% of the samples, and that the previously suggested criteria do not satisfactorily serve to distinguish those materials which give valid ages. Furthermore, with samples for which such comparisons cannot be made there was often evidence that the systems had not remained closed. Szabo and Rosholt (1969) have attempted to obtain ages from measurements on open systems for which there is disagreement between dates obtained by the ^{230}Th growth method and the analogous method using ^{231}Pa. Their model for this purpose has been criticised by Kaufman et al. (1971) on several grounds. They consider that the assumptions made in its derivation require improbable similarities in the behaviour of ^{231}Pa, ^{230}Th and ^{234}U, and point out that the model does not give correct values for samples of independently known age.

Many samples of coral, within the Pleistocene at least, appear to fulfil the requirements of the method. It has been applied extensively to these deposits and has proved especially valuable in providing an absolute chronology for sea-level stands (Thurber et al., 1965; Broecker and Thurber, 1965; Veeh, 1966; Veeh and Valentine, 1967; Labeyrie et al., 1967; Stearns and Thurber,

1967; Broecker *et al.*, 1968b; Mesolella, *et al.*, 1969; Valentine and Veeh, 1969; Osmond *et al.*, 1970; Veeh and Chappell, 1970). Dating of coralline limestones deposited during interglacial periods by this method has been used by Konishi *et al.* (1970) to estimate rates of vertical displacement of islands due to tectonic activity in the Ryuku island arc. Very good agreement has been found between ages based on the ^{14}C and ^{230}Th methods when applied to material fulfilling the basic criteria (Thurber *et al.*, 1965; Veeh and Chappell, 1970). The ^{14}C method is, of course, also subject to error for material in which there has been considerable inversion of aragonite.

The growth of ^{231}Pa from ^{235}U provides an analogous dating method with a range of $1\cdot2 \times 10^5$ years, which has sometimes been used to obtain ages for corals independently of the ^{230}Th/^{234}U method (Sakanoue *et al.*, 1967; Broecker *et al.*, 1968b; Ku, 1968; Kaufman *et al.*, 1971).

The growth methods using ^{230}Th and ^{231}Pa have been applied to obtain ages for iron-rich sediments of the Red Sea geothermal area, the nuclides being present in these deposits at concentrations below those expected for equilibrium with the relatively high concentrations of uranium (Ku, 1969). The values obtained are maximum ages, because of significant initial incorporation of the daughter nuclides. Because of this the ^{230}Th ages are expected to be greater than those based on ^{231}Pa, the latter being the more accurate estimates; these features were confirmed by the experimental measurements. Maximum ages of phosphorites have also been derived by the ^{230}Th growth method (Veeh *et al.*, 1973).

18.3.6.4. *Geochronology using the decay and growth of short-lived nuclides*

Some attention has recently been given to the use of short-lived nuclides in the natural decay series for measuring deposition rates in rapidly accumulating coastal sediments. Koide *et al.* (1972, 1973) used the decrease in concentration of unsupported ^{210}Pb with depth in two cores, for this purpose. One core was from varved sediment and the method gave a deposition rate in good agreement with the independent estimate from the annual varves. There was evidence of mobility of ^{210}Pb in the uppermost few centimetres, probably associated with changing redox conditions, but below this zone the method, which has a range of about 100 years, gave satisfactory results. Koide *et al.* (1973) have also demonstrated the possibility of using ratios of unsupported ^{228}Th to ^{232}Th for dating over a period of about 10 years.

Estimates of the growth rates of corals on relatively short time scales have also been derived from measurements of short-lived nuclides (Moore and Krishnaswamy, 1972; Moore *et al.*, 1973). Use of the ratio of ^{228}Ra to ^{226}Ra has given satisfactory results for growth rates over some 30 years, but attempts to use the growth of ^{210}Pb from ^{226}Ra have given results which are less readily interpreted, perhaps because of variations in the initial ratio for different zones, or because of migration of this daughter product or an intermediate nuclide in the decay chain.

18.4. NUCLIDES PRODUCED BY COSMIC RADIATION

18.4.1. INTRODUCTION

The prediction by Libby (1946) that carbon-14 and tritium should occur in nature as a result of nuclear reactions between atmospheric nitrogen and neutrons produced by cosmic radiation, and its subsequent confirmation, opened up a new field of research. There have been numerous developments in the application of these two nuclides to geophysical problems, and their artificial production by explosions of nuclear weapons has both complicated the picture and afforded further possibilities of tracer applications. Cosmic ray interactions produce a wide range of other nuclides in terrestrial matter, particularly in the atmosphere, and in extraterrestrial material accreted by the earth. The production of these nuclides and their geophysical applications have been reviewed by Lal and Peters (1962, 1967), Lal (1962, 1963a, b, 1969) and Lal and Suess (1968).

Some general information on the characteristics and distributions of those nuclides which are, at present, of interest for marine studies is given in Table 18.5. These nuclides present a range of chemical properties and half-lives which make them useful in different ways. Beryllium-7, with its short half-life, is of interest as an indicator of fallout rates and possibly as a tracer for studying the mixing of surface waters on a time scale of months. Tritium is an ideal tracer for water mixing processes, with a time-scale appropriate for investigating processes in the mixed layer and the thermocline. The longer half-life of ^{14}C makes it of value for studying exchanges between surface and deep waters, the deep water circulation, and the geochronology of recent sediments. Investigations of this nuclide have been greatly assisted by the comparative readiness with which precise measurements can be obtained. Silicon-32 has a usefully intermediate, but rather uncertainly known, half-life; its applications have been restricted by analytical difficulties associated with its low concentration. The longest lived nuclides in the group considered here, ^{26}Al and ^{10}Be, occur dominantly in sediments, and interest in them centres on their geochronological applications and, for ^{26}Al, possible implications for the accretion of cosmic dust; there are, again, analytical difficulties, especially with ^{26}Al, a positron emitting nuclide.

TABLE 18.5

Basic information concerning nuclides produced by cosmic rays[a]

Nuclide	^{3}H	^{7}Be	^{10}Be	^{14}C	^{26}Al	^{32}Si
Half-life (years)	12·3	0·145	$2·5 \times 10^6$	5730[b]	$7·4 \times 10^5$	500[c]
Production rate[d] in total atmosphere (atom cm^{-2} s^{-1})	0·25	0·081	0·045	2·5	$1·4 \times 10^{-4}$	$1·6 \times 10^{-4}$
Fraction of total earth inventory in						
Atmosphere	0·072	0·71	$3·9 \times 10^{-7}$	0·019	$1·4 \times 10^{-6}$	$2·0 \times 10^{-3}$
Land surface[e]	0·27	0·08	0·29[f]	0·04	0·29[f]	0·29[f]
Ocean—mixed layer[g]	0·35	0·2	$5·7 \times 10^{-6}$	0·022	$1·4 \times 10^{-5}$	0·0035
Ocean—excluding mixed layer	0·3	0·002	10^{-4}	0·92	7×10^{-5}	0·68
Oceanic sediments	0	0	0·71	0·004	0·71	0·028
Average concentration in ocean (10^{-3} dpm kg^{-1} water)	36	—	10^{-3}	260	$1·2 \times 10^{-5}$	$2·4 \times 10^{-2}$
Average specific activity in ocean (dpm g^{-1} element)	$3·3 \times 10^{-4}$	$3·2 \times 10^{-3}$	1600	10	0·0012	0·008
Global inventory (kg)	3·5	1·1	$4·3 \times 10^5$	$7·5 \times 10^4$	$1·1 \times 10^3$	1·4
Global inventory (MCi)	35	1·1	64	340	0·020	0·023

[a] Compiled from the data of Lal and Peters (1967). The fractional inventories, in particular, are approximate values.

[b] The earlier value of 5568 years continues to be used conventionally for many dating purposes.

[c] There is considerable uncertainty about this value as recently reported values vary by a factor of two. It seems probable that the true value is nearer to 700 years than to the value given by Lal and Peters.

[d] These production rates were calculated from data on fluxes of cosmic ray particles and cross sections for the reactions concerned ; estimates from global inventories are also available for some nuclides. Additional information is given in the text concerning the production rates of ^{3}H and ^{14}C. Uncertainties in production rates are reflected in some of the other data tabulated, such as the inventory values.

[e] Comprises the topsoil, terrestrial biosphere, and groundwaters.

[f] Some of this fraction may be transported to the ocean before decay.

[g] The oceanic mixed layer is defined here as the upper 75 m approximately.

The applications of these nuclides to geophysical problems involve assumptions regarding the constancy of the rates of production of the nuclides in question, averaged over the relevant time-scales. Detailed evidence on this point is available for ^{14}C and is discussed below. Measurements of various cosmic-ray produced nuclides in meteorites indicate that their production rates, averaged over periods corresponding to their half-lives, have been constant at least within a factor of two, these observations covering periods of up to some millions of years (Arnold et al., 1961). Average production rates of ^{10}Be have been calculated by Higdon and Lingenfelter (1973) from measurements on sediments. Constant sedimentation rates must be assumed so that the interpretation is not unequivocal, but on this basis the results are consistent with a significant increase in the production rate of ^{10}Be between 1·4 and 3×10^6 years ago, which these authors suggest may be related to a supernova explosion. Better records of such variations may be preserved in ferro-manganese concretions, but the studies on these have so far indicated the average values on a time-scale of about 5×10^6 years, over which such variations are not apparent (Bhat et al., 1973).

18.4.2. CARBON-14

Carbon-14 is formed in the atmosphere by an n,p reaction of secondary neutrons, produced by cosmic rays, with nitrogen. Various estimates of the rate of production have given values of about 2 atom cm^{-2} s^{-1}. Lingenfelter (1963) calculated the production rate over the solar cycle from 1946–1958 as $2 \cdot 5 \pm 0 \cdot 5$ atom cm^{-2} s^{-1}; the rate varied by as much as 16% from this value during the cycle. The average production rate over the last three solar cycles has been estimated as $2 \cdot 2 \pm 0 \cdot 4$ atom cm^{-2} s^{-1} (Lingenfelter and Ramaty, 1970). The ^{14}C atoms are oxidized in the atmosphere, and the ^{14}C thus enters the exchangeable system of carbon; this system is discussed in detail in Chapter 9. Of the total exchangeable carbon of about 8 g cm^{-2}, more than 90% occurs in the deep oceanic reservoir; the fractions in the atmosphere and oceanic mixed layer are each about 2%.

The specific activity of ^{14}C (ratio of ^{14}C to stable carbon) in the atmosphere is closely reflected in the terrestrial biosphere so that a record of specific activities is preserved in old wood. The residence time in the atmosphere is sufficiently long for mixing to mask the latitudinal variation in production rate. For wood from the late 19th century the specific activity is close to 14 dpm $g^{-1}C$. Comparison of this value with Lingenfelter's estimate of the more recent natural production rate suggests that the latter may somewhat exceed the decay rate, so that the inventory of natural ^{14}C may presently be increasing.

The use of measurements of ^{14}C in various dating applications depends upon the essential constancy of its production rate and rates of exchange between reservoirs in the exchangeable system over the periods concerned. Very detailed examinations have been made, using tree rings, of variations in the specific activity of atmospheric carbon over several millenia (see, for example, Suess (1965), Bray (1966), Damon *et al.* (1966), Kigoshi and Hasegawa (1966)), and it has been possible, using the bristle-cone pine, *Pinus aristata*, and varved clay sediments, to extend knowledge of these variations back over more than 7×10^3 years (see Ferguson (1968, 1970, 1972) and additional papers in Olsson (1970)). It is apparent that significant variations have occurred on several time-scales. Thus, there has been oscillation on a time-scale of about 10^4 years with an amplitude of about 10% and variations on shorter time-scales of about 200, 400 and 1000 years with amplitudes of up to about 6%. The implications of these variations have been discussed, particularly by Houtermans *et al.* (1973). Because the total ^{14}C in the exchangeable system is about 10^4 times greater than the annual production, variations in the latter on time-scales of up to 1000 years have very little effect on the average concentration in the whole system and on the concentration in the deep oceanic reservoir. The effects of variations on the concentrations in the smaller reservoirs (atmosphere, biosphere, oceanic mixed layer) depend on their time-scales and the residence time of ^{14}C in the reservoir concerned. Unfortunately, the residence time of carbon dioxide in the atmosphere, a most important quantity for an understanding of the system, is poorly known. Estimates range from about 2–50 years, and it is only possible to arrive at an order of magnitude value of about 10 years. The estimates have been derived from box-model calculations, as discussed below, using various observations of steady-state and perturbed situations. Variability arises partly from differences in definition of residence time in the models employed, partly from consideration of processes on different time-scales, and partly from the application of data reflecting mixing processes in particular regions to simplified and generalized models. Houtermans *et al.* (1973) concluded that a simple two-box model (see p. 148) was adequate for considering longer-term changes in the atm. concentration of ^{14}C in relation to variations in its production rate, and they showed that changes that have occurred on time-scales of greater than 100 years are explicable by changes in production rate of up to 50% amplitude.

Several factors may be involved in producing changes in the production rate of ^{14}C (see Lal and Suess (1968) and papers in Olsson (1970)). The flux of galactic cosmic rays alters through solar modulation. This appears to be the most significant cause of short-term variations, but changes in cosmic-ray intensity associated with solar flares must also be considered. Long term changes occur through modulation of the flux of cosmic rays entering the atmosphere, as a result of changes in the earth's magnetic dipole. Additionally, supernova variations and climatic factors influencing the distribution of ^{14}C in the exchangeable system must be considered as possible factors affecting the atmospheric concentration.

Two artificial effects have also disturbed the steady-state condition in the exchangeable carbon system. During this century there has been a significant addition to the atmosphere of carbon dioxide from intensive consumption of fuels. Because this added carbon dioxide is essentially free from ^{14}C this has led to a detectable reduction in the specific activity of atmospheric carbon dioxide as compared with the essentially pre-industrial value of the mid-nineteenth century. Baxter and Walton (1970) have estimated that this effect (the fossil fuel or Suess effect) caused average reductions of 0·5, 3·2 and 5·9% by 1890, 1950 and 1969 respectively. The effect can show pronounced regional variations in industrial areas for which changes of 10% or greater have been reported (Young and Fairhall, 1968; Walton et al., 1970). Baxter and Walton (1970) predict a fossil fuel effect of almost 25% by the year 2000.

The second artificial effect has arisen over the last two decades from the production of ^{14}C by explosions of nuclear weapons. Increases from this source in the specific activity of tropospheric carbon dioxide were significant during the 1950's, but the major increase arose from intensive testing of weapons in the early 1960's. Nydal and Lövseth (1970) have given a detailed description of the resulting trends in the specific activity in the troposphere. Their results show that in the Northern Hemisphere a steep rise in 1963, superimposed on earlier increases, led to peak concentrations during mid-1963 and mid-1964 which were about 100% above the natural level; there were seasonal oscillations associated with the transference of fallout from the stratosphere to the troposphere each spring. Specific activities decreased from these peak levels to about 60% above the natural level by 1969. This reflected the moratorium on testing of weapons from 1963 and the passage of excess ^{14}C into other reservoirs of the exchangeable system. In the Southern Hemisphere, which has received most of its fallout by atmospheric transference rather than direct injection, the peak values were reached somewhat later and were about 70% above the natural level. In 1968–69 the levels in the troposphere of both hemispheres were very similar. Stratospheric concentrations, however, were still higher than those in the troposphere (Walton et al., 1970). The rate at which the atmospheric specific activity will decrease depends not only on the rate of transference to other reservoirs but also on the fossil-fuel effect. Baxter and Walton (1970) predicted that, in the absence of further substantial inputs of ^{14}C, the specific activity would return to the "pre-bomb" level before the end of this century.

F

The inventory of "bomb" ^{14}C has been estimated as 6.5×10^{28} atoms (United Nations Scientific Committee on the Effects of Atomic Radiation, 1964), more than twice the normal atmospheric inventory and about 3% of the total natural ^{14}C.

In the presentation of data for ^{14}C, results are generally expressed as a *per mille* difference from a standard, with normalization on the basis of ^{13}C/^{12}C ratios to eliminate differences due simply to isotopic fractionation, using the scale proposed by Broecker and Olsen (1961). In this, a Δ value is defined by the equation:

$$\Delta = \delta^{14}C - (2\delta^{13}C + 50)\left(1 + \frac{\delta^{14}C}{1000}\right) \tag{18.10}$$

where

$$\delta^{13}C = \frac{(^{13}C/^{12}C)\,\text{sample} - (^{13}C/^{12}C)\,\text{standard}}{(^{13}C/^{12}C)\,\text{standard}} \times 1000 \tag{18.11}$$

and

$$\delta^{14}C = \frac{(^{14}C/^{12}C)\,\text{sample} - 0.95\,(^{14}C/^{12}C)\,\text{standard}}{0.95\,(^{14}C/^{12}C)\,\text{standard}} \times 1000. \tag{18.12}$$

In equation (18.11) the standard is a belemnite. In equation (18.12) it is the National Bureau of Standards oxalic acid standard corrected for decay since 1958. The factor 0.95 in equation (18.12) and the constant 50 in equation (18.10) are introduced to place the zeropoints of both the δ^{14}C and Δ scales close to the values for 19th century wood.*

A Δ^{14}C value may be converted into an age (t) by the expression

$$t = \frac{t_{\frac{1}{2}}}{0.693}\log_e\left(\frac{1}{1 + \Delta^{14}C}\right) \tag{18.14}$$

* The Δ values defined by equation (10) replace Δ^{14}C values as defined in the original scale proposed by Broecker and Olson (1959) which incorporated a logical inconsistency. The relation between the values is:

$$\Delta = \Delta^{14}C - \frac{\delta^{14}C}{20} \tag{18.13}$$

Both Δ^{14}C and Δ values appear in the literature. Throughout the following text the term Δ^{14}C is used since many of the works cited employ these values and the general sense of the discussion is equally well conveyed by either notation. The use of oxalic acid as a standard also led initially to some inconsistency since in converting this compound to the form used for measurement of its radioactivity varying degrees of isotopic fractionation were introduced. This problem is discussed by Craig (1961) and Broecker and Olson (1961). It is overcome by normalizing the measured activities of the standard to correspond to a δ^{13}C value of -19.0.

where $t_{\frac{1}{2}}$ is the half-life of ^{14}C. When material is removed from the exchange-able system, as in carbonate sediments or dead wood, the derived age has an unequivocal significance and represents the time elapsed since the material was part of the exchangeable system. This is exploited in the general techniques of archaeological dating and sedimentary geochronology using ^{14}C. The technique has been of great value in the dating of marine sediments, normally using carbonate material in oozes and corals, but sometimes using organic carbon. Within its range of about 4×10^4 years, it has proved useful as a comparative method for assessing other radiometric techniques over parts of their time scales, and has been invaluable in estab-lishing an absolute chronology for the climatic events of the early Quaternary revealed by palaeotemperature and other stratigraphic studies (see, for example, Ericson *et al.* (1956), Broecker *et al.* (1958, 1960a), Ku *et al.* (1969)).

In selecting sediment material for dating it is important to exclude detrital carbonate material which has an age greater than that corresponding to the time of sedimentation. This can be essentially achieved by suitable selection of the fraction for dating on the basis of particle size; Olsson and Eriksson (1965) have discussed this question with reference to their own and earlier results.

In addition to these differences arising from decay after removal from the exchangeable system, there are significant variations in $\Delta^{14}C$ within the system itself. These variations have provided unique approaches to the study of mixing processes between the atmosphere and the ocean and within the oceanic reservoirs, since they reflect the slow exchange between the atmosphere and the oceanic mixed layer and the mixing time of the deeper oceanic reservoirs. Thus, the carbon in the surface layers and, to a greater degree, the deep ocean, is deficient in ^{14}C compared with sea water in equilibrium with the atmosphere, reflecting decay during the time that the water is isolated from exchange at the surface. The conversion of $\Delta^{14}C$ values into apparent ages for such samples can be an equivocal procedure, however, since the "age" depends upon the mixing history of the water mass. In qualitative terms this is readily understood from the fact that a given "age" may arise by the mixing of different proportions of waters of various older and younger "ages". Thus, if meaningful interpretations are to be made, the $\Delta^{14}C$ values must be incorporated into a model, appropriate to the particular purpose of the investigation. However, before briefly con-sidering this question, some general features of the distribution of $\Delta^{14}C$ values in the ocean are described.

Information on the variations in the specific activity of inorganic carbon in the oceans is available from a number of investigations. Most of the data are given in the papers by Broecker *et al.* (1960b, 1961), Broecker (1963b, c) and Bien *et al.* (1965). In Fig. 18.7, average $\Delta^{14}C$ values are shown for the inorganic carbon of the Atlantic Ocean. It is apparent that the values reflect the distribution of the major water masses. Vertical profiles for the Pacific and Indian Oceans, measured by Bien *et al.* (1965) show increasingly negative $\Delta^{14}C$ values with depth to about 1500 m below which the values

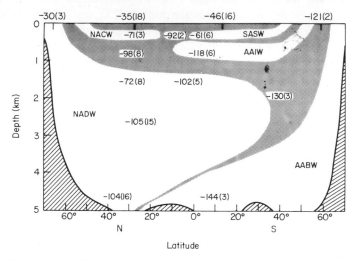

FIG. 18.7. Distribution of ^{14}C in the Atlantic Ocean. Values shown are for $\Delta^{14}C$ (‰) as defined in the text; numbers of observations are given in parentheses. Values refer to the dissolved inorganic carbon. NACW—North Atlantic Central Water; SASW—South Atlantic Surface Water; AAIW—Antarctic Intermediate Water; NADW—North Atlantic Deep Water; AABW—Antarctic Bottom Water. After Broecker (1963).

show little variation in the deep water at a given position. The average values for deep water found by these workers are summarised in Fig. 18.8; the increasing ages (more negative $\Delta^{14}C$ values) reflect the movement of bottom water.

Subsequently to these investigations, measurements on surface waters have been made which clearly reflect the influence of ^{14}C from nuclear weapons (Bien and Suess, 1967; Münnich and Roether, 1967; Rafter, 1968; Young and Fairhall, 1968; Nydal and Lövseth, 1970; Rafter and O'Brien, 1970; Fairhall *et al.*, 1970, 1971). These recent concentrations show a wide scatter, but broadly reflect the changes in atmospheric specific activity. Nydal and Lövseth (1970) have summarized many of the data (see Fig. 18.9) which show a maximum increase in the surface layer in about 1968 of, on average, 10–15%. According to these workers this represented the peak concentration

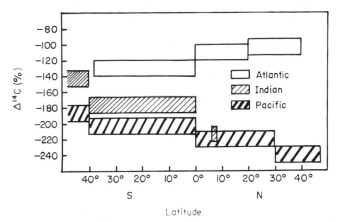

FIG. 18.8. Average values for $\Delta^{14}C$ in ocean waters below 3000 m. After Bien *et al.* (1965).

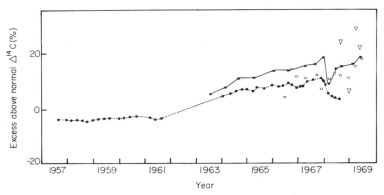

FIG. 18.9. Variations from 1957 to 1969 in the excess carbon-14 introduced to the surface ocean by testing of nuclear weapons. ●—40° S, 174° W; ■—5°–45° N; □—5° S, 30′ W; ▽—*ca.* 33° S, 50° W. After Nydal and Lövseth (1970).

in surface waters. Rafter and O'Brien (1970) found that for nine stations in the South Pacific and Antarctic Oceans there was an average increase in specific activity for surface waters of 14% from 1958 to 1968. There was again evidence that the increase in surface water concentration had then passed its maximum value. When further results are available, the distribution of excess ^{14}C into the deep oceanic reservoirs should give more insight into the time-scale of exchanges with surface waters. The available data suggest a rather rapid transfer of ^{14}C from the mixed layer as generally defined in simple box models, with a residence time relative to transfer to deeper waters of only about 5 years.

The linear box or reservoir models used for the interpretation of values for ^{14}C have been of varying complexity according to their purpose. A great deal of the modelling of the exchangeable carbon system has, in fact, been concerned with the fate of the carbon dioxide added to the atmosphere by man's activities; much information on this aspect is contained in the papers by Bolin and Eriksson (1959), Schell *et al.* (1967), Nydal (1968), Broecker *et al.* (1971), Machta (1972) and Keeling (1973). In the box model approach, transfer from a reservoir is treated as a first-order process. The rate constant or transfer coefficient for transfer between reservoirs is conveniently presented as the reciprocal value, described usually as the residence time, but sometimes as the mean reservoir life time or the turnover time. Very generalized models have proved valuable for a broad understanding of the main features of the exchangeable system; they indicate the dependence of concepts of residence time and exchange rates on the assumptions of the model and have been adequate for the derivation of atmospheric residence times and examination of the implications of variations in the production rate of ^{14}C.

Two such models, introduced by Craig (1957, 1958, 1963) are shown in Fig. 18.10. In the first (chain) model, carbon dioxide can exchange with the deep ocean only through the mixed layer. The atmospheric residence time is defined here by the exchange rate with the mixed layer. The residence

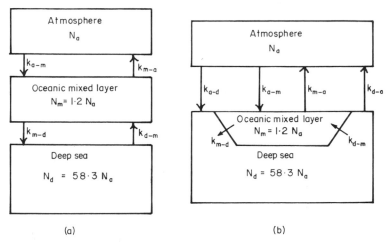

FIG. 18.10. Reservoir models for the exchangeable carbon system. (a) Chain model (Craig, 1957), (b) Cyclic model (Craig, 1958, 1963). Terrestrial reservoirs are not shown. N_a (amount of carbon in the atmosphere) $= 0.126$ g cm^{-2}. Values of k are the rate constants for the transfers indicated. See text for further details.

times of water and carbon dioxide in the deep sea each have identical significance, if the particulate flux derived from the biosphere is neglected. In the second (cyclic) model, the mixed layer occupies about 75% of the ocean surface. Carbon dioxide in the deep sea can be replenished directly from the atmosphere over the outcrop area as well as by transfer through the mixed layer. The total residence time in the atmosphere is defined by reference to both exchanges. In this model, the mean life of a carbon atom in the deep sea is less than the mean life of a water molecule relative to renewal from the mixed layer. The relation between the age of the water from ^{14}C measurements and the renewal time for the water mass thus depends upon the model used, as does the estimate of residence time obtained from given observations of specific activities in the reservoirs.

Extension of the use of models originally intended to relate to steady-state conditions, to interpret distribution of fallout nuclides recently introduced to the surface ocean, has given discrepant results because of the unsuitability of the models used for this purpose. Broecker (1966) has developed a model to represent mixing across the thermocline in which the mixed layer is replaced by four boxes; this gave compatible results. With another model intended for the examination of the removal of carbon dioxide from the atmosphere into the ocean, Broecker et al. (1971) found it necessary to introduce into the two-box ocean model a third reservoir representing the water below the mixed layer to a depth of 1000 m. This took account of the existence of water masses which in exchange characteristics are intermediate between the mixed layer and the deep ocean. Box models incorporating specific major oceanic water masses have been developed by a number of workers (Broecker, 1963b; Keeling and Bolin, 1967, 1968; Broecker and Li, 1970) and have been used in the study of ocean circulation using $\Delta^{14}C$ values in conjunction with other oceanographic measurements such as total carbon dioxide, salinity and temperature. From such models residence times in the deep Atlantic and Pacific Oceans of about 500 and 1000 years have been derived.

In these approaches the distribution of ^{14}C is treated as in a steady-state condition with concentration gradients determined by circulation and mixing rates. Some implications of discontinuous mixing have been discussed by Broecker (1963b). Craig (1969, 1971) has questioned the basic assumption of a closed-system flow in interpreting $\Delta^{14}C$ values for the deep Pacific Ocean. He noted that there is essentially no vertical gradient or latitudinal variation in the concentration of ^{14}C in water below 1300 m in the Pacific, because addition of ^{14}C by particulate transference is in approximate balance with radioactive decay. There is, however, a decrease in total inorganic carbon and he suggested that a downward diffusive flux of stable carbon reduces the specific activity of bottom water; thus, changes in specific activity do not simply reflect elapsed flow time. Suess and Goldberg (1971) have argued that, in interpreting the results for deep-water masses, Bien *et al.* (1960, 1963a, b) have already taken account of the likely maximum correction for any additional input of inorganic carbon into bottom water, but this refers to the increase by oxidation of organic carbon and release from particles rather than to the diffusive flux.

Direct evidence of significant transport of ^{14}C from surface waters to the deep ocean by particulate material has been provided by Somayajulu *et al.* (1969). They found that calcareous particles collected at depths of 50–3500 m showed specific activities of ^{14}C corresponding to recent surface water values enhanced by "bomb" ^{14}C and were able to derive estimates of the sinking rate of the particles from these measurements.

It has been shown by Williams *et al.* (1969) that dissolved organic carbon in deep water of the Pacific Ocean has an apparent age of about 3400 years. This is considerably older than the inorganic carbon present in the same water mass and demonstrates the resistant nature of most of the dissolved organic matter of the deep ocean.

18.4.3. TRITIUM

Tritium is produced in the atmosphere primarily by galactic cosmic rays through the *n,t* reaction of secondary neutrons on ^{14}N, and by spallation of nitrogen and oxygen; production by solar flare accelerated particles is minor (Nir *et al.*, 1966). Estimates of the production rate from cosmic-ray and cross-section data have ranged widely. The critical assessment by Nir *et al.* (1966) suggests a mean production rate of $0.19 + 0.09$ atom $cm^{-2} s^{-1}$. Teegarden (1967) obtained a value of 0.20 ± 0.05 atom $cm^{-2} s^{-1}$ from experiments using balloon flights. The presence of tritium from nuclear weapon explosions has greatly complicated attempts to estimate the production rate from the natural inventory of tritium, since data on natural concentrations are very limited. The most reliable estimate obtained in this way appears to be that of Craig and Lal (1961) who obtained a value of 0.5 ± 0.3 atom $cm^{-2} s^{-1}$. The limits of error are such that these estimates may be compatible. There is a possibility that significant direct accretion of solar tritium occurs, but the magnitude of this process is uncertain (Nir *et al.*, 1966). The world natural inventory corresponding to the higher estimate of production rate by Craig and Lal (1961) is about 7 kg (70M Ci) and nuclear weapon explosions, up to the cessation of widespread atmospheric testing in 1963, introduced an artificial inventory of the order of 100 kg (Martell, 1963; Suess, 1969).

Tritium of natural and artificial origin is transferred through the troposphere to the surface ocean in precipitation and, possibly to a similar extent by molecular exchange. The HTO molecules form an ideal tracer for water mixing processes, and the half-life of tritium is very suitable for studies of vertical mixing above and in the thermocline. The interpretation of distributions of tritium in terms of mixing processes has been examined particularly by Rooth and Östlund (1972). The introduction of artificial tritium occurs with a pronounced latitudinal variation, resulting from its major production in high northern latitudes, and in these latitudes there has also been a marked seasonal variation in the concentration in surface ocean water, reflecting seasonal variations in the input from the troposphere. These seasonal variations are caused by the transference of material from the stratospheric reservoir into the troposphere each spring. Data for the North Pacific Ocean which illustrate the regional and seasonal variations are shown in Fig. 18.11. The amplitudes of seasonal variations in the concentrations of

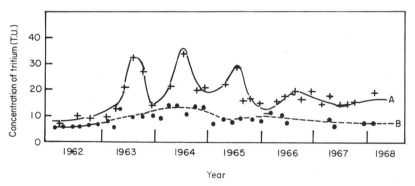

FIG. 18.11. Variations from 1962 to 1968 in the concentration of tritium in surface ocean water at Adak, Alaska (52° N) (curve A) and at Johnston Island (17° N) (curve B). After Suess (1969).

artificial tritium in the atmosphere are much greater than those observed for ^{14}C, reflecting the short tropospheric residence time of tritium of about 7 weeks (Suess, 1970). Because of the marked latitudinal variation in injection of tritium to the surface ocean it is necessary in the study of vertical profiles to consider the influence of horizontal movements of tritium-rich water from higher latitudes as a source of the nuclide in sub-surface waters at lower latitudes.

Measurements of tritium in surface water prior to 1954, when significant wide-scale artificial releases began, are few and somewhat discrepant. The best estimate that can be made of the natural steady-state concentration in surface water is about 1×10^{-18} atom of tritium per atom of hydrogen (abbreviated as 1 tritium unit or 1 T.U.). Libby (1962) estimated the average concentration in surface waters of the Northern Hemisphere in 1961 as 8 T.U., the corresponding value for the Southern Hemisphere being 3·5–4 T.U. Low values were found in areas of upwelling (Bainbridge and O'Brien, 1962; Libby, 1962). Detailed information is available on variations in the North Pacific Ocean over the period 1959 to 1966 (Bainbridge, 1963a, b; Dockins et al., 1967). Following attainment of the maximum value of 13 T.U. in 1963 the average concentration remained essentially constant until 1965 and then began to decrease. Rooth and Östlund (1972) have summarized data for the Atlantic Ocean. In 1967, concentrations in surface waters decreased with latitude from about 25 T.U. at 43° N to about 9 T.U. around 20° N and 5 T.U. at the equator. Temporal variations differed markedly with latitude. There was a four-fold decrease in concentration for the equatorial samples between 1964 and 1967, whereas at 22–26° N concentrations remained quite uniform during 1965 and 1966 and decreased by about 30% in 1967. In 1972, Antarctic and Sub-Antarctic Surface Water masses contained 0·9–2·1 T.U. (Michel and Williams, 1973).

A number of vertical profiles of tritium have been reported. At the North Pacific (1969) GEOSECS station essentially constant values of about 23 T.U. occurred in the upper 150 m below which there was a decrease to 0·2 T.U. or less at 400 m (Roether et al., 1970). This contrasts markedly with the Atlantic (1970) GEOSECS station (Roether and Münnich, 1972). Surface values there were 8·5 T.U., with a decrease to 6·5 T.U. by 600 m and the presence of significant amounts below (1 T.U. down) to about 2000 m. The concentrations in the deep water were attributable to horizontal advection. This effect explains the presence of a maximum concentration of tritium of 17 T.U. at about 50 m in a profile for the tropical North Pacific Ocean, described by Suess (1969) and has also been discussed in relation to the presence of tritium in low temperature water of the Sargasso Sea (Rooth and Östlund, 1972). In the equatorial Atlantic Ocean in 1964 there was no penetration of tritium below the thermocline (Östlund et al., 1969). Profiles measured in 1967 at stations on an Atlantic Ocean section at about 45°N showed contrasting vertical distributions according to location (Bowen and Roether, 1973) although surface values were similar. Concentrations were quite uniform with depth, with values of about 12 T.U. down to some 400 m in the Eastern basin, whereas in the Western basin the uniform zone extended only to about 100 m. Significant concentrations of tritium in the deep water of the Western basin were attributed to a contribution by advection of Arctic Ocean water flowing out from the Greenland Basin. Michel and Williams (1973) have given information on vertical profiles in the Antarctic Ocean during 1972. At two stations north of the Antarctic Convergence the maximum concentrations of 2 T.U. occurred at about 100 m. Significant amounts of tritium (about 1 T.U. in some cases) were found in some deep water samples, but in the transition zone between South Pacific Deep Water and Antarctic Intermediate Water the concentrations were below 0·3 T.U.

18.4.4. BERYLLIUM-7, BERYLLIUM-10 AND ALUMINIUM-26

Beryllium-7 is produced by the spallation of nitrogen and oxygen by cosmic rays. The presence of this short-lived nuclide in surface sea water was first demonstrated by Lal et al. (1964) who found 0·01–0·07 dpm l^{-1} for samples from the Indian Ocean. Further measurements have been made by Silker et al. (1968) and Silker (1972a, b). These show pronounced latitudinal differences in North Pacific Ocean waters, reflecting the latitudinal variation in production rate and rapid transference to the ocean in precipitation. Surface concentrations varied from 0·45–0·64 dpm l^{-1} at about 8–14° N, through a pronounced minimum of 0·04–0·07 dpm l^{-1} between 17–40° N, to values up to 0·6 dpm l^{-1} at higher latitudes. Comparatively uniform concentrations, averaging 0·3 dpm l^{-1}, were found for Atlantic Ocean waters between 5–25° N. Most of the ^7Be in sea water appears to be in a dissolved form (Silker et al., 1968) and it may therefore be a useful tracer for mixing processes in surface waters on the time-scale of several months. Vertical profiles showing pronounced decreases in concentration with depth have been reported by Silker (1972a, b).

The long lived radioactive isotope of beryllium, ^{10}Be, is also produced by spallation of atmospheric nitrogen and oxygen. The nuclide enters the ocean *via* precipitation; the rate of removal is rapid compared with its half-life. Peters (1955) suggested that ^{10}Be might be detectable in pelagic sediments and this was confirmed by Arnold (1956) and Goel et al. (1957). Further measurements on deep-sea clays have been reported by Merrill et al. (1960), Kharkar et al. (1963), Amin et al. (1966) and Tanaka et al. (1968). The concentrations found by these workers are almost entirely in the range 1–10 dpm kg^{-1}. Potentially, the nuclide provides interesting possibilities for geochronology with a unique range of greater than 5×10^6 years. The measurements are laborious however, and those made on core sections from various depths have often shown erratic variations. Some estimates of deposition rates have been made by comparing the measured concentrations in sediments with the rate of introduction of the nuclide to the oceans (Amin et al., 1966).

The application of ^{10}Be to the geochronology of very slowly accumulating ferro-manganese concretions has proved successful. The time-scale of the method is highly advantageous, enabling distinct layers within the concretion to be dated, whereas most techniques cover only the superficial layers. Results on several concretions (Somayajulu, 1967; Bhat et al., 1973; Krishnaswamy and Lal, 1972) have shown extrapolated surface concentrations from about 10 to more than 100 dpm kg^{-1}. Clear decreases in concentration within the concretions correspond to accumulation rates of 1–4 $\times$ 10^{-6} mm year^{-1}. These values agree satisfactorily with those obtained by other methods; because of the time-scale of the method they provide the most unequivocal radiometric evidence for these low rates of accumulation.

Lal (1962) suggested that ratios of ^{26}Al to ^{10}Be in sediment cores could be of value in dating. This suggestion was based on the rather similar geochemical behaviour of the elements, which both have short oceanic residence times, and the fact that regional variations in rates of production and introduction to the ocean should apply equally to both nuclides in so far as they are formed in the atmosphere.

The production rate of ^{26}Al from spallation of atmospheric argon by cosmic rays is, however, estimated as about 1·4 $\times$ 10^{-4} atom cm^{-2} sec^{-1}, about two orders of magnitude below that of ^{10}Be (see Table 18.5). There are, accordingly, great difficulties in measuring ^{26}Al in sediments. The first measurements by Amin et al. (1966) and Wasson et al. (1967) gave concentrations in deep sea clays of 0·46 and 0·81 dpm kg^{-1}, corresponding to an activity ratio of ^{26}Al to ^{10}Be of the order of magnitude of 0·1 rather than the order of 0·01 expected from the production rates. These workers concluded that the excess ^{26}Al represents a dominant contribution by cosmic dust accreted to the earth, in which the nuclide is produced by bombardment with solar-flare particles (see Wasson, 1963). This conclusion is at variance with the observations by McCorkell et al. (1967) on Greenland ice samples in which the concentrations of ^{10}Be and ^{26}Al agreed well with the accepted atmospheric production rates. Analyses by Tanaka et al. (1968) on pelagic sediments also give a lower ratio of ^{26}Al to ^{10}Be than was indicated by the earlier measurements. These workers found a concentration of ^{10}Be very similar to those observed by Amin et al. (1966), but found the concentration of ^{26}Al to be lower, giving an upper limit for the activity ratio of ^{26}Al to ^{10}Be of 0·03. This value is, however, not incompatible with a significant influx of ^{26}Al in cosmic dust.

Additional uncertainties concerning the interpretation of measurements of ^{10}Be and ^{26}Al in sediments arise from a revised estimate of the atmospheric production rate of ^{10}Be, made by Yokoyama (1967) on the basis of recent nuclear cross-section data. This estimate is about an order of magnitude lower than the generally accepted value shown in Table 18.5. On this basis the results of Amin et al. (1966) could be explained without invoking a major contribution of ^{26}Al from accreted cosmic dust. Subsequently, Yokoyama (1968) endeavoured to set an upper limit to the rate of accretion of cosmic dust by comparing the activity ratios of ^{26}Al to ^{10}Be reported for sediments with his calculated atmospheric production ratio. Tanaka *et al.* (1972) have made the converse calculation, estimating the influx of ^{26}Al to the earth in cosmic dust, from nuclear data for the production of the nuclide and an assumed accretion rate for cosmic dust which is substantially higher than the upper limit estimated by Yokoyama (1968). They show that the sum of this estimated influx and the atmospheric production rate agree with the experimental upper limit found by Tanaka *et al.* (1968) for the deposition rate of ^{26}Al in recently deposited sediments. Their calculations suggest that some 60% of the ^{26}Al entering sediments has its origin in cosmic dust.

The information available at present on the production rates of these nuclides appears too uncertain to enable this confusing situation to be clarified, and the status of the measurements of ^{26}Al in sediments also needs clarification by further work. Somayajulu (1967) showed from measurements on a ferro-manganese concretion that the specific activity of beryllium in sea water corresponded reasonably well with that predicted from the production rate of ^{10}Be, as given in Table 18.5. However, this prediction involves the introduction of additional uncertainties regarding stable beryllium in sea water, and the agreement obtained provides only tenuous evidence for the correctness of the production rate used in the prediction.

18.4.5. SILICON-32

Silicon-32 is produced in the atmosphere by cosmic rays, through the spallation of argon. There appears to be no significant contribution from testing of nuclear weapons (Kharkar *et al.*, 1966). The nuclide was first detected in nature in siliceous sponges (Lal *et al.*, 1959, 1960). These and subsequent measurements have given specific activities for sponges and plankton from various regions in the range of 2 to 77 dpm ^{32}Si kg^{-1} SiO$_2$ (Kharkar *et al.*, 1963; Somayajulu, 1969; Lal, 1969; Lal *et al.*, 1970). Measurements on sea water have also been made using *in situ* sampling (Lal *et al.*, 1964; Somayajulu, 1969; Somayajulu *et al.*, 1973); they give a somewhat similar range of specific activities.

The results of measurements on water from the Pacific Ocean are shown in Fig. 18.12. The extrapolated specific activity for surface water of 110 dpm kg^{-1} SiO$_2$ is higher than the values for sponges from the coastal Pacific Ocean, but the specific activities shown by deep-sea sponges agree well with the measurements on sea water. In terms of absolute concentrations, the deep water is an order of magnitude higher in ^{32}Si than is the water of

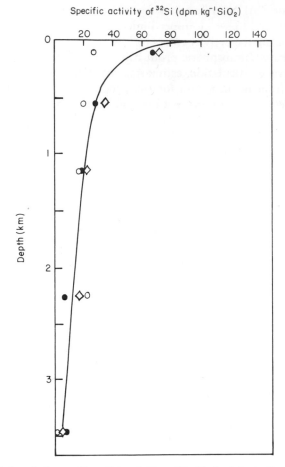

FIG. 18.12. Variations in the specific activity of silicon-32 with depth in the South Pacific Ocean. ◇—31° S 177° W; ●—27° S 175° E; ○—24° S 176° E. Other measurements of the nuclide in surface waters support the shape of the curve fitted. After Somayajulu *et al.* (1973).

the mixed layer. The depletion of ^{32}Si in surface waters is a consequence of transference as skeletal remains. Integrated activities in the water column are broadly consistent with the cosmic-ray production rate of ^{32}Si. Use of specific activities (dpm ^{32}Si kg^{-1} SiO$_2$) in a two-box model provides a means of estimating deep-water residence times (Lal *et al.*, 1960; Lal, 1962, 1969). The half-life of ^{32}Si is in a very useful range for this purpose, but unfortunately there is much uncertainty as to the true value (see Table 18.5).

The application of ^{32}Si to the geochronology of sediments rich in biogenous silica has been examined by Kharkar *et al.* (1963, 1969). The method has a range of only a few thousand years, and the deposition rates thus determined on this time scale are much higher than those derived by measurements on longer time-scales using other methods. Thus, Kharkar *et al.* (1969) obtained deposition rates for a siliceous ooze of 47×10^{-3} cm year^{-1} using ^{32}Si, and 0.6×10^{-3} cm year^{-1} using ^{230}Th. The reasons for this are not clear, but the latter value gives a much more accurate reflection of the long-term deposition rate.

18.5. Artificially Produced Nuclides

18.5.1. INTRODUCTION

The introduction of artificially produced nuclides to the environment on any appreciable scale began in the mid-1940's with the earliest discharges from nuclear plants and the first explosions of nuclear weapons. The testing of nuclear weapons in the atmosphere and the ocean has been the major source of artificial radioactivity released to the environment. Releases from this source were greatest during the major series of weapons tests in the early 1960's; following the cessation of widespread atmospheric testing in 1963, such inputs have been minor. Some explosions of nuclear weapons have taken place within the ocean, but the bulk of the artificial radioactivity entering the seas has been in the form of world-wide fallout from explosions in the atmosphere, mainly at high altitudes, but some near sea or ground level; most of this input occurs directly in precipitation and, to a lesser extent, by run-off of material deposited on land. Between 1945 and 1963, weapon tests produced almost 200 megaton equivalents of fission, corresponding to the fission of about 2.8×10^{28} atoms of uranium or plutonium (data cited by Joseph *et al.*, 1971).

The other main source of artificial radioactivity is the controlled discharge of wastes containing low levels of radioactivity and activated cooling water from nuclear fuel processing plants and nuclear reactors. Although this has contributed much less radioactivity overall than has the fall-out from weapons (see Table 18.6), these discharges have occurred in coastal environments and thus have been important in a few marginal regions of limited extent. With the growth of the nuclear power industry the problems associated with these discharges continue to receive much attention. Estimates summarized in Table 18.6 indicate the possible future increase in environmental radioactivity from this source; these estimates assume that discharges will increase in proportion to the production of nuclear power, but improved management of wastes may reduce the predicted inventory.

TABLE 18.6

Estimated total contributions from artificial sources to the oceanic inventories of radioactive nuclides[a]

Nuclides and source	Estimated radioactivity (Ci)	
	in 1970	in 2000
Nuclear explosions		
Fission products	$2–6 \times 10^8$	$n \times 10^8$
Tritium	10^9	$n \times 10^9$
Reactors and associated plant		
Fission and activation products other than tritium	3×10^5	3×10^7
Tritium	3×10^5	$n \times 10^8$
TOTAL	ca. 10^9	ca. 10^9

[a] From Food and Agriculture Organization of the United Nations (1971). Projections of inventories from nuclear explosions are based on the assumption that tests in the atmosphere will continue at a similar rate to that in the period 1968–70.

Other sources which require consideration are accidental release from nuclear plants and from encapsulated sources used in technology and medicine, discharges from nuclear powered ships, wastes from the many technical and scientific uses of radioactive substances, and the dumping of packaged nuclear wastes. Up to the present, these have made only minor contributions to environmental levels. Accounts of the sources of artificial nuclides in marine environments have been given recently by Joseph *et al.* (1971) and by Rice and Wolfe (1971).

18.5.2. PRINCIPAL ARTIFICIALLY PRODUCED NUCLIDES AND THEIR PHYSICO-CHEMICAL FORMS IN SEA WATER

The fission products comprise more than 200 nuclides of elements ranging in atomic number from 30 (zinc) to 66 (dysprosium). From a knowledge of fission yields, which vary to some extent according to the fissile material and the energy of the neutrons causing fission, it is possible to define fairly precisely the composition of the mixture of fission products formed in a nuclear explosion. However, the composition of the material entering a given oceanic region at any particular time will be very different as a result of radioactive decay between formation and deposition, and the mixing of material of different ages. The compositions of radioactive wastes may vary still more markedly because of their different sources and previous treatments.

The principal fission products with half-lives between two months and about thirty years are listed in Table 18.7; radioactive daughter products of these nuclides are also shown. Shorter-lived nuclides, particularly ^{143}Pr (13·6 day), ^{140}Ba (12·8 day), ^{147}Nd (11·1 day) and ^{131}I (8·05 day) make a substantial contribution to the total activity of mixed fission products in the early stages, but become unimportant several months after fission. Fission products of longer half-life than 30 years, which include ^{151}Sm (about 90 years) and ^{129}I (1·7 × 10^7 years), make little contribution to the artificial radioactivity in the oceans. In a sample of fission products about one year old, more than three-quarters of the activity is attributable to ^{95}Zr, ^{95}Nb and rare earth nuclides; after 20 years, ^{90}Sr, ^{137}Cs and their daughter nuclides account for over 90% of the total activity. At present, with few injections of fresh material from weapons having occurred for over a decade, the composition of world-wide fallout is dominated by the latter nuclides with minor periodic inputs of the shorter-lived nuclides.

Also listed in Table 18.7 are the more important activation products which are formed by capture of neutrons in material such as weapons structures, reactor coolants, and in earth, water or air. Tritium is included here for convenience although its presence is also due to its incorporation in some types of weapon and production in thermonuclear reactions. Tritium and ^{14}C have already been discussed in Section 18.4. The proportions of activation products vary greatly according to the nature of the irradiated material and the magnitude of the neutron flux. In addition to those listed in Table 18.7, several other nuclides were produced in significant amounts in particular series of weapons tests; they include ^{102}Rh (half-life 2·9 years), ^{181}W and ^{185}W (half-lives 140 and 73 days, respectively).

Among the alpha-emitting nuclides of the transuranic elements, the most important from the environmental standpoint is plutonium-239 (half-life 2·4 × 10^4 years), present either as unreacted fissile material or as a product of neutron capture by uranium-238. The shorter lived ^{238}Pu (half-life 86.4 years) produced by weapon explosions constitutes only a small fraction of the total plutonium. A release of about 1 kg (17 kCi) of this nuclide occurred in 1964 from the burn-up in the atmosphere of the Southern Hemisphere of a navigational satellite, with a nuclear auxiliary power system, which failed to enter orbit. This release increased the global fallout of the nuclide about threefold and because of the location of the injection the distribution pattern was quite different from that of weapons fallout (Hardy et al., 1973). The activity ratio of ^{238}Pu to ^{239}Pu prior to this release averaged 0·024. Results reported for ^{239}Pu include any small fractions of ^{240}Pu (half-life 6580 years) which is present, but to avoid repeated qualification they are treated here simply as values for ^{239}Pu. Thermonuclear explosions have been estimated to have introduced some 10^{28} atoms of ^{239}Pu, comparable amounts of ^{237}Np

TABLE 18.7

Principal artificial nuclides

I. Fission products[a]

Nuclide	Half-life	Total chain yield from fission of ^{235}U by thermal neutrons[b]	Radioactive daughter product	Half-life of daughter
Strontium-89	51 day	4·8	—	
Strontium-90	28 year	5·8	Yttrium-90	64 hour
Yttrium-91	58 day	5·8	—	
Zirconium-95	65 day	6.4	Niobium-95	35 day
Ruthenium-103	40 day	2·9	Rhodium-103m	57 min
Ruthenium-106	368 day	0·4	Rhodium-106	35 s
Antimony-125	2·7 year	0·04		
Tellurium-129m	33 day	1·0	{ Tellurium-129 { Iodine-129	70 min $1·7 \times 10^7$ year
Caesium-137	30 year	6·0	Barium-137m	2·6 min
Cerium-141	33 day	5·8	—	
Cerium-144	282 day	5·6	Praseodymium-144 Neodymium-144	17·3 min 2×10^{15} year
Promethium-147	2·6 year	2·6	Samarium-147	1×10^{11} year
Europium-155	1·8 year	0·03		

II. Neutron activation products

Nuclide	Half-life	Nuclide	Half-life
Tritium	12·3 year	Cobalt-57	270 day
Carbon-14	5730 year	Cobalt-58	72 day
Phosphorus-32	14·4 day	Cobalt-60	5·2 year
Sulphur-35	87 day	Nickel-63	ca. 100 year
Scandium-46	84 day	Zinc-65	246 day
Chromium-51	27·8 day	Silver-108m	>5 year
Manganese-54	300 day	Silver-110m	253 day
Iron-55	2·8 year	Cadmium-113m	14 year
Iron-59	45 day	Europium-152	12·5 year
		Mercury-203	47 day

[a] Some of the fission products can also be produced by neutron capture processes.
[b] Values from the Chart of the Nuclides issued by the Federal Minister of Nuclear Energy, Federal Republic of Germany.

(half-life 2.1×10^6 years), and about one-tenth of this quantity of ^{240}Pu (Joseph et al., 1971); two other isotopes of plutonium, ^{241}Pu and ^{242}Pu (half-lives 13 years and 3.8×10^5 years respectively) were produced in significant amounts in a single explosion during 1952.

A major factor influencing the distribution of artificial nuclides entering sea water is the physical and chemical form in which they are introduced. A closely related question is the extent to which they undergo exchange with the stable forms of the element present in sea water. Some short-range fallout from nuclear weapons enters the ocean as large fused particles but long-range material is finely divided with particle sizes mostly below 1 μm. Freiling and Ballou (1962) have reported results for underwater nuclear explosions and experimental simulations. The results, together with additional data summarised by Joseph et al. (1971), indicate that strontium and caesium are present in forms which are almost wholly soluble, whereas cerium, yttrium, zirconium, niobium, ruthenium and tellurium are mainly associated with particulate and colloidal material.

Knowledge of the chemical speciation of trace elements in sea water is, from a practical standpoint, at an underdeveloped stage. It is therefore not surprising that little is known about possible differences in speciation of the stable elements (present at steady-state concentrations which reflect long-term processes in sea water) and the recently introduced radioactive forms, and about the implications of such differences. However, the importance of this question is demonstrated by such work as that of Jones (1960) and Kečkeš et al. (1967), which showed differences in uptake of ^{106}Ru on sediments and by organisms according to the chemical form in which the nuclide was present, and also by the findings of Jenkins (1969) that ^{55}Fe, ^{60}Co and ^{65}Zn, freshly introduced into sea water in fallout, were taken up by marine organisms more readily (by orders of magnitude) than the corresponding stable elements already present.

Changes in the chemical form of cerium ions added to sea water have been studied, using radioactive techniques, by Ivanov et al. (1972). They concluded that within a few hours of addition, cerium was converted from free ions and charged hydrolysed species, to neutral and weakly charged species. Following these changes there was a decrease in the adsorption coefficient of cerium which was attributed to polymerization of the hydrolytic forms, giving an essentially colloidal product. Similar changes in adsorptive behaviour with time were reported for yttrium by Ivanov and Lyubimov (1970) and, according to Lyubimov et al. (1970), these changes in physicochemical state are reflected by variations in the accumulation of yttrium by Ulva sp. and by molluscan shells. The association of elements with precipitates such as hydrous oxides has been studied, using radioactive tracers, by Fukai and Huynh-Ngoc (1968). It was found that Cr (III) was readily adsorbed on ferric hydroxide and that Cr (VI) tended to remain in solution although its behaviour was somewhat variable and depended on the nature of the organic material present in solution.

18.5.3. SOME RADIOLOGICAL AND RADIOECOLOGICAL ASPECTS OF ARTIFICIALLY PRODUCED NUCLIDES

Much of the work concerning artificial nuclides in the environment has been directed to the assessment of possible hazards, especially those to man. There are two different aspects to this problem. First, world-wide fallout has led to some increase in man's exposure to radiation, primarily through the passage of certain nuclides, especially ^{90}Sr, into human diet. In this context, marine food-chains have required much less attention than those on dry land because, in all countries, marine foods have contributed only a small part of the total intake of the more hazardous nuclides in the average diet (United Nations Scientific Committee on the Effects of Atomic Radiation, 1962). The second aspect concerns the controlled discharge of low-level radioactive waste and its regulation within acceptable levels.

Approaches to this second problem rely upon the assumption that, in this context, safety for man implies safety for the ecosystem. There is no evidence that ecosystems have suffered significant damage from artificial changes in radiation levels associated with waste discharges, and the above assumption seems fully justifiable, especially in view of the high standards adopted to protect human beings. Assessments of safe levels of discharge to aquatic environments have been made by one of two approaches, namely the specific activity approach and critical pathway analysis; they have been fully reviewed by Foster *et al.* (1971). Each is an operational approach to relate the level of a discharge into the environment to the resulting radiation dose, above background, incurred by exposed individuals in the human population. It is then possible to control the discharge at a level which will ensure that acceptable dose limits are not exceeded. The criteria of acceptability are based upon internationally accepted recommendations for persons not occupationally exposed to radiation. The acceptable radiation doses correspond to very low probabilities of biological effects. There are, in practice, additional margins of safety in the application of the approaches to determine the capacity of an environment to receive waste. Furthermore, environmental and dietary levels are monitored directly, and efforts are made to maintain discharges at the minimum practicable level so that frequently only a small fraction of the environmental capacity to accept waste is actually utilized.

The approach to radiological protection using specific activities is based upon the assumption that discharged radioactive nuclides and the stable isotopes already present in the environment move along environmental pathways in an essentially identical way. This assumption is valid only if exchange occurs between radioactive and stable forms. From the maximum permissible human body burden for a particular nuclide and a knowledge of the stable element concentrations in human tissues it is possible to express the permissible level as a specific activity (ratio of radioactive nuclide to stable element). If the assumption of exchange is valid then it is only necessary to ensure that this specific activity is not exceeded in the environment to which the discharge is made, for example, in sea water.

The alternative approach of critical pathway analysis is widely used. In this case, an examination is made to determine, for the given circumstances, the critical nuclide and dietary or other pathway which will give rise to the most significant increase in radiation exposure for the critical group in the population. This critical group comprises those people who incur such an increase because of their dietary habits or their exposure to external radiation. From a knowledge of the relationship between the amount of the nuclide discharged and the dose incurred by this group it is possible to determine the capacity of the environment to receive waste without an unacceptable increase in exposure for the individuals or small groups concerned. Other sectors of the population will receive lesser increases in exposure. In the initial stages of identifying a critical pathway, much information is needed on such matters as the physical dispersion of discharged material, concentration factors in organisms, utilization of local resources, and dietary habits in exposed populations. Preoperational predictions are, properly, based on pessimistic assumptions, and postoperational assessments may often enable more realistic use to be made of environmental capacity. The degree of continued monitoring needed in a particular situation depends upon the fraction of the environmental capacity which is utilized.

A much-quoted example of the use of critical pathway analysis has been in the establishment of the capacity of the environment for wastes from the Windscale Chemical Reprocessing Plant which discharges material into the Irish Sea. From 1957 to 1967 the total discharge of beta-radioactivity averaged 6.5×10^4 Ci y^{-1} (Joseph et al., 1971). In this waste the dominant nuclides are ^{95}Zr + ^{95}Nb, ^{106}Ru and ^{144}Ce. In 1965 average concentrations of ^{106}Ru, ^{144}Ce and ^{95}Zr + ^{95}Nb in edible seaweeds of the genus *Porphyra*, collected at eleven stations, were 123, 2·3 and 20·8 pCi g^{-1} wet weight, respectively (Preston and Jefferies, 1967). In surface sediments near the discharge, ^{95}Zr + ^{95}Nb and ^{106}Ru are also the major gamma-emitting nuclides (Jefferies, 1968). In this region, where *Porphyra* is harvested for human consumption, ^{106}Ru is the critical nuclide and the greatest exposure to radiation resulting from the discharge is the dose to the gastro-intestinal tract of persons consuming foodstuffs prepared from the alga. The environmental capacity is determined by this critical pathway, and it has been appropriate to concentrate monitoring on this food chain, although this has been done against a wider background of studies on levels of radioactive nuclides in the Irish Sea (Mauchline and Templeton, 1963; Dunster et al., 1964). In 1964–66, the intestinal doses received by consumers of large amounts of processed seaweed were about 40% of the recommended dose limit. For the Blackwater Estuary, England, which receives waste from the nuclear power station at Bradwell, the critical pathway is through the concentration of ^{65}Zn into oysters consumed by man (Preston, 1968); here the doses to the whole body and the gastro-intestinal tract have been much less than 1% of the recommended limits. Other applications of critical pathway analysis have been discussed by Foster et al. (1971). An example of a critical pathway which does not involve a food-chain is that which arises at Dounreay, Scotland, where the most significant increased exposure arising from discharges from the reactor establishment is the external dose incurred by the handling of contaminated fishing gear.

Interaction of dissolved material with sediments is an important factor in influencing the pathways of radioactive nuclides in estuarine and coastal waters. Because of the often complex patterns of transport, deposition and resuspension of sediment and the processes of sorption and desorption which may occur under different conditions it is difficult to generalize about the role of sediments. Settled sediment may frequently remove a considerable fraction of some nuclides, causing a reduction in concentration in solution and reducing availability to many organisms, but there are circumstances where associations with particulate material may increase uptake in some food chains. A summary of this complex question is given by Duursma and Gross (1971).

These authors give a generalized order of decreasing distribution coefficients, representing a decreasing adsorptive tendency, for a number of nuclides with a variety of oxic sediments, viz.

$$^{147}\text{Pm} > {}^{106}\text{Ru} > {}^{54}\text{Mn}, {}^{95}\text{Zr} + {}^{95}\text{Nb} > {}^{59}\text{Fe} > {}^{65}\text{Zn} > {}^{137}\text{Cs} > {}^{90}\text{Sr}.$$

This order is of only general predictive value, since factors such as the physicochemical state of the element, the mineralogy of the sediment, and the existence of coatings of hydrous oxides, may modify adsorptive behaviour. For some purposes it may be important to know the rates of sorption reactions as well as the quasi-equilibrium distribution coefficients. On anoxic sediments, nuclides such as ^{54}Mn,^{59}Fe, ^{95}Zr + ^{95}Nb, ^{106}Ru and

^{147}Pm, show a lesser tendency to be adsorbed, whereas ^{60}Co, ^{65}Zn and ^{90}Sr are more firmly held. In coastal environments, where reducing conditions commonly develop below an oxidized surface layer, migration of the elements by diffusion in interstitial water may occur through these differences in behaviour.

Diffusion coefficients determined for the movement of nuclides in interstitial waters of sediments are influenced by the holdback of ions by sorption processes and are, accordingly, lower than the values corresponding to diffusive processes alone and, moreover, must be expected to show some differences according to the nature of the solid phases. The value corresponding to diffusive processes alone was found by Duursma and Bosch (1970) using ^{36}Cl to be about 10^{-6} cm^2 s^{-1}, and values for interactive nuclides range from about 10^{-7} cm^2 s^{-1} for ^{90}Sr to between 10^{-10} and 10^{-11} cm^2 s^{-1} for lanthanide nuclides and ^{106}Ru (Duursma and Gross, 1971).

Problems concerning radioactive waste disposal and the exposure to radiation of ecosystems and human populations have given rise to very many studies concerned with environmental monitoring, the uptake of nuclides by organisms and sediments, and their movement in marine food chains. These topics cannot be dealt with here but much of the information has been summarized elsewhere (National Academy of Sciences—National Research Council, 1957, 1959a, b, 1962; National Academy of Sciences 1971; International Atomic Energy Agency, 1960, 1961, 1962, 1966, 1971, 1973; Schultz and Klement, 1963; Mauchline and Templeton, 1964; Baranov and Khitrov, 1964; Polikarpov, 1966; Åberg and Hungate, 1967; European Nuclear Energy Agency, 1968; Nelson and Evans, 1969; Rice and Wolfe, 1971).

18.5.4. OCCURRENCE AND DISTRIBUTION OF SOME ARTIFICIALLY PRODUCED NUCLIDES IN OCEAN WATERS

18.5.4.1. *Strontium-90 and caesium-137*

By far the greatest part of the effort devoted to the study of oceanic distributions of artificially produced nuclides has centred on ^{90}Sr and ^{137}Cs, the most abundant of the longer lived nuclides introduced in fallout. The primary reason for this is that they are readily dissolved and become isotopically diluted with the stable forms of these elements which, particularly in the case of strontium, show conservative behaviour in sea water; there has, accordingly, been great interest in their use as tracers for the mixing of surface waters, particularly with deeper oceanic water masses.

Since the first investigations of this subject were made by Bowen and Sugihara (1957, 1958, 1960), over 2000 analyses of these nuclides have been reported for ocean waters. These extensive data, beginning in 1954 for the Atlantic Ocean and in 1957 for the Pacific Ocean, have recently been summarized by Volchok *et al.* (1971), who have provided a comprehensive bibliography. The average ratio of ^{90}Sr to ^{137}Cs calculated by these authors for oceanic samples is 1.6 ± 0.3, a value which is compatible with that measured in fallout. No significant trends in the ratio with time or location are evident, except in the atypical conditions of the Baltic Sea. The uniformity of the ratio, indicating little geochemical fractionation, supports the view that these nuclides are valid tracers for water movements and enables measurements of ^{137}Cs to be converted to values for ^{90}Sr, for inclusion in an overall survey. Following Volchok *et al.* (1971), this discussion is presented primarily in terms of concentrations of ^{90}Sr, but may be considered as generally applying also to ^{137}Cs.

The spatial and temporal variations in the concentrations of ^{90}Sr in surface ocean waters are summarized in Fig. 18.13, which reproduces the presentation by Duursma (1972) of the data summarized by Volchok *et al.* (1971). The highest concentrations have been recorded in north-western Pacific Ocean waters which received direct inputs from some of the earlier major series of weapon tests. The cumulative deposition of ^{90}Sr in fallout on land areas has shown a very pronounced latitudinal variation, associated with the major releases in high northern latitudes, particularly in explosions during 1961–62. Data for the cumulative deposition on land up to the end of 1966 (Joseph *et al.*, 1971) show a pronounced maximum centred on 50° N, a decrease by about an order of magnitude by 10° S, and a broad maximum between 25–45° S which is lower than that in the Northern Hemisphere by a factor of over 4; the ratio of total deposition in the two hemispheres is also about 4. Seasonal variations associated with transfer of fallout from the stratosphere to the troposphere also occur. They do not appear to be closely reflected in the surface ocean and the long-term temporal and spatial variations in the ocean are considerably modified compared with those on land. As discussed below, there is uncertainty concerning the relationship between deposition over land and sea. It has been suggested that precipitation is not so dominant a mechanism of deposition into the ocean as it is over land, and that a different pattern of transfer across the tropopause applies to oceanic regions (Bowen *et al.*, 1968).

Interpretation of the variations in concentration of ^{90}Sr in surface ocean waters, with both region and time, is complicated by the fact that several processes are involved, namely, the initial input of material and its transfer by surface current movements and by mixing into subsurface waters. Detailed discussions of the distribution of ^{90}Sr in parts of the Atlantic Ocean, in terms of oceanographic features, have been given by Bowen and Sugihara

Fig. 18.13. Distribution of Strontium-90 in the world ocean and its variation with region, depth and time. In the vertical profiles the sub-surface concentrations are expressed as percentages pilation by Volchok *et al.* (1971). Reproduced with permission of the author and George Allen and Unwin, Ltd.

(1965) and by Bowen *et al.* (1968, 1969). Surface concentrations in the North and Equatorial Atlantic Ocean have changed in a manner which is not explicable in terms of the pattern of fallout on land. For example, in 1961 during a period of decreased terrestrial fallout, the amounts of ^{90}Sr in surface waters were quite uniform over the region between the Equator and 20° N. The rapidity with which this uniformity was attained could not be fully explained by ocean current fields. It was necessary to invoke either "un-oriented" mixing processes within water masses or the differences, discussed above, and on p. 171, between the patterns of fallout over land and ocean.

Several composite profiles to about 700 m are also shown in Fig. 18.13; these represent averaged vertical distributions for the regions or stations concerned. They show generally rather smooth distributions with the maximum concentrations at the surface; the secondary maxima at 250 m in the Atlantic Ocean profiles probably arise fortuitously through differences in the numbers of observations available for averaging at different depths. There is a marked contrast in the extent of vertical penetration in the regions of the Pacific Ocean shown in the figure; several hydrographic factors are discussed by Volchok *et al.* (1971) as possible contributory causes to the greater penetration at the station in the north-western Pacific Ocean. In general, the depth of mixing of ^{90}Sr and ^{137}Cs in surface layers of the North Atlantic Ocean is about twice that in the North Pacific Ocean, a feature which is consistent with the vertical stability of the surface layers in these two regions (Pritchard *et al.*, 1971).

Exceptionally, sub-surface maxima in concentration of ^{90}Sr have been found in vertical profiles, as a result of lateral advection along isopycnal surfaces, as described for tritium on p. 152. A consistent inversion with a maximum concentration of ^{90}Sr at about 100 m was observed at the south-eastern approaches to the Caribbean Sea in 1970 by Bowen *et al.* (1972), reflecting the layering of water masses in that region. Bowen *et al.* (1966) attributed increases in concentration at intermediate latitudes in the Atlantic Ocean at depths below 1000 m, between 1961 and 1963, to the southward flow of water which had received higher fallout when at the surface in higher latitudes.

The results for ^{90}Sr and ^{137}Cs obtained by different workers for waters at the surface and to depths of about 700 m are in good general agreement. However, there is controversy about the concentrations in deep waters. Penetration of low but detectable amounts of these nuclides to depths below 1000 m has been found by some workers. In the profiles examined by Miyake *et al.* (1962) in the north-western Pacific Ocean, penetration into deep waters could possibly be explained as an anomaly arising from the direct injections in this region of radioactivity from explosions in the Pacific Ocean testing grounds. Significant concentrations were also reported, however, in Atlantic Ocean deep waters, sampled for the most part between 1957–61, by Bowen and Sugihara (1960, 1963, 1965) and Bowen et al. (1966). On the well-based assumption that such penetration can be validly regarded as an indication of corresponding water movements, the values indicate much shorter residence times for water in these deep ocean water masses than are indicated by measurements of ^{14}C present in steady state concentrations (see Section 18.4.2). Broecker and his co-workers have determined concentrations for a number of vertical profiles (Broecker and Rocco, 1963; Rocco and Broecker, 1963; Broecker et al., 1966b) and their interpretation of the problem is at variance with the conclusions reached by Bowen and his colleagues. For a number of deep-water samples, neither ^{90}Sr nor ^{137}Cs was present in detectable amounts and many, but not all, of their values are significantly lower than those reported by Bowen and his co-workers. They considered that their results gave no unequivocal evidence of penetration into deep water which could be interpreted in terms of water mixing. Folsom and Mohanrao (1960) have also found no significant penetration of ^{137}Cs below 300 m for a profile in the north-eastern Pacific in 1960. Later measurements of ^{137}Cs in this region by Folsom during 1965–1967, as cited by Pritchard et al. (1971) showed a similar pattern. Broecker (1966) has discussed the conflicting results in terms of his model for mixing of surface waters (p. 149) and a summary of the oceanographic implications is presented by Pritchard et al. (1971).

This controversy could not be resolved in the examination and summary of data made by Volchok et al. (1971). In that report, Broecker considers that analyses showing significant amounts of ^{90}Sr and ^{137}Cs at depths below 1500 m up to 1965 are affected by undetected blanks. This view is argued on two main grounds.

(i) Such penetration is incompatible with rates of vertical mixing from other studies and with estimates of the global inventories of ^{90}Sr and ^{137}Cs.

(ii) Reported concentrations in deep water do not show the expected response to increases in cumulative fallout.

It is implicit in Broecker's argument that those measurements which show low or negligible penetration are more accurate because they are unaffected, or less affected, by undetected blanks.

Bowen, in the same report and elsewhere (Bowen et al., 1969), has argued the opposite view, namely that significant penetration in deep waters of the Atlantic Ocean and Caribbean Sea has occurred and that it reflects mixing of water. This argument is based on three main points:

(i) Concentrations of ^{90}Sr decreased in the surface layers of the Atlantic

Ocean in 1961 and 1965, during periods of reduced rate of fallout, in a way which indicates rapid penetration to deep water.

(ii) Transport along isopycnal surfaces, which extend from considerable depths at mid and low latitudes to near the surface at high latitudes, can account for the observed penetration, which is not considered to involve transport across the thermocline in the former regions. Vertical mixing during severe storms may also be a factor.

(iii) Comparisons of results for samples and blanks by various workers support the analytical validity of the findings of significant penetration.

In addition to the data summarized in the report, information on the concentrations of both ^{90}Sr and tritium has been given by Bowen and Roether (1973) for samples from vertical profiles on a section at 45° N in the Atlantic Ocean. The values for tritium are discussed on p. 152; the concentrations of ^{90}Sr showed closely similar variations with depth. Greater concentrations of ^{90}Sr were found in deep water at the stations in the Western Basin than occurred at similar depths in the Eastern Basin; the penetration of these greater amounts was attributed to advection of surface water from the Arctic region.

In another individual comment in the report of Volchok et al. (1971), Schuert accepts the general validity of measurements showing significant concentrations of ^{90}Sr and ^{137}Cs in deep waters. He presents evidence that in situ extraction of ^{137}Cs by ion-exchange showed no detectable amounts between 380 and 1800 m at a station in the East Pacific although concurrent measurements using conventional water sampling showed detectable concentrations down to 1000 m. Schuert's conclusion is that ionic ^{137}Cs had not mixed into the deep water. He suggests that where penetration into the deep ocean has been observed it may be explained by the processes of transfer on particulate material and of lateral advection.

The evaluation and interpretation of measurements of ^{90}Sr and ^{137}Cs in deep waters thus remains unresolved, and these controversial matters require clarification by further investigations. Pritchard et al. (1971) have pointed out that the existence of a mechanism for deep penetration of the nuclides by movement along isopycnals does not resolve the basic discrepancy between the long residence times of deep water indicated by other approaches and the concentrations of ^{90}Sr reported for many profiles, the discrepancy being the unexpectedly high amounts of the nuclide. Clarification of these problems is very important for our understanding of deep water mixing.

The results for deep water also have implications for the estimation of the oceanic inventory of ^{90}Sr. The calculations made by Volchok et al. (1971) show that even if it is assumed that there is no ^{90}Sr in deep water, the oceanic inventory is about twice that which would be predicted by extrapolating the extensive data for cumulative deposition on land. Since there is evidence that some ^{90}Sr has penetrated into deep waters, even though the amounts and the extent of penetration are subjects of controversy, oceanic and land deposition appear to differ by a factor considerably greater than two. The higher the inventory in deep water, the greater is this difference. Even the unrealistic minimum estimate is greater than is compatible with the most reliable estimates of the global inventory of ^{90}Sr, and this discrepancy has not been explained. It does appear, however, that the relative importance of different mechanisms of deposition differs for oceanic and land surfaces.

Some more direct comparisons have been made of rates of fallout for oceanic and land surfaces. Measurements by Broecker *et al.* (1966a) of ^{90}Sr in waters of the Bahama Banks suggested that fallout rates were essentially similar to those expected from the results of fallout sampling on land in the same latitude belt. If these results were typical then oceanic rates of fallout of ^{90}Sr would appear to agree with terrestrial rates within a factor of not more than two. From measurements of shorter lived nuclides, Chesselet *et al.* (1965) obtained evidence of higher rates of fallout over ocean waters near France than over adjoining land areas; the magnitude of the difference was variable, but indicated an average excess significantly greater than twofold. Differences in deposition between oceanic and terrestrial areas at the same latitude are also indicated by the work of Noshkin (1969) on radioactive nuclides in aerosols in the near-surface atmosphere.

The behaviour of ^{90}Sr and ^{137}Cs in the mixing of river and sea waters in the plume of the Columbia River was studied by Park *et al.* (1965). This river received contributions of radioactivity from the nuclear plant at Hanford, leading to a circumstance in which the concentration of ^{90}Sr was slightly higher in the river water than in the sea water with which it mixes in the plume, whereas the gradient for ^{137}Cs was in the opposite direction. Within the scatter of the data, the behaviour of ^{90}Sr appeared to be conservative. For ^{137}Cs, there was some indication that the nuclide behaved less conservatively, but the analyses were too few to enable a firm conclusion to be drawn.

18.5.4.2. *Other fission products*

A number of fission products in addition to ^{90}Sr and ^{137}Cs have been detected in marine environments by analysis of sea water and organisms, but the information is much sparser for these other nuclides, except in relation to waste discharges. A comprehensive bibliography up to the late 1960's has been given by Volchok *et al.* (1971). Discussion of measurements of these nuclides in open ocean waters receiving long-range fallout generally assumes that the observed distributions reflect the input in dissolved or dispersed states which can enter into geochemical processes and that they do not reflect the distribution of unreactive particles. This is supported by those geochemical fractionations which have been observed. For many of these nuclides, the elements concerned are geochemically reactive and can exist in more than one oxidation state. Thus, it cannot necessarily be assumed that the radioactive nuclides act as tracers for the stable elements already present in sea water. The special circumstances associated with waters influenced by waste discharges or local fallout from weapons tests are not considered here.

The distributions of the lanthanide nuclides of moderate half-life, ^{144}Ce and ^{147}Pm, in parts of the Atlantic Ocean between 1956 and 1961 have been reported by Sugihara and Bowen (1962) and Bowen and Sugihara (1965). Surface concentrations were usually in the range of $4–65 \times 10^{-2} \, pCi \, l^{-1}$ for ^{144}Ce and $0.5–14 \times 10^{-2} \, pCi \, l^{-1}$ for ^{147}Pm, but considerably higher concentrations were occasionally recorded. Promethium-147 is more depleted in surface waters, relative to ^{90}Sr, than is ^{144}Ce. Vertical profiles show secondary maxima in which ^{147}Pm is enriched relative to ^{144}Ce. These distributions are attributable to the downward transfer of the lanthanide nuclides with particulate matter, including organic material. It appears that ^{147}Pm is more involved in these processes than is ^{144}Ce, but both nuclides show significant penetration. Evidence for relatively rapid particulate transport of isotopes of cerium has also been obtained by Osterberg et al. (1963b).

In periods following weapon explosions, a number of shorter-lived nuclides, such as ^{95}Zr + ^{95}Nb, ^{103}Ru and ^{141}Ce, have been measured in ocean waters receiving long-range fallout (Kameda, 1963; Chakravarti et al., 1964; Chesselet et al., 1965; Gross et al., 1965; Silker, 1972a, b), and others, ^{140}Ba + ^{140}La and ^{131}I, have been detected in organisms (Palumbo et al., 1966; Jenkins, 1969). The occurrence of ^{125}Sb has been reported for coastal waters (Slowey et al., 1965). Many of the fallout nuclides of relatively short half-life have been detected in sediments of coastal areas, as, for example, by Osterberg et al. (1963a). The appearance of ^{95}Zr + ^{95}Nb in bottom deposits from the north-eastern Pacific Ocean following the weapon tests of 1961–62 was used by Gross (1967) to estimate the sinking rate of particles in that area as about $10 \, m \, day^{-1}$. This value is consistent with the rapid appearance of these nuclides, with ^{141}Ce and ^{144}Ce, in deep benthic organisms off the Oregon coast, as recorded by Osterberg et al. (1963b). Transport with particulate matter is important in determining the vertical distributions of ^{95}Zr + ^{95}Nb and isotopes of ruthenium, as with the longer-lived nuclides discussed above, and, in a similar way, it leads to fractionation of these groups from each other (Chesselet *et al.*, 1965; Slowey *et al.*, 1965) and from the conservative nuclides as represented by ^{90}Sr (Volchok *et al.*, 1971).

The possible use of the long-lived nuclide ^{129}I in studies of water mixing has been discussed by Edwards (1962), but no relevant measurements have been reported.

18.5.4.3. *Neutron activation products*

Investigations of neutron activation products in ocean waters have been undertaken in two different contexts. Firstly, there have been a number of studies of these nuclides in waters affected by discharges from nuclear reactors. The most extensive of these investigations, and those which have given the most useful information in oceanographic terms, were carried out in the plume of the Columbia River which was labelled with a number of activation products. Summaries are available in the work edited by Pruter and Alverson (1972). Secondly studies have been made of a wider range of nuclides present in ocean waters as a result of fallout from explosions of nuclear weapons. Most of the information on these nuclides relates to concentrations in organisms, and for some nuclides the data were obtained in regions directly influenced by local weapon testing.

Reactor operations at Hanford led to the discharge of many radioactive nuclides to the Columbia River over a period of more than two decades prior to the cessation of operations early in 1971; some operation was subsequently resumed. Of those nuclides which reached the Pacific Ocean in significant amounts, ^{51}Cr and ^{65}Zn were the most important. The average annual discharges to the Pacific Ocean over the period 1960–67 were $2 \cdot 7 \times 10^5$ Ci of ^{51}Cr and $1 \cdot 3 \times 10^4$ Ci of ^{65}Zn (Joseph *et al.*, 1971).

Despite the fact that ^{51}Cr has a considerably shorter half-life than does ^{65}Zn its greater input and dominantly conservative behaviour in the river plume enabled it to be detected at distances of 100–350 km from the river mouth, about ten times greater than the distance over which ^{65}Zn could be detected (Gross *et al.*, 1965; Osterberg *et al.*, 1965). Although some ^{51}Cr was associated with particles, possibly indicating that a fraction of the nuclide was present in the tripositive oxidation state (Curl *et al.*, 1965), the dominantly conservative behaviour (Osterberg *et al.*, 1965; Cutshall *et al.*, 1966) suggests that the nuclide mostly remained in the hexapositive oxidation state in which it was introduced.

The interactions of ^{65}Zn in the river plume have been investigated in some detail. Johnson *et al.* (1967) examined the association of the nuclide with river sediment and showed that it is not readily desorbed from river-borne particles when they come into contact with sea water. In the plume, several biological processes have been identified which can accelerate the vertical transport of ^{65}Zn, namely transference by organisms which migrate vertically (Pearcy and Osterberg, 1967) and the sinking of the moulted exoskeletons of euphausiids (Fowler and Small, 1967). The amount of river water present in the plume shows a marked seasonal variation; this was reflected in a seasonal variation in the concentrations of ^{65}Zn in organisms trawled in the upper 150 m and in the livers of albacore; organisms trawled below 500 m in the plume region and albacore collected off Southern California did not show a related variation (Pearcy and Osterberg, 1967, 1968). Concentrations of ^{65}Zn in organisms show the expected marked decrease with distance offshore. Close to the river mouth, during the period when the discharge of the nuclide was comparatively uniform, concentrations in the range of about 40–136 pCi g^{-1} dry weight were reported for *Salpa* spp. and *Euphausia pacifica*, whereas at about 300 km along the plume, concentrations were more typically about 7 pCi g^{-1} dry weight in similar material (Osterberg, 1962; Osterberg *et al.*, 1964). Organisms trawled in the upper 150 m at about 100 km from the river mouth were reported by Pearcy and Osterberg (1967) to contain about 1 pCi g^{-1} wet weight. Some of the highest concentrations occurred in algae and benthic organisms near the river mouth where a value of 1600 pCi g^{-1} dry weight was found in *Mytilus edulis* during 1960 (Watson *et al.*, 1963).

These concentrations contrast with values of from less than 0·01 to 0·1 pCi g^{-1} wet weight in organisms collected by midwater trawling at distances greater than 500 km from the river discharge (Pearcy and Osterberg, 1967) and the range of 0·01–0·2 pCi g^{-1} wet weight found for Californian coastal organisms in 1963 (Folsom *et al.*, 1963). These values essentially reflect weapon fallout, and are compatible with the concentrations of 0·02–0·7 pCi g^{-1} dry weight found by Alexander and Rowlands (1966) in N. American coastal invertebrate organisms, collected in 1964–65, which had been exposed only to world-wide fallout.

Concentrations of ^{54}Mn found by Alexander and Rowlands (1966) in barnacles and oysters were in the range of 0·24–2·2 pCi g^{-1} dry weight, in good agreement with the values of 0·01–0·4 pCi g^{-1} wet weight found slightly earlier for Californian coastal organisms by Folsom *et al.* (1963). At the mouth of the Columbia River in 1959–60, plankton contained about 8 pCi g^{-1} dry weight and *Mytilus edulis* contained 4·6 pCi g^{-1} dry weight, these concentrations being the highest found in a number of organisms analysed (Watson *et al.*, 1963). In benthic organisms from the Mediterranean Sea, Chipman and Thommeret (1970) found concentrations of this nuclide to range from the limits of detection to 400 pCi g^{-1} ash; there was generally a parallel trend with concentrations of stable manganese, the specific activity being usually in the range 7–14 pCi mg^{-1} Mn.

The presence of a wide range of neutron activation products has been established in organisms from the central and north-western Pacific Ocean, following tests of nuclear weapons. Many observations relate to samples exposed to appreciable local fallout, whereas other studies have shown lower concentrations associated with long-range fallout. Measurements in the mid and late 1950s showed the presence of ^{54}Mn, ^{55}Fe, ^{59}Fe, ^{57}Co, ^{58}Co, ^{60}Co, ^{65}Zn and ^{113m}Cd in widely differing proportions in various tissues. Sources for these data were given by Burton (1965); more recent observations on such nuclides include those by Folsom and Young (1965), Palumbo et al. (1966), Pearcy and Osterberg (1968), Jenkins (1969) and Beasley et al. (1969, 1971). Jenkins (1969) identified a number of additional neutron activation products in tissues of Pacific salmon, including ^{203}Hg, ^{152}Eu and ^{46}Sc; the presence of the last two nuclides may reflect the influence of Columbia River effluent rather than fallout. It was found that ^{54}Mn, ^{60}Co and ^{65}Zn were more concentrated in salmon livers than in muscle, by one or two orders of magnitude.

Silver-110m was detected by Folsom and Young (1965) in some organisms collected from the North Pacific Ocean in 1964; the highest concentration was 4200 pCi kg^{-1} wet weight in the digestive gland of a squid. This nuclide was also measured in Pacific salmon by Jenkins (1969) who found the concentration to be greatest in the liver and to show a successively decreasing trend in roe, bone, and muscle. The longer lived isotope, ^{108m}Ag, was subsequently found by Folsom et al. (1970) to be also present. The activity ratio, ^{108m}Ag to ^{110m}Ag, was considered to indicate that these nuclides had been released primarily in the weapon tests of 1961–62 and this suggested possible tracer applications for these nuclides. However, Beasley and Held (1971) have analysed tissues and sediments collected after earlier explosions and have found that ^{108m}Ag was produced then in significant amounts. Organisms exposed to effluent from nuclear plant have been shown to contain ^{110m}Ag (Preston et al., 1968).

The presence in marine organisms of another long-lived activation product, ^{63}Ni, has been reported by Beasley and Held (1969). The highest concentrations were found in kidneys of clams from the testing areas in the Pacific Ocean, where a maximum level of 73 pCi g^{-1} dry weight was found. In shellfish from the eastern coast of North America the concentration was less than 0·01 pCi g^{-1} dry weight.

Concentrations of ^{55}Fe in fish tissues have been measured quite extensively (Rama et al., 1961b; Palmer et al., 1966; Jenkins, 1969; Preston, 1970). Relatively high specific activities of this nuclide can occur in such tissues, probably reflecting the low concentrations of biologically available stable iron in surface waters and the comparatively ready uptake of the freshly introduced radioactive form. Much of the total fallout of ^{55}Fe was produced by explosions during 1961–62 and the total deposition has been greatest at about 40–50° N (Joseph *et al.*, 1971). There have been marked variations in concentrations in fish with time and latitude, as well as with feeding habits. Pronounced latitudinal dependence of levels in North Atlantic cod in 1967–69 was found by Preston (1970); the maximum specific activity of 221 pCi mg^{-1} Fe compared with a peak value of 2·9 × 10^{4} pCi mg^{-1} Fe for Pacific salmon in 1964. Concentrations in tuna and salmon in late 1965 were about two orders of magnitude higher than those measured after earlier series of weapon tests.

18.5.4.4. *Plutonium*

There is good agreement between the various measurements which have been made of concentrations of ^{239}Pu in sea water. The earliest reported values, for Pacific Ocean surface waters in 1964 (Pillai *et al.*, 1964), were about 0·9 × 10^{-3} dpm l^{-1} for coastal waters and 4·5 × 10^{-3} dpm l^{-1} for open waters. The ratio of ^{239}Pu to ^{90}Sr was then similar to that in fallout, reflecting the fact that the bulk of the deposit was relatively recent and had not undergone much geochemical fractional after entry to the surface ocean.

Subsequent investigations (Miyake and Sugimura, 1968; Miyake *et al.*, 1970c; Bowen *et al.*, 1971) have shown surface concentrations in the Atlantic and Pacific Oceans within the range from the limit of detection to 7·7 × 10^{-3} dpm l^{-1}. About 1971, surface coastal waters off California contained 1·3 × 10^{-3} dpm l^{-1} (Wong *et al.*, 1972). Miyake and Sugimura (1968) and Miyake *et al.* (1970c) recorded activity ratios of ^{238}Pu to ^{239}Pu for surface waters collected in 1967–68 of 0·07–0·38, this relatively high range reflecting the release of ^{238}Pu from the satellite burn-up of 1964 (see p. 159). Measurements on algal material indicated a ratio of 0·055 for Californian coastal waters in the early 1970s (Wong *et al.*, 1972), which is similar to that of about 0·03 determined in a similar way by Pillai *et al.* (1964) in 1964; this latter value was similar to that measured for fallout in the U.S.A. during 1961–63. In some of the later work the geochemical separation of ^{239}Pu and the conservative nuclides, ^{90}Sr and ^{137}Cs, had become apparent, with relative depletion of ^{239}Pu in surface waters and penetration as deep as 3000 m in the north-western Pacific Ocean by 1967. In the Sargasso Sea a maximum in the concentration of ^{239}Pu in a vertical profile was found at 500 m. This behaviour of ^{239}Pu resembles that of the lanthanide nuclides (p. 172) and is attributable primarily to transfer by association with organisms and subsequent dissolution of detritus in deeper water.

Concentration factors for uptake by organisms of about 10^{3} on a wet weight basis were reported by Pillai *et al.* (1964) for fixed algae, dino-flagellates, and zooplankton, with a decrease in concentration at higher trophic levels. Concentrations in a number of coastal and oceanic organisms from the Atlantic Ocean in 1970 have been determined by Wong (1971). A high concentration occurred in Sargasso weed (6 pCi kg^{-1} wet weight). Other typical or average values (pCi kg^{-1} wet weight) included 0·1 and 0·013 in shark liver and bone, respectively, 0·002 in fish muscle, 0·13 in fish bone, 0·16 in blue mussels and 1·3 in zooplankton. Samples of sediments from Cape Cod in 1968 contained on average 80 dpm kg^{-1} dry weight. Concentrations of ^{239}Pu in the giant brown alga, *Pelagophycus porra*, collected in 1971 from the Californian coast, showed considerable variations between different parts of the plant (Wong *et al.*, 1972). The bladder surfaces contained 2 pCi kg^{-1} wet weight, but the concentration decreased by two orders of magnitude in the inner tissue of the bladder. The use of such algae as indicator organisms for monitoring plutonium concentrations would, therefore, require the use of carefully controlled sampling procedures.

References

Åberg, B. and Hungate, F. P. (eds) (1967). "Radioecological Concentration Processes", xiv + 1040 pp. Pergamon, Oxford.

Adams, J. A. S. (1954). *Bull. geol. Soc. Am.* **65**, 1225.

Agamirov, S. Sh. (1963). *Geochemistry*, 104.

Alexander, G. V. and Rowlands, R. H. (1966). *Nature, Lond.* **210**, 155.

Amiel, A. J., Miller, D. S. and Friedman, G. M. (1973). *Sedimentology*, **20**, 523.

Amin, B. S., Kharkar, D. P. and Lal, D. (1966). *Deep Sea Res.* **13**, 805.

Antal, P. S. (1966). *Limnol. Oceanogr.* **11**, 278.

Armstrong, R. L. (1971). *Nature, Phys. Sci.* **230**, 132.

Arnold, J. R. (1956). *Science, N.Y.* **124**, 584.

Arnold, J. R., Honda, M. and Lal, D. (1961). *J. geophys. Res.* **66**, 3519.

Arrhenius, G. O. S. (1963). *In* "The Sea" (M. N. Hill, ed.), Vol. 3, pp. 655–727. Interscience, New York.

Arrhenius, G. O. S., Bramlette, M. N. and Picciotto, E. (1957). *Nature, Lond.* **180**, 85.

Arrhenius, G. O. S., Mero, J. L. and Korkisch, J. (1964). *Science, N.Y.* **144**, 170.

Aten, A. H. W., Jr., Dalenberg, J. W. and Bakkum, W. C. M. (1961). *Hlth Phys.* **5**, 225.

Attree, R. W., Cabell, M. J., Cushing, R. L. and Pieroni, J. J. (1962). *Can. J. Phys.* **40**, 194.

Aumento, F. (1969). *Can. J. Earth Sci.* **6**, 1431.

Aumento, F. (1971). *Earth planet. Sci. Lett.* **11**, 90.

Aumento, F., Wanless, R. K. and Stevens, R. D. (1968). *Science, N.Y.* **161**, 1338.

Bainbridge, A. E. (1963a). *In* "Nuclear Geophysics" (P. M. Hurley, G. Faure and C. Schnetzler, eds), Publication 1075, pp. 129–137. National Academy of Sciences—National Research Council, Washington, D.C.

Bainbridge, A. E. (1963b). *J. geophys. Res.* **68**, 3785.

Bainbridge, A. E. and O'Brien, B. J. (1962). *In* "Tritium in the Physical and Biological Sciences", Vol. I, pp. 33–38. International Atomic Energy Agency, Vienna.

Baranov, V. I. and Khitrov, L. M. (eds) (1964). "Radioactive Contamination of the Sea", vi + 194 pp. Izdatel'stvo "Nauka", Moscow. (Translation published 1966 Israel Program for Scientific Translations, Jerusalem).

Baranov, V. I. and Khristianova, L. A. (1959). *Geochemistry*, 765.

Barnes, S. S. and Dymond, J. R. (1967). *Nature, Lond.* **213**, 1218.

Baturin, G. N. (1968). *Geochemistry int.* **5**, 344.

Baturin, G. N. (1969). *Oceanology*, **9**, 828.

Baturin, G. N. and Kochenov, A. V. (1969). *Geokhimiya*, 715.

Baturin, G. N., Kochenov, A. V. and Kovaleva, S. A. (1966). *Dokl. Acad. Sci. U.S.S.R.* **166**, 172.

Baturin, G. N., Kochenov, A. V. and Shimkus, K. M. (1967). *Geochemistry int.* **4**, 29.

Baturin, G. N., Kochenov, A. V. and Senin, Yu. M. (1971). *Geochemistry int.* **8**, 281.

Baturin, G. N., Merkulova, K. I. and Chalov, P. I. (1972). *Mar. Geol.* **13**, M37.

Baxter, M. S. and Walton, A. (1970). *Proc. R. Soc., Ser. A*, **318**, 213.

Beasley, T. M. (1969). *Nature, Lond.* **224**, 573.

Beasley, T. M. and Held, E. E. (1969). *Science, N.Y.* **164**, 1161.

Beasley, T. M. and Held, E. E. (1971). *Nature, Lond.* **230**, 450.

Beasley, T. M., Osterberg, C. L. and Jones, Y. M. (1969). *Nature, Lond.* **221**, 1207.

Beasley, T. M., Jokela, T. A. and Eagle, R. J. (1971). *Hlth. Phys.* **21**, 815.

Bender, M., Broecker, W., Gornitz, V., Middel, U., Kay, R., Sun, S.-S. and Biscaye, P. (1971). *Earth planet. Sci. Lett.* **12**, 425.

Bernat, M. and Goldberg, E. D. (1969). *Earth planet. Sci. Lett.* **5**, 308.

Bernat, M., Bieri, R. H., Koide, M., Griffin, J. J. and Goldberg, E. D. (1970). *Geochim. cosmochim. Acta*, **34**, 1053.

Bertine, K. K., Chan, L. H. and Turekian, K. K. (1970). *Geochim. cosmochim. Acta*, **34**, 641.

Bhat, S. G. and Krishnaswamy, S. (1969). *Proc. Indian Acad. Sci., Section A*, **70**, 1.

Bhat, S. G., Krishnaswamy, S., Lal, D., Rama and Moore, W. S. (1969). *Earth planet. Sci. Lett.* **5**, 483.

Bhat, S. G., Krishnaswamy, S., Lal, D., Rama and Somayajulu, B. L. K. (1973). *In* "Proceedings of Symposium on Hydrogeochemistry and Biogeochemistry" (E. Ingerson, ed.), Vol. 1, pp. 443–462. Clarke, Washington, D.C.

Bien, G. and Suess, H. (1967). *In* "Radioactive Dating and Methods of Low-Level Counting", pp. 105–115. International Atomic Energy Agency, Vienna.

Bien, G. S., Rakestraw, N. W. and Suess, H. E. (1960). *Tellus*, **12**, 436.

Bien, G. S., Rakestraw, N. W. and Suess, H. E. (1963a). *In* "Nuclear Geophysics" (P. M. Hurley, G. Faure and C. Schnetzler, eds), Publication 1075, pp. 152–160. National Academy of Sciences—National Research Council, Washington, D.C.

Bien, G. S., Rakestraw, N. W. and Suess, H. E. (1963b). *In* "Radioactive Dating", pp. 159–172. International Atomic Energy Agency, Vienna.

Bien, G. S., Rakestraw, N. W. and Suess, H. E. (1965). *Limnol. Oceanogr.* **10**, R25.

Biscaye, P. E. and Dasch, E. J. (1971). *J. geophys. Res.* **76**, 5087.

Blackman, A. and Somayajulu, B. L. K. (1966). *Science, N.Y.* **154**, 886.

Blanchard, R. L. (1965). *J. geophys. Res.* **70**, 4055.

Blanchard, R. L. and Oakes, D. (1965). *J. geophys. Res.* **70**, 2911.

Blanchard, R. L., Cheng, M. H. and Potratz, H. A. (1967). *J. geophys. Res.* **72**, 4745.

Bolin, B. and Eriksson, E. (1959). *In* "The Atmosphere and the Sea in Motion" (B.

Bolin, ed.), pp. 130–142. Rockefeller Institute Press, New York.

Bonatti, E., Fisher, D. E., Joensuu, O. and Rydell, H. S. (1971). Geochim. cosmochim. Acta, 35, 189.

Bonatti, E., Fisher, D. E., Joensuu, O., Rydell, H. S. and Beyth, M. (1972). Econ. Geol. 67, 717.

Boström, K. and Fisher, D. E. (1971). Earth planet. Sci. Lett. 11, 95.

Bowen, V. T. and Roether, W. (1973). J. geophys. Res. 78, 6277.

Bowen, V. T. and Sugihara, T. T. (1957). Proc. natn. Acad. Sci. U.S.A. 43, 576.

Bowen, V. T. and Sugihara, T. T. (1958). Proc. second United Nations int. Conf. peaceful Uses atomic Energy, 18, 434.

Bowen, V. T. and Sugihara, T. T. (1960). Nature, Lond. 186, 71.

Bowen, V. T. and Sugihara, T. T. (1963). In "Radioecology" (V. Schultz and A. W. Klement, Jr., eds), pp. 135–139. Reinhold, New York; American Institute of Biological Sciences, Washington, D.C.

Bowen, V. T. and Sugihara, T. T. (1965). J. mar. Res. 23, 123.

Bowen, V. T., Noshkin, V. E. and Sugihara, T. T. (1966). Nature, Lond. 212, 383.

Bowen, V. T., Noshkin, V. E., Volchok, H. L. and Sugihara, T. T. (1968). In "Health and Safety Laboratory, Fallout Program, Quarterly Summary Report". Rept. HASL 197, pp. I-2–I-64. United States Atomic Energy Commission, New York Operations Office.

Bowen, V. T., Noshkin, V. E., Volchok, H. L. and Sugihara, T. T. (1969). Science, N.Y. 164, 825.

Bowen, V. T., Wong, K. M. and Noshkin, V. E. (1971). J. mar. Res. 29, 1.

Bowen, V. T., Metcalf, W. G. and Burke, J. C. (1972). J. mar. Res. 30, 112.

Bowie, S. H. U. and Atkin, D. (1956). Nature, Lond. 177, 487.

Bray, J. R. (1966). Nature, Lond. 209, 1065.

Broecker, W. S. (1963a). J. geophys. Res. 68, 2817.

Broecker, W. S. (1963b). In "The Sea" (M. N. Hill, ed.), Vol. 2, pp. 88–108. Interscience, New York.

Broecker, W. S. (1963c). In "Nuclear Geophysics" (P. M. Hurley, G. Faure and C. Schnetzler, eds.), Publication 1075, pp. 138–149. National Academy of Sciences— National Research Council, Washington, D.C.

Broecker, W. S. (1965). In "Symposium on Diffusion in Oceans and Fresh Waters" (T. Ichiye, ed.), pp. 116–144. Lamont Geological Observatory, Palisades, N.Y.

Broecker, W. S. (1966). J. geophys. Res. 71, 5827.

Broecker, W. S. and Kaufman, A. (1970). J. geophys. Res. 75, 7679.

Broecker, W. S. and Ku, T.-L. (1969). Science, N.Y. 166, 404.

Broecker, W. S. and Li, Y.-H. (1970). J. geophys. Res. 75, 3545.

Broecker, W. S. and Olson, E. A. (1959). Am. J. Sci. Radiocarbon Suppl. 1, 111.

Broecker, W. S. and Olson, E. A. (1961). Am. J. Sci. Radiocarbon Suppl. 3, 176.

Broecker, W. S. and Peng, T. H. (1971). Earth planet. Sci. Lett. 11, 99.

Broecker, W. S. and Rocco, G. C. (1963). In "Nuclear Geophysics" (P. M. Hurley, G. Faure and C. Schnetzler, eds), Publication 1075, pp. 150–151. National Academy of Sciences—National Research Council, Washington, D.C.

Broecker, W. S. and Thurber, D. I.. (1965). Science, N.Y. 149, 58.

Broecker, W. S., Turekian, K. K. and Heezen, B. C. (1958). Am. J. Sci. 256, 503.

Broecker, W. S., Ewing, M. and Heezen, B. C. (1960a). Am. J. Sci. 258, 429.

Broecker, W. S., Gerard, R., Ewing, M. and Heezen, B. C. (1960b). J. geophys. Res. 65, 2903.

Broecker, W. S., Gerard, R. D., Ewing, M. and Heezen, B. C. (1961). *In* "Oceanography" (M. Sears, ed.), pp. 301–322. Publ. 67 American Association for the Advancement of Science, Washington, D.C.

Broecker, W. S., Rocco, G. G. and Volchok, H. L. (1966a). *Science, N.Y.* **152**, 639.

Broecker, W. S., Bonebakker, E. R. and Rocco, G. G. (1966b). *J. geophys. Res.* **71**, 1999.

Broecker, W. S., Li, Y.-H. and Cromwell, J. (1967). *Science, N.Y.* **158**, 1307.

Broecker, W. S., Cromwell, J. and Li, Y.-H. (1968a). *Earth planet. Sci. Lett.* **5**, 101.

Broecker, W. S., Thurber, D. L., Goddard, J., Ku, T.-L., Matthews, R. K. and Mesolella, K. J. (1968b). *Science, N.Y.* **159**, 297.

Broecker, W., Kaufman, A., Ku, T.-L., Chung, Y. and Craig, H. (1970). *J. geophys. Res.* **75**, 7682.

Broecker, W. S., Li, Y.-H. and Peng, T.-H. (1971). *In* "Impingement of Man on the Oceans" (D. W. Hood, ed.), pp. 287–324. Wiley-Interscience, New York.

Broecker, W. S., Kaufman, A. and Trier, R. M. (1973). *Earth planet. Sci. Lett.* **20**, 35.

Burton, J. D. (1965). *In* "Chemical Oceanography" (J. P. Riley and G. Skirrow, eds), Vol. 2, pp. 425–475, Academic, London.

Chakravarti, D., Lewis, G. B., Palumbo, R. F. and Seymour, A. H. (1964). *Nature, Lond.* **203**, 571.

Chalov, P. I., Tuzova, T. V. and Musin, Ye. A. (1964). *Geochemistry int.* **1**, 402.

Cherdyntsev, V. V. (1955). *Trudy Tret'ei sessii Komissii po opredeleniyu absolyutnogo vozrasta geologicheskikh formatsii*, 175. Izdatel'stvo Akad. Nauk SSSR, Moscow.

Cherdyntsev, V. V. (1969). "Uranium-234", v + 234 pp. Atomizdat, Moscow. (Translation published 1971, Israel Program for Scientific Translations, Jerusalem).

Cherry, R. D., Geriche, I. H. and Shannon, L. V. (1969). *Earth planet. Sci. Lett.* **6**, 451.

Chesselet, R., Nordemann, D. and Lalou, C. (1965). *C.r. hebd. Séanc. Acad. Sci. Paris*, **260**, 2875.

Chipman, W. and Thommeret, J. (1970). *Bull. Inst. océanogr. Monaco*, **69**, No. 1402, 15pp.

Chung, Y. (1973). *Earth planet. Sci. Lett.* **17**, 319.

Chung, Y. and Craig, H. (1972). *Earth planet. Sci. Lett.* **14**, 55.

Chung, Y. and Craig, H. (1973). *Earth planet. Sci. Lett.* **17**, 306.

Church, T. M. and Bernat, M. (1972). *Earth planet. Sci. Lett.* **14**, 139.

Clarke, R. S. Jr. and Altschuler, Z. S. (1958). *Geochim. cosmochim. Acta*, **13**, 127.

Craig, H. (1957). *Tellus*, **9**, 1.

Craig, H. (1958). *Proc. second United Nations int. Conf. peaceful Uses atomic Energy*, **18**, 358.

Craig, H. (1961). *Am. J. Sci. Radiocarbon Suppl.* **3**, 1.

Craig, H. (1963). *In* "Earth Science and Meteoritics" (J. Geiss and E. D. Goldberg, eds), pp. 103–114. North-Holland, Amsterdam.

Craig, H. (1969). *J. geophys. Res.* **74**, 5491.

Craig, H. (1971). *J. geophys. Res.* **76**, 5133.

Craig, H. and Lal, D. (1961). *Tellus*, **13**, 85.

Craig, H., Krishnaswami, S. and Somayajulu, B. L. K. (1973). *Earth planet Sci. Lett.* **17**, 295.

Curl, H., Jr., Cutshall, N. and Osterberg, C. (1965). *Nature, Lond.* **205**, 275.

Cutshall, N., Johnson, V. and Osterberg, C. (1966). *Science, N.Y.* **152**, 202.

Dalrymple, G. B. and Moore, J. G. (1968). *Science, N.Y.* **161**, 1132.

Damon, P. E., Long, A. and Grey, D. C. (1966). *J. geophys. Res.* **71**, 1055.
Dasch, E. J. (1969). *Geochim. cosmochim. Acta,* **33**, 1521.
Dockins, K. O., Bainbridge, A. E., Houtermans, J. C. and Suess, H. E. (1967). *In* "Radioactive Dating and Methods of Low-Level Counting", pp. 129–140. International Atomic Energy Agency, Vienna.
Dorta, C. C. and Rona, E. (1971). *Bull. mar. Sci.* **21**, 754.
Dunster, H. J., Garner, R. J., Howells, H. and Wix, L. F. U. (1964). *Hlth Phys.* **10**, 353.
Duursma, E. K. (1972). *Oceanogr. Mar. Biol. ann. Rev.* **10**, 137.
Duursma, E. K. and Bosch, C. J. (1970). *Neth. J. Sea Res.* **4**, 395.
Duursma, E. K. and Gross, M. G. (1971). *In* "Radioactivity in the Marine Environment", pp. 147–160. National Academy of Sciences, Washington, D.C.
Dymond, J. R. (1966). *Science, N.Y.* **152**, 1239.
Dymond, J. (1969). *Earth planet. Sci. Lett.* **6**, 9.
Dymond, J. (1970). *Bull. geol. Soc. Am.* **81**, 1229.
Dymond, J. and Windom, H. L. (1968). *Earth planet. Sci. Lett.* **4**, 47.
Edgington, D. N., Gordon, S. A., Thommes, M. M. and Almodovar, L. R. (1970). *Limnol. Oceanogr.* **15**, 945.
Edmond, J. M. (1970). *J. geophys. Res.* **75**, 6878.
Edwards, R. R. (1962). *Science, N.Y.* **137**, 851.
Emiliani, C. and Rona, E. (1969). *Science, N.Y.* **166**, 1551.
Ericson, D. B., Broecker, W. S., Kulp, J. I. and Wollin, G. (1956). *Science, N.Y.* **124**, 385.
European Nuclear Energy Agency (1968). "Radioactive Waste Disposal Operation into the Atlantic 1967", 74 pp. European Nuclear Energy Agency, Paris.
Fairhall, A. W., Ostenson, A. T., Yang, I. C. and Young, A. W. (1970). *Antarctic J.U.S.* **5**, 190.
Fairhall, A. W., Bradford, P., Yang, I. C. and Young, A. W. (1971). *Antarctic J.U.S.* **6**, 163.
Fanale, F. P. and Schaeffer, O. A. (1965). *Science, N.Y.* **149**, 312.
Faure, G. and Powell, J. L. (1972). "Strontium Isotope Geology", 188 pp. Springer-Verlag, Berlin.
Faure, G., Hurley, P. M. and Powell, J. L. (1965). *Geochim. cosmochim. Acta,* **29**, 209.
Faure, G., Crockett, J. H. and Hurley, P. M. (1967). *Geochim. cosmochim. Acta,* **31**, 451.
Ferguson, C. W. (1958). *Science, N.Y.* **159**, 839.
Ferguson, C. W. (1970). *In* "Radiocarbon Variations and Absolute Chronology" (I. U. Olsson, ed.), pp. 237–259. Wiley Interscience, New York: Almqvist and Wiksell, Stockholm.
Ferguson, C. W. (1972). *In* "Proceedings of the 8th International Conference on Radiocarbon Dating" (T. A. Rafter and T. L. Grant-Taylor, eds), pp. A1–A10. R. and D. Associates, Santa Monica.
Fisher, D. E. (1971). *Earth planet. Sci. Lett.* **12**, 321.
Fisher, D. E. (1972). *Earth planet. Sci. Lett.* **14**, 255.
Fisher, D. E. and Boström, K. (1969). *Nature, Lond.* **224**, 64.
Fisher, D. E., Bonatti, E., Joensuu, O. and Funkhouser, J. (1968). *Science, N.Y.* **160**, 1106.
Folsom, T. R. and Mohanrao, G. J. (1960). *J. Radiat. Res.* **1**, 150.
Folsom, T. R. and Young, D. R. (1965). *Nature, Lond.* **206**, 803.

Folsom, T. R., Young, D. R., Johnson, J. N. and Pillai, K. C. (1963). *Nature, Lond.* **200**, 327.

Folsom, T. R., Grismore, R. and Young, D. R. (1970). *Nature, Lond.* **227**, 941.

Food and Agriculture Organization of the United Nations (1971). "Report of the Seminar on Methods of Detection, Measurement and Monitoring of Pollutants in the Marine Environment", FAO Fisheries Rep. No. 99, Suppl. 1. Food and Agriculture Organization of the United Nations, Rome.

Foster, R. F., Ophel, I. L. and Preston, A. (1971). *In* "Radioactivity in the Marine Environment", pp. 240–260. National Academy of Sciences, Washington, D.C.

Fowler, S. W. and Small, L. F. (1967). *Int. J. Oceanol. Limnol.* **1**, 237.

Freiling, E. C. and Ballou, N. E. (1962). *Nature, Lond.* **195**, 1283.

Fukai, R. and Huynh-Ngoc, I. (1968). *Radioact. Sea Ser.* No. 22, 26 pp.

Funkhouser, J. G., Fisher, D. E. and Bonatti, E. (1968). *Earth planet. Sci. Lett.* **5**, 95.

Goel, P. S., Kharkar, D. P., Lal, D., Narsappaya, N., Peters, B. and Yatirajam, V. (1957). *Deep Sea Res.* **4**, 202.

Goldberg, E. D. (1961). *In* "Oceanography" (M. Sears, ed.), pp. 583–597. American Association for the Advancement of Science, Washington, D.C.

Goldberg, E. D. (1963). *In* "Radioactive Dating", pp. 121–130. International Atomic Energy Agency, Vienna.

Goldberg, E. D. (1965). *Limnol. Oceanogr.* **10**, R125.

Goldberg, E. D. (1968). *Earth planet. Sci. Lett.* **4**, 17.

Goldberg, E. D. and Griffin, J. J. (1964). *J. geophys. Res.* **69**, 4293.

Goldberg, E. D. and Koide, M. (1958). *Science, N.Y.* **128**, 1003.

Goldberg, E. D. and Koide, M. (1962). *Geochim. cosmochim. Acta,* **26**, 417.

Goldberg, E. D. and Koide, M. (1963). *In* "Earth Science and Meteoritics" (J. Geiss and E. D. Goldberg, eds), pp. 91–102. North-Holland, Amsterdam.

Goldberg, E. D., Koide, M., Griffin, J. J. and Peterson, M. N. A. (1964a). *In* "Isotopic and Cosmic Chemistry" (H. Craig, S. Miller and G. J. Wasserburg, eds), pp. 211–232. North-Holland, Amsterdam.

Goldberg, E. D., Koide, M. and Griffin, J. J. (1964b). *In* "Recent Researches in the Fields of Hydrosphere, Atmosphere and Nuclear Geochemistry" (Y. Miyake and T. Koyoma, eds), pp. 117–126. Maruzen, Tokyo.

Goldberg, E. D., Somayajulu, B. L. K., Galloway, J., Kaplan, I. R. and Faure, G. (1969). *Geochim cosmochim Acta,* **33**, 287.

Grashchenko, S. M., Nikolaev, D. S., Kolyadin, L. B., Kuznetsov, Yu. V. and Lazarev, K. F. (1960). *Dokl. Akad. Nauk SSSR,* **132**, 1171.

Griffin, J. J., Koide, M., Höhndorf, A., Hawkins, J. W. and Goldberg, E. D. (1972). *Deep Sea Res.* **19**, 139.

Gross, M. G. (1967). *Nature, Lond.* **216**, 670.

Gross, M. G., Barnes, C. A. and Riel, G. K. (1965). *Science, N.Y.* **149**, 1088.

Hardy, E. P., Krey, P. W. and Volchok, H. L. (1973). *Nature, Lond.* **241**, 444.

Hart, R. A. (1973). *Nature, Lond.* **243**, 76.

Hashimoto, T. (1971). *Analytica chim. Acta,* **56**, 347.

Heath, G. R., Moore, T. C., Jr. and Somayajulu, B. L. K. (1970). *J. mar. Res.* **28**, 225.

Heye, D. (1969). *Earth planet. Sci. Lett.* **6**, 112.

Higdon, J. C. and Lingenfelter, R. E. (1973). *Nature, Lond.* **246**, 403.

Holmes, C. W., Osmond, J. K. and Goodell, H. G. (1968). *Earth planet. Sci. Lett.* **4**, 368.

Holtzmann, R. B. (1967). *In* "Symposium on Radioecology" (D. J. Nelson and F. C Evans, eds), Rep. CONF-670503, pp. 535–546. United States Atomic Energy Commission, Washington, D.C.

Houtermans, J. C., Suess, H. E. and Oeschger, H. (1973). *J. geophys. Res.* **78**, 1897.

Hurley, P. M., Heezen, B. C., Pinson, W. H. and Fairbairn, H. W. (1963). *Geochim. cosmochim. Acta,* **27**, 393.

International Atomic Energy Agency (1960). "Disposal of Radioactive Wastes", Vol. I, 603 pp., Vol. II, 575 pp. International Atomic Energy Agency, Vienna.

International Atomic Energy Agency (1961). "Radioactive Waste Disposal into the Sea", 168 pp. International Atomic Energy Agency, Vienna.

International Atomic Energy Agency (1962). "Disposal of Radioactive Wastes into Marine and Fresh Waters", 365 pp. International Atomic Energy Agency, Vienna.

International Atomic Energy Agency (1966). "Disposal of Radioactive Wastes into Seas, Oceans and Surface Waters", 898 pp. International Atomic Energy Agency, Vienna.

International Atomic Energy Agency (1971). "Disposal of Radioactive Wastes into Rivers, Lakes and Estuaries", 77 pp. International Atomic Energy Agency, Vienna.

International Atomic Energy Agency (1973). "Radioactive Contamination of the Marine Environment", 786 pp. International Atomic Energy Agency, Vienna.

International Union of Geodesy and Geophysics (1963). "Radioactive Tracers in Oceanography", iii + 41 pp. International Union of Geodesy and Geophysics, Paris.

Ivanov, V. N. and Lyubimov, A. A. (1970). *Oceanology,* **10**, 419.

Ivanov, V. N., Leshchenko, L. N. and Shparber, N. Ya. (1972). *Oceanology,* **12**, 46.

Jefferies, D. F. (1968). *Helgoländer wiss. Meeresunters.* **17**, 280.

Jenkins, C. E. (1969). *Hlth Phys.* **17**, 507.

Johnson, V., Cutshall, N. and Osterberg, C. (1967). *Wat. Resour. Res.* **3**, 99.

Jones, R. F. (1960). *Limnol. Oceanogr.* **5**, 312.

Joseph, A. B., Gustafson, P. F., Russell, I. R., Schuert, E. A., Volchok, H. L. and Tamplin, A. (1971). *In* "Radioactivity in the Marine Environment", pp. 6–41. National Academy of Sciences, Washington, D.C.

Kameda, K. (1963). *J. oceanogr. Soc. Japan,* **18**, 230.

Kaufman, A. (1969a). *Trans. Am. geophys. Un.* **50**, 349.

Kaufman, A. (1969b). *Geochim. cosmochim. Acta,* **33**, 717.

Kaufman, A., Broecker, W. S., Ku, T.-L. and Thurber, D. L. (1971). *Geochim. cosmochim, Acta,* **35**, 1155.

Kaufman, A., Trier, R. M., Broecker, W. S. and Feely, H. W. (1973). *J. geophys. Res.* **78**, 8827.

Kazachevskii, I. V., Cherdyntsev, V. V., Kua'mina, E. A., Sulerzhitskii, L. D., Mochalova, V. F. and Kyuregyan, T. N. (1964). *Geochemistry int.* **1**, 1068.

Kečkeš, S., Pučar, Z. and Marazović, L. (1967). *Int. J. Oceanol. Limnol.* **1**, 246.

Keeling, C. D. (1973). *In* "Chemistry of the Lower Atmosphere" (S. I. Rasool, ed.), pp. 251–329. Plenum, New York.

Keeling, C. D. and Bolin, B. (1967). *Tellus,* **19**, 566.

Keeling, C. D. and Bolin, B. (1968). *Tellus,* **20**, 17.

Kharkar, D. P., Lal, D. and Somayajulu, B. L. K. (1963). *In* "Radioactive Dating", pp. 175–186. International Atomic Energy Agency, Vienna.

Kharkar, D. P., Nijampurkar, V. N. and Lal, D. (1966). *Geochim cosmochim. Acta,* **30**, 621.

Kharkar, D. P., Turekian, K. K. and Scott, M. R. (1969). *Earth planet. Sci. Lett.* **6**, 61.

Kigoshi, K. (1971). *Science, N.Y.* **173**, 47.

Kigoshi, K. and Hasegawa, H. (1966). *J. geophys. Res.* **71**, 1065.

Kochenov, A. V. and Baturin, G. N. (1967). *Oceanology*, **7**, 484.

Kochenov, A. V., Baturin, G. N., Kovaleva, S. A., Yemel'yanov, Ye. Me and Shimkus, K. M. (1965). *Geochemistry int.* **2**, 212.

Koczy, F. F. (1958). *Proc. second United Nations int. Conf. peaceful uses atomic Energy*, **18**, 351.

Koczy, F. F. (1961). *Science, N.Y.* **134**, 1978.

Koczy, F. F. (1965). *Prog. Oceanogr.* **3**, 155.

Koczy, F. F. and Rosholt, J. N. (1962). *In* "Nuclear Radiation in Geophysics" (H. Israël and A. Krebs, eds), pp. 18–46. Springer-Verlag, Berlin.

Koczy, F. F. and Szabo, B. J. (1962). *J. oceanogr. Soc. Japan*, 20th *Anniversary Vol.* 590.

Koczy, F. F. and Titze, H. (1958). *J. mar. Res.* **17**, 302.

Koczy, F. F., Tomic, E. and Hecht, F. (1957). *Geochim. cosmochim. Acta*, **11**, 86.

Koide, M. and Goldberg, E. D. (1965). *Prog. Oceanogr.* **3**, 173.

Koide, M., Soutar, A. and Goldberg, E. D. (1972). *Earth planet. Sci. Lett.* **14**, 442.

Koide, M., Bruland, K. W. and Goldberg, E. D. (1973). *Geochim. cosmochim. Acta*, **37**, 1171.

Kolodny, Y. and Kaplan, I. R. (1970). *Geochim. cosmochim. Acta*, **34**, 3.

Kolodny, Y. and Kaplan, I. R. (1973). *In* "Proceedings of Symposium on Hydrogeochemistry and Biogeochemistry" (E. Ingerson, ed.), Vol. 1, pp. 418–442. Clarke, Washington, D.C.

Kolyadin, L. B., Nikolaev, D. S., Grashchenko, S. M., Kuznetsov, Yu. V. and Lazarev, K. F. (1960). *Dokl. Akad. Nauk. SSSR*, **132**, 915.

Konishi, K., Schlanger, S. O. and Omura, A. (1970). *Mar. Geol.* **9**, 225.

Krishnamoorthy, T. M., Sastry, V. N. and Sarma, T. P. (1971). *Curr. Sci.* **40**, 279.

Krishnaswamy, S. and Lal, D. (1972). *In* "The Changing Chemistry of the Oceans" (D. Dyrssen and D. Jagner, eds), pp. 307–320. Wiley Interscience, New York; Almqvist and Wiksell, Stockholm.

Krishnaswamy, S., Lal, D. and Somayajulu, B. L. K. (1970). *Proc. Indian Acad. Sci., Section A*, **71**, 238.

Krishnaswamy, S., Lal, D., Somayajulu, B. L. K., Dixon, F. S., Stonecipher, S. A. and Craig, H. (1972). *Earth planet. Sci. Lett.* **16**, 84.

Krylov, A. Y., Lisin, A. P. and Silin, V. I. (1961). *Izv. Acad. Sci. USSR geol. Ser.*, No. 3, 66.

Ku, T.-L. (1965). *J. geophys. Res.* **70**, 3457.

Ku, T.-L. (1966). "Uranium Series Disequilibrium in Deep-Sea Sediments", Thesis, Columbia University, New York.

Ku, T.-L. (1968). *J. geophys. Res.* **73**, 2271.

Ku, T.-L. (1969). *In* "Hot Brines and Recent Heavy Metal Deposits in the Red Sea" (E. T. Degens and D. A. Ross, eds), pp. 512–524. Springer-Verlag, Berlin.

Ku, T.-L. and Broecker, W. S. (1966). *Science, N.Y.* **151**, 448.

Ku, T.-L. and Broecker, W. S. (1967a). *Earth planet. Sci. Lett.* **2**, 317.

Ku, T.-L. and Broecker, W. S. (1967b). *Prog. Oceanogr.* **4**, 95.

Ku, T.-L. and Broecker, W. S. (1969). *Deep Sea Res.* **16**, 625.

Ku, T.-L. and Glasby, G. P. (1972). *Geochim. cosmochim. Acta*, **36**, 699.

Ku, T.-L., Broecker, W. S. and Opdyke, N. (1968). *Earth planet. Sci. Lett.* **4**, 1.

Ku, T.-L., Thurber, D. L. and Mathieu, G. G. (1969). *In* "Hot Brines and Recent Heavy Metal Deposits in the Red Sea" (E. T. Degens and D. A. Ross, eds), pp. 348–359. Springer-Verlag, Berlin.

Ku, T. L., Li, Y. H. Mathieu, G. G. and Wong, H. K. (1970). *J. geophys. Res.* **75**, 5286.

Ku, T.-L., Bischoff, J. L. and Boersma, A. (1972). *Deep Sea Res.* **19**, 233.

Kuznetsov, Yu. V., Legin, V. K., Lisitsyn, A. P. and Simonyak, Z. N. (1964). *Soviet Radiochemistry*, **6**, 233.

Kuznetsov, Yu. V., Simonyak, Z. N., Elizarova, A. N. and Lisitsyn, A. P. (1966). *Soviet Radiochemistry*, **8**, 421.

Labeyrie, J., Lalou, C. and Delibrias, G. (1967). *In* "Radioactive Dating and Methods of Low-Level Counting", pp. 349–356. International Atomic Energy Agency, Vienna.

Lahoud, J. A., Miller, D. S. and Friedman, G. M. (1966). *J. sedim. Petrol.* **36**, 541.

Lal, D. (1962). *J. oceanogr. Soc. Japan, 20th Anniversary Vol.* 600.

Lal, D. (1963a). *In* "Earth Science and Meteoritics" (J. Geiss and E. D. Goldberg, eds), pp. 115–142, North-Holland, Amsterdam.

Lal, D. (1963b). *In* "Radioactive Dating", pp. 149–156. International Atomic Energy Agency, Vienna.

Lal, D. (1969). *In* "Morning Review Lectures of the Second International Oceanographic Congress", pp. 29–48. United Nations Educational, Scientific and Cultural Organization, Paris.

Lal, D. and Peters, B. (1962). *Prog. elem. Part. cosm. Ray Phys.* **6**, 1.

Lal, D. and Peters, B. (1967). *Handb. Phys.* **46/2**, 551.

Lal, D. and Suess, H. E. (1968). *A. Rev. nucl. Sci.* **18**, 407.

Lal, D., Goldberg, E. D. and Koide, M. (1959). *Phys. Rev. Lett.* **3**, 380.

Lal, D., Goldberg, E. D. and Koide, M. (1960). *Science, N.Y.* **131**, 332.

Lal, D., Arnold, J. R. and Somayajulu, B. L. K. (1964). *Geochim. cosmochim. Acta,* **28**, 1111.

Lal, D., Nijampurkar, V. N. and Somayajulu, B. L. K. (1970). *Galathea Rep.* **11**, 247.

Lazarev, K. F., Nikolaev, D. S. and Grashchenko, S. M. (1961). *Radiokhimiya*, **3**, 623.

Lazarev, K. F., Grashchenko, S. M., Nikolayev, D. S. and Drozhzhin, V. M. (1965). *Dokl. Acad. Sci. U.S.S.R.*, **164**, 189.

Lederer, C. M., Hollander, J. M. and Perlman, I. (1967). "Table of Isotopes", xii + 594 pp. Wiley, New York.

Leung, G., Kim, Y. S. and Zeitlin, H. (1972). *Analytica chim. Acta*, **60**, 229.

Libby, W. F. (1946). *Phys. Rev.* **69**, 671.

Libby, W. F. (1962). *In* "Tritium in the Physical and Biological Sciences", Vol. I, pp. 5–28. International Atomic Energy Agency, Vienna.

Lingenfelter, R. E. (1963). *Rev. Geophys.* **1**, 35.

Lingenfelter, R. E. and Ramaty, R. (1970). *In* "Radiocarbon Variations and Absolute Chronology" (I. U. Olsson, ed.), pp. 513–537. Wiley Interscience, New York; Almqvist and Wiksell, Stockholm.

Livingstone, D. A. (1963). *Prof. Pap. U.S. Geol. Surv.*, **440-G**, vii + 64 pp.

Lyubimov, A. A., Zesenko, A. Ya. and Leshchenko, L. N. (1970). *Oceanology*, **10**, 808.

Machta, L. (1972). *In* "The Changing Chemistry of the Oceans" (D. Dyrssen and D. Jagner, eds), pp. 121–145. Wiley Interscience, New York: Almqvist and Wiksell, Stockholm.

Manheim, F. (1961). *Geochim. cosmochim. Acta*, **25**, 52.

Martell, E. A. (1963). *J. geophys. Res.* **68**, 3759.

Mauchline, J. and Templeton, W. L. (1963). *Nature, Lond.* **198**, 623.
Mauchline, J. and Templeton, W. L. (1964). *Oceanogr. mar. Biol. ann. Rev.* **2**, 229.
McCorkell, R., Fireman, E. L. and Langway, C. C., Jr. (1967). *Science, N.Y.* **158**, 1690.
Merrill, J. R., Lyden, E. F. X., Honda, M. and Arnold, J. R. (1960). *Geochim. cosmochim. Acta,* **18**, 108.
Mesolella, K. J., Matthews, R. K., Broecker, W. S. and Thurber, D. L. (1969). *J. Geol.* **77**, 250.
Michel, R. and Williams, P. M. (1973). *Earth planet. Sci. Lett.* **20**, 381.
Miyake, Y. and Sugimura, Y. (1961). *Science, N.Y.* **133**, 1823.
Miyake, Y. and Sugimura, Y. (1964). *In* "Studies on Oceanography", pp. 274–278. University of Tokyo Press, Tokyo.
Miyake, Y. and Sugimura, Y. (1965). *J. Geogr., Tokyo,* **74**, 95.
Miyake, Y. and Sugimura, Y. (1968). *Pap. Met. Geophys., Tokyo,* **19**, 481.
Miyake, Y., Saruhashi, K., Katsuragi, Y. and Kanazawa, T. (1962). *J. Radiat. Res.* **3**, 141.
Miyake, Y., Saruhashi, K., Katsuragi, Y., Kanazawa, T. and Sugimura, Y. (1964a). *In* "Recent Researches in the Fields of Hydrosphere, Atmosphere and Nuclear Geochemistry" (Y. Miyake and T. Koyoma, eds), pp. 127–141. Maruzen, Tokyo.
Miyake, Y., Sugimura, Y. and Tsubota, H. (1964b). *In* "The Natural Radiation Environment" (J. A. S. Adams and W. M. Lowder, eds), pp. 219–225. University of Chicago Press, Chicago.
Miyake, Y., Sugimura, Y. and Uchida, T. (1966). *J. geophys. Res.* **71**, 3083.
Miyake, Y., Sugimura, Y. and Matsumoto, E. (1968). *Rec. oceanogr. Wks Japan.* **9**, 189.
Miyake, Y., Sugimura, Y. and Mayeda, M. (1970a). *J. oceanogr. Soc. Japan,* **26**, 123.
Miyake, Y., Sugimura, Y. and Yasujima, T. (1970b). *J. oceanogr. Soc. Japan,* **26**, 130.
Miyake, Y., Katsuragi, Y. and Sugimura, Y. (1970c). *J. geophys. Res.* **75**, 2329.
Miyake, Y., Sugimura, Y. and Uchida, T. (1972). *Rec. oceanogr. Wks. Japan,* **11**, 53.
Mo, T., O'Brien, B. C. and Suttle, A. D., Jr. (1971). *Earth planet. Sci. Lett.* **10**, 175.
Mo, T., Suttle, A. D. and Sackett, W. M. (1973). *Geochim. cosmochim. Acta,* **37**, 35.
Moore, W. S. (1967). *Earth planet. Sci. Lett.* **2**, 231.
Moore, W. S. (1969a). *J. geophys. Res.* **74**, 694.
Moore, W. S. (1969b). *Earth planet. Sci. Lett.* **6**, 437.
Moore, W. S. (1972). *Earth. planet. Sci. Lett.* **16**, 421.
Moore, W. S. and Krishnaswami, S. (1972). *Earth planet. Sci. Lett.* **15**, 187.
Moore, W. S. and Sackett, W. M. (1964). *J. geophys. Res.* **69**, 5401.
Moore, W. S., Krishnaswami, S. and Bhat, S. G. (1973). *Bull. mar. Sci.* **23**, 157.
Münnich, K. O. and Roether, W. (1967). *In* "Radioactive Dating and Methods of Low-Level Counting", pp. 93–104. International Atomic Energy Agency, Vienna.
Murthy, V. R. and Beiser, E. (1968). *Geochim. cosmochim. Acta,* **32**, 1121.
National Academy of Sciences—National Research Council (1957). "The Effects of Atomic Radiation on Oceanography and Fisheries", Publication No. 551, 137 pp. National Academy of Sciences—National Research Council, Washington, D.C.
National Academy of Sciences—National Research Council (1959a). "Radioactive Waste Disposal into Atlantic and Gulf Coastal Waters", Publication No. 655, 37 pp. National Academy of Sciences—National Research Council, Washington, D.C.
National Academy of Sciences—National Research Council (1959b). "Considerations on the Disposal of Radioactive Wastes from Nuclear-powered Ships into the

Marine Environment", Publication No. 658, 52 pp. National Academy of Sciences—National Research Council, Washington, D.C.

National Academy of Sciences—National Research Council (1962). "Disposal of Radioactive Waste into Pacific Coastal Waters", Publication No. 985, 87 pp. National Academy of Sciences—National Research Council, Washington, D.C.

National Academy of Sciences (1971). "Radioactivity in the Marine Environment", ix + 272 pp. National Academy of Sciences, Washington, D.C.

Nelson, D. J. and Evans, F. C. (eds) (1969). "Symposium on Radioecology", Rep. CONF-670503, xii + 774 pp. United States Atomic Energy Commission, Washington, D.C.

Nikolayev, P. S. and Yefimova, E. I. (1963). Geochemistry, 703.

Nikolayev, D. S., Korn, O. P., Lazarev, K. F., Kolyadin, L. B., Kuznetsov, Yu. V. and Grashchenko, S. M. (1960). Dokl. Akad. Nauk SSSR, 132, 1411.

Nikolayev, D. S., Lazarev, K. F. and Grashchenko, S. M. (1961). Dokl. Acad. Sci. U.S.S.R. 138, 489.

Nikolaev, D. S., Lazarev, K. F., Korn, O. P., Yakunin, M. L., Drozhzhin, V. M. and Samartseva, A. G. (1965). Dokl. Akad. Nauk SSSR, 165, 175.

Nir, A., Kruger, S. T., Lingenfelter, R. E. and Flamm, E. J. (1966). Rev. Geophys. 4, 441.

Noakes, J. E., Kim, S. M. and Supernaw, I. R. (1967a). Trans. Am. geophys. Un. 48, 237.

Noakes, J. E., Supernaw, I. R. and Akers, L. K. (1967b). J. geophys. Res. 72, 2679.

Noble, C. S. and Naughton, J. J. (1968). Science, N.Y. 162, 265.

Noshkin, V. E. (1969). Tellus, 21, 414.

Nozaki, Y. and Tsunogai, S. (1973). Earth planet. Sci. Lett. 20, 88.

Nydal, R. (1968). J. geophys. Res. 73, 3617.

Nydal, R. and Lövseth, K. (1970). J. geophys. Res. 75, 2271.

Olsson, I. U. (ed.) (1970). "Radiocarbon Variations and Absolute Chronology", 652 pp. Wiley Interscience, New York; Almqvist and Wiksell, Stockholm.

Olsson, I. U. and Eriksson, K. G. (1965). Prog. Oceanogr. 3, 253.

Osmond, J. K. and Pollard, L. D. (1967). Earth planet. Sci. Lett. 3, 476.

Osmond, J. K., Carpenter, J. R. and Windom, H. L. (1965). J. geophys. Res. 70, 1843.

Osmond, J. K., May, J. P. and Tanna, W. F. (1970). J. geophys. Res. 75, 469.

Osterberg, C. (1962). Limnol. Oceanogr. 7, 478.

Osterberg, C., Kulm, L. D. and Byrne, J. V. (1963a). Science, N.Y. 139, 916.

Osterberg, C., Carey, A. G. and Curl, H., Jr. (1963b). Nature, Lond. 200, 1276.

Osterberg, C. L., Pattullo, J. and Pearcy, W. (1964). Limnol. Oceanogr. 9, 249.

Osterberg, C., Cutshall, N. and Cronin, J. (1965). Science, N.Y. 150, 1585.

Östlund, H. G., Rinkel, M. and Rooth, C. G. (1969). J. geophys. Res. 74, 4535.

Ozima, M., Ozima, M. and Kaneoka, I. (1968). J. geophys. Res. 73, 711.

Ozima, M., Kaneoka, I. and Aramaki, S. (1970). Earth planet. Sci. Lett. 8, 237.

Palmer, H. E., Beasley, T. M. and Folsom, T. R. (1966). Nature, Lond. 211, 1253.

Palumbo, R. F., Seymour, A. H. and Welander, A. D. (1966). Nature, Lond. 209, 1190.

Park, K., George, M. J., Miyake, Y., Saruhashi, K., Katsauragi, Y. and Kanazawa, T. (1965). Nature, Lond. 208, 1084.

Pearcy, W. G. and Osterberg, C. L. (1967). Int. J. Oceanol. Limnol. 1, 103.

Pearcy, W. G. and Osterberg, C. L. (1968). Limnol. Oceanogr. 13, 490.

Peterman, Z. E., Hedge, C. E. and Tourtelot, H. A. (1970). Geochim. cosmochim. Acta, 34, 105.

Peters, B. (1955). Proc. Indian Acad. Sci., Section A, 41, 67.

Pettersson, H. (1955). Pap. mar. Biol. Oceanogr., Deep Sea Res. Suppl. 3, 335.

Picciotto, E. and Wilgain, S. (1954). *Nature, Lond.* **173**, 632.

Pillai, K. C., Smith, R. C. and Folsom, T. R. (1964). *Nature, Lond.* **203**, 568.

Polikarpov, G. G. (1966). "Radioecology of Aquatic Organisms", xxviii + 314 pp. North-Holland, Amsterdam; Reinhold, New York.

Popov, N. I. and Grekov, A. S. (1971). *Oceanology,* **11**, 433.

Potratz, H. A. and Sackett, W. M. (1959). *In* "International Oceanographic Congress. Preprints" (M. Sears, ed.), pp. 502–503. American Association for the Advancement of Science, Washington, D.C.

Preston, A. (1968). *Helgoländer wiss. Meeresunters.* **17**, 269.

Preston, A. (1970). *Mar. Biol.* **6**, 345.

Preston, A. and Jefferies, D. F. (1967). *Hlth Phys.* **13**, 477.

Preston, A., Dutton, J. W. R. and Harvey, B. R. (1968). *Nature, Lond.* **218**, 689.

Pritchard, D. W., Reid, R. O., Okubo, A. and Carter, H. H. (1971). *In* "Radioactivity in the Marine Environment", pp. 90–136. National Academy of Sciences, Washington, D.C.

Pruter, A. T. and Alverson, D. L. (eds) (1972). "The Columbia River Estuary and Adjacent Ocean Waters", xiii + 868 pp. University of Washington Press, Seattle.

Pushkar, P. and Peterson, M. N. A. (1967). *Earth planet. Sci. Lett.* **2**, 349.

Rafter, T. A. (1968). *N.Z. Jl Sci.* **11**, 551.

Rafter, T. A. and O'Brien, B. J. (1970). *In* "Radiocarbon Variations and Absolute Chronology" (I. U. Olsson, ed.), pp. 355–377. Wiley Interscience, New York; Almqvist and Wiksell, Stockholm.

Rama, Koide, M. and Goldberg, E. D. (1961a). *Science, N.Y.* **134**, 98.

Rama, Koide, M. and Goldberg, E. D. (1961b). *Nature, Lond.* **191**, 162.

Rice, T. R. and Wolfe, D. (1971). *In* "Impingement of Man on the Oceans" (D. W. Hood, ed.), pp. 325–379. Wiley-Interscience, New York.

Rocco, G. G. and Broecker, W. S. (1963). *J. geophys. Res.* **68**, 4501.

Roether, W. and Münnich, K. O. (1972). *Earth planet. Sci. Lett.* **16**, 127.

Roether, W., Münnich, K. O. and Östlund, H. G. (1970). *J. geophys. Res.* **75**, 7672.

Rona, E. and Emiliani, C. (1969). *Science, N.Y.* **163**, 66.

Rona, E. and Urry, W. D. (1952). *Am. J. Sci.* **250**, 241.

Rona, E., Gilpatrick, L. O. and Jeffrey, L. M. (1956). *Trans. Am. geophys. Un.* **37**, 697.

Rona, E., Akers, L. K., Noakes, J. E. and Supernaw, I. (1965). *Prog. Oceanogr.* **3**, 289.

Rooth, C. G. and Östlund, H. G. (1972). *Deep-Sea Res.* **19**, 481.

Rosholt, J. N., Emiliani, C., Geiss, J., Koczy, F. F. and Wangersky, P. J. (1961). *J. Geol.* **69**, 162.

Rosholt, J. N., Emiliani, C., Geiss, J., Koczy, F. F. and Wangersky, P. J. (1962). *J. geophys. Res.* **67**, 2907.

Rydell, H. S. and Bonatti, E. (1973). *Geochim. cosmochim. Acta,* **37**, 2557.

Rydell, H. and Fisher, D. E. (1971). *Bull. mar. Sci.* **21**, 787.

Sackett, W. M. (1958). "Ionium–Uranium Ratios in Marine Deposited Calcium Carbonates and Related Materials", Thesis, Washington University.

Sackett, W. M. (1960). *Science, N.Y.* **132**, 1761.

Sackett, W. M. (1964). *Ann. N.Y. Acad. Sci.* **119**, 339.

Sackett, W. M. (1965). *In* "Marine Geochemistry" (D. R. Schink and J. T. Corless, eds), pp. 29–40. University of Rhode Island, Kingston.

Sackett, W. M. (1966). *Science, N.Y.* **154**, 646.

Sackett, W. M. and Cook, G. (1969). *Trans. Gulf-Cst Ass. geol. Socs,* **19**, 233.

Sackett, W. M. and Potratz, H. A. (1963). *Prof. Pap. U.S. Geol. Surv.* **260-BB**, 1053

Sakanoue, M., Konishi, K. and Komura, K. (1967). *In* "Radioactive Dating and Methods of Low-Level Counting", pp. 313–329. International Atomic Energy Agency, Vienna.

Sarma, T. P. and Krishnamoorthy, T. M. (1968). *Curr. Sci.* **37**, 422.

Schaeffer, O. A. and Zähringer, J. (eds) (1966). "Potassium Argon Dating," xi + 234 pp. Springer-Verlag, Berlin.

Schell, W. R., Fairhall, A. W., and Harp, G. D. (1967). *In* "Radioactive Dating and Methods of Low-Level Counting", pp. 79–91. International Atomic Energy Agency, Vienna.

Schroeder, J. H., Miller, D. S. and Friedman, G. M. (1970). *J. sedim. Petrol.* **40**, 672.

Schultz, V. and Klement, A. W., Jr. (eds) (1963). "Radioecology", xvii + 746 pp. Reinhold, New York; American Institute of Biological Sciences, Washington, D.C.

Scott, M. R. (1968). *Earth planet. Sci. Lett.* **4**, 245.

Scott, M. R., Osmond, J. K. and Cochran, J. K. (1972). *In* "Antarctic Oceanology II The Australian-New Zealand Sector" (D. E. Hayes, ed.), pp. 317–334. American Geophysical Union, Washington, D.C.

Shannon, L. V. (1972). *Investl Rep. Div. Sea Fish. S. Afr.* No. 99, 20 pp.

Shannon, L. V. (1973). *Investl Rep. Div. Sea Fish. S. Afr.* No. 100, 34 pp.

Shannon, L. V. and Cherry, R. D. (1967). *Nature, Lond.* **216**, 352.

Shannon, L. V., Cherry, R. D. and Orren, M. J. (1970). *Geochim. cosmochim. Acta*, **34**, 701.

Silker, W. B. (1972a). *Earth planet Sci. Lett.* **16**, 131.

Silker, W. B. (1972b). *J. geophys. Res.* **77**, 1061.

Silker, W. B., Robertson, D. E., Rieck, H. G., Jr., Perkins, R. W. and Prospero, J. M. (1968). *Science, N.Y.* **161**, 879.

Slowey, J. F., Hayes, D., Dixon, B. and Hood, D. W. (1965). *In* "Marine Geochemistry" (D. R. Schink and J. T. Corless, eds), pp. 109–129. University of Rhode Island, Kingston.

Somayajulu, B. L. K. (1967). *Science, N.Y.* **156**, 1219.

Somayajulu, B. L. K. (1969). *Proc. Indian Acad. Sci., Section A*, **64**, 338.

Somayajulu, B. L. K. and Goldberg, E. D. (1966). *Earth planet. Sci. Lett.* **1**, 102.

Somayajulu, B. L. K., Lal, D. and Kusumgar, S. (1969). *Science, N.Y.* **166**, 1397.

Somayajulu, B. L. K., Heath, G. R., Moore, T. C., Jr. and Cronan, D. S. (1971). *Geochim. cosmochim. Acta*, **35**, 621.

Somayajulu, B. L. K., Lal, D. and Craig, H. (1973). *Earth planet. Sci. Lett.* **18**, 181.

Spalding, R. F. and Sackett, W. M. (1972). *Science, N.Y.* **175**, 629.

Spencer, D. W., Robertson, D. E., Turekian, K. K. and Folsom, T. R. (1970). *J. geophys. Res.* **75**, 7688.

Starik, I. E. and Kolyadin, L. B. (1957). *Geochemistry*, 245.

Starik, I. Ye., Lazarev, K. F., Nikolayev, D. S., Grashchenko, S. M., Kolyadin, L. B. and Kuznetsov, Yu. V. (1959). *Dokl. (Proc.) Acad. Sci. U.S.S.R.* **129**, 1041.

Stearns, C. E. and Thurber, D. L. (1967). *Prog. Oceanogr.* **4**, 293.

Strohal, P. and Pinter, T. (1973). *Limnol. Oceanogr.* **18**, 250.

Strøm, K. M. (1948). *Nature, Lond.* **162**, 922.

Suess, H. E. (1958). *A. Rev. nucl. Sci.* **8**, 243.

Suess, H. E. (1965). *J. geophys. Res.* **70**, 5937.

Suess, H. E. (1969). *Science, N.Y.* **163**, 1405.

Suess, H. E. (1970). *J. geophys. Res.* **75**, 2363.

Suess, H. E. and Goldberg, E. D. (1971). *J. geophys. Res.* **76**, 5131.

Sugihara, T. T. and Bowen, V. T. (1962). *In* "Radioisotopes in the Physical Sciences and Industry", pp. 57–65. International Atomic Energy Agency, Vienna.

Sugimura, Y. and Miyake, Y. (1968). *Bull. natn. Sci. Mus., Tokyo,* **11**, 327.

Swallow, A. J. (1969). *Nature, Lond.* **222**, 369.

Swanson, V. E. (1961). *Prof. Pap. U.S. Geol. Surv.* **356-C**, 112 pp.

Szabo, B. J. (1967). *Geochim. cosmochim. Acta,* **31**, 1321.

Szabo, B. J. (1971). *Bull. mar. Sci.* **21**, 748.

Szabo, B. J. and Rosholt, J. N. (1969). *J. geophys. Res.* **74**, 3253.

Szabo, B. J., Koczy, F. F. and Östlund, G. (1967). *Earth planet. Sci. Lett.* **3**, 51.

Tanaka, S., Sakamoto, K., Takagi, J. and Tsuchimoto, M. (1968). *Science, N.Y.* **160**, 1348.

Tanaka, S., Sakamoto, K. and Komura, K. (1972). *J. geophys. Res.* **77**, 4281.

Tatsumoto, M. T. and Goldberg, E. D. (1959). *Geochim. cosmochim. Acta,* **17**, 201.

Teegarden, B. J. (1967). *J. geophys. Res.* **72**, 4863.

Thompson, G. and Livingston, H. D. (1970). *Earth planet. Sci. Lett.* **8**, 439.

Thurber, D. L. (1962). *J. geophys. Res.* **67**, 4518

Thurber, D. L. (1963). *In* "Radioactive Dating", pp. 113–119. International Atomic Energy Agency, Vienna.

Thurber, D. L. (1964). *Trans. Am. geophys. Un.* **45**, 119.

Thurber, D. L., Broecker, W. S., Blanchard, R. L. and Potratz, H. A. (1965). *Science, N.Y.* **149**, 55.

Torii, T. and Murata, S. (1964). *In* "Recent Researches in the Fields of Hydrosphere, Atmosphere and Nuclear Geochemistry" (Y. Miyake and T. Koyama, eds), pp. 321–334. Maruzen, Tokyo.

Trier, R. M., Broecker, W. S. and Feely, H. W. (1972). *Earth planet. Sci. Lett.* **16**, 141.

Tsunogai, S. and Nozaki, Y. (1971). *Geochemical J.* **5**, 165.

Turekian, K. K. and Bertine, K. K. (1971). *Nature, Lond.* **229**, 250.

Turekian, K. K. and Chan, L. H. (1971). *In* "Activation Analysis in Geochemistry and Cosmochemistry" (A. O. Brunfelt and E. Steinnes, eds), pp. 311–320. Universitetsforlaget, Oslo.

Turner, D. L., Forbes, R. B. and Naeser, C. W. (1973). *Science, N.Y.* **182**, 579.

Umemoto, S. (1965). *J. geophys. Res.* **70**, 5326.

United Nations Scientific Committee on the Effects of Atomic Radiation (1962). General Assembly, Official Records, Seventeenth Session, Supplement No. 16 (A/5216), 442 pp. United Nations, New York.

United Nations Scientific Committee on the Effects of Atomic Radiation (1964). General Assembly, Official Records, Nineteenth Session, Supplement No. 14 (A/5814), 117 pp. United Nations, New York.

Valentine, J. W. and Veeh, H. H. (1969). *Bull. geol. Soc. Am.* **80**, 1415.

Veeh, H. H. (1966). *J. geophys. Res.* **71**, 3379.

Veeh, H. H. (1967). *Earth planet. Sci. Lett.* **3**, 145.

Veeh, H. H. (1968). *Geochim. Cosmochim. Acta,* **32**, 117.

Veeh, H. H. and Boström, K. (1971). *Earth planet. Sci. Lett.* **10**, 372.

Veeh, H. H. and Chappell, J. (1970). *Science, N.Y.* **167**, 862.

Veeh, H. H. and Turekian, K. K. (1968). *Limnol. Oceanogr.* **13**, 304.

Veeh, H. H. and Valentine, J. W. (1967). *Bull. geol. Soc. Am.* **78**, 547.

Veeh, H. H., Burnett, W. C. and Soutar, A. (1973). *Science, N.Y.* **181**, 844.

Vinogradov, A. P. (1953). "The Elementary Chemical Composition of Marine Organisms", 647 pp. Sears Foundation for Marine Research, New Haven.

Volchok, H. L., Bowen, V. T., Folsom, T. R., Broecker, W. S., Schuert, E. A. and Bien, G. S. (1971). *In* "Radioactivity in the Marine Environment", pp. 42–89. National Academy of Sciences, Washington, D.C.

Walton, A., Ergin, M. and Harkness, D. D. (1970). *J. geophys. Res.* **75**, 3089.

Wasson, J. T. (1963). *Icarus*, **2**, 54.

Wasson, J. T., Alder, B. and Oeschger, H. (1967). *Science, N.Y.* **155**, 446.

Watson, D. G., Davis, J. J. and Hanson, W. C. (1963). *Limnol. Oceanogr.* **8**, 305.

Williams, P. M., Oeschger, H. and Kinney, P. (1969). *Nature, Lond.* **224**, 256.

Wilson, J. D., Webster, R. K., Milner, G. W. C., Barnett, G. A. and Smales, A. A. (1960). *Analyt. chim. Acta*, **23**, 505.

Wolgemuth, K. (1970). *J. geophys. Res.* **75**, 7686.

Wolgemuth, K. and Broecker, W. S. (1970). *Earth planet. Sci. Lett.* **8**, 372.

Wong, K. M. (1971). *Analyt. chim. Acta*, **56**, 355.

Wong, K. M., Hodge, V. F. and Folsom, T. R. (1972). *Nature, Lond.* **237**, 460.

Young, J. A. and Fairhall, A. W. (1968). *J. geophys. Res.* **73**, 1185.

Yokoyama, Y. (1967). *Nature, Lond.* **216**, 569.

Yokoyama, Y. (1968). *Nature, Lond.* **220**, 1016.

Yokoyama, Y., Tobailem, J., Grjebine, T. and Labeyrie, J. (1968). *Geochim. cosmochim. Acta*, **32**, 347.

Chapter 19

Analytical Chemistry of Sea Water

J. P. RILEY

Department of Oceanography, The University of Liverpool, England

With contributions by

D. E. ROBERTSON

Pacific Northwest Laboratories, Richland, Washington 99352, U.S.A.
(Section 19.10.4 on Neutron activation analysis)

J. W. R. DUTTON AND N. T. MITCHELL

Fisheries Radiobiological Laboratory, Lowestoft, Suffolk, England
(Section 19.13 on Determination of radionuclides other than those of
the uranium and thorium series)

P. J. LE B. WILLIAMS

Department of Oceanography, The University, Southampton, England
(Section 19.14 on Determination of organic compounds)

SECTION 19.1. INTRODUCTION 193

19.2. ANALYTICAL ERROR 197

19.3. WATER SAMPLING 199
    19.3.1. Criteria for water sampling bottles 202
    19.3.2. Water samplers 203
    19.3.3. Sampling by pumping 209

19.4. FILTRATION OF WATER SAMPLES.. 211

19.5. STORAGE OF WATER SAMPLES 218

19.6. DETERMINATION OF SALINITY AND CHLORINITY 223

19.7. DETERMINATION OF TOTAL CATIONS BY ION EXCHANGE 225

19.8. DETERMINATION OF MAJOR COMPONENTS 225
    19.8.1. Introduction 225
    19.8.2. Determination of major cations 226
    19.8.3. Determination of major anions 243

19.9.   DETERMINATION OF DISSOLVED GASES 253
         19.9.1.   Oxygen 253
         19.9.2.   Nitrogen 263
         19.9.3.   Argon 263
         19.9.4.   Other noble gases 264
         19.9.5.   Total carbon dioxide, P_{CO_2}, carbonate and bicarbonate 265
         19.9.6.   Hydrogen sulphide 265
         19.9.7.   Nitrous oxide 266
         19.9.8.   Hydrogen 267
         19.9.9.   Carbon monoxide 267
         19.9.10. Aliphatic hydrocarbons and chlorinated hydrocarbons 268

19.10.   DETERMINATION OF MINOR INORGANIC CONSTITUENTS 269
         19.10.1. Introduction 269
         19.10.2. Methods for concentration of trace elements from sea water .. 278
         19.10.3. Separation of elements after concentration 289
         19.10.4. Physicochemical techniques for the determination of trace
                   amounts of elements * 290
         19.10.5. Methods for the determination of individual trace elements 335

19.11.   DETERMINATION OF MICRONUTRIENT ELEMENTS 408
         19.11.1. Nitrogen 408
         19.11.2. Phosphorous 419
         19.11.3. Silicon 427

19.12.   DETERMINATION OF THE RADIOELEMENTS, URANIUM-238, URANIUM-235
         AND THORIUM AND THEIR DECAY PRODUCTS 431
         19.12.1. Introduction 431
         19.12.2. Determination of radionuclides 433

19.13.   DETERMINATION OF RADIONUCLIDES OTHER THAN THOSE OF THE URANIUM
         AND THORIUM SERIES† 437
         19.13.1. Introduction 437
         19.13.2. Sample storage 438
         19.13.3. Direct measurement of radioactivity 439
         19.13.4. Radiochemical separations 439
         19.13.5. Determination of individual radionuclides 443

19.14.   DETERMINATION OF ORGANIC COMPONENTS‡ 443
         19.14.1. Introduction 443
         19.14.2. Filtration 454
         19.14.3. Particulate organic matter 455
         19.14.4. Dissolved organic matter 460
         19.14.5. Conclusion 476

REFERENCES 477

19.1. INTRODUCTION

Analytical chemistry occupies a key position in marine chemistry, and it is true to say that there are few facets of the subject in which its aid is not in-

---

* Section 19.10.4.10. on neutron activation techniques by D. E. Robertson.
† By J. Dutton and N. Mitchell.
‡ By P. J. le B. Williams.

voked. Because of the interdisciplinary nature of oceanography the analyst may also be presented with a wide variety of problems by physical oceanographers, marine biologists, biochemists and geochemists. The samples to be examined may include not only sea water itself but also surface films, marine organisms, and sediments and their interstitial waters. The elements which he may be required to determine may be drawn from the whole periodic table, including the transuranic elements, and may cover concentrations ranging from grams per litre for major elements down to $10^{-18}\,g\,l^{-1}$ for ^{210}Pb and other radionuclides. In addition, determinations of physiologically important organic compounds are being increasingly required. Monitoring and control of marine and estuarine pollution is another aspect of the work of the marine chemist which is of growing importance, and one which can stretch his analytical ability to the full.

Until about 15 years ago practically the only analytical procedures which were used by the marine chemist were gravimetric, titrimetric and colorimetric ones. Since that time there have been great advances in the development of physico-chemical techniques, e.g. atomic absorption spectrophotometry, anodic stripping voltammetry and neutron activation analysis, many of them specific and of very high sensitivity. The application of these to the analysis of sea water has, in many instances, both simplified the problem of analysis, and enabled elements to be determined at concentrations which are orders of magnitude lower than those detectable by previous methods. In addition, there has been considerable success in the development of automatic methods of analysis. Important advances have been made in the development of sensors which can be used for making *in situ* measurements in the sea, but the range of useful sensors is still very limited (e.g. to salinity, pH, P_{O_2} a_{F^-}). The automation of colorimetric methods of determining micro-nutrient elements has done much to relieve the analyst of the tedium of examining the large number of samples collected during physical or biological oceanographic cruises. Concurrently with these advances in instrumental analysis there have been considerable developments in methods for the concentration and separation of elements, particularly those using solvent extraction and ion exchange techniques. It is probable that future progress in inorganic aspects of sea water analysis will centre around three principal topics. Firstly, the attainment of greater sensitivity by the examination of some potentially useful instrument techniques which have not so far been exploited (e.g. gas liquid chromatography and mass spectrometry). Secondly, the production of sensitive sensors which can be used in probes for making *in situ* determinations. Thirdly, the development of techniques for the study of the speciation of elements in sea water, not only those in true solution, but also those present in colloidal and particulate forms.

In comparison with the advances which have occurred in the inorganic field, development of techniques for the determination of dissolved organic compounds has been slow. This probably mainly reflects the difficulty of concentrating the compounds (particularly the hydrophilic ones) from the preponderating mass of inorganic ions, since the actual determination and identification of sub-microgram amounts of the compounds by combinations of chromatographic and mass spectrometric techniques is often a fairly simple matter. Much effort is being devoted to the solution of this problem, and it seems likely that there will be considerable progress in this field within the next ten years.

Because of the rapidity with which changes can occur in the concentration of many components within a short time of sampling it is often desirable that the analysis be carried out, or at least commenced, at sea. Unfortunately, because of vibration, rolling and pitching, it is not possible to use many of the more sophisticated types of analytical instruments on board ship. Indeed, even precision weighing is out of the question because of vertical accelerations, and if any standard solutions are required they must either be pre-prepared or made up from pre-weighed material. The techniques which can be used on shipboard include: titrimetry, colorimetry, conductimetric salinometry, electrometric pH measurement, gas and thin-layer chromatography, fluorimetry and flame emission and atomic absorption spectrophotometry, although the inflammable gases necessary for the latter two techniques may present an unacceptable fire hazard on smaller ships. Fortunately they enable a considerable range of the more important analyses to be carried out, including those required by physical oceanographers and marine biologists. If the analysis requires instruments which it is not possible to use on board ship, it may be possible to preconcentrate the desired component at sea and bring the concentrate back to the land based laboratory. This has several advantages as it not only saves time and space, but also obviates the need for storing and transporting bulky samples.

Analyses carried out afloat, even under favourable conditions, are inherently less precise than those made ashore, owing to the effect of the motion of the ship on the analyst and his instruments. In this respect, there are conspicuous advantages in the adoption of automatic or semi-automatic methods of analysis (e.g. for the determination of micro-nutrients) since their precisions are scarcely affected by ship motion, and indeed they can often be used in conditions so rough that the use of manual methods would be out of the question.

In common with most other branches of marine chemistry, the analytical chemistry of sea water has developed enormously since the previous edition of this book appeared ten years ago. Although some of the techniques which were described in this earlier volume have stood the test of time, many others are now only of historical interest as they have been superseded by simpler and more precise methods, often using instruments which were not available in 1965. It is the purpose of this chapter to review these newer procedures and to discuss the older techniques if they are still relevant (the subject has also been reviewed by Spencer and Brewer, 1970). The rapidly expanding field of the application of electroanalytical techniques to the analysis of sea water will be discussed in Chapter 20. For practical details of procedures the original papers should be consulted. At present, the only books devoted to practical aspects of sea water analysis are those primarily concerned with measurements required by marine biologists (Strickland and Parsons, 1968; Barnes, 1959; Genovese and Magazzu, 1969; Atlas *et al.*, 1971), or with determination of radionuclides (International Atomic Energy Agency, 1970). There is a great need for a compendium of reliable and well tested methods for major and trace inorganic constituents, both dissolved and particulate, as well as for the more important organic components including pollutants.

19.2. ANALYTICAL ERROR

All analytical measurements are subject to errors which are of three types:

(i) *Gross errors* which originate from mistakes in the analytical process or from carelessness of the analyst (e.g. from spillage of the sample while it is being analyzed, from mistakes in weighing or in reading of an instrument, or even from miscalculation of results). Such errors can almost always be eliminated by careful attention to detail.

(ii) *Systematic errors* are constant in character for any given method and lead to bias in the results. They thus affect the *accuracy* of the method (i.e. the closeness with which the analytical result approaches the actual content of the component in question). They may result, for example, from inaccuracy of a measuring instrument, non-stoichiometry of a reaction, interference from some other element, or systematic losses occurring during the analysis (e.g. loss of iodine by volatilization during the Winkler determination of oxygen). The causes and magnitudes of these errors are often difficult to determine. Accuracy can be achieved only by careful design of the analytical procedure accompanied by stringent tests. Accuracy can be tested in two ways (1) by analyzing a synthetic sample, similar in composition to the samples being analyzed, but containing a known amount of the constituent being determined, (2) by analysing a sample in which the constituent has already been determined by some other reliable (referee) method. In the development of methods for sea water analysis, the former of these approaches is usually the more convenient one.

(iii) *Random errors* which result from a combination of a variety of small irregular errors arising in the individual operations in the analysis. These may be associated with, for example variability of the measuring instruments, imperfect manipulation and irreproducibility of the chemical reaction used. The total effect of these errors in any determination governs its *precision*, and results are regarded as precise if those of a number of replicate analyses agree well with one another. The scatter of such results about their mean usually shows a Gaussian (normal) distribution, and under these circumstances, the error can be characterized by the standard deviation,* σ, which can be assessed from the results of a number of replicate determinations n using the expression

$$\sigma = \sqrt{\frac{\Sigma(x_i - \bar{x})^2}{n - 1}}.$$

where $\bar{x}$ is the arithmetical mean of n determinations and x_i is a determined value, if n is large, the probability, λ, of obtaining a value of x_i in the range $x_i + dx$ in a determination is given approximately by

$$\lambda = dx\, f(x_i) = \frac{1}{\sigma\sqrt{2\pi}} \exp\left[-\frac{1}{2}\left(\frac{x_i - \bar{x}}{\sigma}\right)^2 \right] dx$$

The width of the range between this individual determination and the mean will thus depend on σ and on the acceptable probability. For most oceanographic purposes a 95% confidence level, which corresponds to 1.96σ is acceptable. That is to say, for 95% of the determinations the determined value will lie within ca. $\pm 2\sigma$ of the mean value. Strickland and Parsons (1968) have accepted this confidence level in their monograph on sea water analysis, and quote the precision (P) of each method in a form which is such that there is a 95% probability that the mean of n determinations will lie within $\pm Pn^{-\frac{1}{2}}$. A discussion of the use of statistical methods in analysis is beyond the scope of the present chapter, and the reader is referred to the texts by Youden (1959) and Eckschlager (1969) for a detailed treatment of the subject.

The precision required in chemical oceanography varies considerably, and is greatest in the determination of salinity for which a coefficient of variation of $\pm 0.03\%$ or better is essential. For the major ions, which vary only slightly, if at all, relative to chlorinity, a coefficient of variation of $\pm 0.1\%$ is desirable, but not always attainable. The limitations of the analytical methods

* The standard deviation expressed as a percentage of the determined value (i.e. $100\,\sigma/\bar{x}$) is termed the coefficient of variation or percentage standard deviation.

make it difficult to determine most trace metals with a precision of better than $\pm 2\%$, and for those occurring at extreme dilutions, a coefficient of variation of $\pm 10\%$ may be the best which can be achieved.

Analyses of the same sea water sample made in two or more laboratories may unfortunately give quite different results, even when the same method is used. A number of collaborative intercalibration exercises have been carried out, both nationally and internationally, in order to investigate the magnitudes and causes of such variations. The components which have been studied include, salinity (Cox, 1968) dissolved oxygen (see p. 259), phosphate (see p. 425) nitrate, silicate (see p. 431) various radionuclides (Section 19.13.4) and a number of trace elements (see Vol. 1 p. 437). Many of these studies have revealed a disquieting lack of consistency between the results obtained in different laboratories*. The causes of these inconsistencies are uncertain, and obviously further investigation of the problem is necessary. The use of Standard Sea Water from Copenhagen has greatly improved the consistency of the results of chlorinity determinations. Cooper (1958) and Palmork (1969) have suggested that, if possible, sealed ampoules containing different standards for other determinations should also be made available. The widespread use of these should then lead to a corresponding improvement in the general reliability of analyses. However, it is much more difficult to prepared stable standards for these determinations than for chlorinity estimation. An initial approach to this problem has been made by Sugawara (1969) who has succeeded in preparing stable standards for micro-nutrient determinations.

It should be borne in mind that precision as considered above is that of the analytical procedure only, and it takes no account of errors of sampling. Many of the minor components (e.g. micronutrients, chlorophyll) are very patchily distributed in the sea both horizontally and vertically. It is therefore important to plan the sampling programme with great care so that a truly representative picture of the distribution of a particular parameter shall be obtained.

19.3. WATER SAMPLING

The work of the marine chemist embraces not only the examination of water drawn from the water column itself, but also on occasions, analysis of the interstitial waters of the sediments and of the surface film. Ideally, it would be desirable if these analyses could be carried out *in situ* by means of a probe

---

* For a general discussion of the rationale of collaborative tests see Youden (1967) and Palmork (1969).

which could be lowered from the surface to give a profile of the water column. This can in fact be done with satisfactory precision for a few parameters e.g. salinity (see Chapter 6), pH (Section 9.2.3.1), P_{O_2} (p. 640), and perhaps P_{CO_2}, $a_{S_2^-}$ and a_{F^-}. However, there seems to be little likelihood that it will be possible to greatly extend this list in the near future because of the lack of adequate sensors. Thus, although sensors for the major cations are available, inter-element interferences and their Nernstian (logarithmic) responses, makes their precision far too poor for them to be of any value for the determination of these elements. In addition, the concentrations of trace elements lie well below the levels of detection with any electrodes available at present. Because of this, the chemical oceanographer is still almost entirely dependent on water samples collected at the depth of interest. For the collection of these samples a very wide range of samplers is available. In choosing one for a particular purpose it should be borne in mind that its essential function is to capture an appropriate volume of water from the desired environment, and to convey it unchanged on board ship. Since collection of water from the water column is the most important type of sampling it will be discussed in some detail below (see also Herdman, 1963; Grasshoff, 1969); sampling of the surface layer and interstitial waters will be discussed briefly subsequently.

Sampling of the water column may be carried out in three ways: (i) by use of a water sampling bottle which is lowered open to the appropriate depth and then closed by a signal from the surface; (ii) by lowering a hose to the depth required and pumping the water to the surface; (iii) by adsorbing the required element, or compound, onto a mass of an appropriate adsorbent which is towed through the water column at the desired depth (this technique is of particular value for processing the large volumes of water needed for the determination of radionuclides, such as ^{210}Pb and ^{32}Si; see e.g. Krishnaswami et al., 1972).* Of these, the first approach is undoubtedly the more important, although pumping has advantages if large samples are to be collected and is essential if continuous profiling is to be carried out.

In sampling, the marine chemist is confronted with two important problems; at what depth(s) to collect the sample and how to ensure that it is representative of the water mass being sampled and free from contamination. For many purposes sampling is carried out at the "standard depths" laid down in 1936 by the International Association on Physical Oceanography. (Table 19.1). However, such a sampling schedule will obviously need modification in the light of any special requirements (e.g. if it is desired to investigate

---

* Pumping of water through an in situ tube containing an adsorbent such as Amberlite XAD-2 resin is a potentially valuable means of isolating hydrophobic organic compounds without risk of the losses that occur through adsorption onto the walls of samplers or storage containers (see e.g. Dawson and Riley, 1975).

TABLE 19.1

Standard sampling depths (m)

0	300	2,000
10	400	2,500
20	500	3,000
30	600	4,000
50	(700)	5,000
75	800	6,000
100	1,000	7,000
150	1,200	
200	1,500	etc.
(250)		

the fine detail of any micro-structural feature, or if only the photosynthetic zone is being studied). Indeed, the very high cost of ship time* may enforce some simplification of the depth sampling schedule in many investigations.

Avoidance of contamination or alteration of the sample during sampling is of paramount importance. Such changes may be produced not only through the influence of the water sampler, but also from wastes discharged into the sea from the ship. Contamination from the latter source can usually be minimized by carrying out sampling to windward; however, pollution of the surface water may become significant under calm conditions, or when sampling is prolonged. Although surface samples are the ones most likely to suffer in this way, it should be borne in mind that most water bottles are open when they are lowered through the surface layer, and there is thus the possibility of them adsorbing impurities (particularly organic ones). Sometimes it may be preferable to carry out sampling of the surface water from a small boat at a distance from the research ship. Serious contamination may sometimes also result from the hydrographic wire and new wires must be cleaned before use to remove oil and grease which may even prevent the messenger falling freely. Grasshoff (1969) has indicated that the hydrowire may be a significant cause of contamination with trace metals, especially iron. In order to prevent this he suggests that the sampling bottle should be suspended from a nylon rope spliced to the wire; the splice being bridged with another sampler to allow transmission of the messenger. The adoption of this precaution does, however, preclude multiple sampling, and practical experience shows that it is only rarely necessary.

* This may amount to $3,000 per day for a large research vessel. Each sample from a deep-sea 10 bottle cast down to 5000 m, which takes 3–4 hours, may thus cost $50.

19.3.1. CRITERIA FOR WATER SAMPLING BOTTLES

There is no single type of water sampler which is suitable for all applications, and bearing in mind the nature and concentration of the component or components to be determined, the most appropriate type of sampler must be selected from among the many available. In making this selection attention must be paid to the following criteria.

(i) Exchange between the surrounding water and that within the sampler should be rapid and complete. In this connection it should be noted that many of the earlier types of sampler (e.g. the Nansen and Knudsen types— see Fjarlie, 1953), and even some of the more recent ones, are deficient in this respect.

(ii) The closing mechanism should be reliable and positive acting, and after it has been actuated the sampler must seal completely to ensure that no exchange of water occurs as it is brought to the surface (failure to do this is a common fault of many types of bottle, including the Nansen and Knudsen patterns).

(iii) The sampler should be constructed of materials which are resistant to corrosion, and which will neither contaminate the sample nor adsorb material from it. Until about 15 years ago water bottles were nearly all constructed from brass or other copper based alloys which has been heavily coated with nickel, or more recently with PTFE or epoxide resin, to protect them from corrosion. Unfortunately, in the course of time the protective plating becomes pitted or worn. This exposes the brass to the corrosive action of sea water which can be represented by the equation:

$$2(Cu, Zn) + O_2 + 4HCO_3^- \rightarrow 2(Cu, Zn)^{2+} + 4CO_3^{2-} + 2H_2O$$

When water is enclosed in a corroded bottle for a period of an hour or two (as it could be when a deep station is being worked) this reaction may cause the dissolved oxygen concentration to decrease by as much as $0.16 \, \text{ml} \, l^{-1}$ (Park, 1968; see also Rotschi, 1963; Bogoyavlenskii, 1965b). Similarly, the transformation of bicarbonate into carbonate ion will lead to an increase of both pH and alkalinity and to a decrease of the electrical conductivity and hence of the salinity if determined by means of a salinometer (Park, 1968). Plastic coated metal bottles are also not completely satisfactory, as Reed and Ryan (1965) have found that sacs which develop under the film can entrap highly oxygenated surface water and lead to oxygen concentrations which are $0.1–0.3 \, \text{ml} \, l^{-1}$ too high when oxygen-impoverished deep waters are sampled. Metallic water samplers should be avoided in trace element work; however, they are suitable for collection of samples to be analyzed for dissolved organic carbon or for individual organic compounds. In recent years there has been a move away from the use of metallic samplers to ones made of plastics. Usually the preferred constructional materials have been polypropylene polycarbonate or polyvinyl chloride, although polymethyl methacrylate (Plexiglass, Perspex) has also been employed, but it is rather too brittle for many purposes. Because of their lightness, incorrodibility and resilience plastic samplers have much to recommend them, and indeed their use is mandatory for trace element work. Even with these samplers trace element contamination (e.g. Ba, Sb and Zn) may arise from the metal fittings (unless these are Teflon coated) and especially from fibres present in the resilient sealing caps. It is desirable that the latter should be constructed of neoprene low in fillers. The use of plastic samplers is not to be recommended for collection of water to be analyzed for organic compounds.

(iv) The sampler should be light and easy to handle. Plastic bottles are generally only about one third of the weight of metal ones of the same capacity, but even with these, those which contain more than 10 litres are awkward and too heavy to handle. Although large samplers are available, they are not suitable for use in a bottle cast. The breaking strain of the hydrographic wire limits the number of samples which can be safely attached to it. The breaking strain of normal 4 mm hydrographic wire in good condition is ~ 1000 kg. In order to make allowance for strain produced by rolling of the ship it is usual to allow a safety factor of three, i.e. the safe load should not exceed ~ 350 kg. Since the weight of wire which may have to be paid out in deep sampling together with the weight of the sinker may amount to ~ 300 kg, this only leaves a comparatively small margin for the weight of the bottles. In practice, it is usual to limit the number of the 1–1·5 1 samplers on any cast to 12, or to 16 if plastic ones are used. Obviously, the number will have to be reduced if bottles of larger capacity are being used, or if the weather is rough.

19.3.2. WATER SAMPLERS

19.3.2.1 *General purpose small volume samplers*
In the limited space available it is only possible to describe some of the more important types in use. Even today the reversing samplers designed by Nansen (Wüst, 1932), Knudsen (Knudsen, 1929) and Ekman (Knudsen, 1923) or modifications of them are still in use, although they are being increasingly superseded by plastic ones of more recent design. The Ekman sampler consists of a brass tube closed by spring operated lids; this is pivoted about its midpoint within a brass frame. It is so arranged that a brass weight (messenger) which is allowed to slide down the suspension cable releases a catch allowing the bottle to rotate through 180° in the frame, simultaneously closing the end

caps. Although it has relatively good flushing characteristics and seals satisfactorily, it is heavy and awkward to use. The Nansen sampler consists of a tube having a capacity of ∼1·2 litre which is closed at each end with conical stopcocks. The sampler is attached to the hydrographic wire at its lower end by means of a screw clamp and at its upper by means of a releasable catch. The device is triggered by means of a messenger. This releases the catch allowing the sampling tube to pivot on the wire; as it does so the stopcocks are closed. The sampling tubes of both of these samplers are fitted with frames bearing reversing thermometers which are set when the instruments are triggered. The temperature of the water and the depth of sampling are determined from the readings of protected and unprotected reversing thermometers respectively (see e.g. Sverdrup et al., 1942). Although both the Nansen and Knudsen samplers are mechanically reliable they have very poor flushing characteristics and have a tendency not to seal perfectly when closed. Because of problems arising from corrosion, many modern versions of the Nansen bottle are made of, or coated with, plastic and the greased stopcock is replaced by one made of plastic. However, the flushing characteristics are still poor because of the restricted bore of the stopcock. This difficulty has been overcome by Fischer (1968) who has designed a 1·7 litre Nansen type sampler* made of transparent polycarbonate which is sealed by means of wide aperture plastic ball valves. This instrument is only about one third of the weight of the conventional metal Nansen bottle.

Fjarlie (1953) has developed a metallic sampler having excellent flushing and sealing characteristics. In this, the sampling tube is attached at both ends to the hydrographic wire and only the thermometer frame rotates. When the sampler is triggered with a messenger, hinged lids fitted with rubber gaskets are snapped shut at each end of the body by means of strong springs, and the thermometer frame rotates. A highly successful water sampler based on this principle is produced by the Institute of Oceanographic Sciences, Wormley, Surrey, England. (Herdman, 1963). The body of this instrument is constructed of polypropylene, and practically all the other fitments, including most of the release mechanism are plastic. This instrument has a capacity of 1·3 litre and since it weighs only ca. 1·4 kg in water it is possible to use up to 20 water bottles per hoist. A larger (ca. 9 litre) version of this bottle is also available. Fischer (1964) has designed a 5 litre non-metallic sampler based on a similar principle which can be triggered electrically from the surface, and Krey and Zeitschel (1968) have employed a deck command multiple sampler using 10 of these samplers in conjunction with temperature, salinity and turbidity sensors.

Van Dorn (1956) has designed a further type of sampler which, because of

* Available from Hydrobios, Kiel-Holtenau, West Germany.

its simplicity and very good flushing and sealing characteristics, has been widely adopted. This sampler is available in a range of sizes (3–60 litre). Closure in the original model was effected by means of two plumber's rubber plungers, which on release, were drawn together by a strong rubber band thus sealing both ends of the wide plastic sampling tube. Modifications of this apparatus have been described by Finucane and May (1961) and by Stephens (1962). In more recent versions the rubber plungers have been replaced by contoured conical P.V.C. stoppers*. For samples collected for trace element analysis the rubber band is a possible source of contamination and Spencer et al. (1972) have recommended its replacement by a P.T.F.E. coated spring. Niskin (1968) has described a multiple sampler using ten 1·7 litre Van Dorn water samplers which are triggered electrically from the surface as required. This type of instrument can be used with benefit in conjunction with an STD or CTD probe† (see also Gerard, 1968).

One further type of water sampler is worthy of mention because modifications of it have been used for the collection of very large samples. This is the bag sampler which was developed by Niskin (1962) for the collection of 2 litre samples of sterile water. It consists of a hinged metal frame which, by means of springs, opens a detachable sterile polythene bag when triggered by a messenger. Water enters via a rubber tube with a glass tip which is fractured by the action of the messenger; the rubber tube is subsequently sealed by means of a U clamp to prevent exchange of water as the apparatus is brought to the surface.

19.3.2.2. *Large volume samplers*

There is an increasing need for the collection of very large samples from all depths, particularly for the determination of organic compounds and both natural and man-made radio-nuclides. Several different types of sampler have been used for this purpose, most of them operating on the same principles as those described in the previous section. Samples of up to 220 l have been collected with Van Dorn-type samplers (see e.g. Van Dorn, 1956; Duursma, 1967; Bien et al., 1960). Young et al. (1969) have described a modified Van Dorn sampler in which, for cheapness, a 60 l stainless steel beer barrel is used as the body; they have used this apparatus successfully to depths in excess of 400 m. Bag samplers, working on the principle of the Niskin sampler (see Section 19.3.2.1) are probably the most suitable ones for the collection of very large

---

* Van Dorn samplers are available from a number of suppliers including Hydro Products San Diego, the Kahl Scientific Instrument Corp. San Diego, Calif, and General Oceanics Inc. Miami, Florida.

† Multiple samplers of this type are obtainable from General Oceanics, Inc., 5535 Northwest Seventh Avenue, Florida 33127, U.S.A.

samples and ones with capacities of as much as 30 tons have been used (see e.g. Schink and Anderson, 1969; Moore, 1969b)*. Although this type has been used down to depths of 2500 m, it is very fragile. For this reason it is impossible to lift the full sampler on board and instead the water must be pumped out while the sampler is still 20 m or so beneath the surface.

Bodman *et al.* (1961) have described a very satisfactory large volume sampler (up to 140 l) which has excellent flushing characteristics. It is a modified Kammerer sampler and is constructed of stainless steel or aluminium and lined with plastic; closure is effected by means of neoprene or polyethylene valves. Broecker *et al.* (1960) have developed a sampler having a capacity of 400 l (see also Gerard and Ewing, 1961). This consists of a rigid steel barrel divided longitudinally by a partition reaching within ca. 15 cm of the top and bottom. At the top of the barrel there is a 25 cm circular orifice which can be closed by means of a hinged door operated by a messenger. As the sampler is lowered a hood-like baffle directs the water flow down on one side of the partition to the bottom, it then passes under the partition and up and out on the other side. The efficiency of the flushing is such that only 0·1 % of the original water remains within the sampler when it is lowered 100 m.

19.3.2.3. *Sampling for organic compounds*

In sampling water for analysis for organic carbon or organic compounds it is essential to keep contact with plastic to a minimum. For many purposes conventional metallic samplers can be used; however, for special purposes the 15 l sampler designed by Clark *et al.* (1967) can be used down to depths as great as 4500 m. The body of this instrument consists of a thick-walled aluminium tube which is lined with a borosilicate glass tube, The outer tube is sealed with a rupture disc which will break at any selected pressure between 1 and 2000 atm. When the assembly is lowered to the desired depth the disc ruptures. Water then enters the sampler via a flow restrictor and is filtered by means of a membrane filter before entering the body of the sampler. A check valve allows excess air to escape during retrieval of the sampler.

i9.3.2.4. *Sampling for gas analysis*

Because of the ease with which gases diffuse, sampling of water for gas analysis presents a number of problems which do not arise when collecting samples for the determination of other components. The criteria for samplers for this

---

* Bag samplers ranging in size from 100–2000 l are available from General Oceanics, Miami, Florida, who also make a 1000 l segmented water bottle, in which the sample is collected, for ease of handling in a series of 90 l segmented cylinders.

purpose have been discussed by Weiss (1968) who has described a simple sampler which is used in conjunction with a standard Nansen bottle. It consists of a length of thin walled soft copper tubing which is connected between high pressure ball valves using standard couplings. The stems of the ball valves are connected to the rotating plugs of the Nansen bottle so that these valves also close when the bottle is closed. Polyethylene funnels are attached to the ball valves to ensure flushing of the copper pipe as the assembly is lowered. Once aboard the research ship the gas sampler is detached and the copper tube is pinched off with a clamp so as to give a gas-tight seal. A more complicated approach has been adopted by Bieri (1965) who has developed a battery operated sampler which pinches off the water sample in a length of aluminium tubing *in situ*.

Werner and Waldichuk (1967) have described a simple apparatus for sampling gases in bottom sediments; it is capable of operating down to depths of 200 m.

19.3.2.5. *Sampling close to the sea floor*

The composition of the water in the vicinity of the sea floor is often modified by reactions occurring with the sediments. There are two important extra criteria which must be fulfilled by any satisfactory device for the collection of water samples to be used in studies of this process. Firstly, it must be possible to collect samples at known distances above the sediment-water interface. Secondly, it must be possible to ensure that the water captured is truly representative of that at the particular depth sampled, i.e. the lowering of the device should cause minimal disturbance of the water, or it should be possible to leave it in position for sufficiently long for the current to normalize the water composition again. Samplers operating on several different principles have been used for sampling close to the bottom, these include:

(i) water bottles attached to a sediment sampler (Carey and Paul, 1968);

(ii) horizontal samplers having a variety of different closure mechanisms (Joeris, 1964; Wunderlich, 1969; Sholkovitz, 1970; Joyce, 1973). The apparatus developed by Sholkovitz for sampling within 180 cm of the sea floor is a free floating device consisting of 10 horizontal 600 ml Van Dorn bottles attached to a float with radar reflector. The train of bottles is connected to a concrete sinker by means of a corrodible linkage which releases it from the bottom after 3–4 hours;

(iii) instruments in which the water is pumped to the surface through intakes mounted either on a frame which stands on the bottom (Scheimen and Schubel, 1970), or on a bottom sled (Smith, 1971). These devices are suitable only for use in shallow water.

H

19.3.2.6. *Sampling while under way*

Spilhaus and Miller (1948) have designed an automatic sampler which can be used from a moving ship. It consists of a standard bathythermograph which has been modified by the addition of twelve 50 ml sampling bottles. The pressure sensor of the bathythermograph triggers the closing valves at preset depths down to 100 m. Good (1968) has described the water sampling equipment on the deep submersible DEEPSTAR 4000, which can be operated at depths of up to 1200 m. Two different types of solenoid-operated bottles are used—1·3 l Fjarlie bottles and 2·0 l Van Dorn bottles—these are mounted in pairs at 8 different depths from the bottom.

19.3.2.7. *Sampling under sterile conditions*

A number of different types of sampler have been designed for the collection of water under sterile conditions for bacteriological purposes. The first of these, which was designed by Zobell (1941), consists of a sterilized compressed 150 ml rubber bulb closed with a sealed glass capillary. At the appropriate depth the capillary is fractured by means of a messenger and the elasticity of the rubber causes the bulb to fill. This simple sampling apparatus suffers from a number of drawbacks (see Niskin, 1962), and can only be used down to shallow depths because the rubber loses its elasticity under high pressure. The Niskin bag samplers (Section 19.3.2.1) are free from most of these disadvantages, and have proved very satisfactory in practice. A further sterile water sampler has been described by Jannasch and Maddox (1967).

19.3.2.8. *Sampling of the sea surface film*

The chemical composition of the aqueous film at the atmosphere–sea boundary has been the subject of a number of investigations in the last ten years. These have shown that many non-polar organic compounds (see Jarvis *et al.*, 1967) and trace metals (Duce *et al.*, 1972) are considerably enriched in the surface layer relative to the underlying water (see Chapter 10). The surface film is probably usually a monomolecular one. Methods of sampling it are discussed in Section 10.4; see also the review by Baier (1972).

19.3.2.9. *Sampling of sediment pore waters*

Until recently interstitial water has usually been recovered from sediments by taking a conventional core and then extracting the pore water by means of a type of filter press called a squeezer (see e.g. Siever, 1962; Siever *et al.*, 1965; Reburgh, 1967; Brookes *et al.*, 1968). However, it has been shown (Mangelsdorf *et al.*, 1969; see also Chapter 32) that the composition of the pore water may change significantly unless the core is kept at its original temperature until it is squeezed. As it is difficult to maintain such constant temperature conditions during recovery of a deep-sea core it would obviously be preferable to carry out the sampling of the interstitial water *in situ*. Only two workers have described apparatus by which this can be achieved. Barnes (1973) has designed a sampler which is fitted as an outrigger on a corer. In this, the pore water is filtered under the *in situ* hydrostatic pressure through 1 µm stainless steel mesh and collected in an evacuated stainless steel container. Wilson *et al.*, (1972) have given preliminary details of a more elaborate interstitial water sampler which draws in and filters water at various depths in the upper 2 m of the sediment column. Sampling is carried out by means of spring loaded syringes, mounted within the cylindrical body of the instrument; these are actuated when a collar at the top of the body comes into contact with the sediment surface.

The sampling of the interstitial waters of intertidal mud flats poses a much simpler problem than sampling in deep water. For this purpose Creaser (1971) has developed an apparatus consisting of a length of P.V.C. tubing sealed at its lower end to a porous (20 µm) alundum cone. In practice, the tube is pushed into the mud to the desired depth and the interstitial water is drawn into it by slight suction.

19.3.3. SAMPLING BY PUMPING

Sampling of the water column by means of water samplers has two major limitations. Firstly, it is difficult to collect very large samples, such as those required for work on radionuclides. Secondly, as a result of profiling carried out by means of STD and other probes it is now being generally recognized that there is much fine structure present in the water column; much of this will be missed if observations are carried out using samples collected at the standard depths. Collection of water by pumping is free from these objections, and this technique has been used for collection of large (> 20 ton) water samples from depths down to 3000 m (see e.g. Jeffrey *et al.*, 1973) and also, in conjunction with continuous analyzers, for vertical and horizontal profiling (see Sigalove and Pearlman, 1972). The equipment required for these two tasks is basically similar. It consists of a reinforced hose of sufficient length to reach to the desired depth and a submersible pump which is connected to a sampling manifold on the ship.

A number of different types of pumping systems have been used for sampling. Thus, Hood (1963) has used a pump operating on the jet pump principle which functioned by cycling a large volume of water under high pressure through a jet situated 100 m below the surface; the pressure drop which this caused drew water from the sampling depth and delivered it to the pump through the return line. Some of this water was then recycled to the jet. This type of pump has the advantage that all the electrical components are on board the ship, but it is difficult to prime. Freudenthal et al. (1968) have described a pneumatic ejector pump for the collection of discrete 3 litre water samples from shallow depths. Most recent workers have favoured the use of multistage submersible electric pumps (see e.g. Strickland, 1968; Sigalove and Pearlman, 1972; Jeffrey et al. 1973) which are more versatile, can be used at greater depths and can have faster delivery rates. The number of stages in the pump and the requisite power input to it depend on the flow rate required and the length and diameter of the sampling tubing (because of the power necessary to overcome friction with the wall of the pipe). In the development of the pump system care must be taken to ensure that it does not contaminate the water passing through it or damage particulate matter or plankton. Thus, the multistage axial-flow pump developed by Sigalove and Pearlman (1972) is water lubricated and parts which come into contact with the water are either made of, or coated with, non-contaminating material. The acceleration and deceleration stages have been designed in such a way that even particles as large as 7 mm can pass through without damage. The pump can be used down to depths of 75 m and has a delivery rate of ca. $20 \, l \, min^{-1}$.

For shallow water sampling, pumping may be carried out through a polyethylene tube attached to a strong electric cable, both of which are fed out from a winch having a rotary seal for the water, and slip rings for the electric supply to the pump. It is now possible to obtain an armoured suspension cable which incorporates within it a flow pipe and also conductors for the electric supply and various types of sensor (see Fig. 19.1). When profiling is to be carried out it is important to keep the flushing time of the tubing to a minimum, and to maintain a sharply defined boundary within the tube by keeping the flow rate sufficiently high to ensure turbulent flow. If horizontal profiling is to be carried out underway the use of faired tubing will, by reducing drag, enable a greater depth to be sampled with a given length of tubing.

In the continuous ocean sampling and analysis system described by Sigalove and Perlman (1972), the pump is mounted adjacent to a S.T.D. probe. The deck unit can produce continuous automated analyses for a wide range of chemical parameters by means of colorimetric, gas chromatographic, and potentiometric techniques. Fluorimetry can be used for determination of chlorophyll (Strickland, 1968).

Collapse of the tubing on the winch drum under its own weight presents considerable problems if attempts are made to use pumping to sample deep water masses, and for this reason it does not appear that it is feasible at present to carry out profiling to depths of much below 100 m. For deep sampling Jeffrey, et al. (1973) found it necessary to keep the tubing (2·04 cm

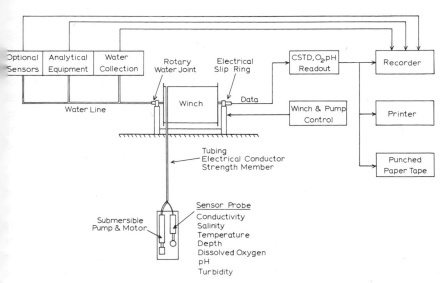

FIG. 19.1. Block diagram of Inter-Ocean ocean sampling and environmental analysis system (after Sigalove and Pearlman, 1972).

ID; 2·5 mm wall) on wooden reels and to clip it onto the suspension cable every 30 m as the latter was paid out. The submersible pump was then fitted into the line so that it would finally be 20–50 m below the surface, depending on the depth to be sampled. In this way it was possible to sample at depths down to 3000 m. At this depth it was necessary to flush the system for 4 hours before collecting the sample at a rate of $624 \, l \, min^{-1}$.

19.4. Filtration of Water Samples (see also Section 19.14.2)

Sea water contains two principal types of suspended matter, (i) inorganic particulate material resulting from rock weathering and authigenic precipitates, (ii) marine organisms, their decay products and other organically-derived detritus. The total amount of suspended matter varies greatly according to the environment from tens of milligrams per litre in estuaries, to a few micrograms or less per litre in the deep waters of the oceans remote from land. Similarly, the ratio of inorganic to organic particulate matter shows wide geographical and seasonal variations, the organic component predominating in highly productive waters. The size distribution of the particulate matter varies considerably with environment, the largest particles often being silt-sized in estuaries and inshore waters, and small clay-sized in deep oceanic

waters. The lower limit is difficult to define since there is probably a con
tinuous spectrum of sizes merging at the lower end with colloidal micelles
Both the inorganic and organic fractions of the suspended matter are capable
of adsorbing trace elements from the water and sometimes of desorbing
them.* In addition, bacterial action may result in the mineralization of micro-
nutrient elements associated with the organic particulate matter. For these
various reasons it is often desirable to separate the particulate fraction before
analyzing the water itself. Although this separation can be carried out by
high speed centrifugation, it can be more efficiently and conveniently achieved
by filtration.

When filtering sea water it is customary to use a filter with pores having
an average diameter of 0·45 μm—the fraction retained by the filter is termed
the particulate material, elements present in the filtrate are said to be
"dissolved". It should, however, be realized that this definition is a purely
arbitrary one. This is so for two reasons. Firstly, particulate matter finer
than 0·45 μm in size is present in sea water; this is demonstrated by the fact
that much less iron (III) can be detected if a sample is filtered through a 0·2 μm
filter rather than a 0·45 μm one.† Secondly, during the course of a filtration
build-up of particulate matter on the surface of the filter or in its pores tends
to reduce its effective pore size (see Sheldon, 1972).

An ideal filtering medium would satisfy the following criteria.

(1) It should have a reproducible and uniform pore size. The pore size most
    commonly used in oceanography is 0·45 μm; this will retain all phyto-
    plankton and most bacteria.
(2) The rate of filtration should be high, and the filter should not clog easily.
(3) It should be easily brought to constant weight so that the concentration
    of particulate material can be determined gravimetrically.
(4) The suspended material should not penetrate into the filter, but should
    be retained on its surface, and thus be easily accessible for microscopic
    examination or for removal.
(5) It should have a low ash content, in order to prevent contamination of the
    water sample and to enable the inorganic components of the particulate
    material to be recovered without contamination by ignition or wet ashing.
(6) It should not absorb trace elements or dissolve organic material.
(7) It should have reasonable mechanical strength.
(8) It should not shed fibres since these might contaminate the filtrate or the

---

* Adsorbed trace elements may be desorbed during the preconcentration stage used in the
analysis of the dissolved fraction.
† Kholina et al., (1973) have described an "electro-filtration" technique which they claim is
capable of separating ionic components of sea water from colloidal ones.

suspended matter, or give erroneous results when the concentration of particulates is determined gravimetrically.

(9) Their use should not lead to rupture of phytoplankton cells.

Although many different types of filter are available none of them satisfy all the above criteria. Cellulose filter papers are unsuitable for most marine work and they should only be used if samples are to be analyzed for major components. Sintered glass filters, which are available in nominal pore sizes down to $0.5 \, \mu m$, are suitable for filtration of samples in which soluble organic compounds are to be determined. They are unsuitable for work in which heavy metals are to be determined. The suspended matter penetrates into the interstices of the filter and cannot usually be recovered without chemical treatment. Glass fibre filters, which are now available in a form capable of retaining $1 \, \mu m$ particles,* are particularly suitable for use in the determination of particulate organic carbon. Since sea water tends to leach heavy metals, such as copper (Marvin et al., 1970) and zinc, from them, they are unsuitable for filtration of samples to be analyzed for most of these elements (except mercury). They are also useful for the filtration of water samples to be analyzed for dissolved organic carbon, particularly since the filter can be cleaned initially by ignition at 450°C. They are also of value for filtering samples for chlorophyll determination. Because of their friable nature they are unsuitable for the gravimetric determination of total particulate matter.

The type of filter most extensively used for the filtration of sea water for analytical purposes is the membrane or micro-filter developed by Goetz and Tsuneishi (1951). These filters are composed of incompletely cross-linked polymers of partially substituted cellulose acetate and nitrate; they have a thickness of ca. $0.15 \, mm$. Those manufactured by the Millipore Corporation Inc., of Bedford, Mass., U.S.A., are available with pore sizes $10 \, nm$–$5.0 \, \mu m$†. The ash weight of this type of filter is very small, amounting to only ca. $0.0001 \, \%$ of the original weight. There is often some penetration of the particulate material into the body of the filter when these filters are used; however, the filter can be rendered transparent for microscopical examination of the particulates by treating it with a liquid with a similar refractive index to its own. Sheldon (1972) considered that this type of filter is probably the best for total particle retention. Particulate material collected on such filters can be examined by pelletizing the filter in a press and analyzing the pellet by instrumental neutron activation analysis (see e.g. Spencer et al., 1972).

---

* Gelman, A. filters ($0.3 \, \mu m$) manufactured by the Gelman Instrument Co., 600 S. Wagner Road, Ann Arbor, Mich., 48103, U.S.A.; Whatman GF/F filters ($0.7 \, \mu m$) manufactured by W. & R. Balston Ltd., Springfield Mill, Maidstone, Kent, England.
† A cheaper British equivalent is available from the Oxoid Division of Oxo Ltd., London, E.C.4, but only with a pore size of $0.5 \, \mu m$.

An important, but expensive new type of filtering medium is the Nuclepore filter.* These filters consist of very thin ($\sim 10\,\mu m$) polycarbonate films which have been perforated by means of electrically charged particles; they are available in a range of pore sizes down to $0.5\,\mu m$. According to Cranston and Buckley (1972a) the $0.5\,\mu m$ filter is extremely retentive for sub-micrometre particles, but is easily overloaded as these particles do not penetrate into the body of the filter. Its weighing characteristics are very reliable, and it has a very small salt retention. There is one further type of filter which is worthy of mention, this is the Flotronic silver membrane filter.† This is available with a pore size of $0.8\,\mu m$, and has a low retention for dissolved humic acids, and salts.

Several workers have examined the performances of some of these types of filter. Sheldon and Sutcliffe (1969) and Sheldon (1972) have used a Coulter Counter to observe changes occurring in the particle size distributions in sea water as a result of passage through filters of various porosities. Their results are shown diagrammatically in Fig. 19.2. They assumed that the median retention diameter (that at which 50% of the particulate material was retained) is a measure of the effective pore size and that the slope of the curve is a measure of the uniformity of the pores; the steeper the slope, the greater is the degree of uniformity. They concluded that there were considerable differences between the nominal and median pore sizes for the Millipore filters, all having similar retention sizes; however, for both the Nuclepore and the Flotronics membranes the median and nominal pore sizes were similar (Table 19.2). Millipore filters were found to have the greatest uniformity. Because the particulates did not penetrate into the body of the Nucleopore filters their median retention size did not vary with the volume filtered, but remained constant until the filter blocked. There is thus the possibility of using them for particle sizing; however the volume which can be filtered with them is very limited. In contrast, with other types of filter the median retention size decreased during the filtration; such filters could not therefore be used for this purpose, but were satisfactory for total particle retention.

A further investigation of filter performance has been made by Cranston and Buckley (1972b) who examined filters by scanning electron microscopy and also studied their power to retain, from suspension, polystyrene beads of a variety of uniform sizes. Their findings are summarized in Table 19.2. They concluded that the order of retention power for $<1\,\mu m$ particles was Nuclepore > Millipore > silver > glass fibre. However, if the filter was

---

* Manufactured by the General Electric Co., Pleasanton, California, 94566, U.S.A.
† Manufactured by Selas Flotronics, Box 300, Springhouse, Penn, 19477, U.S.A.

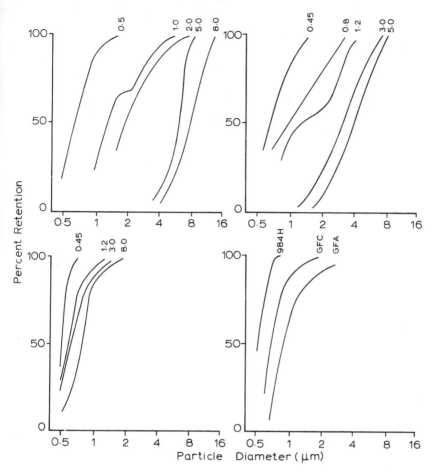

FIG. 19.2. Retention curves of different filtering media. Upper left—Nuclepore; Upper right—Flotronics; Lower left—Millipore; Lower right—Whatman GF/C and GF/A and Reeve Angel 984H. The numbers over the curves are the pore sizes (μm) quoted by the makers (after Sheldon, 1972).

overloaded the efficiency was increased because of the filtering action of the material building in and around the orifices in the filter. The same workers also examined the weighing characteristics of the filters and found that the Nuclepore ones showed the least change in weight on washing, followed by the silver, glass fibre and Millipore filters.

The process of filtration may alter the composition of the water being filtered either by removing dissolved components by adsorption onto the

TABLE 19.2

Properties of some filters used in oceanographic work (all measurements expressed in μm) (from data by Cranston and Buckley, 1972)

Pore diameter	Millipore MF	Nuclepore	Gelman A glass fibre	Flotronics silver membrane
Manufacturer's specification	0·45	0·4	0·3	0·8
Measured by electron microscopy	0·4	0·4	1	2
Measured by retention experiments*	1	0·3	2	2
Thickness				
Manufacturer's specification	150	10	—	51
Measured	180	15	800	45
Pore area as percentage of total area	10	10	5	9
Retention of humic acid (%)†	15 ± 2	44 ± 2	5 ± 3	3 ± 1
Retention of <4 μm clay (%)‡	94 ± 1	92 ± 1	57 ± 7	79 ± 14
Filtration time for <4 μm clay (min)‡	41 ± 2	99 ± 8	6 ± 1	130 ± 34
Mean weight change (mg) original weight of filter (mg) in parentheses§	−0·25 (94·44)	+0·02 (17·65)	−0·16 (140·50)	−0·34 (481·84)

 * Defined as that size of polystyrene sphere which is removed with an efficiency of 90% from 1 l of a 1 mg l^{-1} suspension (Sheldon and Sutcliffe, 1969).
 † From filtration of 100 ml of a 10 mg l^{-1} solution.
 ‡ Filtration of 100 ml of a 50 mg l^{-1} suspension.
 § Dried (60°C). Weighed filters were used to filter 1 l of deionized water they were then redried and reweighed. A radioactive ionizing source was placed in the balance case to minimize effects caused by static electricity.

filter by Van der Waal's forces, or by contaminating it with substances dissolved from the material of the filter. Although such effects may be sufficiently serious to invalidate the subsequent analyses, there have been few systematic studies of the effect of filtration on sea water samples. Marvin *et al.* (1970) have made a statistical study of the influence of filtration through six different filtering media on the copper content of sea water. Unfortunately, to make the analysis easier, most of their work was carried out using unrealistically high levels of copper, and it is therefore difficult to know what extent their findings can be applied to natural sea water. They concluded that filtration was a major source of error in copper analysis; effects due to both contamination and adsorption were observed. At concentrations such as those normally encountered in sea water membrane filters altered the composition of the waters more than did any others. In contrast, glass fibre filters, which they found overall to be the most objectionable ones, affected the composition of normal waters least. Similarly, work in the author's laboratory (Doris Gardner, unpublished manuscript) has shown that Millipore filters strongly adsorb inorganic mercury from sea water, whereas glass fibre filters* (which can be freed from mercury by ignition at 500°C) cause only slight loss ($<7\%$). Figures published by Spencer and Manheim (1969) for the trace element levels in Millipore HA filters (Table 19.4) suggest that such filters should cause negligible contamination of water samples for most trace metals. However, when particulate trace metals are being determined after ashing the filters significant errors may result when the concentration of particulate matter is low.

The effect of filtration on the nutrient content of sea water has been investigated by Marvin et al. (1972). Contamination with ammonium and nitrate ions occurred with many of the six types of filter tested if these were not washed before use; most filters were acceptable if they were first washed. There was some evidence for adsorption of phosphate by many types of filter, and for this nutrient filtration through an unwashed glass fibre filter was recommended.†

Little is known about the influence of filtration on the concentration of dissolved organic materials in sea water (see also Section 19.14.2). Contamination may arise from the organic phosphates used as a plasticizer in certain types of filters. Losses may occur as the result of adsorption and this will probably be most severe for species of high molecular weight. In this connection it is interesting to note that Cranston and Buckley (1972) found that humic acid was strongly taken up by Nuclepore filters (see Table 19.2). Glass fibre filters are probably the most suitable type for filtration of samples to be analysed for dissolved organic matter (DOM), particularly since they can be purified by ignition; however Cranston and Buckley (loc. cit.) have pointed out even these may retain sufficient DOM to affect particulate organic carbon analyses.

---

* Whatman GF/F.
† Jenkins (1968) has found that 47 mm Milipore HA filters contain ca. 1·3 μg of phosphorus of which 1 μg can be removed by washing.

From the above discussion it will be evident that there is no single filter type which is suitable for all purposes. Much further work is required to find the best medium for filtration of samples required for a particular determination. It is recommended that, unless it is known that a particular type of filter is suitable for use with a certain element or compound, tests should be made to select an appropriate filtering medium. In general, these tests need not be as complex as those used by Marvin *et al.* (1970, 1972). They need only embrace a very restricted range of filtering media, and should aim to examine their behaviour as both adsorbents or pollutants. Their former role can often be checked using radio-isotopes. In general, there is probably considerable merit in rejecting the first 250 ml of the filtrate from a sample in case the filter contains soluble impurities.

Filtration is normally carried out using suction, and to prevent the filter tearing it must be supported on a sintered glass disc or perforated plate. The glass filtration apparatus described by Goldberg *et al.* (1952) is suitable for many purposes, but for work in which contamination must be kept to a minimum it is advisable to employ filter units constructed of polytetrafluoro-ethylene (see e.g. Preston *et al.*, 1972). Filtration is usually performed using a vacuum of 5–15 cm of mercury. However, in determinations in which plant cells are to be separated (e.g. for plant pigment analysis) it is probably desirable to use a lower vacuum (> 30 cm of Hg) in order to avoid disintegration of delicate organisms (UNESCO, 1966). There are some advantages in filtering under pressure rather than with suction, and an apparatus for this purpose for use in trace element analysis has been described by Spencer and Brewer (1969) (see also Spencer and Sachs, 1970).

19.5. Storage of Water Samples

Although it is obviously preferable to analyse sea water samples immediately after collection, this is not always practicable, either because of lack of time or because it is not feasible to carry out the analysis on board ship. Unless steps are taken to stabilize them, and to prevent contamination by the containers, many components may undergo rapid changes in concentration; this is particularly true for the micro-nutrients and many trace elements.

It is a simple matter to store without change samples for determination of salinity or major ions, and all that is necessary is to transfer the samples to hard glass bottles with air-tight closures (Thompson, 1940)*. Containers

---

* The use of borosilicate bottles should be avoided if boron is to be determined; Charpiot (1969) recommends storage of samples at 4°C in the presence of $HgCl_2$ for this element.

constructed of soft glass should be avoided if salinity is to be determined by means of a salinometer since, on standing, sufficient alkali may dissolve from the glass to change the electrical conductivity appreciably. Bakelite screw stoppers with polyethylene inserts or screw stoppers with rubber washers are satisfactory closures, but ground glass stoppers should not be used as evaporation of water past them tends to be excessive. Bottles made of high density polyethylene are suitable for long-term storage of bulk samples. However, low density polyethylene bottles are only suitable for short-term storage (< 7 days) since they are permeable to water vapour (Cox, 1953; Hermann et al., 1959). Johnston (1964) has reviewed the literature on storage of sea water samples for salinity determination.

The oxygen content of water samples can change very rapidly after collection. Loss of oxygen can occur as cold well-oxygenated deep water samples warm up to the ambient temperature. Conversely, poorly oxygenated samples quickly absorb oxygen if exposed to the atmosphere. In addition, rapid changes can occur as a result of the biological processes of respiration and photosynthesis. It is therefore extremely important that samples to be used for the determination of dissolved oxygen should be the first to be drawn from the sampling bottle. Care must be taken to avoid turbulence and bubble formation during the transfer of the sample to the analysis bottle, and the Winkler reagents must be added immediately. Acid should not be added to the samples until a few minutes before the titration is to be commenced in order to minimize photochemical oxidation of iodide ion. Storage of the treated samples, if necessary, should be carried out in the alkaline state, and it is advantageous to cover the bottles with water to prevent air being drawn into them as their contents expand and contract in response to change in the ambient temperature.

Hydrogen sulphide is very rapidly oxidized by atmospheric oxygen, and no attempt should be made to store samples of euxinic water which are to be analysed for this gas. Such samples should be immediately transferred to a stoppered bottle and, using a technique similar to that used in the Winkler method, treated with a fixative such as zinc or cadmium acetate (see also Section 19.6.6).

For most other gases, if the complete analysis cannot be carried out at sea it is preferable to carry out the extraction on shipboard and to transfer the gas to sealed ampoules for transport to the laboratory. If this cannot be done, the water should be stored in glass vessels fitted with mercury seals (see e.g. Rakestraw and Emmel, 1937) or in the special piston storage tubes devised by Benson and Parker (1961) and Weimer and Lee (1971). For the storage of samples for analysis for noble gases Clarke et al. (1969) have developed bakeable stainless steel flasks which are fitted with steel stem tip valves.

Polyethylene bottles should not be used for storage of such samples because of their permeability to gases (see e.g. Kerr et al., 1962).

The concentrations of micro-nutrient elements in sea water may alter rapidly when samples are stored, as a result of biological or enzymatic processes, or, with phosphorus also because of adsorption. Since appreciable changes may occur in few hours it is advisable that, if possible, analyses for these elements should be commenced as soon as possible after the samples have been collected. Unfortunately, these changes are very difficult to inhibit, and although many attempts have been made to find satisfactory ways of storing samples, there is no general agreement about how this can be done. Both freezing and the use of chemical preservatives have been recommended. Although the latter techniques have the virtue of simplicity, most workers at present appear to favour rapid freezing of the samples at $-20°C$ (often in a glycol bath), followed by storage at $-20°C$ in a deep-freeze It seems likely that this procedure brings about reasonable stabilization of most micro-nutrient forms, particularly if samples are filtered before freezing. The literature on the preservation of samples is voluminous and often contradictory, and can only be reviewed briefly below; for a more complete treatment of the subject the reader is referred to Charpiot (1969) and Jenkins (1968).

The concentration of orthophosphate in sea water may change significantly within as little as 1 hour of collection (Collier and Marvin, 1953). Losses are accentuated if the particulate matter is not removed (Heron, 1962). Losses are particularly pronounced if the samples are stored in containers made of polyethylene or polyvinyl chloride (see e.g. Murphy and Riley, 1956; Hassenteufel et al., 1963) and, in fact, it is possible to reduce phosphate to undetectable levels by storing samples in such containers for 2–3 weeks. The cause of these losses is uncertain. It seems unlikely that the mechanism involves adsorption of orthophosphate ions, since Ryan (1966) has shown that, under sterile conditions, phosphate is adsorbed to a negligible extent from sea water onto beads or granules of glass or plastics having a large specific surface area. A more likely explanation appears to be that the phosphorus is taken up by proliferating micro-organisms which attach themselves to the walls of the containers. This is supported by the observation by Ryan that losses of phosphate from sea water which had been filtered through a $0·2\ \mu m$ membrane filter varied according to the container material. Losses were greatest and most rapid with polyvinyl chloride and polyethylene,* those with polypolypropylene and polystyrene were slower and more regular; negligible changes occurred with glass. These results suggest that the small residual

---

* Jones (1968) has reported that bacteria produced during storage of sea water samples in polyethylene containers cause interference in the photometric determination of phosphate.

population of micro-organisms proliferated more slowly on the last three of these materials and that they would be suitable for storage containers. According to Heron (1962) polyethylene vessels can be used for storage if they are first treated with a solution of iodine in potassium iodide; this presumably inhibits the growth of micro-organisms. Hassenteufel et al. (1963) recommend that samples for phosphate determination should be stored in bottles made of borosilicate glass which have been treated for several days with a dilute solution of hydrofluoric acid. They found that polytetrafluoroethylene containers were also suitable.

Freezing has been frequently used for preservation of samples for phosphate determination (see e.g. Collier and Marvin, 1953; Strickland and Parsons, 1968; Thayer, 1970). However, the pretreatment before freezing varies. Thus, Fitzgerald and Faust (1967) recommended that samples should be filtered before freezing as, under some circumstances, phosphate may be released from algal cells on freezing. Gilmartin (1967) suggested that 0·7% v/v of chloroform should be added before freezing to stabilize the phosphate as changes occurred more rapidly after thawing than initially. However, Thayer (1970) has found that freezing in the presence of chloroform caused significant changes in phosphate levels, whereas in its absence the samples were stable for at least 10 days. This accords with the observation by Fitzgerald and Faust (1967) that phytoplankton cells, on freezing or treatment with chloroform, release labile phosphorus as orthophosphate. On balance, it would thus seem that freezing at -10 to $-20°C$ is an effective means of stabilizing samples particularly if they are first filtered to remove algal cells (Henriksen, 1965b). Rapid increases in phosphate concentration have been reported for samples which have been merely refrigerated at 5°C (Thayer, 1970).

Several workers have investigated the possibility of stabilizing samples by purely chemical means. In addition to chloroform and other organic solvents which have proved unreliable (*vide supra*), mercury (II) chloride has been used as a preservative (Jenkins, 1967; Charpiot, 1969). Acidification of the sample has been recommended by Charpiot (1969), but is likely to be unsatisfactory because it will lead to hydrolysis of labile organic phosphates (see Vogler, 1965a; Jenkins, 1968) and also of inorganic polyphosphates* which may be present in estuarine samples.

Very little work has been carried out to find the best way of stabilizing samples to be used for the determination of dissolved organic phosphorus. Many of the compounds concerned are labile and it is recommended that samples should be filtered and rapidly frozen as soon as possible after collection. If they are then kept at $-20°C$ they will undergo little change

---

* The hydrolysis of condensed phosphates in lake waters under both sterile and non-sterile conditions has been studied by Clesceri and Lee (1965).

within 2–3 weeks (see e.g. Thayer, 1970; Strickland and Parsons, 1968).

The concentrations of the various dissolved fixed nitrogen species in samples may change significantly even within an hour of collection. (Harvey, 1955; Strickland and Parsons, 1968; Jenkins, 1967)*. If the analysis cannot be commenced immediately, it is therefore essential that steps be taken to minimize such changes. Of the methods which have been suggested for preservation, rapid freezing followed by storage at $-20°C$ is probably the best (see e.g. Varlet, 1958; Proctor, 1962; Strickland and Parsons, 1968; Jenkins, 1968; Thayer, 1970). However, even under these conditions prolonged storage (> 20 days) may lead to appreciable changes in the concentrations of nitrate and particularly ammonia (Strickland and Parsons, 1968; Truesdale, 1971). According to Thayer (1970) refrigeration at $5°C$ is inadequate to inhibit alteration of the concentrations of NO_3^-, NO_2^- and NH_3 in sea water. Chemical methods of preservation have been recommended from time to time. Chloroform has been found (Thayer, 1970) to be unsatisfactory as a stabilizer with both refrigerated and deep frozen samples, probably because it disrupts the algal cells and releases nitrogen compounds (thus, Degobbis, 1973 found that ammonia is rapidly released when chloroform is added). The use of mercury (II) chloride (ca. $40 \text{ mg HgCl}_2 \text{ l}^{-1}$) as a preservative for samples to be analysed for nitrite and nitrate has been recommended by several workers (Mullin and Riley, 1955b; Henriksen, 1965a; Jenkins, 1967; Charpiot, 1969). This technique, particularly if used in conjunction with refrigeration, is of considerable value for short-term preservation of samples when deep-freeze facilities are not available.

For storage of samples to be analysed for dissolved silicate the use of plastic (e.g. polyethylene) bottles is essential. Variations in silicate concentration are considerably slower than those of other micro-nutrient elements and samples can be kept for a few days at room temperature without significant change. However, on prolonged storage alteration can occur, and to prevent this Strickland and Parsons (1968) have suggested that samples should be kept at $-20°C$. However, Charpiot (1969) has found that this is liable to produce anomalous results. In this connection, it is interesting to note that Kobayashi (1968) has observed that the dissolved silicate in fresh water polymerizes on freezing and only slowly depolymerizes on thawing. However, this does not appear to happen in sea water medium (Leatherland, 1969), but may do so with water of intermediate salinity. Charpiot has also observed that samples stored in the dark at $4°C$ are stable for at least a month, but prolonged storage (5 months) led to a significant decrease in the silicate

---

* Ostrowski et al. (1968) who have rejected chemical preserving agents, have claimed that nitrate in filtered (0·5 μm) samples is stabilized for 8 days at room temperature.

concentration; this change could be prevented by the addition of 40 mg $HgCl_2$ l^{-1} to the sample before storage. Mullin and Riley (1955c) have recommended acidification of the sample to pH 2·5 followed by storage in the dark.

Care in the choice of storage conditions is particularly important when trace elements are to be determined. Changes in the concentrations of such elements may occur as a result of contamination arising from desorption of the element from particulate matter, its liberation during breakdown of marine organisms, or its dissolution or desorption from container walls. Contamination from the first two of these sources can be prevented by filtration of the sample, although even with this, difficulties may arise because of leaching of the filters (see Section 19.10.1). In addition, elements may be lost from solution by adsorption onto the walls of the sample bottle. For many trace elements these changes may be quite rapid. Since preservation of samples for analysis for such elements is often difficult, there is much to be said for carrying out at least the pre-concentration stage of the analysis within a few hours of sampling. This can be readily achieved, even on shipboard, by the use of ion-exchange methods, particularly those employing chelating ion exchangers (see Section 19.10.2.3). Such procedures have the added advantage that they obviate the necessity for transporting large samples back to the land-based laboratory. Notwithstanding this, it is occasionally necessary to bring samples back to the laboratory before commencing the analysis. Unfortunately, there is no single method of storage which is suitable for all elements; probably the most generally applicable one involves filtration of the sample through a pre-washed membrane filter, acidification to pH 2·0–2·5 and storage in the frozen state at $-20°C$ in a polyethylene bottle. The reader is referred to the papers by Robertson (1968a) and Bowen et al. (1970) for discussions of the storage of samples for trace element analysis (see also Section 10.10.1).

19.6. DETERMINATION OF SALINITY AND CHLORINITY

A knowledge of the salinity of sea water is of fundamental importance in physical oceanography since, if the in situ temperature is also known, the in situ density of the water can be computed and used in dynamical calculations of currents and salt transport. Since the horizontal and vertical salinity gradients in oceanic regions are generally very small it is essential to be able to determine salinity with an accuracy of better than $\pm 0·01\%_0$ or, in extreme instances to $\pm 0·002\%_0$. Considerably greater variations in salinity occur in coastal waters and particularly in estuaries, and for these waters a considerably lower accuracy is usually acceptable.

DETERMINATION OF SALINITY

When sea water is evaporated to dryness, hydrogen chloride is lost through hydrolysis of magnesium chloride. This, and the fact that the dried salts retain water tenaciously, makes it difficult to determine salinity directly. Until comparatively recently, salinity has usually been calculated from measurements of chlorinity by application of the well-known equation

$$\text{Salinity} = 0.03 + 1.8050\ \text{Cl}\%_0$$

derived by Knudsen (1901) from the salinity data of Sørensen (1902) for mine water samples. These data were obtained using an elaborate and time-consuming gravimetric technique in which compensation was made for loss of hydrogen chloride.

The use of chlorinity for the estimation of salinity has, in recent years, been largely superseded by methods based on physical measurements, principally electrical conductivity. These procedures, which have been discussed in Chapter 6, have the advantages of cheapness, precision, speed, and convenience for use on board ship. However, it should be pointed out that they do require calibration with Standard Sea Water which, of course, has itself been volumetrically standardized against pure silver using the Volhard method (Jacobsen and Knudsen, 1940). These developments have led to a reappraisal of the relationships between the chlorinity, salinity and electrical conductivity of sea water. As a result of this, a new equation has been adopted relating chlorinity to salinity (Wooster *et al.*, 1969)

$$\text{Salinity} = 1.80655\ \text{Cl}\%_{oo}.$$

When saline waters of anomalous composition are being examined it may be useful to make a direct measurement of salinity. This can be carried out by Morris and Riley's (1964) modification of the gravimetric method of Guntz and Kocher (1952). In this, loss of hydrogen chloride by hydrolysis of magnesium chloride is prevented by precipitating the magnesium by adding sodium fluoride before the evaporation. The decrepitation which occurs when sea salts are dried is inhibited by precipitating the salts as fine crystals by addition of alcohol to the concentrated brine. After completion of the evaporation, the salts must be heated to a temperature of 650°C in order to remove the last traces of water. If precautions are taken to prevent uptake of water by the very hygroscopic salts a coefficient of variation of ca. $\pm 0.03\%$ can be achieved. With normal sea waters, the results obtained agree with the salinity calculated from the equation of Wooster *et al.*, (1969) to within $\pm 0.02\%$, provided that correction is made for the failure to convert bromide to the equivalent amount of chloride, as required by the definition of salinity.

This method has been used for the examination of the waters of the Baltic (Kremling, 1969) and the Suez Canal (Morcos and Riley, 1966).

DETERMINATION OF CHLORINITY

The determination of chlorinity is now of considerably less importance than it was ten years ago as a result of the development of reliable conducti-metric salinometers. These are not only several times more rapid, but enable salinity to be determined by relatively unskilled personnel with a precision of $\pm 0.003\%$. For this reason, it has not been thought appropriate to review the literature on the determination of chlorinity here. Earlier work on this subject has been summarized by Riley (1965b) and by Johnston (1969), and recent developments are described in Chapter 6 of the present book; electrochemical techniques are also discussed in Chapter 20.

19.7. DETERMINATION OF TOTAL CATIONS BY ION EXCHANGE

Carritt (1962) has described a method for the determination of the total cation concentration of sea water (as meq kg^{-1}) (see also Culkin and Cox, 1966). In this, a known weight of the sample is passed through a bed of the hydrogen form of a strongly acidic cation exchange resin. The resin is washed with water and the liberated acid is titrated with standard sodium hydroxide solution (carbonate free). Carritt has suggested that this may be a better method of assessing the property which salinity represents than a measurement of chlorinity.

19.8. DETERMINATION OF MAJOR COMPONENTS

19.8.1. INTRODUCTION

The major components of sea water have been defined by Culkin (1965) as those which contribute significantly to salinity. This therefore implies those components having concentrations of 1 mg kg^{-1} or greater, i.e., the cations of sodium, potassium, calcium, magnesium and strontium and the anions chloride, sulphate, bromide, bicarbonate and fluoride, together with boric acid.* These constituents† are generally considered to be con-

* Except of course bicarbonate.
† Silicate is not included among the major components of sea water, although it may sometimes reach concentrations of 4 mg Si kg^{-1}, as it does not exhibit conservative behaviour. Because of its involvement in the biosphere it is more appropriately included among the micronutrient elements.

servative, or almost so, in behaviour. In order to make meaningful studies of the variations of the ratios of these ions to chlorinity in the sea it is therefore essentially to use analytical methods capable of high precision and accuracy. For the more abundant of the ions a precision of at least $\pm 0.15\%$ is essential. However, for boron and fluorine a precision of $\pm 1\%$ is probably acceptable. The determination of the major ions with precisions lying within these ranges is probably one of the more difficult tasks confronting the marine chemist.

The widespread adoption of conductimetric salinometers has led to a revival of interest in the salinity–chlorinity–conductivity relationship of sea water (see Chapter 6). This, in its turn, has given a fresh impetus to the investigation of the constancy of the ratios of the major ions to chlorinity. Very few complete analyses of sea water for all its major components have been published since the classical work of Dittmar (1884). Most workers in this field have, instead, been content to study the ratios of single elements to chlorinity. The literature on the determination of the major components of sea water, much of which is only of historical interest, has been comprehensively surveyed by Culkin (1965). In the present review stress will be laid on the considerable developments which have occurred over the last 10–15 years, as a result of the development of new reagents and improvements in instrumental techniques. Older procedures will, however, be discussed where they offer advantages.

19.8.2. DETERMINATION OF MAJOR CATIONS

Before schemes for the complete analysis for the major cations are described, methods for the determination of the individual cations will be discussed. It should be pointed out that the precision attainable by flame photometric and atomic absorption procedures (at best $+1\%$) and with specific ion electrodes (see Chapter 20) falls far short of that required for the determination of most of the major cations, and even for strontium it is barely adequate. It seems highly probable that the very large variations in the major cation/ chlorinity ratios reported by some recent workers (e.g. Billings et al., 1969) result from failure to appreciate the limits of these techniques, particularly when applied directly to sea water.

Ion exchange separation of major cations. Because of mutual interferences there are considerable advantages in separating the major cations from one another before determining them. Greenhalgh et al. (1966) have described an ion exchange scheme for this purpose. (see also Riley and Tongudai, 1966a). The sample (30 ml) was adsorbed on a 18×0.6 cm column of

Amberlite CG.120 (100–200 mesh). Sodium was eluted together with potassium using 200 ml of 0·15 M ammonium chloride. Elution of magnesium was carried out with 450 ml of 0·35 M ammonium chloride, after which calcium was removed with 75 ml of 1 M ammonium acetyl acetonate as suggested by Carpenter (1957). Finally, the column was washed with water and strontium was eluted with 100 ml of 2 M nitric acid. Separations were clear-cut and recoveries were better than 99·9 %. A slightly modified version of this scheme has been adopted by Morcos (1968) for the analysis of high salinity waters.

19.8.2.1. *Sodium*

Because of the lack of accurate and specific methods, the determination of sodium in sea water has usually been carried out by difference (the gravimetric technique used by Webb (1939) and by Knapman and Robinson (1941), in which sodium is weighed as sodium zinc uranyl acetate, has little to recommend it becase of the high solubility of the precipitate, $58·5$ g kg^{-1}). Direct methods of analysis are usually to be preferred to those based on differences as the precision of the latter will be influenced by the errors in the analyses of the other relevant components. However, since the concentration of sodium in sea water is so much greater than that of the other major cations ($Mg/Na = 0·1201$, $Ca/N = 0·038$, $K/Na = 0·037$) this is not a serious objection in this instance.

Dittmar (1884) determined the combined sulphates of the alkali and alkaline earth metals gravimetrically. Calcium, magnesium and potassium were estimated individually, and sodium was determined by difference. Other workers have determined total sodium + potassium as sulphates or chlorides after removal of the alkaline earth metals. After potassium has been determined in the sample, or in the sulphate or chloride mixture, sodium was estimated by difference. Usually the separation of the alkaline earth metals has been carried out by precipitating calcium as CaC_2O_4 and magnesium as NH_4MgPO_4. In both instances considerable coprecipitation of the alkali metals occur. Although this difficulty can be overcome by reprecipitating the alkaline earth metals, this is at the expense of the contamination of the alkali metal fraction with calcium and magnesium as the precipitates are appreciably soluble. A much simpler and more clear cut separation of the alkalis from the alkaline earth metals can be achieved by the use of ion exchange procedures (see e.g. Bhavnagary and Krishnaswamy, 1967). In their scheme for the determination of the major cations in sea water Greenhalgh et al. (1966) concentrated sodium and potassium by ion exchange prior to weighing them as sulphates; sodium was then determined by difference after potassium had been estimated gravimetrically (see also p. 229).

A rather different approach to the use of ion exchange for the indirect determination of sodium has been employed by Culkin and Cox (1966) in their definitive work on the ratios of the major cations to chlorinity in sea water. Total cations were determined by passing the sample through a column of cation exchanger in its hydrogen form and titrating the liberated acid. Potassium, calcium and magnesium were then determined separately and sodium was estimated by difference.

19.8.2.2. Potassium

Potassium has been determined in sea water by chemical, spectrochemical and difference chromatographic methods*. The latter technique, which has been described on p. 240, is rapid and is particularly suitable for studies of variations in the $K/Cl\%_0$ ratio in the oceans as it is capable of a precision of better than $\pm 0.1\%$. (Mangelsdorf and Wilson, 1971).

19.8.2.2.1. *Gravimetric and titrimetric techniques.* The choice of reagents for the quantitative precipitation of potassium is very limited. Their application to the determination of the element in sea water is complicated by its comparatively low concentration and the very adverse sodium to potassium ratio (26.8:1). Before 1956 the favoured precipitation forms were potassium hexachloroplatinate (K_2PtCl_6), potassium perchlorate, and the cobaltinitrites $K_2NaCo(NO_2)_6$ or $K_2AgCo(NO_2)_6$.

All of them have considerable disadvantages (see Culkin, 1965) and the advent of sodium tetraphenyl boron ($NaB(C_6H_5)_4$) as a precipitant for potassium (Wittig, *et al.*, 1949, Wittig, 1950) has rendered them obsolete. Gloss (1953) has summarized the advantages of this excellent reagent:

(i) sodium tetraphenyl boron is very soluble, whereas the solubility of the potassium salt (54 mg l^{-1}) is very low in comparison with those of other potassium salts,

(ii) the composition of the potassium salt corresponds exactly with $KB(C_6H_5)_4$, and the gravimetric factor is high,

(iii) the potassium salt is stable up to at least $200°C$ and it can therefore be dried at $120°C$ without risk of decomposition,

(iv) the precipitate is easy to filter,

(v) if the precipitation is carried out from a dilute acid solution, none of the common cations or anions interfere.

---

* Although potassium-selective valinomycin electrodes are available (Stefanac and Simon, 1967) they are too sensitive to sodium to be of any value for the determination of potassium in sea water. Even with a computer processed standard addition technique T. Anfält and D. Jagner (personal communication) were unable to obtain a coefficient of variation of better than 2.6% using this electrode.

Normally, precipitation is carried out from dilute mineral acid medium (<0.1 N) to produce a precipitate which has good filtering characteristics and to minimize interference from other cations. Unfortunately, the reagent tends to decompose under acid conditions at a rate which increases with the acidity. There is some difference of opinion about the optimum acidity for the precipitation (Kallmann, 1961). However, a normality of ca. 0.03–0.1 N appears to be satisfactory, particularly if the solution is cooled to $<3°C$ to retard decomposition of the reagent.

Sodium tetraphenyl boron was first applied to the gravimetric determination of potassium in sea water by Sporek (1956). He carried out the precipitation at $0°C$ from a medium 0.44 N with respect to hydrochloric acid. A coefficient of variation of $\pm0.6\%$ was claimed for the method which gave results which were ca. 1% too high, either because of interference from other cations, or because of contamination of the precipitate with decomposition products of the reagent. Most subsequent workers have favoured lower acidities for the precipitation in order to minimize decomposition of the reagent. Thus, Culkin and Cox (1966) and Riley and Tongudai (1966b) who carried out the precipitation from 0.03 N and 0.11 N hydrochloric acid medium respectively, obtained quantitative recoveries of potassium from synthetic sea waters. The gravimetric tetraphenyl boron method is probably second only to the difference chromatographic technique (see p. 240) in its suitability for use in studies of the potassium/chlorinity ratio in sea water. In skilled hands a precision of ca. $\pm0.2\%$ can be attained. The technique has been used for the analysis of sea water, both directly (Culkin and Cox, 1966; Bauman and Tagliatti, 1964; Kremling 1969) or using concentrates of the alkali metals obtained by ion exchange (Riley and Tongudai, 1966b, 1967; Morcos, 1968).

Sodium tetraphenylboron has also been used for the volumetric determination of potasium in sea water. Viswanathan et al. (1965) quantitatively precipitated the potassium derivative. The precipitate was taken up in acetone, silver nitrate was added and the excess silver was determined by a Volhard titration with ammonium thiocyanate. A coefficient of variation of $\pm0.36\%$ was claimed

$$KB(C_6H_5)_4 + Ag^+ \rightleftharpoons K^+ + AgB(C_6H_5)_4$$
$$Ag^+ + CNS^- \rightleftharpoons Ag\,CNS \downarrow$$

for this procedure. The same workers have also described an alternative, but less precise, procedure in which the solution of the potassium tetraphenyl boron was passed through a column of a cation exchanger in its hydrogen form. The hydrogen ion liberated was titrated with standard alkali solution. Murakami (1965) has described a procedure based on that of Schall (1957) in which potassium was precipitated from sea water using a slight excess of sodium tetraphenylboron. After removal of the precipitate, the excess reagent was determined by titration with a solution of benzene alkonium chloride with which it forms an insoluble compound. Sunchromine Blue FBG or Chromazurol S were recommended as indicators. Anfält and Jagner (1973) have employed a potassium selective electrode for the detection of the end point in the direct titration of K in sea water with Na B(C_6H_5)$_4$. A coefficient of variation of ca. 0·27% was claimed (see Section 20.3.3.2).

19.8.2.2.2. *Spectrophotometric methods.* Several attempts have been made to use flame and atomic absorption spectrophotometric methods for the determination of potassium in sea water (see e.g. Chow, 1964; Fabricand *et al.*, 1966; Buyanov and Anisimov, 1966; Viswanathan *et al.*, 1969; Romanov and Eremeeva, 1970). The irreproducible nature of the flame and of the atomization process limits their precision to ±1%, at best, (see p. 300) which is nearly an order of magnitude greater than the expected range of variations in the potassium/chlorinity ratios in sea waters. The application of these techniques to the determination of potassium in sea water is therefore not to be recommended.

19.8.2.3. *Calcium and magnesium*
Since the determinations of calcium and magnesium in sea water are closely related, they will for convenience be discussed under the same heading.

19.8.2.3.1. *Gravimetric determination* In most of the early work on the determination of calcium and magnesium in sea water, (see Culkin, 1965), calcium was precipitated from sea water as calcium oxalate and determined either gravimetrically by ignition to calcium oxide, or volumetrically by titration with permanganate. Magnesium was subsequently precipitated as magnesium ammonium phosphate and ignited to magnesium pyrophosphate for weighing. Considerable coprecipitation of other ions occurs during the precipitation of both elements. The calcium oxalate carries sodium and oxalate ions and the magnesium ammonium phosphate coprecipitates potassium, ammonium and phosphate ions. Although these adsorbed ions can be eliminated by carrying out time-consuming multiple precipitations, this introduces serious errors arising from the solubility of the precipitate and transference losses. An added uncertainty is introduced into the determination of calcium owing to the fact that strontium is also partially precipitated with the calcium oxalate (Carpenter, 1957). Because of their tediousness and the difficulty of attaining acceptable precision, these gravimetric procedures have not been used in recent years for the analysis of sea water, and they have been replaced by much more satisfactory methods.

19.8.2.3.2. *Volumetric determination* The accurate determination of calcium and magnesium in sea water has been much simplified following the development by Schwarzenbach and his co-workers of methods in which these elements are titrated with a solution of a chelating agent (*complexone*) containing the characteristic grouping —$N(CH_2COOH)_2$ (Kampitsch, *et al.*, 1945). The most important of these compounds is ethylenediamine-N,N,N',N-tetra-acetic acid (EDTA) which is usually used as its more soluble di-sodium salt. This complexing agent forms strong chelates with many polyvalent cations. In an alkaline medium it gives rise to stable complexes with magnesium, calcium and strontium having log stability constants of 8·69, 10·70 and 8·63 respectively at 20°C and $\mu = 0\cdot1$ (Schwarzenbach, 1957). When one of these elements is titrated in alkaline medium with the di-sodium salt of EDTA, the concentration of the free ions of the element will decrease as they are chelated. This change in concentration can be observed by using a mercury electrode as an indicator for pM ($= -\log$ [Metal ion] or, strictly speaking $= -\log a_{metal}$). A plot of pM against EDTA titre is sigmoidal in shape (Fig. 19.3) and resembles a pH titration curve.

Although the EDTA titration can be carried out potentiometrically, the end point can be more satisfactorily detected by means of metal indicators. These are dyestuffs which form coloured chelates with metal cations and serve as indicators for pM values, in an analogous fashion to the response of acid-based indicators to changes of pH. Metal indicators to be used in EDTA titrations must possess the following features (Reilly and Schmid, 1959); (i) the colour of the complex formed between the metal and the indicator must differ from that of the free indicator, (ii) the metal complex must have a lower stability than the metal–EDTA complex, (iii) both of the complexes must be formed rapidly.

Most of the metal-indicators which have been used in the titration of calcium and magnesium with EDTA are *o, o'*-dihydroxyazo derivatives of naphthalene. Of these, Eriochrome Black T (EBT) has been most extensively used because of the suitability of the stability constants of the complexes which it forms with these elements. In this compound, I, the sulphonic acid group is completely ionized over the whole pH range of analytical interest and only the dissociation of the two phenolic acid groups needs to be considered. The pK values for the ionization of these are 6.3 and 11.55 (Schwarzenbach and Biedermann, 1948). Over the pH range 7–11 the indicator therefore exists principally as the blue double charged anion HE^{2-}, in which only one of these groups is ionized. Addition of calcium or magnesium to a solution of

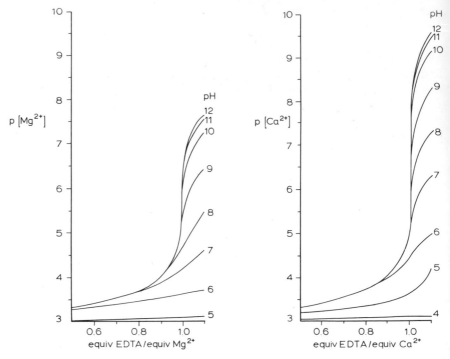

FIG. 19.3. Titration curves of magnesium and calcium with EDTA (after Schwarzenbach, 1957).

the indicator, buffered within the pH range 8–11, yields relatively weak chelate complexes of the formula CaE^- and MgE^-. These complexes have an intense red colour (the molar absorbance of the magnesium complex is ca. 28,100 at 545 nm.). The stability of the calcium complex is much lower than that of the magnesium complex (log stability constants at pH 9 and μ − ·0·1 of 2·85 and 4·45 respectively at 20°C).

Eriochrome Black T (H_2E^-)

$$H_2E^- \rightleftharpoons HE^{2-} + H^+ \rightleftharpoons E^{3-} + H^+$$

During the initial stages of the titration of magnesium or calcium at pH 9 with *di*-sodium EDTA, using EBT as indicator, the solution has a red colour imparted by the EBT complexes of these metals. As the titration continues the ions of the metals are chelated by the EDTA. When the equivalence point of the titration is approached the concentration of the free metal ions will tend to zero. The stability constants of the EDTA complexes of these metals are several orders of magnitude greater than those of their EBT complexes. Further addition of EDTA to the solution will, therefore, lead to the breakdown of the EBT-metal complex and to the liberation of the free indicator ion. When the end-point of the titration is reached the colour of the solution will change from red to blue. Although this colour change is reasonably sharp to the eye, for the greatest precision the titration should be carried out photometrically (see e.g. Headridge, 1961). It should be noted that the stability constant of the Ca-EBT complex is so low that the indicator is of little use for the titration of calcium alone. For this reason, if calcium is to be titrated in the absence of magnesium, a small amount of the Mg–EDTA chelate must be added to the solution. During the titration the EDTA complexes preferentially, and the end-point is much sharper as it is virtually that obtained with magnesium alone.

There is at present no procedure for the direct EDTA titration of magnesium in sea water.* A few workers (Carpenter, 1957; Riley and Tongudai, 1966b; Morcos, 1968) have preferred to separate the element by ion exchange before titrating it. However, the majority have estimated it, probably less satisfactorily, by difference. Total magnesium + calcium + strontium are determined in the sample by EDTA titration at pH 10 using Eriochrome Black T as indicator. Calcium + strontium are determined by titration with EDTA after precipitating magnesium as its hydroxide at a $pH > 12$.

The EDTA titration of the total alkaline earths, or of magnesium after ion exchange separation, is normally carried out in a medium buffered with ammonia–ammonium chloride to pH 10.0–10.5. Eriochrome Black T is probably the most suitable indicator. Visual detection of the end point is reasonably satisfactory and has been widely used, (Skopintsev and Kobanov, 1958; Pate and Robinson, 1961; Okuda, 1964; Rohde, 1966; Traganza and Szabo, 1967; Mameli and Mosetti, 1966; Kremling 1969; Shah and Gogate, 1969; Lyakhin, 1971; Atwood *et al.* 1973). However, the precision is considerably improved if the titration is carried out photometrically (Carpenter, 1957; Culkin and Cox, 1966; Riley and Tongudai, 1966b, 1967; Morcos, 1968; Kremling, 1969). Jagner and Årén (1971) have described a photometric

* Cheng and Cheng (1971) have given preliminary details of a complexometric method using EGTA for the selective titration of magnesium in sea water, in which the end-point is detected with a specific ion electrode (p. 236).

procedure in which the absorbance–titre data is recorded automatically and the end point is evaluated by means of a computer processed Gran plot method. For this they claim a coefficient of variation of 0·01%.

Before calcium can be determined in sea water by EDTA titration it is necessary to remove magnesium. Many workers have carried out the titration at a pH of greater than 12 (sodium hydroxide, or diethylamine buffer). Under these conditions magnesium is precipitated as its hydroxide which does not react with EDTA. This method suffers from a number of drawbacks:

(i) calcium is coprecipitated to an extent of up to ca. 1% by the magnesium hydroxide (Voipio, 1959). Pate and Robinson (1958) have claimed that this loss can be reduced by addition of 95% of the theoretical amount of EDTA before making alkaline. However, there is considerable doubt whether this does more than slightly reduced the proportion of calcium coprecipitated. Thus, Lyakhin (1971) has found that more accurate results can be obtained if the sample is buffered precisely to pH 12 than if Pate and Robinson's procedure is used,

(ii) strontium is not removed during the precipitation, and in precise work it is necessary to correct the calcium figure to allow for this,

(iii) Eriochrome Black T cannot be used as indicator in the titration, not only because of the low stability of its calcium complex, but also because at pH 12 and above it exists in the completely ionized form (E^{3-}, see p. 232). The yellow-brown colour of this species is similar to that of the Ca-EBT complex. A number of other metal indicators have been used in the titration (Table 19.3). However, it is probably true to say that none of them gives as sharp an end point as can be achieved with Eriochrome Black T at pH 10 in the presence of magnesium (see p. 233),

(iv) adsorption of the indicator by the magnesium hydroxide causes the indicator colour to fade and makes it difficult to detect the end-point of the titration accurately either visually or photometrically.

Because of the difficulties inherent in the above technique, a number of workers have separated calcium from magnesium by other methods before titrating it with EDTA. Shah and Gogate (1969) carried out a double precipitation of calcium as its oxalate, dissolved the precipitate in excess EDTA and back-titrated with standard magnesium solution. However, this method has little to recommend it because of the risk of loss of calcium during the manipulation and through the appreciable solubility of the precipitate. A much more satisfactory separation can be achieved by cation exchange procedures. They have the advantage not only of giving quantitative separation of calcium from magnesium, but also of providing a separation of the former from strontium. It is then possible to use Eriochrome Black T as the indicator for the EDTA titration of calcium as well as magnesium.

TABLE 19.3

Indicators used for the EDTA titration of calcium in sea and estuarine waters

Metal indicator	Reference
Sea waters	
Murexide	de Sousa (1954); Skopintsev et al. (1957);
(ammonium purpurate)	Skopintsev and Kobanov (1958);
	Carpenter (1957)[1]; Okuda (1964); May (1964);
	Corless (1965)[2]; Kato (1966); Kremling (1969);
	Lyakhin (1971).
Calcon	Voipio (1959); Atwood et al. (1973);
Calver II	McIntyre and Platford (1964);
Calred	Pate and Robinson (1958); Rohde (1966)[3];
	Szabo (1967); Traganza and Szabo (1967);
	Mameli and Mosetti (1966, 1967);
	Bojanowski and Ostrowski (1968).
Estuarine waters	
Calcein	Rial and Molins (1962).
Eriochrome Blue S.E.	Price and Priddy (1961).
Sea waters after separation from magnesium	
Eriochrome Black T	Riley and Tongudai (1966b, 1967);[1]
	Shah and Gogate (1969)[4]; Morcos (1968)[1];
	Carpenter and Manella (1973)[5]

[1] Used in photometric titration after separation of magnesium by ion exchange.
[2] Calcium separated from magnesium by ion exchange prior to titration.
[3] Photometric titration.
[4] Calcium and strontium precipitated as oxalate, precipitate dissolved in excess EDTA, the excess of which was titrated with standard magnesium solution.
[5] Photometric titration carried out with 1,2-diamino-cyclohexane tetraacetic acid after separation of magnesium by ion exchange.

Ion exchange was first applied to the analytical separation of the alkaline earth elements in sea water by Carpenter (1957) (see also Carpenter and Manella, 1973). The cations were adsorbed onto a cross-linked sulphonated polystyrene ion exchange resin. Magnesium was eluted with 0·2 m ammonium acetylacetonate, and calcium was then removed with 1·0 m ammonium acetylacetonate. Titration of calcium with EDTA was carried out after decomposition of the acetylacetonate by heating. The same method of separating calcium from strontium was used by Riley and Tongudai (1966b) in their scheme for the determination of the major cations in sea water (see p. 226). However, magnesium was eluted by means of 0·35 M ammonium chloride solution instead of ammonium acetylacetonate solution. These workers experienced difficulty in decomposing the latter reagent and found it necessary to evaporate the calcium eluate to dryness with nitric acid before titrating with EDTA. Coefficients of variation of $\pm 0.08\%$ and $\pm 0.04\%$ were claimed for calcium and magnesium respectively. A slightly modified version of this scheme has been used by Morcos (1968) for the analysis of high salinity waters. Tsubota and Kitano (1960) have used an ammonium formate–formic acid buffer (pH4) for the successive elution of calcium and magnesium in sea water analysis; however, the separation is not as satisfactory as it is with the above procedures. Corless (1965) has described a student experiment for the determination of calcium in sea water based on cation exchange sepation and EDTA titration.

A considerable amount of research has been devoted to the development of selective chelating agents based on the nitrilodiacetic acid grouping. One of the results of this was the discovery of a complexone (1,2-*bis*-[2-*di*-(carboxymethyl)-aminoethoxy]-ethane = EGTA), which forms a strong complex with calcium (log $K = 10.0$), but only a weak one with magnesium (log $K = 5.2$) (Schwarzenbach, *et al.*, 1957). This enables this compound to be used for the titration of calcium at pH9–10 in the presence of magnesium. The end-point of the titration is difficult to detect directly because of the lack of a specific colour indicator for calcium. Dyrssen *et al.* (1969) attempted to detect the end-point by means of a calcium selective membrane electrode, but were unsuccessful because of interference from sodium and magnesium; later attempts to use a variety calcium selective electrodes in the titration have also been unsuccessful (see Section 20.3.2.2). Ringbom *et al.* (1958) have suggested an indirect indicator consisting of zincon (HIn^{3-} which forms a weak blue coloured complex ($ZnIn^{2-}$) with zinc) and the zinc-EGTA complex. The reaction occurring at the end-point can be represented by the equation:

$$Ca^{2+} + HIn^{3-} + Zn\text{–}EGTA^{2-} \rightleftharpoons Ca\text{–}EGTA^{2-} + ZnIn^{2-} + H^{+}$$

orange   blue

This procedure has been adapted for the photometric titration of calcium in sea water by Cox and Culkin (1966). These workers found that magnesium and strontium at the concentrations at which they occur in sea water increased the calcium titre by 0.39% and 0.73% respectively. They therefore corrected for this slight interference by multiplying their results by an empirical factor of 0.989.

Chelatometric titration with EGTA has been used in a novel method for the determination of calcium in sea water with a coefficient of variation of $\pm 0.06\%$ (Tsunogai *et al.*, 1968a, b) In this calcium is extracted at pH 11.7 as its red complex with glyoxal-*bis*-(2-hydroxy-anil) (GHA) using *n*-amyl alcohol. The two phases are then titrated, without separation, using alkaline

EGTA solution. The disappearance of the red colour of the Ca-GHA complex from the organic layer is used to indicate the end-point. Some interference is caused by magnesium and strontium; correction can be made for this by multiplication of the calcium results by 0·9946.

Carpenter and Manella (1973) have pointed out that considerably sharper end points in the titration of magnesium can be achieved by the substitution of 1,2-diamino-*cyclo*-hexane tetraacetic acid (CDTA) for EDTA. This is because the stability constant of the MgCDTA complex is appreciably greater than that of the corresponding EDTA complex. They have used this reagent for the photometric titration of magnesium preconcentrated from sea water by ion exchange; a coefficient of variation of $\pm0·03\%$ was claimed.

Jagner (1971) has attempted to determine magnesium in sea water by a direct volumetric method. The sample is acidified and boiled to remove carbon dioxide. After cooling and addition of mannitol to complex boric acid it is brought to pH7 and then titrated with sodium hydroxide. The titration is followed by means of a glass electrode

$$Mg^{2+} + 2OH^- \rightarrow Mg(OH)_2$$

Although the reproducibility of the method was satisfactory (coefficient of variation $= \pm0·16\%$), results with standard sea water were ca. 1% lower than those generally accepted.

19.8.2.3.3. *Determination by difference chromatography.* The speed and relatively high precision of difference chromatography (see p. 240) makes it potentially a very useful technique for use in an investigation of the variations of the Ca/Cl‰ and Mg/Cl‰ ratios in the oceans. Mangelsdorf and Wilson (1969) have claimed to be able to detect differences in concentration of as little as 0·15% magnesium and 0·05% calcium relative to a standard sea water.

19.8.2.3.4. *Flame and atomic spectrophotometric determination.* Flame and atomic absorption spectrophotometric methods (see p. 300) are of little value in studies of the variations of the Ca/Cl‰ and Mg/Cl‰ ratios in the oceans since their precisions (at best $\pm1\%$) are similar to the maximum differences likely to be encountered. The very large spreads for these ratios found by some workers using these techniques (e.g. Billings *et al.*, 1969; Viswanathan *et al.*, 1969) are almost certainly due to analytical errors. They emphasise the undesirabilty of using these methods for the determination of these elements in sea water.

19.8.2.4. *Strontium*

Strontium is difficult to determine in sea water by classical analytical techniques because of its relatively low concentration, its resemblance to calcium, and its lack of specific reactions. For this reason, most of the determinations using these techniques are mainly of historical interest (see Culkin, 1965). The advent of specific physico-chemical methods has much simplified the estimation of the element in sea water. However, there must be some doubt about the reliability of the results obtained with some of them since some workers have claimed that strontium behaves more or less conservatively, whereas others have stated that the $Sr/Cl‰$ ratio varies considerably. Although some of the more extreme variations can be ascribed to the use of faulty analytical techniques, there are still grounds for controversy, and further efforts are required to develop methods capable of an accuracy of better than $\pm 0.2\%$.

In the limited space available it is not possible to give more than a brief survey of physico-chemical methods for the determination of strontium in sea water. The reader requiring a more thorough treatment of the subject should consult the review by Bowen (1970).

19.8.2.4.1. *Spectrochemical methods.* Both atomic absorption and flame emission spectrophotometric techniques have been used for the determination of strontium in sea water. When they are used for the direct determination of strontium in sea water, interference is caused by some of the major ions present and by silicon (Brass and Turekian, 1972). Correction can be made in two ways for the combined effect of these interferences (i) by calibrating the method with artificial standards matching the salinity and composition of the samples (Angino *et al.,* 1966; Carr, 1970; Brass and Turekian, 1972) (ii) by a standard addition method in which the instrumental response of the sample is measured both alone and spiked with known amounts of strontium. The strontium content is determined from the intercept of the graph relating instrumental response to strontium added (see Fig. 19.9). The latter procedure, which was first applied to the analysis of sea water by Chow and Thompson (1955), has been favoured by most workers both for flame emission (Culkin and Cox, 1966; Tamont'ev and Brujewicz, 1964; Bojanowski and Ostrowski, 1968, Osmolovskaya, 1964) and atomic absorption spectrophotometry (Nagaya, *et al.,* 1971; Viswanathan, *et al.,* 1969; Atwood *et al.* 1973). Problems associated with the atomization of sea water (arising from its high ionic strength) are a major source of irreproducibility in these methods. In order to overcome this difficulty many workers prefer to separate strontium from the other major cations before determining it. Kawasaki and Sugawara (1958) and Odum (1951, 1957) concentrated strontium by coprecipitating it with calcium oxalate. However, serious errors arise from incompleteness of the coprecipitation. A much more satisfactory separation can be achieved by the use of catio exchange procedures (Riley and Tongudai, 1966b, 1967; Angino et al., 1966; Andersen and Hume, 1968a, b, c; Anderson et al., 1970; Noshkin and Mott, 1967; Sednev et al., 1966). In most methods based on this principle, complexones (e.g. EDTA, or cyclohexane dinitrilotetraacetic acid) are used for elution of strontium after calcium has been eluted previously. Considerably better precision can be attained when the ion exchange concentrates are sprayed into the flame of the spectrophotometer rather than sea water itself; thus, Andersen and Hume (1968a) have obtained a coefficient of variation of $\pm$ 0·5% for the flame emission determination of strontium following ion exchange separation. However, since the results which they obtained are appreciably lower than those reported by most recent workers there may be some source of systematic error in their procedure.

19.8.2.4.2. *Neutron activation procedures.* Neutron activation was first applied to the determination of strontium in sea water by Hummel and Smales (1956) who irradiated sea water with thermal neutrons. Inert strontium was then added. Strontium was separated as sulphate and weighed to determine the chemical yield. Finally, the radiation from ^{87}Sr or ^{89}Sr was counted and compared with that from irradiated standards. A coefficient of variation of $\pm 4\%$ was claimed for the method. Bowen (1956) and Rao et al. (1964) have used neutron activation to determine strontium in concentrates prepared from sea water by ion exchange and oxalate coprecipitation respectively. Strontium has also been estimated in sea water by counting ^{87m}Sr produced by irradiation of the freeze-dried samples with 30 MeV bremstrahlung; a coefficient of variation of ± 4–5% was claimed (Gordon and Larsen, 1970).

19.8.2.4.3. *Isotope dilution methods.* Isotope dilution techniques using both stable and radioactive nuclides of strontium have been used for the analysis of sea water. In the first of these (Aleksandruk and Stepanov, 1968), sea water was enriched with ^{87}Sr and strontium was separated by ion exchange. The concentrate was transferred to the hollow cathode of a discharge tube and the spectrum was excited. Strontium was determined by comparison of the intensities of the two hyperfine lines at 4078 Å arising from the internal standard ^{87}Sr and ^{88}Sr which constitutes 82·6% of natural strontium. More recently, mass spectrometric isotope dilution techniques have been applied to the determination of strontium in sea water. Small samples of sea water (2 ml) are spiked with known amounts of either ^{84}Sr or ^{85}Sr, and a small volume of the mixture is applied to the filament of a mass spectrometer either directly (Bernat et al., 1972), or after separation of the strontium by ion exchange (Brass and Turekian, 1972). The strontium content of the sample

I

is evaluated from the ratio of the peak intensity of the spiking isotope to that of ^{88}Sr. These techniques offer the greatest precision at present available for the determination of strontium in sea water, and indeed Brass and Turekian (1972) have claimed a coefficient of variation of $\pm 0.2\%$ for their method.

Hummel and Smales (1956) spiked a large sample of sea water with a radioactive strontium nuclide (^{89}Sr); strontium was coprecipitated with calcium carbonate, and after separation from calcium as the nitrate, it was precipitated and weighed as strontium sulphate. Comparison of the activity of the precipitate with that of the ^{89}Sr spike added enabled correction to be made for losses of strontium occurring during processing. A coefficient of variation of $\pm 1\%$ was quoted for this technique.

19.8.2.4.4. *Colorimetric determination*. Uesugi *et al.* (1964) have described a colorimetric procedure for the determination of strontium in sea water. The element is coprecipitated at pH 4 with calcium oxalate. After separation from calcium using ion exchange, strontium is determined photometrically with *o*-cresolphthalein.

19.8.2.5. *Difference chromatography for the determination of major cations*

Studies of possible variations in the major cation/chlorinity ratios in sea water are hampered by the fact that differences are so small that they can only be observed with difficulty even by the most refined chemical techniques. A very important advance has been made by Mangelsdorf (1966) who suggested the application of difference chromatography to this problem. This ion exchange technique was subsequently refined to provide a rapid method of unparalleled sensitivity for the observation of small differences in relative composition between the unknown sample and a reference standard sea water (Mangelsdorf and Wilson, 1969, 1971). The apparatus used is shown diagrammatically in Fig. 19.4. A stream of the reference standard sea water of salinity 27·3‰* is pumped under pressure at a rate of 2 ml/min^{-1} through a 15×0.4 cm column of cation exchanger, such as Amberlite CG 120 (400–600 mesh). The sample, (2 or 4 ml) adjusted to salinity 27·3‰, is injected into this flow. The effluent from the column passes to the detector consisting of an ion exchange membrane cell (Fig. 19.5) of the form

Ag/AgCl Electrode	standard sea water	Anion exchange membrane	Column effluent	Cation exchange membrane	standard sea water	Ag/AgCl electrode

* Sargasso Sea water was employed because it is low in nutrients and therefore relatively sterile. Dilution to salinity 27·3‰ was carried out with deionized water.

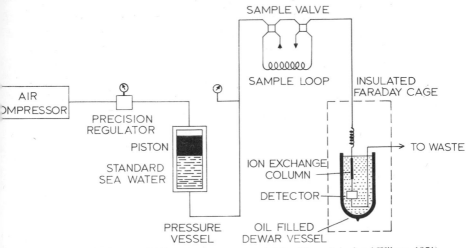

FIG. 19.4. Schematic diagram of difference chromatograph (after Mangelsdorf and Wilson, 1971).

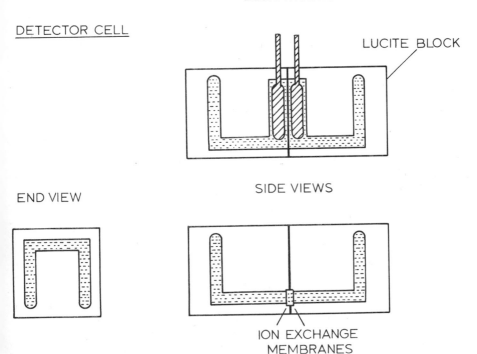

FIG. 19.5. Schematic diagram of membrane cell detector (after Mangelsdorf and Wilson, 1971).

The highly amplified output from the detector cell is fed to a chart recorder. The method is a null one, and if the sample has the same composition as the reference standard, the recorder will give a steady undeflected reading. However, if slight differences do exist, these will be manifested as a set

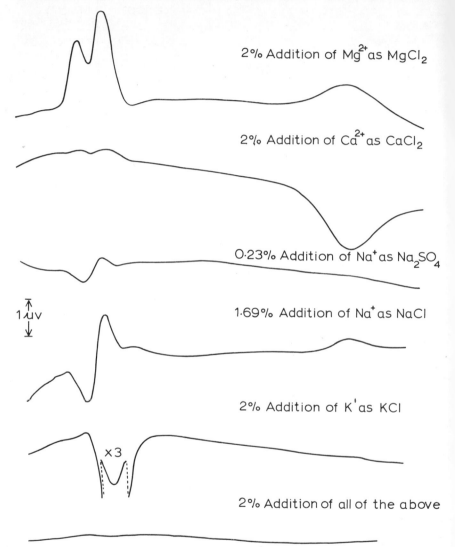

2% Addition of Mg^{2+} as $MgCl_2$

2% Addition of Ca^{2+} as $CaCl_2$

0.23% Addition of Na^+ as Na_2SO_4

1.69% Addition of Na^+ as $NaCl$

2% Addition of K^+ as KCl

2% Addition of all of the above

$1 \mu v$

$\times 3$

FIG. 19.6. Difference chromatograms resulting from the addition of various salts to sea water. The large initial anion peaks have been omitted. However, some residual tailing is present. The time span of each trace is *ca.* 20 min. (after Mangelsdorf and Wilson, 1971).

of composition pulses. These will travel down the column at their characteristic speeds, and are detected as a series of peaks on the recorder. The method is calibrated by injecting samples of the reference standard sea water which have been spiked with the small increments of the chlorides of the various cations. A least squares regression is plotted between the amounts of Ca, Mg and K used as spikes and the corresponding peak heights (Fig. 19.6). Correction must be applied for the anion wave which travels through the column marginally ahead of the cation peaks.

Using this equipment Mangelsdorf and Wilson were able to detect changes in the K/Cl‰, Ca/Cl‰ and Mg/Cl‰ ratios in sea water of $< 0.1 \%$, 0.25% and 0.1% respectively. The sensitivity for strontium was too low for natural variations to be detected. The method has been used by Mangelsdorf *et al.* (1969) to study changes occurring in sea water when it is equilibrated with montmorillonite clay.

19.8.3. DETERMINATION OF MAJOR ANIONS

19.8.3.1. *Sulphate*

The classical gravimetric technique in which the sulphate is precipitated and weighed as barium sulphate has been extensively used for the analysis of sea water since the early part of the 19th century (see e.g. Culkin, 1965). However, it is extremely tedious and subject to a number of serious errors caused by coprecipitation of other ions, principally those of calcium and strontium. Although the coprecipitation of the latter elements can be reduced by carrying out the coprecipitation in the presence of hydrochloric acid, this is at the expense of increasing the solubility of the barium sulphate. Bather and Riley (1954) have found that the most accurate results are obtained if the precipitation is performed from an almost boiling medium in which the hydrochloric acid coprecipitation is $0.06 \, \text{M}$ and which contains picric acid to inhibit coprecipitation of alkali metals. Under these conditions they were able to obtain sulphate recoveries of $100.00 \pm 0.03 \%$ from synthetic sea waters, probably by the counterbalancing of positive and negative errors. This procedure was used subsequently by Morris and Riley (1966) in a study of the $SO_4^{2-}/Cl\%$ ratios of a total of 345 samples of sea water collected from the world oceans which showed an average ratio of 0.1400 ± 0.00022. It has been used subsequently for studies of the ratio in Baltic water by Kremling (1969) who found a coefficient of variation of $\pm 0.16 \%$.

Although the classical gravimetric method can yield results of high accuracy in expert hands, it is very time consuming, and a number of attempts have been made to find alternative procedures mainly based on the precipitation of barium sulphate or the more soluble lead or benzidine sulphates. Volumetric methods, both direct and indirect, have been described by several workers. Macchi *et al.* (1969) titrated sea water with 0·02 M barium chloride solution using thorin as indicator, after first removing cations by ion exchange, diluting with alcohol and adjusting to pH 2·5–4·0. Recoveries of 100·0 ± 0·1% were found if the barium chloride solution was standardized with known amounts of sulphate ion in the presence of the same amount of chloride as would be present during the analysis of samples. A coefficient of variation of ±0·13% was claimed for the method. Several other indicators, e.g. carboxyarsenazo, are potentially useful in this titration in place of thorin (see, for example Basargin and Nogina, 1969). Attempts to directly titrate sea water potentiometrically with either barium or lead ions appear to have been unsuccessful. Thus, Jagner (1970) was unable to detect the end point in the titration of sea water with barium nitrate solution by means of a silver sulphide membrane electrode which had been specially treated to make it responsive to sulphate ion activity. Chloride interferes seriously in the potentiometric titration of sulphate with lead nitrate using a lead elective electrode (Ross and Frant, 1969). However, Mascini (1973) has been able to use this technique with a precision of ca. ±4% after first removing chloride by means of a cation exchanger in its silver form (see Section 20.3.2.4). Thermometric techniques have been successfully used by Millero *et al.* (1974) and Stainton (1974) for detection of the end point in direct titration of sea water with barium chloride. Precisions of ±0·3% were claimed. Since other major cations are coprecipitated by the barium sulphate produced, results will be ca. 1% low unless the titrant is standardized against standard sea water.

The indirect determination of sulphate in sea water has been carried out by precipitating it as an insoluble derivative, and either dissolving the precipitate and titrating it, or titrating the excess of the precipitating agent after removal of the precipitate. The precision of these techniques is usually inferior to that of the gravimetric method, errors due to coprecipitation and solubility of the precipitate being augmented by those arising from the titration, the end point of which is often not very sharp.

The first of these approaches was used by del Riego (1965) who after removal of chloride ion and addition of alcohol treated the sample with lead acetate. The resultant precipitate was dissolved in a known amount of ethylene diaminetetraacetic acid solution, and the excess of the latter remaining was determined by titration with a standard zinc solution at pH 10 using Eriochrome Black T as indicator. Page and Spurlock (1965) precipitated the sulphate from the cation-free sample as the moderately insoluble benzidine sulphate which, after separation by filtration, was titrated with standard sodium hydroxide solution. Coefficients of variation of ±0·5–2·8% and ±1·4% were claimed for these procedures respectively.

In the alternative indirect procedure a known, (excess) amount of barium ion is added to the sample. After removal of the precipitated barium sulphate, the barium remaining in the solution is determined by titration with EDTA using Eriochrome Black T as indicator. The amount of barium used in precipitating the sulphate in the sample is assessed by difference. Since calcium and magnesium are also chelated by EDTA, both of these elements must be removed from the sample by cation exchange before carrying out the determination (Page and Spurlock, 1965), or less satisfactory, the sample must be titrated directly with EDTA in order to make a correction for their presence (Kwiecinski, 1965).

An indirect polagraphic technique for the determination of sulphate in sea water has been described by Berge and Brügmann (1970) which it is claimed has a coefficient of variation of $\pm 0.5\%$—only slightly inferior to that of the gravimetric method. The sea water sample is treated with a known excess amount of a solution of barium chromate in perchloric acid. After making alkaline, the mixed precipitate of barium sulphate and barium chromate is filtered off. The liberated chromate ion which is equivalent to the sulphate originally present is determined polarographically in the filtrate.

An entirely new approach to the determination of sulphate in sea water has been made by Jagner (1970), who has developed rapid manual and automatic titrimetric methods with precisions superior to that of the gravimetric procedure. It is based on the observation that, because of hydrogen bonding, the hydrogen sulphate ion has a very high stability constant ($pK_a = 14.5$; Kolthoff and Chantooni, 1968) in dimethyl sulphoxide ($(CH_3)_2SO$), a solvent which itself is practically devoid of acidic properties ($pK_a = 1.04$). This enables sulphate to be titrated in this medium using hydrochloric acid $H^+ + SO_4^{2-} \rightarrow HSO_4^-$. In the Jagner (1970) procedure the titration is carried out photometrically using bromocresol green.

A novel direct titrimetric method for the determination of sulphate in sea water has been suggested by Dollman (1968). The diluted sample, containing 5–10 mg SO_4^{2-} is washed through a column of the hydrogen form of the strongly acidic cation exchanger Dowex 50. The percolate, in which the sea water cations have been substituted by an equivalent amount of hydrogen ion, is evaporated at 75°C until all the volatile acids (e.g. HCl and HBr) have been driven off. The sulphuric acid that remains is titrated with standard sodium hydroxide using bromocresol green as indicator. No data are given from which the precision can be assessed.

Jasinski and Trachtenberg (1973) have described a procedure for the determination of sulphate in sea water in which advantage was taken of the strong complexes formed between Fe(III) and sulphate. The sample was titrated with barium chloride in the presence of Fe^{3+}. The endpoint of the titration was detected using a special glass electrode sensitive to Fe(III) (See Section 20.3.2.4).

19.8.3.2. *Bromine*

Bromine occurs in sea water in the form of the bromide ion. The earlier literature on the distribution of the element in sea water has been summarized by Culkin (1965). The main problem in its determination is the presence of ca. 300 times its own weight of the closely related element chlorine. After a critical survey of published methods for the determination of traces of bromide, Haslam and Moses (1950) concluded that the most accurate procedure for the analysis of sea water is that developed by Kolthoff and Yutzy (1937). In this method, bromide ion is oxidized to bromate ion, in a medium buffered to pH 6·0, by means of hypochlorite solution. Excess of the oxidizing agent is reduced with sodium formate. Potassium iodide and a catalytic amount of ammonium molybdate are added and the solution is then acidified. The iodine liberated (6 equivalents per equivalent of bromide) is titrated with standard thiosulphate solution using starch as indicator.

$$Br^- + 3ClO^- \rightarrow BrO_3^- + 3Cl^-$$

$$BrO_3^- + 9I^- + 6H^+ \rightarrow Br^- + 3I_3^- + 3H_2O$$

This technique was first employed for the analysis of sea waters by Thompson and Korpi (1942). The optimum conditions for oxidation of bromide to bromate in sea water medium have been investigated by Haslam and Moses (1950) and by Morris and Riley (1966). The latter workers found that the oxidation is stoichiometric in the pH range 4·0–7·0 and recommended that the reaction be carried out at pH 6·2 (phosphate buffer). The procedure which they developed gave a bromine recovery of $100·00 \pm 0·06\%$ for synthetic sea waters containing known amounts of bromide and a coefficient of variation of $\pm 0·15\%$. This method has been used by Kremling (1969) in an extensive survey of the $Br/Cl\%_0$ ratio in the Baltic Sea.

A spectrophotometric method for the determination of bromine in sea water has been developed by Saenger (1972). The acidified sample is treated with chloramine T. The elemental bromine liberated is extracted with chloroform and the absorbance of the extract is measured at 400 nm. No data are given by which the precision can be judged.

Berge and Brügmann (1971) have described two rapid polarographic procedures for the determination of bromide in sea water. In the first of these the bromide is oxidized to bromate by means of hypochlorite, the excess of which is subsequently reduced by means of formic acid. The solution is then brought to neutrality by addition of a slight excess of calcium carbonate, and the polarographic wave at the dropping mercury electrode corresponding to the reduction of bromate to bromide ($BrO_3^- + 6H^+ + 6e^- \rightarrow Br^- + 3H_2O$) is recorded. A coefficient of variation of $\pm 0.6\%$ was claimed for this procedure which has a negligible salt error. In the second, bromide ion is oxidized by means of permangate at pH 1·6 to elemental bromine. After reduction of the excess oxidant by disproportionation with Mn^{2+} and centrifugation of the precipitated manganese dioxide, bromine is determined polarographically using a rotating platinum electrode ($Br_2 + 2e^- \rightarrow 2Br^-$ for which $E_{\frac{1}{2}} = +0.725$ V). Perhaps because of the volatility of bromine the precision of this technique is only about one-half of that of the first procedure.

19.8.3.3. *Carbonic acid system*
The determination of the various parameters of the carbon dioxide system has been discussed in Chapters 9 and 20.

19.8.3.4. *Fluorine*
In most sea waters fluorine behaves conservatively, with a F/Cl‰ ratio of ca. $6.7 \pm 0.1 \times 10^{-5}$. However, in some very localized bottom waters the ratio may reach 9.0×10^{-5} (Riley, 1965; see also Section 6.3.2.4). The element occurs in sea water as the fluoride ion, a considerable proportion of which is associated as ion pairs, principally MgF^+ (50%, Brewer, *et al.*, 1970) and CaF^+ (2% Elgquist, 1970). The earliest methods for the determination of fluoride in sea water have been reviewed by Taylor and Thompson (1933). Most of them were at best, only semiquantitative, and it was not until the development of reliable photometric procedures that the estimation of the element was placed on a quantitative basis.

The earliest photometric methods for the determination of fluoride in sea water were based on the ability of the ion to break down the coloured complexes (lakes) formed between some organic dyes, e.g. alizarin red S and thorin, and certain metals, such as Zr and Th, with which fluoride forms much more stable complexes.* The fluoride concentration is assessed from the resultant decrease in the light absorption of the complex. Sulphate and chloride also form complexes with these metals, albeit much weaker ones than those with fluoride. As their concentrations in sea water are several orders of magnitude greater, they interfere by bleaching the lakes; this effect is most pronounced for sulphate. For this reason the method must be calibrated with synthetic sea water containing known concentrations of fluoride (Taylor and Thompson,

---

* Berge and Brügman (1972) have described a procedure for the determination of fluoride in sea water in which the sample is treated with an acidic solution of the Zr-alizarin red S and the alizarin red S which is liberated is determined polarographically (see Section 20.4.1).

1933; Riva, 1966). Alternatively, both the determination and calibration may be carried out in the presence of a swamping excess of sulphuric acid (Anselm and Robinson, 1951).

In 1959 Belcher *et al.* described the first specific colour reaction for fluoride and used it as the basis for a spectrophotometric procedure. In the presence of fluoride the red chelate formed between alizarin complexone (3-amino-methylalizarin-N,N-diacetic acid) and cerium and certain other lanthanides is converted to a blue ternary complex. The nature of the complexes involved has been investigated recently by Langmyhr *et al.* (1971) who showed that at pH 4·5 a dimeric chelate of the formula $(LaL)_2$ is formed between the lanthanide (La) and the complexone (L). In the presence of fluoride this is converted into a blue ternary complex of the formula $(LaL)_2F_2$ ($K = 6·3 \times 10^{13}$). These workers suggested that this complex has the structure shown in Fig. 19.7. The reaction was first applied to the analysis of sea water by Green-

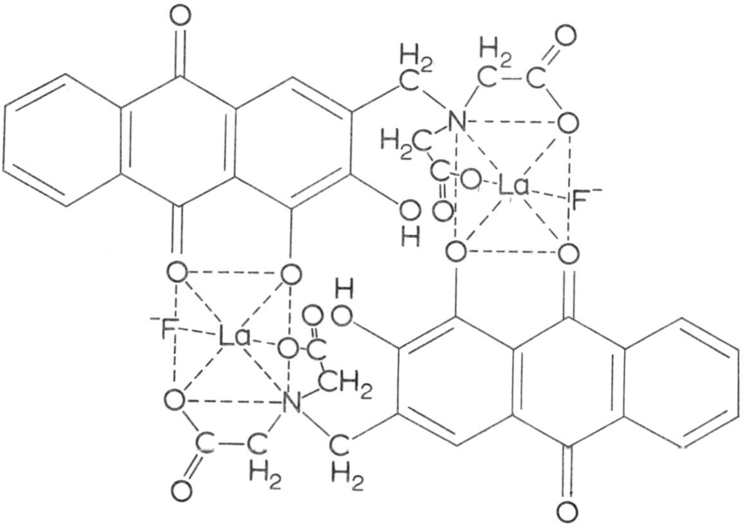

FIG. 19.7. Probable structure of lanthanum-alizarin complexone-fluorine tertiary complex (after Langmyhr *et al.*, 1971).

halgh and Riley (1961) who used the lanthanum chelate and carried out the reaction in the presence of acetone which not only doubled the sensitivity, but gave a much more stable colour. Although the major anions of sea water do not interfere, there is a small salt error mainly caused by magnesium. The method shows a coefficient of variation of less than $\pm 1\%$. Kletsch and Richards (1970) have described a very similar technique, which differs mainly in the substitution of cerium for lanthanum in the chelate, this leads to a significant reduction in sensitivity. Greenhalgh and Riley's manual method has been extensively used in studies of the $F/Cl\%_0$ ratio in oceanic and inshore waters (Riley, 1965a, Kremling, 1969; Brewer et al., 1970; Bewers, 1971; Kester, 1971; Windom, 1971; Kitano and Furukawa, 1972; Kullenberg and Sen Gupta, 1973). This procedure has also been successfully adapted for automatic analysis using the Technicon AutoAnalyzer flow system (Grasshoff, 1966; Chan and Riley, 1966a).

The observation by Frant and Ross (1966) that electrodes can be made from the fluorides of several of the lanthanides, notably lanthanum, has led to the commercial development of fluoride-specific electrodes (by Orion Research Inc., see Table 20.2). Such electrodes show a Nernstian response to fluoride ion activity extending from ca. 1 M down to 10^{-6} M. It should be stressed that the electrode system responds to the *activity* of fluoride ion and not to its *concentration*. If the electrode is to be applied to the determination of fluoride in sea water, it would therefore be necessary to modify direct potentiometric measurements to allow for (i) the fact that about half the total fluoride is bound as ion pairs with magnesium and calcium, (ii) changes in activity coefficients and liquid junction potentials caused by the high ionic strength of the medium. Although fluoride ion could be liberated from the ion pairs by the addition of a stronger complexing agent, e.g. EDTA, this is not feasible in this instance as the latter would interfere with the working of the electrode (Anfält and Jagner, 1970). For this reason, it is usual to calibrate the electrode either with standards having the same salinity as the samples and containing known amounts of fluoride (see e.g. Brewer et al., 1970), or with samples to which known increments of fluoride have been added (Bauman, 1968). Mainly because of the logarithmic response of electrode, the direct potentiometric method shows a variance of ca. $\pm 3\%$; (Durst and Taylor, 1967) about 3 times that of the lanthanum-alizarin complexone method. However, because of its speed and simplicity it has been widely used in surveys of fluoride distribution in both inshore and oceanic waters (Warner, 1969a, b; 1971; Brewer et al., 1970; Windom, 1971).

In an attempt to improve the precision of the potentiometric method for fluoride, Durst (1968) has developed a differential null point potentiometric technique for the analysis of 10 µl samples of natural waters; including interstitial waters. Two fluoride electrodes are employed and these are connected by a 0·1 M potassium nitrate salt bridge. One of the electrodes is in contact with the sample and the other is immersed in a fluoride-free synthetic sea water of similar salinity to the sample. Standard fluoride solution is added to the latter half-cell until a null potential on the valve voltmeter shows that the concentrations of fluoride in the two half cells are equal. It is claimed that the procedure shows a variance of $\pm 0·5-1·0\%$. An alternative approach has been used by Anfält and Jagner (1971) who employed a modified standard addition technique. In this, the sample itself is titrated potentiometrically with standard fluoride solution. The titration data are evaluated in terms of fluoride concentration using a computer program. The procedure gave an average mean deviation of ca. 0.6% with water of salinity $35\%_0$, but the precision deteriorated considerably at salinities below $5\%_0$. For a more detailed account of the use of fluoride electrodes in the determination of fluoride in sea water see Sections 20.3.1.3 and 20.3.3.2.

Some differences have been observed between the fluoride concentrations found in sea water by the alizarin complexone and the direct potentiometric techniques. Thus, Brewer et al., (1970) have found that the bottom waters from certain restricted areas of the North Atlantic which show anomalously high F/Cl ratios by the photometric method give normal ratios when the electrode is used. The excess fluoride is, therefore, not present in the free ionic state, and it was suggested that it might occur in a colloidal form (see also Warner, 1971).

Wilkniss and Linnenbom (1968) have determined fluoride with a precision of $\pm 5\%$ in freeze-dried sea water by photon activation analysis. However, because of its complexity the method seems to have very little to recommend it.

19.8.3.5. Boron

Boron occurs in sea water as free boric acid and as the borate ion $B(OH)_4^-$. The latter, which at the pH of sea water amounts to ca. 10% of the total boron, is probably partly associated with calcium and magnesium as the ion pairs, $CaB(OH)_4^+$ and $MgB(OH)_4^+$ (Dyrssen and Hansson, 1973) and also as $NaB(OH)_4$ (Byrne and Kester, 1974). Calculations by Hansson (1973) suggest that polynuclear species of boric acid amount to less than 0.001% of the total boron present in sea water. It has been postulated (Gast and Thompson, 1958; Hood and Noakes, 1961) that appreciable amounts of the borate ion of sea water occur as complexes with cis type carbohydrates. However, Williams and Strack (1966) have demonstrated from a consideration of the stability constants of these complexes that they will be present only in negligible proportions at the very low levels of carbohydrate prevailing in the sea (see also Kanamori, 1971).

19.8.3.5.1. *Titrimetric determination of boron.* Since boric acid is a weak acid ($pK_B = 8.67$ at $20°C$ and $S = 35\%_0$) it cannot be titrated directly with acceptable precision at the low concentration at which it occurs in sea water. However, in the presence of a massive excess of organic polyhydroxy compounds having 1,2 cis-hydroxy groups, e.g. mannitol, it is converted into complexes, which are strong acids and therefore can be titrated readily.

$$H_3BO_3 + \quad \begin{array}{c} | \\ HC-OH \\ | \\ HC-OH \\ | \end{array} \quad \rightleftarrows \quad \begin{array}{c} | \\ HC-O \\ | \\ HC-O \\ | \end{array} \diagdown B-OH + 2H_2O \rightleftarrows \begin{array}{c} | \\ HC-O \\ | \\ HC-O \\ | \end{array} \diagdown B-O^- + H^+$$

When this technique is used for the analysis of sea water samples it is necessary first to remove carbon dioxide by addition of excess hydrochloric acid and boiling under reflux (boric acid is volatile in steam). After titration to pH 7·0 with carbonate-free sodium hydroxide, a very large excess of mannitol is added and the resultant mannito-boric acid is titrated with standard carbonate-free sodium hydroxide to a final pH value of 7·0. In most early, and some recent, work on the boron content of sea water the titration was carried out using an indicator (see e.g. Buch, 1933; Moberg and Harding, 1933; Rakestraw and Mahncke, 1935; Miyake, 1939; Miyake and Sakurai, 1952; Belyaev and Ovsyanyi, 1968). However, because of the low concentration of boron in sea water (ca. 0·4 mM) the end point is not easy to observe visually, and it is difficult to obtain a coefficient of variation of better than $\pm 2\%$. The precision can be considerably improved by carrying out the titrations potentiometrically using a glass electrode (see, for example Gast and Thompson, 1958; Hood and Noakes, 1961; Fossato, 1968/69; Kremling, 1970; Sheremet'eva, 1970; Ryabinin, 1972). It seems likely that the precision of the potentiometric procedure could be further improved by using Gran plot methods (Anfält and Jagner, 1971; Hansson and Jagner, 1972) to evaluate the equivalence point of the titration.

19.8.3.5.2. *Photometric determination of boron.* Several reagents have been used for the spectrophotometric determination of boron in sea water. Of these, curcumin shows the highest sensitivity (ca. $0·00015 \ \mu g \ cm^{-2}$). This reagent reacts with boric acid in the dry state, or in certain non-aqueous media, to yield the intensely coloured complex rosocyanin, or if oxalic acid is present the somewhat less strongly coloured rubrocurcumin (the chemistry of the boron-curcumin reaction has been studied by Dyrssen *et al.* 1972). The determination in fresh waters is often carried out by treating the sample with hydrochloric and oxalic acids and a solution of the reagent in alcohol. The mixture is evaporated in a silica crucible on a water bath at 55°C. The colour forming reaction takes place after the water has evaporated and the reaction product is taken up in aqueous alcohol for spectrophotometry at 550 nm. The precision of this method tends to be poor because of the difficulty of controlling the many variables in the evaporation and colour development processes. This is particularly the case when sea water samples are analysed, and for these Greenhalgh and Riley (1962) have suggested that the reaction should be carried out at 120° in a homogeneous medium of phenol and acetic acid. Under these conditions they were able to determine boron in sea water with a coefficient of variation of $\pm 0.5\%$. A rather simpler approach to the problem was used by Uppström (1968, 1974) who eliminated the water from the sample by means of propionic anhydride in the presence of catalytic amounts of oxalyl chloride. Sulphuric acid and a solution of curcumin in acetone were then added and spectrophotometry was carried out on the rosocyanin thus produced. Hulthe et al. (1970) have successfully adapted this method to batchwise automatic analysis using the Autolab system, and their system has been used manually by Liss and Pointon (1973).

A number of organic compounds form coloured complexes with boric acid in strong sulphuric acid medium. Two of these have been employed for the photometric determination of boron in sea water. Gassaway (1967) has used dianthrimide and Charpiot (1969) has automated the carminic acid method of Demmitt (1965).

In the presence of hydrofluoric acid boric acid is converted into the fluoroborate ion $(H_3BO_3 + 4HF \rightleftarrows H_3O^+ + BF_4^- + 2H_2O)$. With certain cationic dyes the latter forms intensely coloured ion-pairs which can be extracted with suitable non-polar solvents. This principle has been adopted for the photometric analysis of sea water by Nicholson (1971) who extracted the blue ion-pair formed between the fluoroborate ion and Nile Blue A using o-dichlorobenzene. Since > 100 p.p.m. of chloride interferes it must be removed by precipitation as silver chloride as a preliminary to the analysis proper. A coefficient of variation of ca. $\pm 3.5\%$ was claimed for the method.

A fluorometric method for the determination of boron in sea water using benzoin has been described by Barnes and Parker (1960). A preliminary ion exchange treatment of the sample is necessary.

Spectrographic methods have been used for the direct determination of boron in sea water by Frederickson and Reynolds (1960) and by Lotrian and Johannin-Gilles (1969) These procedures differ principally in the way in which the boron emission spectrum is excited. The former workers employed spark excitation using a rotating disc electrode, whereas the latter, after an examination of a number of D.C. arc methods, selected a tube electrode-spray technique. Although the spectrographic methods have the advantage of speed, their precision is too low for them to be of value in examining possible variation of the $B/Cl\%_0$ ratio in sea water.

19.9. DETERMINATION OF DISSOLVED GASES

A knowledge of the distribution of the dissolved gases in sea water, in particular oxygen, and carbon dioxide, is of considerable value in hydrographical and biological studies. In recent years, differences in the relative concentrations of the noble gases have been used to study air-sea interactions (see 8.5.3). The concentrations of gases in sea water may be expressed in several ways. Conventionally, they are given either in terms of the weight or volume (at S.T.P.) present in 1 l of the water, or in terms of percentage saturation. However, there are considerable advantages to be gained in expressing the amount of sea water in kg since it is then independent of temperature and pressure (Weiss, 1970). In studies of oxygen utilization etc. it is of value to give the oxygen concentration in units similar to those used for the nutrients i.e. μg-at l^{-1}.

Considerable effort has been devoted in the last decade to the accurate determination of the solubilities of atmospheric gases in sea water. For oxygen and nitrogen there is satisfactory agreement between the results obtained by recent workers (Carpenter, 1966; Murray and Riley, 1969; Murray et al. 1969; Douglas, 1964, 1965; see also Weiss, 1970). Detailed oxygen solubility tables based on the work of Carpenter and Murray and Riley have been published by UNESCO (1973), see Appendix Tables 6 and 7. However, Weiss (1971a) has pointed out that discrepancies exist at intermediate salinities between the argon solubility data of Murray and Riley (1970) and those interpolated from those of Douglas (1965). For helium and neon, the data of Weiss (1971b) are probably the best available. (Tabulated solubility values are presented in Chapter 8 and the Appendix).

19.9.1. OXYGEN

Water sampling for oxygen determination should only be carried out with plastic sampling bottles, since corrosion of the metal can cause significant loss of oxygen in samplers made of metal, particularly if the plating has been corroded (Rotschi, 1963; Rochford, 1963; Bogoyavlenskii, 1965b; Strickland and Parsons, 1968). Reed and Ryan (1965) have reported that sacs which may develop under the coatings of plastic lined sampling bottles can entrap sufficient oxygen-rich surface water to cause results for deep waters to be $0.1–0.3$ ml $O_2 l^{-1}$ too high. The general problem of sampling water for dissolved oxygen analysis has been discussed by Montgomery and Cockburn (1964). Owing to the risk of loss of oxygen as the water warms, and to changes produced by microbiological respiration, samples for oxygen analysis should be removed from the sampler as soon as it comes aboard. For the same reasons, analysis of the samples should be commenced as soon as possible, and certainly within 15 min.

Several different techniques are available for the determination of oxygen in sea water; these include titrimetric, electrometric, colorimetric and radio-metric procedures. Of these probably only the titrimetric (Winkler) and electrometric (potentiometric and amperometric) techniques are in general use. All aspects of the determination of oxygen in natural waters have been reviewed by Mancy and Jaffe (1966).

19.9.1.1. *Winkler method*

The Winkler (1888) method is still the most precise and accurate procedure for the determination of dissolved oxygen in sea water. In it the sample is treated with a concentrated solution of sodium hydroxide and sodium iodide and then with a strong solution of manganous chloride or sulphate.* The gelatinous precipitate of manganous hydroxide which is formed reacts with the dissolved oxygen, and manganese is oxidized to a higher oxidation state†. The precipitate is allowed to settle for a few hours and then dissolved in hydrochloric or sulphuric acid. The iodine liberated by reaction of the oxidized manganese with the iodide ion is titrated with standard thiosulphate solution using starch as an indicator.

$$Mn^{2+} + 2OH^- \rightarrow Mn(OH)_2$$

$$Mn(OH)_2 + O \rightarrow MnO(OH)_2$$

$$MnO(OH)_2 + 4H^+ + 3I^- \rightarrow Mn^{2+} + I_3^- + 3H_2O$$

$$I_3^- + 2S_2O_3^{2-} \rightarrow 3I^- + S_4O_6^{2-}$$

The method was first applied to the analysis of sea water by Natterer in 1892, and many minor modifications of it have been described (see e.g. Kalle, 1939; Thompson and Robinson, 1939; Jacobsen et al., 1950; Grasshoff, 1968; Andersen and Føyn, 1969). However, it is only comparatively recently that the Winkler method has been studied systematically (Grasshoff, 1962a, b; Carpenter, 1965a, Carritt and Carpenter, 1966). These investigations have revealed that most of the techniques which have been suggested are subject

---

* Syringes are normally used for the addition of these reagents. However, Öström (1973) has suggested that there are some advantages in using disposable cylindrical ampoules for adding the reagents.

† Carritt and Carpenter (1966) have shown that the overall oxidation state of the oxidized manganese lies between $+3$ and $+4$. However, the actual nature of the product is immaterial since the subsequent oxidation of iodide is stoichiometric to the amount of oxygen which has reacted. For convenience the oxidized form of manganese is represented in the equations by $MnO(OH)_2$.

to serious systematic errors, and Carpenter (1966) has pointed out that the use of data produced by such methods in conjunction with accurate oxygen solubility values may lead to misleading deductions. The principal sources of these errors are:

(i) Loss of iodine by volatilization from the acidified solution during, or immediately prior to, titration. Error from this source can be reduced, but by no means eliminated, by using an alkaline iodide reagent which is almost saturated with respect to sodium iodide (e.g. that proposed by Pomeroy and Kirschman, 1945). This serves to lower the vapour pressure of iodine by complexing it as the tri-iodide ion. Even when this precaution is taken, 1–2% of the iodine can be lost when the solution is transferred to a flask for titration, as it is in most published methods. The loss of iodine in this way can only be obviated if transference is avoided by using the reaction vessel also for the titration. Green and Carritt (1966) have described a simple apparatus with which this can be achieved. It consists of a 125 ml conical flask fitted with an elongated 19/38 stopper, which serves to displace sufficient of the sample to allow the titration to be carried out with a microburette. With this apparatus, which is essential for accurate work, iodine loss can be kept to $\sim 0.05\%$, provided that the flask is covered with a loosely fitting perspex lid during the titration. If such equipment is not available, the entire acidified sample should be transferred rapidly to the titration vessel.

(ii) Photo-oxidation of iodide may occur if the acidified test solutions are exposed to strong light and this will lead to high results. The rate of oxidation increases with pH, and a pH value of about 2 is probably optimal (Carpenter, 1965a); the final pH must be less than 2·7 for the dissolution and reaction of the precipitated manganese. Acidification may be carried out with sulphuric, hydrochloric, phosphoric or sulphamic acids.

(iii) Serious errors can be produced by impurities in the reagents. The presence of icdate in the iodide used for the preparation of the alkaline iodide reagent will cause the blank to be positive. Reducing impurities in the reagents will give rise to negative blanks. Carpenter (1965a) has shown that the reducing impurities are present as dark particles in the sodium iodide and can be removed by hand-picking.

(iv) A slight systematic error is introduced by failure to make allowance for the oxygen present in the Winkler reagents. Murray et al. (1968) have estimated that the error produced in this way, using a sample volume of 140 ml and the reagents described by Carpenter (1965b), amounts to 0.012 ml l^{-1} at 20°C. In the analysis of oxygen depleted waters it is advantageous to degas the Winkler reagents with a current of nitrogen.

(v) Diffusion of oxygen into, or out of, the sample either during its transfer to the test bottle, or while reagents are being added, does not appear to be a significant source of error except when moderately anoxic water is being analysed. Thus, Grasshoff (1962a) found that when a water sample was left in an unstoppered oxygen bottle for 10 minutes its oxygen saturation only increased from 25 to 26%. Since the relative error arising from this cause increases as the oxygen concentration decreases, considerable care must be taken to minimize ingress of oxygen when nearly anoxic samples are analysed. For such waters, it is advantageous to use a technique, such as that of Broenkow and Cline (1969), in which the sample is drawn into a syringe, and treated by injection with the Winkler reagents.

(vi) Errors may arise from faulty standardization of the standard thio-sulphate solution. Dichromate is an undesirable standard since its reaction with iodide is slow, except in the presence of relatively high concentrations of acid and iodide. Under these conditions there is a serious risk of photo-chemical oxidation of the iodide. (See Carritt and Carpenter, 1966.) Potassium iodate, which reacts with iodide ion at low acidity, is probably the most satisfactory reagent for the standardization of thiosulphate for the Winkler method, and has been adopted by most recent workers. Sugawara (1969) has investigated the stability of the 0·01 N potassium iodate solution used for the standardization and has found that the normality of the sterilized solution did not change appreciably in 17 months. Carritt (1963) has pointed out that there are advantages to be gained by using air-saturated water to standardize the method since this should enable systematic errors to be cancelled out. Grasshoff (1964c) has described a simple apparatus for the continuous production of such water. In it a stream of water-saturated air passes counter-current wise over a film of water flowing down a thermostatically controlled glass spiral (see also Murray, et al. 1968). The temperature of the water must be controlled to $\pm 0·05°C$, and it is necessary to allow for differences between the ambient pressure and the 760 torr standard pressure.

(vii) The cleanliness of the test bottles and titration flasks is very important if errors are to be avoided, since any contamination with residual amounts of manganese will lead to high results (Carritt and Carpenter, 1966)

In an attempt to minimize the errors arising from these various sources Carpenter (1965b) has developed a dissolved oxygen procedure which is probably the most accurate and precise routine method available at present (see also Carritt and Carpenter, 1966).

End-point detection. Starch is a satisfactory visual indicator for routine Winkler titrations. It should be prepared freshly each day as it soon loses its sensitivity through biologically induced hydrolysis. Van Landingham (1960) who has reviewed the literature on the stabilization of starch solutions has concluded that it is difficult to stabilize its aqueous solution, and has suggested the use of a reagent prepared by heating starch to 190°C with glycerol.

Although the visual end-point with starch indicator is satisfactory for routine determinations of dissolved oxygen in most sea water, it is insufficiently sensitive (10μ equiv l^{-1}) if high precision is required, or if nearly anoxic waters are to be analyzed. The sensitivity of the detection of the end-point can be enhanced approximately five-fold by carrying out the titration photo-metrically. (Bradbury and Hambly, 1952). However, the indicated end-point is significantly different from the equivalence point in titrations of iodine solutions having concentrations similar to those obtained when Winkler titrations are carried out on air-saturated water. A further ten-fold increase in sensitivity (to 0.02μ equiv l^{-1}) can be achieved if the photometric titration is carried out by making use of the intense near-ultraviolet absorption band of the tri-iodide ion. Using this principle Carpenter (1965b, 1966) has been able to attain a coefficient of variation of better than $\pm 0.06\%$ in his work on the solubility of oxygen in pure water and sea water.

Electrochemical methods have been used for the detection of the end-point in Winkler determination by several workers. Knowles and Lowden (1953) have compared amperometric, potentiometric and dead stop methods and have recommended amperometric titration using a platinum and standard calomel electrode system. Amperometric titration has also been employed by Bradbury and Hambly (1952) who found that the sensitivity of end point detection is about twelve times better than could be achieved visually with starch. In practice, the variability of the sensitivity of the platinum electrode is a drawback in this method. Grasshoff (1962a, b) has adopted a modified dead stop technique for automatic titration in the Winkler method. Excess of standard thiosulphate is added and the solution is then titrated with standard potassium iodate solution. The end-point is detected by means of two platinum electrodes polarized with a potential of ca. 100 mV. Iodine, which begins to be liberated at the equivalence point in the titration, causes depolarization of the anode, and the resultant current actuates the automatic titrator. The precision is claimed to be similar to that of careful manual Winkler titration.

Spectrophotometric modifications of the Winkler method

A number of modifications of the Winkler method have been described in which the determination is completed by spectrophotometry instead of titration. Thus, Trotti and Sacks (1962) measure the absorption of the liberated iodine. The precision ($\sigma = \pm 0.037$ ml l^{-1}) which they claim for their method is somewhat inferior to that of the normal Winkler method with oxygenated waters. However, the photometric procedures can offer a worthwhile gain in precision in the analysis of nearly anoxic waters. Although the absorption band of the molecular iodine itself has been used for this purpose (Ivanoff, 1962), its absorbance is too low to give adequate sensitivity. Much greater sensitivity can be achieved using the near ultra-violet maximum arising from the tri-iodide ion, and in this way Broenkow and Cline (1969) have been able to determine oxygen at a level of ca. 0.2 ml l^{-1} with a coefficient of variation of ca. $\pm 1.2\%$.* If a spectrophotometer covering the ultra violet range is not available the approach used by Ostrowski et al. (1965) may be of value. In this, the sample is treated with manganese chloride and sodium hydroxide. After the precipitate has settled a solution of N, N-*bis* (2-hydroxy-propyl)-*o*-phenylenediamine in hydrochloric acid is added. The oxidized manganese oxidizes the organic base to an intensely coloured compound, the absorbance of which is measured at 533 nm. It should be pointed out that the presence of colour or suspended matter in the sample can cause interference in the spectrophotometric methods. In methods depending on the formation of iodine correction can be made by subsequently reducing the iodine with a drop of sodium sulphite solution and re-measuring.

Interferences in the Winkler method

Both reducing and oxidizing agents interfere in the Winkler method. Although the concentrations of these substances in ocean waters are too low to cause appreciable interference, they may be sufficiently high in badly polluted estuarine or coastal waters to cause serious difficulties with the normal Winkler procedure. However, many of these interferences can be overcome by suitable modification of the method.

Oxidizing agents lead to high results because of the oxidation of the iodide ion. The most important of these is the nitrite ion. Interference from this source can be prevented by incorporating sodium azide in the alkaline iodide reagent (Alsterberg, 1925; Montgomery et al., 1964), or by using sulphamic acid either as a preliminary reagent (Cohen and Ruchhoft, 1941) or to acidify the sample before titration (Nusbaum, 1958). The presence of ferric iron also leads to positive errors; its interference can be prevented by the addition of phosphoric acid which complexes the iron and lowers its redox potential (Montgomery et al., 1964).

---

* Cline and Richards (1972) have found that for oxygen depleted waters this method gives results systematically ~ 0.1 ml l^{-1} lower than those obtained by the Winkler method, perhaps because of a systematic error in the latter.

Reducing agents cause negative errors in the Winkler method since they bring about reduction of the oxidized form of manganese. The most important of them are ferrous iron, hydrogen sulphide and organic matter. Their interference can frequently be prevented by carrying out a preliminary oxidation with either bromine (Ross, 1964) or acidic permanganate (Rideal and Stewart, 1901). For the analysis of sea water samples containing traces of hydrogen sulphide Mor and Beccaria (1971) preferred to measure the sulphide photometrically by the methylene blue method and to correct for its effect on the Winkler determination.

Accuracy and precision of the Winkler technique. The overall accuracy of the Winkler method is determined not only by the accuracies by which the various manipulations are carried out (including the determination of the reagent blank and standardization of the thiosulphate), but also by the stoichiometry of the reaction sequence. With very careful work Carpenter (1965a) and Murray and Riley (1969) have been able to obtain an accuracy of better than 0·1%, relative to accurately prepared solutions of oxygen and to physical measurements of oxygen solubility respectively. However, the accuracy is likely to be considerably poorer in routine measurements, particularly if, as is the usual practice, the solution is transferred to a flask for titration.

Numerous interlaboratory comparisons of the Winkler method have been carried out over the last decade (see e.g. Rochford, 1963, 1964; Bogoyavlenskii, 1965a; Carritt and Carpenter, 1966). The results obtained have shown a disquieting range of variations, and it is apparent that although the technique is simple, small variations in the techniques used can produce significant differences in the results. The most comprehensive of these intercalibration experiments has been described by Carritt and Carpenter *loc. cit.* who noted that faulty standardization of the thiosulphate solution could be a major source of error in the Winkler method. Thus, when eleven analysts analyzed a standard thiosulphate solution, known to be accurately 0·01040 N, their mean results ranged from 0·01034 N to 0·01067 N. However, the results of the 10 replicate standardizations by each participant showed acceptable precision and ranged among themselves ca. by only 0·1–0·5%. Six out of the eleven mean results were significantly higher than the true value whereas only two were significantly lower. This suggests that iodine is being lost from the solution, probably by volatilization before or during the titration. Low results were found if dichromate was used for the standardization, since the high acidity which is necessary to speed up the reaction of the dichromate with the iodide ion also favours the photochemical oxidation of the latter. In testing the Winkler method itself six participants carried out ten replicate analyses on water samples equilibrated with three levels of oxygen, using four different

procedures. These showed standard deviations for three of the four methods*
of $\pm 0.02-0.09$, $\pm 0.11-0.13$ and $\pm 0.06-0.09$ ml $O_2 l^{-1}$ at mean oxygen
concentrations of 4·79, 2·53 and 0·44 ml $O_2 l^{-1}$ respectively. There appeared
to be no significant differences between the results yielded by the three methods.

19.9.1.2. Other chemical methods

Roskam and de Langen (1963) have described a complexometric procedure
for the determination of oxygen in polluted waters. In this, the water sample
is treated with a solution of ferrous ethylenediamine sulphate and tris-
(hydroxymethyl)-amino methane. Under these mildly alkaline conditions
(pH 7·5) the dissolved oxygen oxidizes the ferrous iron stoichiometrically
to the ferric state. After acidification to pH 2·4 the ferric iron is determined by
titration with EDTA using salicyclic acid as indicator. The method, which
showed a coefficient of variation of $\pm 0.6\%$, gave results which agreed well
with those obtained by the Winkler method.

The oxidation of the leuco-bases of certain dyes to the dyes themselves has
been used in several methods for the photometric determination of low
concentrations of oxygen in boiler feed waters (see e.g. Loomis, 1954; Alcock
and Coates, 1958). Although these techniques do not appear to have been
applied to sea water, they might prove of value for the analysis of oxygen-
depleted waters.

19.9.1.3. Gasometric method

Wheatland and Smith (1955) have developed a gasometric method for the
determination of oxygen in fresh and saline waters. In it, the dissolved gases
are pumped from the sample under vacuum. They are transferred to a Bone
and Wheeler constant volume gas analysis unit and their volume and composition is then measured. The technique, for which a total maximum error of
0·04 p.p.m. of oxygen was claimed, has the advantage that it can be used for the
the analysis of samples heavily polluted with organic matter.

19.9.1.4. Radiochemical methods

A radiochemical procedure for monitoring low concentrations of oxygen
in sea water has been developed by Richter and Gillespie (1962) and Gillespie
and Richter (1965). The sample is allowed to flow through a column of
molybdenum filings coated electrolytically with thallium containing ^{204}Tl
(specific activity 2 mCi g^{-1}). The dissolved oxygen reacts rapidly and stoichiometrically with the thallium and the amount of thallous ion liberated to the

---

* The fourth method gave results ca. 12% lower than the mean of those obtained by the other
three, perhaps because of the use of dichromate for standardization.

water is measured by counting the ^{204}Tl present in the water using a liquid G.M. counter and rate meter. The ^{204}Tl counting rate varies linearly with oxygen concentration up to at least 100 p.p.m. The precision was $\pm 5\%$ at an oxygen level of 1 p.p.m., but could be improved by use of thallium with a higher specific activity.

19.9.1.5. *Gas chromatographic methods*

The application by Swinnerton *et al.* (1962a) of gas chromatography to the determination of dissolved gases in aqueous solutions gave for the first time the prospect of a rapid method for the simultaneous estimation of the major dissolved atmospheric gases in sea water. In their method, the sample to be analyzed is injected by means of a gas-tight syringe into an all-glass stripping tube, fitted at its lower end with a coarse sintered glass filter. A current of helium which is passed through the filter rapidly strips the dissolved gases from the water sample. The helium is passed through a drying tube and then into a chromatographic column packed with 30% hexamethyl phosphor-amide on an inorganic support phase. The emergent gas is passed through a thermal conductivity cell and next to another chromatograph tube packed with layers of diatomaceous support phase and molecular sieve. Finally, the effluent gas is passed through another thermal conductivity cell. The first column serves to separate carbon dioxide from the other major atmospheric gases and its amount is measured with the first conductivity cell. Those gases of the mixture which are not resolved in the first column are separated from one another in the second column and their amounts are measured with the second thermal conductivity cell. Unfortunately, argon and oxygen are not separated on the molecular sieve column at room temperature. With air-saturated waters this will lead to an over-estimation of ca. 5% in the observed dissolved oxygen concentration; allowance can be made for this without appreciable error. However, as the oxygen concentration becomes lower the relative contribution of argon to the oxygen peak will be much greater. Under these conditions it is essential to prevent interference from the argon. This can be achieved in a number of ways:

(i) by substitution of argon for the helium carrier gas (Park and Catalfomo, 1964). This reduces the sensitivity by a factor of approximately ten because the difference between the thermal conductivities of oxygen ($5 \cdot 9 \times 10^{-5}$ cal s^{-1} ($^{\circ}$C/cm)$^{-1}$) and argon ($4 \cdot 0 \times 10^{-5}$ cal s^{-1} cm^{-1} ($^{\circ}$C/cm)$^{-1}$) is only small relative to their differences from that of helium. However, it is possible to compensate for some of the loss of sensitivity by using the stripping chamber described by Swinnerton *et al.* (1962b) which permits the use of a larger sample. Using this technique and a sample volume of 13·5 ml Park and Catalfomo *loc cit.* were able to analyze sea water for both oxygen and nitrogen in 5·5 min. A coefficient of variation of $\pm 1\cdot4\%$ was obtained with air-saturated water. A close correlation was found between oxygen concentrations determined by this method and by the Winkler procedure.

(ii) Separation of argon and oxygen can be achieved if the length of the molecular sieve is increased to 10 m and hydrogen is used as a carrier gas (Vizard and Wynne, 1959). However, the sensitivity is low and retention times are excessively long. The same separation can also be achieved on a 2 m column of molecular sieve at $-72°C$ (Lard and Horn, 1960). If nitrogen is to be determined it is necessary to carry out a second run at room temperature as it is adsorbed irreversibly at low temperatures. By striking a balance between the length of the column (5 m) and its operating temperature $(-9°C)$ Gunter and Musgrave (1966) have succeeded in obtaining a rapid and satisfactory separation of oxygen, nitrogen and argon on Molecular Sieve 5A. If it is not desired to determine oxygen as well as argon, the former can be removed by adsorption on activated coconut charcoal (Cooke. 1973).

(iii) If hydrogen is used as carrier gas oxygen can be catalytically converted to water at room temperature by means of palladium. The water can then be separated from argon on a column of Triton X305 on a PTFE support at 55°C and determined using a thermal conductivity detector. After drying, the gas flow can then be passed through a column of molecular sieve to separate argon and nitrogen which are determined using the reference cell of the thermal conductivity detector (Swinnerton, et al. 1964). Alternatively, two runs can be made on the sample one with, and one without the palladium pre-column; the first of these determines the argon alone, and the second the argon plus oxygen (Swinnerton, 1962a).

Gas chromatography was first applied to the determination of the major atmospheric gases in sea water by Swinnerton and Sullivan (1962), and yielded a precision of ca. $\pm 1\cdot3\%$ for both oxygen and nitrogen. Their technique has been subsequently adopted with some modification by a number of workers (e.g. Park and Catalfomo, 1964; Park, 1965; Craig et al., 1967; Hosokawa and Oshima, 1969; Solov'ev and Doroshenko, 1970), and Gary (1968) has described equipment for carrying out the determination at sea. The gas chromatographic procedure has been automated by Reusmann (1968) and used for the determination of nitrogen, total carbon dioxide and oxygen + argon in pumped samples of sea water. The use of an electronic integrator instead of a planimeter for the measurement of the chromatographic peaks more than doubled the precision (from ca. $\pm 1\cdot5\%$ to $\pm 0\cdot5\%$).

19.9.1.6. *Electrometric methods*

Electrometric methods for the determination of dissolved oxygen are described in Section 20.7). Although they are not as accurate as carefully performed Winkler determinations, they have the advantages not only of speed, but also usually of freedom from interference by many of the pollutants commonly found in estuarine environments. They are particularly valuable when continuous monitoring is required. Commercially made dissolved oxygen probes compatible with the Bissett–Berman 9040 STD system (see Section 6.2.2.3) are now available which permit continuous oxygen profiling down to depths of 2000 m (see e.g. Lambert *et al.*, 1973 and Section 8.6.4). A free fall electrochemical oxygen meter analogous to an expendable bathy-thermograph with which oxygen profiles to 500 m can be made in 80 s has been described by Jeter *et al.* (1972).

19.9.2. NITROGEN

The accurate and precise determination of dissolved nitrogen in sea water has proved very difficult. Most of the earlier methods, which have been reviewed by Riley (1965b) and Craig *et al.* (1967), are time-consuming and often imprecise; and in many of them it is the total nitrogen + argon which is measured. However, the development, within the last twelve years, of gas chromatographic techniques for the determination of the major atmospheric gases has revolutionized this field. The application of these methods to the analysis of sea water has been discussed in Section 19.9.1.5. Among their advantages are rapidity, precision, suitability for use on ship-board and their ability to determine several gases simultaneously. A typical gas chromatographic system described by Atkinson (1972) permits on-line monitoring of nitrogen and argon in sea water with a precision of $\pm 0.5\%$. In it the gases are sparged from the sample by bubbling with helium. Nitrogen is separated from oxygen and argon by means of a 1·9 m column of Molecular Sieve 5A at 100°C. Oxygen is adsorbed on activated charcoal and the peaks of nitrogen and argon are detected with a helium ionization detector coupled to an electronic integrator. Calibration of the instrument for nitrogen is carried out using molecular nitrogen generated by electrolytic break-down of hydrazine (Page and Lingane, 1957).

Mass spectrometric techniques are available for the estimation of the $^{15}N/^{14}N$ ratio of the dissolved nitrogen in sea water (Benson and Parker, 1961; Miyake and Wada, 1967).

19.9.3. ARGON

Although the other noble gases can only be determined in sea water by mass spectrometric analysis, argon because of its abundance can be readily determined by shipboard gas chromatography using samples of only 10–20 ml. As the element is chemically unreactive it behaves in a conservative manner and can be used for the investigation of oceanic diffusive processes. It is also of considerable value as an "internal standard" in studying the variations in the oceans of the biologically involved gases, particularly as the latter can be measured in the same chromatographic run as the argon itself.

Although some gasometric techniques have been described for the determination of argon in sea water (see e.g. Rakestraw and Emmel, 1937; Oana, 1957), they have been rendered obsolete by the development of rapid and more accurate chromatographic methods. The latter have been reviewed above in Section 19.9.15. It should be borne in mind that the retention times of argon and oxygen on columns of molecular sieve are quite similar. For this reason it is necessary either to operate the column at reduced temperatures, or to remove the oxygen e.g. by adsorption on activated charcoal (Atkinson, 1972). Under optimum conditions a coefficient of variation of $\pm 0.5\%$ can be achieved if the chromatographic peaks are assessed with an electronic integrator (see e.g. Craig et al., 1967; Atkinson, 1972). Craig and Weiss (1968) have compared the results obtained for dissolved argon by gas chromatography with those found by an isotope dilution-mass spectrometric procedure (Section 19.9.4) and found that the two methods gave a similar accuracy and precision. Benson and Parker (1961) have described a mass spectrometric technique for the determination of the N_2/Ar ratio in sea water.

19.9.4. OTHER NOBLE GASES

The use of noble gas saturation anomalies in studying gas fluxes into the ocean (Section 8.5.3) has necessitated the development of high precision techniques for the determination of these elements. All the analytical procedures which have been used for this purpose involve stripping the dissolved gases from the water, removing O_2, N_2, CO_2 and water vapour, and in some instances fractionating the noble gases before determining them by direct (Bieri et al., 1966) or isotope dilution mass spectrometry (see e.g. Mazor et al., 1964; Hintenberger et al., 1964; Craig et al., 1967; Bieri et al., 1968); the latter leads to considerably better precision.

Special precautions must be taken in collection and storage of samples for analysis for noble gases in order to avoid losses due to diffusion of helium. Water samples must, therefore, be transferred immediately after collection to well baked, evacuated or nitrogen filled, stainless steel storage flasks (see e.g. Bieri, 1965; Clarke et al., 1969). Vacuum trains must be constructed of glass having a low permeability to helium (e.g. Corning 1723).

The procedure described by Craig *et al.* (1967) is typical of the analytical methods used for determination of noble gases in sea water. The sample is acidified with phosphorus pentoxide and boiled under reflux. The evolved gases are continuously pumped into a sample tube using a Toepler pump. The sample tube is closed with its stopcock and removed from the degassing line. After returning to the laboratory, the samples are injected with small amounts of concentrates of the less abundant isotopes of the elements to be determined (e.g. ^{3}He and ^{22}Ne), for isotope dilution purposes. Carbon dioxide and water vapour are eliminated by means of a cold-trap, and nitrogen and oxygen are removed with a titanium "getter". Argon and krypton are separated from the neon and helium fraction by adsorption onto charcoal at $-187°$C. If krypton is to be measured it can be separated from argon by a two-stage adsorption on charcoal at $-115°$C (Bieri *et al.*, 1968). The separated gases are then submitted to mass spectrometry and the response of the major isotope of the element being determined is measured relative to that of the corresponding isotopic spike. An overall reproducibility of $\pm1·4\%$ for neon and $\pm2\%$ for helium can be attained.

19.9.5. TOTAL CARBON DIOXIDE, P_{CO_2}, CARBONATE AND BICARBONATE

The determination of total carbon dioxide and the components of the carbonic acid equilibrium system in sea water are discussed in Chapter 9. Electro-analytical procedures of the measurement of pH and both total alkalinity and ΣCO_2 are described in Sections 20.3.1.1 and 20.3.2.2 respectively. The electrometric determination of P_{CO_2} is discussed in Section 20.3.3.1.

19.9.6. HYDROGEN SULPHIDE

Hydrogen sulphide is a frequent component of anoxic waters, attaining concentrations as high as 70 mg l^{-1} under extreme conditions. It behaves as a weak acid, and is present in natural waters as both the undissociated compound and the HS^- ion (below pH 12 the concentration of the S^{2-} ion is negligible). Because of the volatility of hydrogen sulphide and the rapidity with which it reacts with oxygen great care must be taken during sampling to prevent the water coming into contact with the atmosphere. To reduce the risk of loss of hydrogen sulphide the addition of reagents to samples should preferably be carried out using techniques similar to those used in the Winkler method for dissolved oxygen. If it is not possible to analyze the samples immediately after collection they should be treated with a solution of zinc or cadmium acetate (Kudo, 1964); this fixes the hydrogen sulphide as the insoluble metal sulphide.

Concentrations of hydrogen sulphide greater than ca. $4 \, mg \, l^{-1}$ can be determined volumetrically by addition of excess iodine solution and back titration with standard thiosulphate solution. This method is somewhat unsatisfactory since, if the sample is added to the iodine solution in order to avoid side-reactions, it is difficult to avoid oxidation of the sulphide by air. Armstrong (personal communication) has suggested treating the acidified sample with excess standard arsenite solution, filtering off the arsenious sulphide and then back titrating the excess arsenite with standard iodine solution. Both sulphite and thiosulphate ions interfere in these methods. However, their interference can be overcome by treating the sample with zinc acetate, filtering off the precipitated zinc sulphide, and determining it iodometrically (Johnson et al., 1964).

The photometric methylene blue method of Fischer (1883) was first adopted for the determination of hydrogen sulphide in sea water by Fonselius (1962). In this procedure the sample is treated, in acidic solution, with N, N-dimethyl-p-phenylenediamine and ferric iron. In the presence of sulphide, methylene blue is produced, the absorbance of which is measured at 667 nm. The extensive literature on the method has been reviewed by Cline (1969) who has studied its application to natural waters, and has developed a method employing a stable single reagent for the determination of 0.03–$35 \, mg \, H_2S \, l^{-1}$. A coefficient of variation of ± 0.8–1.5% was found over this range. The method is free from salt error, and no interference is caused by up to $100 \, \mu$ moles l^{-1} of thiosulphate or sulphite. Grasshoff and Chan (1969, 1971) have adapted the methylene blue technique for automatic analysis using the Technicon AutoAnalyzer system. Samplers are stabilized by addition of cadmium chloride and gelatine. Beer's Law is obeyed over the range 0.007–10 $mg \, H_2S \, l^{-1}$. Rather similar procedures using p-phenylenediamine itself and its ethoxyethyl derivative have been described by Strickland and Parsons (1968) and Smirnov (1971) respectively. However, by comparison with the method of Cline (loc. cit.) they suffer from a number of drawbacks, including lower precision, and the need to use two reagents.

The spectrofluorometric method of Bark and Rixon (1970), may be of value for the determination of low concentrations of sulphide in sea water (1–$30 \, \mu g \, S^{2-} \, l^{-1}$). It depends on the liberation of fluorescent 2, 2'-pyridyl-benzimidazole when its mercury derivative reacts with sulphide ion. Specific ion electrodes are available for the determination of sulphide and can be used successfully in sea water medium. (See Section 20.3.1.2).

19.9.7. NITROUS OXIDE

The pioneering work of Junge and Hahn (1971) (see also Junge et al., 1971) has revealed that very low but significant concentrations (<0.1–$0.6 \, \mu g \, l^{-1}$)

of nitrous oxide are present in the sea (see Chapter 8). Hahn (1972) has described a method by which the determination of the gas in sea water can be carried out on board ship. The sample (5 litre) is stripped of nitrous oxide by boiling under reflux while passing a current purified nitrogen through it. The nitrogen stream is passed through an absorption train to remove carbon dioxide and water vapour. Nitrous oxide is adsorbed using a column of Molecular Sieve 5A. After collection of the nitrous oxide the pre-column is sealed and can be stored until required. To continue the analysis the pre-column is connected to the gas sampling valve of a gas chromatograph and both are evacuated. After switching the valve and heating the pre-column to ca. 300–350° the desorbed nitrous oxide is swept into the gas chromatograph column (0·9 m of Molecular Sieve 5A) with a current of helium. After 10 minutes the sampling valve is again switched. The gas chromatography is then commenced by raising the column temperature at a rate of $20°C\,min^{-1}$ to 250°C and continuing isothermally. Helium is used as the carrier gas and the separated nitrous oxide is determined with a katharometer. A coefficient of variation of $\pm 10\%$ can be attained.

19.9.8. HYDROGEN

Williams and Bainbridge (1973) have given a preliminary account of a gas chromatographic procedure for the determination of hydrogen in sea water. In this method air is circulated through the sample. After equilibrium has been attained a 10 ml aliquot of the air is injected into a chromatograph column packed with molecular sieve. Nitrogen is used as carrier gas and the separated gases are detected by allowing the column effluent to flow between the electrodes of a low pressure glow discharge tube operating at a constant current of 1 mA. The voltage changes occurring across the discharge tube are a measure of the partial pressures of the gases in the air and thus in the water sample.

Schmidt (1974) has described a simple technique for the continuous determination of hydrogen in sea water. In this, sea water is equilibrated with cycled hydrogen-free air. After equilibration, the air is dried and then passed through columns containing molecular sieve and silver oxide (the latter to remove carbon monoxide). Finally, the air is passed through a heated tube (220°C) packed with yellow mercury (II) oxide. Hydrogen reacts with the mercury oxide liberating an equivalent amount of elemental mercury which is determined by flameless atomic absorption spectrophotometry (Schmidt and Seiler, 1970). A precision of $\pm 15\%$ was claimed (see also Seiler and Schmidt, 1974).

19.9.9. CARBON MONOXIDE

The determination of carbon monoxide in sea water has been described by

Swinnerton *et al.* (1968). The procedure used is similar to that employed for hydrocarbons by Swinnerton and Linnenbom (1967a, b) (see below). Carbon monoxide is trapped along with methane in the second cold trap which contains a mixture of activated charcoal and Molecular Sieve. After stripping of the gas from the water is complete, the temperature of the trap is raised to 90°C and the desorbed gases are swept into the gas chromatograph with a current of helium. Separation of carbon monoxide from methane and a small amount of residual oxygen is carried out on a column of Molecular Sieve. The gas emerging from the chromatograph column is mixed with hydrogen and passed over a heated nickel catalyst. In this way carbon monoxide is converted to methane which is then determined using a flame ionization detector. The time required for an analysis is approximately 45 min. The method, which has a sensitivity of ca. 10^{-8} ml CO l^{-1}, has been used in a number of studies of the distribution of carbon monoxide in the sea (see e.g. Elmer and Robbins, 1968; Lamontagne *et al.*, 1971; Junge, *et al.*, 1971).

A different approach has been adopted by Williams and Bainbridge (1973) in their method for the determination of carbon monoxide in sea water. In this, air is circulated through the sample. After equilibration the partial pressure of carbon monoxide in the air is measured gas chromatographically using a column of Molecular Sieve for the separation and a d.c. glow discharge tube as a detector.

19.9.10. ALIPHATIC HYDROCARBONS AND CHLORINATED HYDROCARBONS

The gas chromatographic determination of C_1–C_4 hydrocarbons in aqueous media, including sea water, has been described by Swinnerton and Linnenbom (1967a, b). The hydrocarbons are stripped from the sample by bubbling with helium. The purged gases are dried and then passed via sampling valves through two cold traps cooled to $-78°C$. The first of these, which is filled with alumina, traps all the hydrocarbons except methane. The second trap contains activated charcoal which serves to adsorb the latter and separate it from the major atmospheric gases. After stripping has been completed, the traps are isolated by means of the sampling valves, they are then heated to 90° to desorb the hydrocarbons. Each trap is back flushed separately with helium, and the contents of each are analyzed by gas chromatography using a flame ionization detector. Methane which may be 1000 times as abundant as the other hydrocarbons is separated from any residual air using a silica gel column. The C_2–C_4 saturated and olefinic hydrocarbons are separated on a column of activated alumina containing 10% of paraffin oil. For higher molecular weight hydrocarbons it is advantageous to use a stationary phase consisting of 20% silcone oil on Chromosorb. A sensitivity of 1 in 10^{13} (by weight) is claimed for the method.

A different principle has been used by Williams and Bainbridge (1973) in their method for the determination of methane in sea water. In this, air is circulated through the sample. After equilibrium has been attained an aliquot of the air is analyzed gas chromatographically on a column of Molecular Sieve using for detection the voltage changes occurring in a constant current d.c. glow discharge tube.

A somewhat similar procedure has been used by Murray and Riley (1973) for the determination of very low concentrations of chlorinated aliphatic hydrocarbons (e.g. CCl_4, $CHCl_3$, $CHCl=CCl_2$) in sea water. The chloro-compounds are stripped from the water with a stream of nitrogen and absorbed in a cold-trap packed with silicone oil on Chromosorb-W. They are subsequently removed from the trap with a current of argon and determined by gas chromatography using an electron capture detector (see also Lovelock et al., 1973).

19.10. Determination of Minor Inorganic Constituents

19.10.1. introduction

Many trace elements in the sea do not bear the more or less constant relationship to chlorinity which is found for the major components. This is the result of either their geochemical reactivity or their uptake by marine organisms. The determination of the biologically essential micronutrient elements nitrogen, phosphorus and silicon will not be considered in the present section as they are of sufficient importance to merit extensive separate treatment. Discussion in this section will be confined to other elements occurring at concentrations below 200 µg l^{-1}, but excluding the noble gases and artificial radio-nuclides, the determinations of which are described in Sections 19.9.4 and 19.12 respectively.

Owing to the extremely low concentrations of the trace elements in the sea it is, as yet, only possible in a very few instances to determine the dissolved species present directly (for an account of the use of polarographic techniques in the study of speciation see Sections 20.4.1 and 20.2.5.3). Most of our knowledge of the speciation of these elements has been obtained from calculations made using stability constants derived from measurements made at much higher concentrations in simpler media than sea water, the high ionic strength of the sea water being usually simulated by the use of an alkali metal perchlorate, the anion of which has minimal complexing powers. Such studies have shown that dissolved elements in sea water occur as a variety of different species, including simple (hydrated) ions (e.g. Li^+, Ba^{2+}), oxyanions (e.g. MoO_4^{2-}, WO_4^{2-}), co-ordination compounds (e.g. $HgCl_4^{2-}$, $CdCl^+$ [UO_2-

$(CO_3)_3]^{4-}$) and ion pairs (e.g. $BaSO_4^0$) (see Chapter 3). Calculations suggest that, because the concentrations of dissolved organic compounds are very low, only insignificant amounts of the trace elements will exist in the form of chelates (see Chapter 3). However, several workers (e.g. Morris, 1974; Slowey and Hood, 1971; Muzzarelli and Rochetti, 1974) have claimed that significant proportions of certain elements, such as copper and zinc, do occur in the form of organic complexes of uncertain nature. The possible existence of such organic species poses considerable problems in the determination of the inorganic forms of the elements present in the sea. Thus, ion exchange or solvent extraction procedures which are often used for the preconcentration of the element will tend to partly break down metal chelates by competitive complexation. When concentration is performed by coprecipitation, organic complexes may also be carried down. It should, therefore, be understood that the analyses for the inorganic forms of a trace metal in sea water will also embrace much of the metal associated with labile chelates. The proportion of the organically complexed metal which will be included in the analysis will depend on the relative stability constants of the complexes which the metal forms with the extracting reagent and the natural chelator. If a signifi-cant proportion of a metal is present in water as a chelate* the use of different methods of preconcentration may, therefore, produce somewhat different analytical results. At present, the only additional parameter which can be determined for a metal is its total concentration. Although this can be deter-mined in a few instances by neutron activation analysis of the freeze dried sample, it is more frequently estimated by conventional analytical techniques following a preliminary oxidation stage. In early work oxidation of organic matter was carried out by heating the sample with peroxydisulphate ion $(S_2O_8^{2-})$ (Gast and Thompson, 1958; see also Holm-Hansen et al. 1970; Slowey and Hood, 1971; Muzzarelli and Rochetti, 1974). However, the photo-oxidation procedure of Armstrong and Tibbitts (1968), which has been used for example by Fitzgerald (1970), Morris (1974) and by Paus (1973), is probably more satisfactory because of the smaller risk of contamination.

Some elements (particularly iron, and those of Groups III and IV of the periodic table) may exist in sea water as colloidal, or sometimes more coarsely particulate, hydroxy compounds. These may be removed to a greater or lesser extent, during filtration through a 0·45 μm membrane filter. In addition, appreciable amounts of trace metals are often present in adsorbed form on the suspended inorganic and organic detritus in sea water. A considerable number of elements, such as Fe, Cu, Zn, Cd, Pb and Mn are strongly con-

---

* Work by Duursma and Sevenhuysen (1966) suggests that the levels of chelates are low for most elements.

centrated by living organisms. It is, therefore, important before analysis to filter samples which contain significant proportions of suspended matter (i.e. all samples from neritic and entrophic waters). As in nutrient analysis it is customary to use a 0·45 μm membrane filter for this purpose (e.g. a Millipore® HA filter). It should, of course, be realized that this choice of pore size is a purely arbitrary one, and that the "dissolved" fraction which passes through the filter will contain at least some of the colloidal species present in the sample (see Section 19.4). Problems arising from contamination of the filtrate by soluble impurities present in certain membrane filters have been reported by several workers (see e.g. Robertson, et al., 1968; see also Section 19.4). However, determinations by Spencer and Manheim (1969) of trace elements in Millipore® HA filters (Table 19.4) suggest that these filters should cause little contamination of samples even if, as is very unlikely, all of the metal is leached from the filter, provided that a sample of > 10 l is filtered through a single filter. In practice it is preferable to discard the first 1–2 l which passes through the filter (Preston et al., 1972).

Because of the extremely low concentrations at which many trace elements are present in sea water it is very important to guard against contamination during collection, storage, and analysis. This is particularly true for the common industrial metals such as iron, copper, zinc, lead and nickel. Thus, it is essential to use plastic water sampling bottles; even with these, care must be taken to ensure that rubber and neoprene used as closures are free from fillers which are often rich in zinc, barium and other heavy metals. In addition, any metal fittings on the bottles should preferably be well coated with plastic (see Section 19.3 for a discussion of water sampling). In order to minimize changes in the concentrations of trace elements which often occur on storage, as a result of adsorption or contamination by the container, it is advisable to commence the analysis as soon as possible after sampling. If the analysis cannot be completed immediately it is possible, even on shipboard, to carry out a preconcentration (e.g. by chelating ion exchange) and to store the concentration for analysis later.

The storage, without change, of samples for trace metal analysis is a matter of considerable difficulty. There appears to be no one method which is applicable to all trace metals. However, little alteration occurs for many trace elements (e.g. Cu, Zn, Fe, Ni, Co, Mn, In, U, Rb, Re, Cr, La) over 2–3 weeks if the samples are filtered, acidified to pH 2·0–2·5 with purified hydrochloric acid and stored in acid cleaned polyethylene bottles (Robertson, 1968a). Unfiltered samples should not be acidified because of risk of desorption of trace elements from the particulate matter (see e.g. Carr and Wilkniss, 1973). It is perhaps advantageous to freeze the acidified samples (Robertson, 1968a). Changes on storage are particularly pronounced for mercury (see e.g. Carr

K

TABLE 19.4

Maximum possible contamination by 47 mm Millipore® HA (0·45 µm) filters using a 10 l water sample (after Spencer and Manheim, 1969)*

Element	Fe	Cr	Mn	Cu	Zn	Ba	Ni	Ag	Mo	Co
Concentration in filter ash (%)	1·4	0·75	0·12	1·2	0·77	0·05	0·08	0·01	0·02	0·006
Sea water concentration (µg l^{-1})	1	0·5	2	3	3	20	7	0·03	10	0·01
Maximum contamination as percentage of sea water concentration	6	6	0·25	1·7	1	0·01	0·04	1·7	0·008	2

* For data on Sc, Hg, Sb, Au, La see Spencer et al. (1972).

and Wilkniss, 1973; Coyne and Collins, 1972; Rosain and Wai, 1973), silver and gold (see e.g. Chao *et al.*, 1968) because of the ease with which these elements are adsorbed onto the surfaces of polythene containers, even under acid conditions. Storage tests should be carried out to determine the optimum conditions if data are not available for a particular element. It is often convenient to make such tests radio-chemically using a radionuclide of the element concerned; however, care must be taken to ensure that equilibrium is attained between the added nuclide and that in the sample. For a general discussion of problems of sample storage the papers by Robertson (1968a) and Bowen *et al.* (1970) should be consulted.

Although it is obviously preferable to have a laboratory for trace element work air-conditioned, for many trace elements this is not normally essential provided that it is not in an industrialized area, and that it is well separated from workshops in which welding or soldering are being performed. Pollution from air-borne dust can usually be kept at a negligible level by restricting the laboratory to trace element work and cleaning it regularly and thoroughly. Under adverse circumstances it may be necessary to work in positive pressure laminar flow hoods (Fig. 19.12). Extreme cleanliness in working will do much to reduce contamination. Problems concerned with contamination occurring during analysis for trace metals are further discussed in Section 19.10.4.11.5. Such problems are particularly severe for lead (see Section 19.10.5.27).

In addition to contamination arising from air-borne pollution, contamination may occur during processing and analysis from impurities leached from the walls of the apparatus or present in the reagents. Bowen *et al.* (1970) have reviewed the properties of the various materials available for the construction of apparatus for trace element analysis and sample storage. From the data which they have assembled (Table 19.5A) it will be seen that few materials are completely satisfactory. Probably the best is polytetrafluoroethylene, which is chemically very inert and contains very low levels of trace elements (Robertson 1968b; Table 19.5B). A considerable variety of laboratory equipment made of this material is available; however, it is too expensive to use for any but special purposes. The much cheaper polyethylene is only slightly inferior in purity, and is probably the most suitable material for use in applications in which heating above 90°C is not required. According to Bowen *et al.* borosilicate glass apparatus is unsafe to use in trace element analysis. However, in the author's experience it is possible to use it in the determination of many elements (except Zn and Pb) provided that it is not used under strongly alkaline conditions. Before use it should be first treated with a mixture of concentrated nitric and sulphuric acids or with one of the proprietary chelating detergents, such as Decon. Following this it should be well washed with water; it should not be allowed to become dry as this

TABLE 19.5A

Properties and suitabilities of materials for use in trace metal work (after Bowen et al., 1970)

	Comments	Recommendation
Polyvinyl chloride	Often seriously contaminated with trace metals; may adsorb others.	Do not use; but needs more research because of its use in samplers, etc.
Polyethylene A. cross-linked (conventional)	May be a source of Sb, Co, Cr, Fe; may sorb Zn and some other trace metals. Batches vary widely.	Can be used, but needs checking before use in any application.
B. linear	Probably better than cross-linked	Can be used, but needs further study
Polypropylene	Removed Ba (and Sr?) at pH > 3·5, also lanthanides and Ag.	Can be used, but needs more study
Polystyrene	Removed Ag.	Needs more study in view of its use in disposable apparatus.
Polytetrafluoroethylene	Probably the best general purpose plastic; but expensive, soft and subject to pinholing when used as coatings. Sorbed Ag.	Recommended.
Acrylate	Levels of impurities very low; sorbed Zn and Ag.	Needs more study in view of use in samplers.
Silica	Composition variable; subject to attack under alkaline conditions.	Recommended, but needs further study
Borosilicate glass	Attacked chemically at sea water pH	Unsafe in many applications.
Vycor glass	Attacked chemically at sea water pH; trace metal content variable.	Unsafe.
Soft glass		Rarely safe for trace metal work.

TABLE 19.5B

Trace element levels (ng g^{-1}) of various materials used in laboratory equipment (from data from instrumental neutron activation analyses by Robertson, 1968b). (The figures should be regarded only as indications of the general levels of trace metals since variations of orders of magnitude may occur from batch to batch and from products of one manufacturer to another).

Material	Zn	Fe	Sb	Co	Cr	Sc	Cs	Ag	Cu	Hf
Polyvinyl chloride	7×10^3	$2 \cdot 7 \times 10^5$	$2 \cdot 7 \times 10^3$	45	2	4·5	<1	<5	630	nm
Polyethylene										
hose	55	7·4	9×10^3	140	254	11	<100	<200	nm	<100
container (Nalgene)	28	10^4	0·2	7×10^{-2}	76	8×10^{-2}	5×10^{-2}	1·1	6·6	<0·5
Polytetrafluoroethylene	9·3	35	0·4	1·7	<30	$<4 \times 10^{-3}$	$<10^{-2}$	<0·3	22	nm
Polymethyl methacrylate	<10	<140	$<10^{-2}$	$<5 \times 10^{-2}$	<10	$<2 \times 10^{-3}$	$<6 \times 10^{-2}$	$<3 \times 10^{-2}$	<9·5	nm
Surgical rubber tubing*	3×10^6 to 4×10^7	<100	<100 to 360	<30 to 7×10^3	4×10^5	<8 to 185	<100 to 580	<700 to 1240	<6	nm
Neoprene rubber	$1 \cdot 8 \times 10^7$	um	290	2×10^3	um	3×10^3	um	<1000	um	nm
Silica tubing										
(Suprasil)	<1	nm	$<10^{-2}$	12	2·5	0·4	<0·1	$<10^{-2}$	4×10^2	$<5 \times 10^{-3}$
(Spectrosil)	1·5	400	5×10^{-2}	0·44	6·5	3×10^{-2}	1	5×10^{-2}	2	$<5 \times 10^{-3}$
Borosilicate glass	730	3×10^3	3×10^3	81	um	106	<100	$<10^{-3}$	nm	600
Vycor glass	um	um	10^6	um	um	um	um	um	um	um
Millipore filter	2×10^3	330	39	13	$1 \cdot 7 \times 10^4$	0·8	1·5	5×10^{-2}	nm	<0·5
Kimwipe tissue	5×10^4	10^3	16	24	500	14	<0·1	~0·8	nm	nm
Steel hydrographic wire	um	—	5×10^4	6×10^4	um	<50	um	um	2×10^4	um

* Range of values of samples from 3 different suppliers.
nm = not measured. um = could not be measured because of interference from other radio nuclides.

appears to make it soluble again. Fused silica apparatus is eminently suitable for trace metal work but is fragile and costly.

Contamination arising from impurities present in reagents is often a serious problem in the determination of trace elements in sea water. Analytical grade chemicals and even "electronic" grade ones are frequently not sufficiently pure to be used, without purification, in the determination of the more ubiquitous elements, such as Cu, Zn, Pb and As. Robertson (1968b) has used neutron activation analysis to investigate the levels of 10 trace metals (Sc, Cr, Fe, Co, Cu, Zn, Ag, Sb, Cs and Hf) in a wide range of acids, solvents and reagents used in trace element analysis (Table 19.6). Contamination of reagents varies considerably from batch to batch, and Robertson's data should only be taken as a general indication of the order of magnitude of the contamination by a particular element. Nevertheless, his figures do show the severity of the problem which can arise from impurities present in reagents. Impurity levels are often particularly high in organic complexing reagents, such as dithizone. The method to be used for the purification of any particular reagent depends on the reagent itself and the contaminant to be removed. Mineral acids* and solvents can usually be purified by distillation in a well aged silica still. Heavy metals can often be removed from reagents in several ways (i) by extraction with a solution of dithizone in carbon tetrachloride, (ii) by use of a chelating ion exchange resin, (iii) by controlled potential electrolysis into a mercury pool cathode (commercially produced apparatus which is now available for this purpose is capable of reducing levels of certain trace elements to $<0.1 \, \mu g \, l^{-1}$ in litre volumes of reagents). Water to be used for the preparation of reagents should be redistilled from a silica still. It is inadvisable to use de-ionized water since the small quantities of the exchange resin which always dissolve during the de-ionization process may complex the metals to be determined.

Except for neutron activation and anodic stripping procedures no analytical techniques are available for the direct determination of elements in sea water at concentrations below $5 \, \mu g \, l^{-1}$. Although it has been suggested (Arnon, 1953; Bernhard, 1955) that it might be possible to develop specific bio-assay procedures for this purpose, this line of approach does not seem to have been further pursued. It is usually essential therefore to concentrate the element from a large volume of water before determining it. After concentration it may be necessary to separate it from other elements which have also been removed by the concentration step. Finally, the determination is carried out by a specific and sensitive physico-chemical technique. It is extremely important that blank determinations should be carried through

---

* Hydrochloric acid and nitric acid can be satisfactorily purified by iso-thermal distillation.

TABLE 19.6

Typical trace metal concentrations (ng g^{-1}) in some reagents and solvents (from instrumental neutron activation analyses by Robertson, 1968b). (It should be appreciated that there may be large variations between one batch and another, particularly with organic reagents —see e.g. 8-hydroxyquinoline)

Sample	Zn	Fe	Sb	Co	Cr	Sc	Cs	Ag	Cu	Hf
Quartz distilled water	1–10	<0.2–1	0.06–0.10	0.04–0.20	2–10	0.002	<0.01–0.1	<0.02	nd	<0.005
Double distilled water	~1	<0.2	<0.01	<0.02	~2	<10^{-4}	<0.01	<0.02	nd	<0.001
Triple distilled water	~0.5	~1	<0.02	<0.02	12	~2 × 10^{-4}	<0.01	<0.02	nd	<0.001
Nitric acid[1]	13	~2	~0.03	0.02	72	7 × 10^{-4}	<0.01	~0.24	1.3	<0.005
Hydrochloric acid[1]	22	~1	0.20	0.09	1.1	0.002	<0.002	<0.1	82	<0.005
Sodium hydroxide[1]	<20	<900	0.32	5.5	60	0.3	0.7	<0.2	nd	nd
Ammonia solution[1]	2	<0.1	<0.006	~0.009	<0.04	<3 × 10^{-4}	<0.002	<0.1	6.0	—
Carbon tetrachloride[2]	1.2	10	0.3	~0.003	<50	~0.002	<0.1	<0.005	0.12	<0.005
Chloroform[3]	2.1	1.6	0.05	~0.003	<100	~3 × 10^{-5}	<0.02	<0.005	0.29	<0.005
Dithizone	1100	<7000	0.8	1.2	<2000	0.15	10	<10	420	<0.1
Thionalide	120	<300	3.7	5.1	um	0.29	<10	<2	0.4	<100
Ammonium pyrrolidine dithiocarbamate	1970	~5000	1.9	1.3	um	0.11	<1	<1	4	<10
8-hydroxyquinoline[4]	<40–370	<100–5700	<0.2–1210	<0.2–1.8	<50	<0.02–0.14	<0.1–0.4	0.6	290	nm

Key to Table 19.6
[1] Baker and Adamson CP reagent.
[2] Baker Analyzed Reagent, doubly distilled.
[3] Mallinckrodt Analytical Reagent, doubly distilled.
[4] Ranges for 4 samples from different manufacturers.
nd = not determined.
um = could not be determined because of interference by other radionuclides.

the whole analytical process at the same time as the determinations, in order to permit correction to be made for contamination.

When developing analytical methods for the determination of trace elements in sea water it is essential to check the efficiency of the proposed concentration and separation processes. This can be done chemically by spiking stripped sea water samples with known amounts of the appropriate species of the element in question and then checking the recovery by a physico-chemical method. However, if a suitable radio-nuclide of the element is available the recovery can be determined more simply radiochemically (it is important that the radio-nuclide added should have a high specific activity so that the amount of the inert form of the element added is kept to a minimum). In both the chemical and radio-chemical methods of checking the separation process it is, of course, vital that the spike should be equilibrated with that present naturally in the sample.* In practice it is usually possible to devise analytical schemes which will give 98–99% recoveries of trace elements at levels down to $0.1 \mu g \, l^{-1}$, or less. If the yield is significantly lower than this, ca. $0.2 \mu Ci$ of a carrier-free radio-nuclide of the element being determined should be equilibrated with the water sample. After concentration and separation, but before completing the determination, the radioactivity of the product is compared with that of the spike. The chemical yield is the ratio of these two activities and can be used to correct the determined amount of the element for losses occurring during the analytical processes.

19.10.2. METHODS FOR CONCENTRATION OF TRACE ELEMENTS FROM SEA WATER

There are comparatively few methods which can be used to concentrate trace elements efficiently from sea water in which they are present at levels of less than 1 in 10^9. Although excellent recoveries of a wide range of trace elements (other than the rare alkali metals) can be achieved by coprecipitation, or the closely related cocrystallization, these techniques are not used as much as formerly. This is because equally efficient concentrations can be attained by solvent extraction and ion exchange procedures which have the advantages of greater speed and selectivity. They also provide the trace elements in a suitable form for direct analysis by atomic absorption spectro-photometry; multi-element analysis is also possible. Techniques for the pre-concentration of trace elements from sea water have been reviewed by Joyner et al. (1967), and Rottschafer et al. (1972) has given a general account of preconcentration techniques as applied to neutron activation analysis.

19.10.2.1. *Coprecipitation*

All precipitates tend to carry down substances which would normally be

---

* There is some evidence that for many elements equilibration with thin stable counterparts in sea water may be a slow process.

soluble under the conditions of the precipitation, i.e. the solubility product of the least soluble compound which could be present is not exceeded. This phenomenon which is termed coprecipitation is of considerable value in the concentration of trace elements from sea water and enables many of them to be recovered almost quantitatively even at levels $< 100 \, ng \, l^{-1}$. Analytically useful coprecipitation processes are of two types.

(i) That in which the coprecipitated element is incorporated within the lattice of the precipitate. Hahn (1936) has stated that "when the separation of the unweighable amounts takes place within the precipitate and is practically independent of the conditions of precipitation—such as speed and excess of precipitate—the process is called true coprecipitation with the mass of the precipitate, and is accompanied by the formation of mixed crystals or of systems resembling mixed crystals". Because of the requirement for mixed crystal formation this type of cocrystallization tends to be specific in its action. Its use in the analysis of sea water has been restricted to the concentration of radium by coprecipitation with barium sulphate (see p. 434) and of caesium and rubidium with potassium cobaltinitrite. Hermann and Suttle (1961) have given a general review of coprecipitation by this mechanism.

(ii) That in which the coprecipitation occurs by a surface adsorption mechanism. Coprecipitation of this type is important with those precipitates which have large specific surface areas, particularly colloids in which the charge borne by the micelle also contributes to the adsorption process. Over a limited range of concentrations the adsorption can be represented by the Freundlich isotherm

$$A = KC^{1/n}$$

where, at equilibrium, A is the amount of impurity element adsorbed by a given weight of adsorbent, C is its concentration in the solution, and K and n are constants. If C is small, only a small fraction of the surface of the adsorbent will be covered by the impurity element and n will approach unity. Under these conditions, the amount of the impurity adsorbed per unit weight of adsorbent is proportional to the equilibrium solute concentration. With high concentrations of impurity, the surface of the adsorbent approaches saturation with the impurity with the result that the efficiency of coprecipitation decreases. If the precipitate is allowed to stand too long in contact with the solution it may undergo recrystallization (ageing). This may reduce the surface area available for adsorption. If the surface is almost saturated, some of the adsorbed impurity will be released unless it can be incorporated in the crystal lattice of the adsorbent. When adsorptive coprecipitation is used for concentration of trace metals from sea water, desorption is prevented by the use of sufficient adsorbent to ensure that the surface will not become saturated.

The factors determining the adsorption of various ions by a particular lattice have been detailed by Fajans and Erdey-Grúz (1932) as follows:

(i) An ion in aqueous solution is readily adsorbed on the unchanged surface of a salt only if it forms a compound of low solubility or weak dissociability with the oppositely charged ion of the lattice. For example, bismuth is coprecipitated with $BaCO_3$ and $Fe(OH)_3$, but not with $BaSO_4$ or $PbSO_4$ in acid solutions.

(ii) The adsorption of a cation is increased (or that of an anion is decreased) in the presence of adsorbed anions, i.e. by a negative charge on the surface of the adsorbent, and decreased (or for anions increased) in the presence of other adsorbed cations, i.e. by a positive charge on the surface. In both instances the effect is greater the greater the adsorption of other ions.

(iii) The tendency towards adsorption increases with increasing polarizing influence of the ions which are being adsorbed upon the oppositely charged ions of the lattice. Thus, if the ions involved have strong polarization, the foreign ions can be adsorbed even when the precipitate has a charge of the same sign.

Coprecipitation by adsorption is a versatile but unspecific method for the concentration of many trace elements from sea water. Among the adsorbents which have been used for this purpose are the hydrous oxides of iron (III) (for e.g. Al, As, Be, Co, Cr^{3+}, Ga, Ge, lanthanides, Mo^{6+}, Nb^{5+}, Ni, Sc, Se, Si, Th, U^{6+}, V^{5+} and W, (see Burrell, 1967)), aluminium (for Cr, Mn, Ti and P), and manganese (IV) (for Mo, W, Sb and Bi), the sulphides of copper (e.g. for Cd, Pb and Hg) cobalt (for radioactive nuclides of Ru), lead (e.g. for Au and Ag), and iron (for Ni), and the mixed precipitate of calcium and magnesium compounds obtained on addition of alkali to sea water (see e.g. Joyner et al., 1966). West (1961) has provided data on the elements coprecipitated with variety of commonly used adsorbents.

Hydrous iron (III) oxide, which forms a negatively charged colloid, is probably the most efficient coprecipitating agent agent available, and has been very widely used for the preconcentration of trace elements from sea water. The coprecipitation is usually carried out by treatment of the sample with iron (III) chloride (equivalent to ca. $5 \, mg \, Fe \, l^{-1}$). Owing to the alkalinity of the water, hydrolysis soon occurs, and the colloid which is initially produced subsequently coagulates under the influence of the major ions present in the water. After the precipitate has settled, it is often advisable to perform a second precipitation using the same amount of iron (III) chloride. After settling, the combined precipitate is separated by siphoning off most of the supernatant liquid. It is then recovered by centrifugation (filtration is not satisfactory because of its gelatinous nature), and washed with 0·5% ammonium nitrate solution by decantation. The major drawback to the use of hydrous iron oxide as a coprecipitating agent is the difficulty of separating iron from the analyte element. Many workers have attempted to carry out this separation by extracting it as its chloride using ether; however, the efficiency of this process falls off rapidly as the iron concentration decreases. Methods based on anion exchange are probably more satisfactory. When the double coprecipitation technique is employed it is frequently possible to recover elements occurring even at levels as low as 20 ng l^{-1} with an efficiency in excess of 97%. In addition to its use for the recovery of non-radioactive elements, coprecipitation with hydrous iron (III) oxide has been frequently used for the isolation of radio nuclides, the inert form of the element being used as carrier. It has also been used as a scavenging agent for the removal of unwanted activities before separation of, for example, ^{137}Cs and ^{90}Sr.

Hydrous manganese dioxide, which can be produced for example by the reduction of permanganate ion with ethanol, is an efficient coprecipitating agent for certain trace elements. It is more selective in its action than is hydrous iron (III) oxide. Fukai (1969) has used adsorption of ions on preformed manganese dioxide to differentiate between the ionic and chelated species of certain elements in sea water.

Coprecipitation with sulphides has been widely used for the collection of trace elements in the examination of fresh waters, particularly as a preliminary to emission spectroscopy. However, it does not seem to have been extensively adopted for sea water analysis.

Attempts have been made to carry out the preconcentration of trace elements by passing the sample through columns containing (preformed) adsorbent incorporated in a suitable stationary phase. Thus, Merrill et al. (1960a, b) have employed short columns of cation exchange resin in which hydrous iron (III) oxide or hydrous manganese dioxide were dispersed for the adsorption of beryllium (1 ng l^{-1}) from 100 l of sea water. They also suggested their use for the recovery of thorium and the lanthanides. Lal et al. (1964) have claimed that the cosmic ray-produced nuclides ^{32}Si and ^{7}Be can be efficiently recovered from sea water by towing through the water perforated stainless steel buckets containing a porous matrix of jute or sponge impregnated with hydrous iron (III) oxide. However, they found that the versatility of this collecting agent was limited by the relatively high trace element content of the support medium; for this reason, in later work, Lal and his co-workers (Krishnaswami et al., 1972) have used hydrolyzed acrilan fibre as the support phase in the determination of thorium isotopes, silicon-32 and lead-210 (see also Tera et al., 1965).

19.10.2.2. *Co-crystallization*

Although coprecipitation has been used very effectively for the preconcentration of trace elements from sea water it suffers from several drawbacks. (i) It is usually unspecific in its action, (ii) the precipitates are often gelatinous and difficult to handle, (iii) it is often difficult to separate the desired element from 10^4 or more times its own weight of the carrier element, (iv) it is sometimes difficult to obtain the carriers sufficiently free from the element being concentrated. If the coprecipitation could be brought about by the use of an organic complexing agent it might be possible to achieve a more selective recovery of the element and to remove the carrier it would only be necessary to wet or dry ash the precipitate.

The first attempt to use this principle was made by Black and Mitchell (1951) who coprecipitated a number of trace elements using a mixture of thionalide, 8-hydroxyquinoline and tannin; a small amount of iron (III) was used as a carrier. The precipitate was examined by emission spectroscopy. Similar procedures have also been adopted by Young *et al.* (1959) and Silvey and Brennan (1962).

Kuznetsov (1954) has employed a different approach in which the sample is treated with a reagent, such as methyl violet, which forms a complex with the analyte element. A precipitant for the complexing agent is then added (e.g. methyl violet is precipitated by means of thiocyanate ion, tannin, or an azo compound, such as arsenazo I or stilbazo II). The precipitate will act as an efficient carrier for a particular element if the compound which it forms with the complexing agent is only sparingly soluble or slightly ionized. This technique has been used for the preconcentration of uranium (Kuznetsov and Akimova, 1958), lanthanides (Balashov and Khitrov, 1961) and thorium (Kuznetsov *et al.*, 1962; but see p. 436).

Weiss and co-workers (Weiss and Lai, 1960; 1961; Weiss and Reid, 1960; Weiss *et al.*, 1961) have found that when slightly soluble organic complexing agents crystallize they tend to carry down any insoluble complexes which they form with metals, and they have devised methods based on this principle for the efficient co-crystallization of trace metals from sea water. The process is simpler than that used by Kuznetsov (*vide supra*) and can be used with a range of different complexing agents. Fajan's rule suggests that the primary requirement for effective cocrystallization is that the reagent should form a complex which is more insoluble that it is itself. In the procedures which they developed the precipitate is produced rapidly by treating the sample with a solution of the reagent in a water-soluble organic solvent. Under such conditions, as recrystallization and diffusion of the trace element within the crystals are negligible, the distribution of the trace element should follow the logarithmic distribution law of Doerner and Hoskins (1925)

$$\lambda = \frac{\log\left[1 - \text{fraction of trace element co-precipitated}\right]}{\log\left[1 - \text{fraction of carrier precipitated}\right]}$$

where λ is the logarithmic distribution coefficient. However, in practice there may be significant deviations from this distribution law (Weiss et al., 1961), perhaps because the concentration of organic solvent increases during the crystallization process.

The co-crystallization process is usually carried out by treating the water sample, adjusted to an appropriate pH value, with a 3–10% solution of the complexing agent in a water-soluble volatile organic solvent. With some reagents it is necessary to boil the solution to remove the solvent. After cooling, the crystalline precipitate is removed by filtration; the precipitate is then wet-ashed (usually with nitric acid) to recover the trace metals.

Potassium rhodizonate has been found by Weiss and Lai (1960) to be an efficient co-crystallizing agent for barium, plutonium, cerium and zirconium from sea water. 1-Nitroso-2-naphthol has been shown (Weiss et al., 1961) to carry cobalt and iron quantitatively over a wide pH range; cerium and uranium are co-crystallized quantitatively at pH 7. Zinc and zirconium could be recovered with an efficiency of ca. 97% at pH 7 and 5 respectively.

Thionalide is a versatile reagent for the co-crystallization of trace metals from sea water. Lai and Weiss (1962) have shown that under the optimum pH conditions the following elements can be recovered with an efficiency of > 90%: Ag, Au, Co, Ga, Hf, Hg, In, Ir, Mn, Os, Ru, Sn, Ta, Tl, W, Zn. For several elements (e.g. Au and Ag) neither the logarithmic distribution law nor Fajan's rule were obeyed. They used the reagent for the co-crystallization of silver from sea water and obtained a recovery of $90 \pm 2\%$ at a concentration level of $150 \, \text{ng Ag} \, l^{-1}$. The same reagent has also been used by Portmann and Riley (1964) for the co-crystallization of arsenic (III) from sea water; in this instance a recovery of $\sim 98\%$ was achieved. Other complexing agents which have been employed for co-crystallization of trace elements from sea water include 2-mercaptobenzimidazole (Weiss and Lai, 1963) and 5,7-dibromo-8-hydroxyquinoline (used by Riley and Topping (1969) for co-crystallization of copper, zinc and iron).

19.10.2.3. Ion exchange

For convenience of discussion ion exchange procedures may be divided into three principal categories depending on the type of exchanger used, viz. cation exchange, anion exchange and chelating ion exchange. It is only possible here to review the application of these techniques to the concentration of trace elements from sea water; for general information on ion exchange, the textbooks by Samuelson (1963) and Rieman and Walton (1970) should be consulted.

Although cation exchange procedures are very useful for the separation and determination of the major cations in sea water (see section 19.8.2), they have proved to be of little value for the preconcentration of trace elements. Indeed, they have only been used for the concentration of lithium (Chow and Goldberg, 1962; Riley and Tongudai, 1964) and barium (see e.g. Andersen and Hume, 1968a and p. 341). Because of its very low affinity for cation exchange resins lithium is extremely easily displaced from such resins by hydrogen ion. Quantitative separation of the element from ca. 50,000 its own weight of sodium can be readily achieved. In contrast, barium is strongly retained by cation exchangers, and after elution of the major elements, can be eluted by means of chelating agents. Novikov et al. (1972) have investigated the uptake of 25 elements from sea water by a variety of cation and anion exchangers. Marchand (1974) has used cation exchange chromatography in an investigation of the speciation of Co, Mn, Zn, Cr and Fe in both a natural sea water and one enriched with marine algal organic matter.

Anion exchange techniques provide a very useful means for concentrating those elements which, in sea water medium acidified with hydrochloric acid, exist as chloro-anions. Krause and Nelson (1956) have studied the uptake of many elements by the strongly basic anion exchanger Dowex I as a function of hydrochloric acid molarity. The volume distribution coefficients (D_v) which they obtained indicated that none of the geochemically abundant elements form strong chloro-complexes in 2M hydrochloric acid medium. Elements, which in their highest oxidation states form strong chloro-anions under these conditions are, in order of decreasing D_v (i.e. decreasing affinity for the exchanger):

$$Au\,(D_v = 10^6), Tl, Hg, Bi, Sn, Cd, Zn, Re*Ag\,(D_v = 100)$$

Iron, which forms a weak chloro-anion $(D_v = 10)$ can be readily separated from the more strongly adsorbed elements.

Only a few attempts have been made to employ anion exchange techniques for the determination of trace elements in sea water. Brooks (1960) achieved enrichment factors of up to 2×10^7 when taking up Tl, Cd, Bi and Au onto $0.5\,cm^2 \times 13.2\,cm$ columns of Amberlite 1R 400 resin from 250 l samples of sea water which had been made 0.1 M with respect to hydrochloric acid. Bromine $(10\,\mu g\,ml^{-1})$ was added to oxidize thallium to the 3 + oxidation state. Only cadmium and bismuth could be detected spectrographically after elution with 0.25 M nitric acid. Gold could not be eluted, as it was

* It seems probable that rhenium does not in fact exist in sea water as a chloro-anion, but rather as the perrhenate ion (ReO_4^-).

perhaps reduced to the elemental state; it was recovered by ignition of the resin. Matthews and Riley (1969a) have recovered thallium (~ 10 ng l^{-1}) quantitatively from sea water by taking it up as a Tl^{3+} chloro-anion on an anion exchanger; after elution of all other adsorbed elements (except gold), thallium was eluted with a reducing agent (H_2SO_3) for determination by neutron activation analysis. Portmann (1965) has found that bismuth (~ 20 ng l^{-1}) can be recovered from acidified sea water by anion exchange with an efficiency of $> 98\%$. Carritt (1962) has given a preliminary account of the use of anion exchange methods for the concentration of zinc, copper and lead from sea water. Rhenium, (~ 10 ng l^{-1}) which probably occurs in sea water as the ReO_4^- ion, can be recovered quantitatively from sea water by anion exchange methods (Matthews and Riley 1970c; see also Section 9.10.5.39).

In recent years, considerable efforts have been made to develop column filling materials with complexing or chelating groups which could be used for the selective uptake of particular trace metals. For such materials to be of value in sea water analysis they must have a very low affinity for the major cations as well as a high affinity for the metal(s) which it is desired to concentrate. Two basic approaches have been adopted in developing such chelating materials. In the more widely used of these, the complexing or chelating groups are integral with the support phase. In the second approach a complexing reagent is sorbed onto an ion exchange resin or other suitable solid phase.

A number of attempts have been made to produce ion exchange media having chelating groups in place of the usual sulphonic acid or quaternary ammonium groups, (see Blasius and Brozio, 1967). The most successful of these media is a copolymer of vinylbenzyliminodiacetic acid, styrene, and divinylbenzene. This is available commercially under the trade name Dowex A1 (the analytical grade of which is known as Chelex-100). The value of this material for the analytical concentration of heavy metals from waters was first realized by Biechler (1965) who employed it in the analysis of waste waters. Subsequently its use was extended to sea waters by Callahan et al. (1966) and by Riley and Taylor (1968, 1968a, b) who found that it was possible to obtain quantitative recoveries of many trace metals (including Cu, Co, Ni, Cd, Zn, Mn, Pd, Lanthanides, In, Mo and V. Several other metals (e.g. Au, Ag, Hg, Bi) are quantitatively adsorbed by the resin, but can only be partially eluted. Unfortunately, the stability constants of the individual trace metal – Dowex A1 chelates differ too little from one another for there to be any prospect of the exchanger being of value for carrying out separations of the metals (van Willigen and Schonebaum, 1966). It should be pointed out that the chelating exchange resin will break down any weak metal–organic complexes present in the water, and the metal recovered will therefore include this fraction. In practice, a $1.5 \text{ cm}^2 \times 12 \text{ cm}$ column of the resin in its ammonium form is used, and this will suffice for the quantitative uptake of trace elements from at least 50 litres of sea water. Elution is carried out by means of 2 M nitric acid. Immediately after elution, the resin must be reconverted to its ammonium form by means of 2 N ammonia solution (purified by iso-thermal distillation) as it is unstable in its hydrogen form. The chelating ion exchange procedure has been adopted for the analysis of sea water by several groups of workers (see e.g. Abdullah and Royle, 1972; Morris, 1974; Knauer and Martin, 1973; Stoner and Chester, 1974). It has the great advantages over solvent extraction procedures that it is easy to use at sea and that it is possible to achieve concentration factors of > 1000. The general use of chelating ion exchangers in water analysis has been reviewed by Hering (1971).

Muzzarelli and his co-workers (Muzzarelli and Tubertini, 1969; Muzzarelli et al., 1970; Muzzarelli and Marinelli, 1972; Muzzarelli and Rochetti, 1974) have suggested the use of chitosan (de-acetylated chitin which is a natural polymer containing glucosamine and N-acetyl glucosamine present in shells of crustacea) as a cheap and possibly superior alternative to the expensive Chelex resin. They have used it for the preconcentration of Au, Sb, Fe, Cu, Ni, Mo, Zn from sea water.

More specific organic complexing agents immobilized on a variety of media have also been used for preconcentration of trace metals. Thus, dithizone has been used for this purpose after immobilization on cellulose acetate (Carritt, 1953), or anion exchange resin (Carritt, 1965)*, or by diazo coupling with carboxy methyl cellulose (Burrell, 1968). A number of complexing stationary phases are available commercially (e.g. the phase consisting of 8-hydroxyquinoline azo-coupled to an arylamine coated on porous glass beads, produced by the Pierce Chemical Co., Rockford, Ill); however, although they appear to be potentially valuable for preconcentration of trace metals, they do not seem to have yet been used in the analysis of sea water.

19.10.2.4. Solvent extraction

Solvent extraction procedures have been used extensively for the preconcentration of trace metals from sea water as a preliminary to their determination by colorimetry and by both flame emission and atomic absorption spectrophotometry. In the first of these applications the possibility of making the extraction selective or even specific is often of advantage. When the extraction is used in conjunction with the two latter methods of determination the enhanced sensitivity resulting from the spraying of the organic

---

* Using this stationary phase Topping (1969) was able to obtain recoveries for cobalt and zinc of only 60% and 40% respectively. However, he was able to achieve quantitative recovery of cobalt, by means of 5,7-dibromo-8-hydroxyquinoline immobilized on an anion exchange column.

solvent is frequently beneficial and multi-element analysis is possible. A limitation in the use of solvent extraction procedures in the determination of very low concentrations of trace elements is that it is not practicable to obtain a concentration factor greater than 100.

In order to make an ion extractable with an organic solvent it is necessary to replace its co-ordinated water molecules with other groups (ligands), so as to produce complexes which are essentially non-polar. Ligands used for this purpose are usually multi-dentate and the complexes which are formed are most stable if the resultant ring has 5 or 6 members.

If the formation of a chelate between a metal M and a protonated ligand is represented by the equation

$$M^{n+} + n\,HR \rightleftarrows MR_n + nH^+,$$

then if hydrolysis and other competing reactions and the formation of intermediate complexes are ignored, it can be shown that

$$D = \frac{\Sigma[M]_{org}}{\Sigma[M]_{aq}} = \frac{K_C K_{CD} K_R^n}{K_{RD}^n} \frac{[HR]_{org}^n}{[H^+]_{aq}}$$

Where D is the *distribution ratio* of the total metal species between the organic and aqueous phases, K_C is the instability constant of the extracted species MR_n, K_{CD} is the partition coefficient of this extracted species, K_R is the partition coefficient of the ligand R^-, $[HR]_{org}$ is the concentration of HR in the organic phase, $[H^+]_{aq}$ is the hydrogen ion concentration in the aqueous phase, and n is the number of ligand groups co-ordinated with each metal ion. From the equation it will be seen that in order to obtain efficient extraction of a metal (i.e. a high value of D) a reagent should be selected which gives high values of K_{CD} and K_R, and as low a value of K_{RD} as possible. For any particular reagent the concentration of HR in the organic phase should be as high as possible and $[H^+]_{aq}$ should be as low as possible consistent with prevention of hydrolysis of M^{n+}. For discussions of theoretical and practical aspects of the solvent extraction of metals the reader is referred to the reference texts by Morrison and Frieser (1957), Stary (1964), De et al. (1970) and to the review by Lyle (1973).

Specific complexing agents for use in extraction are available for only a few elements (e.g. diphenylphenanthroline for iron (II) and biquinolyl for copper (I)). However, it is frequently possible to make specific extractions for particular elements using reagents which are merely selective. This can sometimes be done by carrying out the extraction at a suitable pH (e.g. only Cu, Ag and Hg are extracted by dithizone at pH < 1). Alternatively, interfering elements can be masked by complexing them with ligands so as to produce strong complexes which do not extract under the conditions used, for example, only mercury (II) is extracted by dithizone in the presence of ethylenediamine tetra-acetic acid (EDTA) and thiocyanate, and only mercury (II) and copper (II) are extracted from a medium containing cyanide and thiocyanate. Back extraction with a selective complexing agent, such as EDTA or cyanide can often be used to strip a specific element from the organic phase.

Dithizone (diphenylthiocarbazone) is one of the most versatile chelating agents available for the extraction of metals with which it forms complexes extractable with a variety of organic solvents (usually CCl_4 or $CHCl_3$ are used). It is only moderately selective in its action, and forms extractable complexes with 17 elements—mainly those forming insoluble sulphides. However, by suitable adjustment of the pH of the aqueous phase and by the use of masking agents in either the extraction or back-extraction stages, it can, in many instances, be made fairly specific. Because the mainly red coloured metal dithizonates have very high molar absorptivities, the reagent was formerly much used, not only for concentration of particular trace elements, but also for their colorimetric determination. To do this the element was usually extracted from the sample (at an appropropriate pH value in the presence of suitable masking agents) using a 0·02% solution of dithizone in chloroform or carbon tetrachloride. The element was then back extracted from the organic phase and re-extracted with more diluted dithizone solution in the presence of masking agents. Photometric measurement was made at the wavelength of maximum absorption of the metal dithizonates. Iwantscheff (1958) has comprehensively reviewed the analytical applications of dithizone. Koroleff (1950) has made a systematic study of its use for the extraction of Ag, Bi, Cd, Co, Cu, Hg, Ni, Tl and Zn from sea water. With the exception of thallium, all these metals could be quantitatively extracted at pH 8·5. Selective back-extraction procedures were described for the separation of the elements into groups for further treatment. Methods employing dithizone have been described for the colorimetric determination of a number of elements in sea water including Ag, Bi, Cd, Cu, Hg and Zn (see section 19.10.5).

Derivatives of dithiocarbamic acid ($NH_2-C{\overset{\displaystyle\!\!\nearrow S}{\underset{\searrow SH}{}}}$) have been used extensively for the chelation and extraction of metals for analytical purposes. The diethyl derivative, which forms CCl_4 – extractable complexes with many heavy metals has been employed as a reagent for the extraction and subsequent photometric determination of copper in sea water (see p. 354). However, this reagent is not completely satisfactory for the extraction of many other metals as it decomposes rapidly at pH values below 6. Malissa and Schoffmann (1955) have demonstrated that the pyrrolidine derivative, which is much more stable at low pH values, chelates a large number of metals, excepting the alkali metals and alkaline earths, over broad pH ranges.

Following work by Allan (1961) this derivative has been widely employed as a chelating agent for the extraction of metals as a preliminary to their determination by atomic absorption spectrophotometry (AAS), advantage being taken of the enhancement in sensitivity resulting from the spraying of the metal in an organic solvent (p. 307). The technique was first adopted for the analysis of sea water by Brooks and his co-workers initially for the determination of copper (Brooks et al., 1967b) and later additionally for the estimation of Co, Fe, Pb, Ni and Zn (Brooks et al., 1967a). In this work the pH of the sample (700 ml) was adjusted to a value of 4–5 and a dilute aqueous solution of ammonium pyrrolidine dithiocarbamate (APDC) was then added. Extraction was carried out by shaking mechanically for 30 min with a 35 ml aliquot of methyl iso-butyl ketone. Finally, the extract was examined by conventional AAS. Subsequently, this technique has been used with minor modifications by many other workers (e.g. Burrell, 1967; Brewer et al., 1969; Spencer and Brewer, 1969; Orren, 1971; Preston et al., 1972; Tsalev et al., 1972), not only in conjunction with conventional atomization procedures, but also with the carbon tube furnace (Segar, 1971; Paus, 1973). For elements such as Ag, Cd and Pb, which are undetectable in the extracts using flame atomization, Preston et al. (1972) evaporated a relatively large aliquot of the extract and employed the tantalum boat technique (p. 305). The relatively high solubility of methyl iso-butyl ketone imposes a severe restriction on the concentration factor which can be obtained in these methods. This difficulty can be overcome to some extent by the use of less soluble ketones such as methyl or ethyl iso-amyl ketone; however their cost is high. The solutions of the APDC complexes of most of the extractable metals are fairly stable if kept in the dark; however the complex formed by manganese decomposes readily.

Dithizone and APDC are undoubtedly the most important complexing agents at present used in the extractive preconcentration of trace elements from sea water. However, a variety of other chelating agents, some of them specific for particular metals, have also been used, (see Section 19.10.5). It seems likely that future advances in this field will include the exploitation of new and more selective organic complexing agents (see e.g. Burger, 1973), and also an increase in the number of metals extracted by the simultaneous use of several chelating agents as has been suggested by Sachdev and West (1970).

19.10.3. SEPARATION OF ELEMENTS AFTER CONCENTRATION

As a result of the widespread adoption of specific methods for the determination of trace elements (such as atomic absorption spectrophotometry) it is now less necessary than it was formerly to separate the element from other elements also recovered during the preconcentration stage. Indeed, it is an advantage for it to be often possible to determine several elements in the concentrate. However, there are occasions when less specific techniques are employed (e.g. in neutron activation analysis or colorimetry) when it will be essential to carry out such a separation in order to prevent interference. For this purpose solvent extraction or ion exchange procedures usually provide the simplest and most efficient means of obtaining the analyte element in the necessary degree of purity. The excellent book by Korkisch (1969) gives much valuable information on the techniques available for the separation of the rarer elements.

19.10.4. PHYSICOCHEMICAL TECHNIQUES FOR THE DETERMINATION OF TRACE AMOUNTS OF ELEMENTS

A wide range of physicochemical techniques of high sensitivity is now available for the determination of trace inorganic components in sea water. In the limited space available in the present Section it is not possible to do more than outline their principles, to mention their particular fields of application and to summarize their advantages and disadvantages. Despite many advances in techniques in the last decade, the determination of most inorganic trace components, with the fortunate exception of the micro-nutrient elements, still necessitates a preconcentration stage. However, there has been some encouraging progress in the application of instrumental neutron activation analysis and anodic stripping voltammetry to the direct analysis of sea water.

19.10.4.1. *Absorption spectrophotometry*

Absorption spectrophotometry, often more loosely referred to as colorimetry, is nowadays most widely applied in sea water analysis to the determination of the micro-nutrient elements. For the estimation of trace metals it has been largely displaced by atomic absorption spectrophotometry and other more rapid and specific multielement techniques. However, since the instrumentation required is comparatively cheap, the technique still finds application for the determination of trace metals in less well equipped laboratories.

In colorimetric analysis the element to be determined is converted into a derivative which absorbs radiation. Although the absorption is usually in the visible region of the spectrum it may sometimes be in the ultra-violet or near infra-red regions. The absorption of monochromatic radiation by an absorbing solution is governed by Lambert–Beer's Law.

$$I_\lambda = I_{0\lambda} 10^{-abc} \qquad \text{or} \qquad \log \frac{I_{0\lambda}}{I_\lambda} = abc = A$$

where $I_{0\lambda}$ is the intensity of the light of wavelength λ incident upon the solution, I_λ is its intensity after attenuation by a thickness b (cm) of the solution, and c is the concentration of the absorbing species (usually given in $g\,l^{-1}$). The absorptivity (a) is characteristic of the absorbing species and of the wavelength used. Most spectrophotometers are calibrated in terms of absorbance A (sometimes known as optical density) which should thus be directly proportional to the concentration of the absorbing entity. Because the absorbing molecules may interact among themselves, or with other species in the solution, deviations from this proportionality may occur, particularly at high concentrations of the absorbing species. Fortunately, however, Beer's Law is obeyed in most methods used for the determination of micronutrients and trace metals in sea water as the coloured complexes are present only at great dilutions at which such interactions are minimal.

Colorimetric analysis was formerly carried out using methods in which the colour of the solution was compared visually with those of a series of standards. However, visual methods of comparison have now been almost completely superseded by photoelectric ones which are much less subjective and more precise, and which are useful over a considerably wider spectral range.

In order to increase the selectivity and sensitivity of photometric measurements it is desirable to use monochromatic radiation having the same wavelength as that of the absorption maximum of the coloured species to be determined. Photoelectric instruments are of two principal types. In the cheaper instruments (absorptiometers and colorimeters) which have a range of 400–700 nm, gelatine filters with spectral band passes of ca. 100 nm are used. Although these instruments are generally satisfactory for micro-nutrient analysis, they are not suitable for use in problems in which there is any appreciable overlapping of absorption bands (e.g. in the determination of chlorophylls). The other class of instrument—the spectrophotometer—is considerably more versatile as it employs a prism or grating monochromator to cover a much wider spectral range (often 190–1000 nm) with a spectral bandwidth of a few nanometres. A discussion of the instruments and techniques used in spectrophotometry is beyond the scope of the present chapter and for information on these subjects the reader is referred to the texts by Bauman (1962), Mellon (1950), Edisbury (1966) and Lothian (1969).

The precision of spectrophotometric methods of analysis is limited by both physical and chemical factors. The precision of photoelectric spectrophotometers appears to be limited by electronic noise rather than by the linearity of the amplifier or by the stability of the amplifier or light source. Under noise limiting conditions with instruments using photo-emissive detectors the minimum photometric error will occur at an absorbance of 0·869 (see Fig. 19.8a). If noise is not the limiting factor the minimum error will be found at an absorbance of 0·434 (Fig. 19.8b). In practice it is probable that the minimum

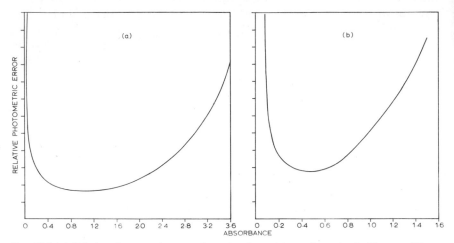

FIG. 19.8. (a) Relative photometric error of spectrophotometer under noise-limiting conditions. (b) Relative photometric errors of spectrophotometer under conditions in which noise is not the limiting factor (Bauman, 1962; reproduced by courtesy of John Wiley & Sons, Inc).

error will occur at some intermediate value depending on instrumental factors such as voltage stability and the level of the photoelectric current. The absorbance corresponding to minimum error cannot be easily calculated, but must be determined empirically for any particular spectrophotometer. In general, for most instruments optimum precision can be attained by working within the absorbance range 0·3–1·0; this can usually be achieved by adjustment of either the cuvette length or the volume of sample used. If, as often occurs in micronutrient studies, the absorbance lies below this optimum range even with cells of the maximum feasible length, some improvement in precision can be obtained by use of the scale expansion facility incorporated in many modern spectrophotometers.

It is not usually necessary that the absorbance scale of a spectrophotometer should be accurate in the absolute sense, since a calibration curve should always be run before using a spectrophotometric method in order to check the linearity of response. In addition, both standards and blanks should be run with each batch of samples. In most analyses it is good practice to carry out the calibration by analysing samples of a typical sea water both alone, and after spiking with a known amount of the test element. In this way allowance can be made for the fact that when colour forming reactions, such as those used in the determination of micronutrients, are carried out in sea water the absorbance produced per unit weight of test element is often less than it would be with distilled water. This *salt error*, which may amount to 20%, is probably produced by changes in the activities of the ions participating in reactions resulting from the high ionic strength of sea water. As an alternative to the above method of calibration by standard additions, a series of stable standard solutions of nutrient elements in sea water is now available and can be used for calibration purposes (see Sugawara, 1969). If, at any time, absolute absorbances are required, as, for example, in the determination of plant pigments, the spectrophotometer should be calibrated with standard glass filters or with solutions of accurately known absorbance, such as 0·05 M potassium dichromate or chromate. From time to time the accuracy of the wavelength scale of the spectrometer should be checked using the emission lines from the deuterium lamp normally employed in the ultra-violet region of the spectrum, or with a didymium glass filter.

Great care must be taken to ensure that the windows of spectrophotometer cells are free from scratches, smears and fingermarks, since these may give rise to appreciable errors. Since pairs of cuvettes rarely balance exactly (particularly in the ultra-violet) it is advisable before commencing an analysis to fill both cells with distilled water and to compare one against the other at the appropriate wavelength. Cells which become out of balance during use can be cleaned by filling with laboratory detergent or with a mixture of equal volumes of concentrated nitric and sulphuric acids.

Samples of inshore waters, or of oceanic surface waters from areas of high productivity, are often appreciably turbid. In carrying out analyses on such waters, or on ones which are discoloured, correction must be made for attenuation of light caused by these factors. This correction can be conveniently assessed by measuring the absorbance of a sample which has been acidified to the same extent as in the determination. A cause of error which is more insidious is the concealed blank; this is a reagent blank which differs from that associated with the determination itself. Such errors may occur, when as is often the case, the blanks are carried out in a different medium from that of sample. As has been pointed out by Jones and Spencer (1963) these concealed blanks will cause particular difficulties in methods in which the sensitivity varies from time to time, since the concealed blank will also vary. Observations by Kuenzler *et al.* (1963) on the determination of low concentrations of phosphate in sea water by two molybdenum blue methods may be used to illustrate effects of the concealed blank. Results obtained using methods employing stannous chloride and ascorbic acid as reductants agreed well provided that blanks were measured using phosphate-depleted sea water. However, when blanks were determined with distilled water the stannous chloride procedure gave anomalously high results.

Sensitivity of photometric procedures. The sensitivity of colorimetric procedures is limited by the molar absorptivity (ε) of the absorbing species formed by the element being determined ($\varepsilon = A/mb$, where A is the absorbance at the wavelength of maximum absorption, m is the molar concentration of the element in the solution (in moles l^{-1}) and b is the thickness (in cm) of the solution). The molar absorptivities of the coloured species used in trace element analysis vary considerably e.g. 6,300 for copper biquinolyl, 78,000 for cadmium dithizonate and 52,000 for the azo-dye formed in Bendschneider and Robinson's (1952) method for the determination of nitrite. In micronutrient analysis, unless the concentrations are extremely low, sufficient sensitivity can usually be achieved by using 10–15 cm cuvettes. However, in the determination of trace metals it is almost always necessary to concentrate the element from a large volume of sample before forming the coloured complex in a small volume of medium. The absorbance of the complex can then be measured in micro-cuvettes in which a light path of 4 cm can be attained with only 1·5–2 ml of liquid.

Automatic photometric analysis. The tedium of routine photometric analysis can be much reduced by the use of automated techniques which were first developed for clinical analysis. Two different approaches have been adapted in the automation of photometric procedures. One is based on the analysis of discrete samples and the other on a flow principle. The first of these techniques effectively automates the various stage of the manual analysis procedure. The desired volume of the sample is transferred by means of a mechanically actuated syringe to a reaction tube mounted on an intermittently moving conveyor belt. Appropriate amounts of the various reagents are then added by means of syringes and heating is applied if necessary to promote the reaction. Finally, the coloured solution is pumped into the flowcell of an automatic spectrophotometer and its absorbance is measured and printed out. Reliable commercial instruments working on this principle have been in use in clinical laboratories for several years; in many instances analysis rates of > 300 samples h^{-1} can be achieved with them. Although there appears to be no intrinsic reason why such instruments, or simple modifications of them, should not be used for the analysis of sea water, there seems to have been a reluctance to adopt them for this purpose, and the only published account of their use appears to be that by Perez and Guillot (1971).

All automated procedures for the photometric analysis of sea waters

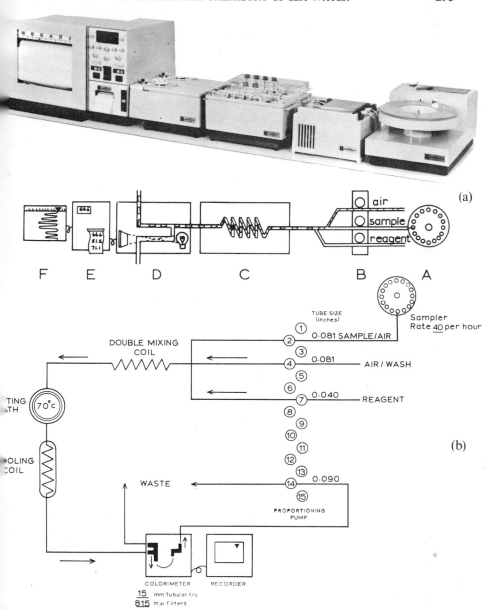

FIG. 19.9 (a) Technicon AutoAnalyzer assemblage consisting of sampler (A), proportioning pump (B), analytical cartridge (C) in which colour forming reactions take place, colorimeter D, digital printer (E), chart recorder F. Photograph by courtesy of Technicon Instruments Co. Ltd., Basingstoke, Hants. (b) AutoAnalyzer flow diagram for determination of phosphate in sea water using a single solution reagent consisting of acid, ammonium molybdate and ascorbic acid.

currently used are based on the flow system patented by the Technicon Instruments Corporation. In the Technicon AutoAnalyser® (see Fig. 19.9) a multichannel peristaltic pump serves to pump the sample from cups mounted on the plate of a rotary sampler and to inject bubbles of air into the sample line. The purpose of these bubbles is to segment the sample stream and so prevent the "tailing" of the sample which would otherwise occur as a result of the liquid at the walls of the tubing used in the flow system flowing more slowly, because of viscous drag, than that at the centre of the tube. Reagents are injected into the flow line at appropriate stages, by means of the peristaltic pump. The correct ratio of volume of sample to that of reagents is achieved by selection of pump tubes of appropriate diameters. Coils of glass tubing of suitable length are used in the flow line to provide sufficient delay for the completion of the reactions (slow reactions can be speeded up by the use of a long coil immersed in a heating bath). After development of the coloured complex, the segmenting air is removed and the coloured liquid is passed through the flow cell of a double beam colorimeter. Light of the appropriate wavelength is selected by means of interference filters. Finally, the signal from the colorimeter is amplified and fed to a logarithmic recorder or data logging system. For the measurement of low absorbances, such as those encountered in the determination of phosphate or nitrite, it is essential that the photometer be fitted with a facility permitting range expansion of up to 10 times. Suspended matter can cause errors in the analysis of water samples from inshore areas or regions of high productivity. However, this turbidity can be removed by means of the Technicon automatic filtration unit or, as Armstrong et al. (1967) have suggested, by use of the Technicon dialysis unit in conjunction with a membrane filter having a pore size of 3 μm.

Although the AutoAnalyzer is considerably slower than automatic analysers working on the batch principle, it is considerably more robust. It is suitable for use on shipboard, and is capable of functioning well even in bad sea conditions (see Coote et al., 1971). As it uses a flow system it can be readily adapted for continuous profiling, both horizontally (see e.g. Park et al., 1968) or vertically (Armstrong et al., 1967) using water pumped inboard. However, in the latter instance problems arise because of the time required for flushing of the long hose. In order to eliminate difficulties of this kind Bernhard (1968) has developed a submersible pressure housing for the Auto-Analyser capable of working at depths of up to 300 m (see also Piro and Rossi, 1969).

Automatic methods employing the AutoAnalyzer have been extensively used for the determination of micronutrient elements in sea water. Most of the procedures which have been developed for this purpose are simple adaptations of manual techniques. When used in land-based laboratories the best of them usually have precisions similar to those of their manual counterparts. However, at sea their precision may be considerably superior to that of the latter as the performance of the instrument is little affected by ship motion. For practical details of the application of the AutoAnalyzer to marine nutrient analysis, the reader is referred to the monographs by Strickland and Parsons (1968) and Atlas *et al.* (1971). Grasshoff (1970b) has described the problems associated with the use of a multichannel system for nutrient analysis at sea with both digital and analogue output of data.

Ammonia has been determined using automated versions of both the indophenol blue method (Grasshoff, 1970a, 1970b; Head, 1971; Slawyk and MacIsaac, 1972; Benesch and Mangelsdorf, 1972; see also p. 412) and the rubazoic acid procedure (MacIsaac and Olund, 1971). Slawyk and MacIsaac (1972) have found a very close correlation ($R = 0.96$) between results obtained by the latter procedure and those found by a simpler indophenol blue technique. No techniques appear yet to have been developed for the automatic determination of organic nitrogen in sea water. However, McDaniel *et al.* (1967) have described an automated Kjeldahl procedure for the examination of estuarine waters.

The method of Bendschneider and Robinson (1952) has been used as the basis for most automated procedures for the determination of nitrite (see e.g. Strickland and Parsons, 1968; Grasshoff, 1969; Atlas *et al.*, 1971; Watanuki, 1969). Mechanized nitrate methods have been described by several workers. These are based on reaction with brucine (Kahn and Brezenski, 1967b), or on reduction to nitrite either homogeneously (Henriksen, 1965a) or heterogeneously using Cd/Hg or Cd/Cu couples (Brewer and Riley, 1965; Armstrong *et al.*, 1967; Strickland and Parsons, 1968; Watanuki, 1969; Grasshoff, 1970b; Hager *et al.*, 1969).

AutoAnalyzer systems for the determination of phosphate have been developed by many investigators (see e.g. Henriksen, 1965b; Chan and Riley, 1966b; Grasshoff, 1967, 1970b; Armstrong *et al.*, 1967; Atlas *et al.*, 1971; Coote *et al.*, 1971). Hager *et al.* (1972) have made a comparison of a manual and an automated phosphate method and found that they gave similar results and had similar precisions. In the automatic determination of total phosphorus, organic phosphorus compounds can be broken down by circulating the sample through a helix of silica tubing surrounding a powerful source of ultra-violet radiation.

Several different automatic molybdenum blue procedures have been described for the determination of dissolved silicon in sea water (see e.g. Wilson, 1965; Brewer and Riley, 1966; Grasshoff, 1966, 1969; Armstrong *et al.*, 1967; Strickland and Parsons, 1968; Watanuki, 1969; Lopez-Benito, 1970; Atlas *et al.*, 1971). Comparative studies by Brewer and Riley (1967) and by Hager *et al.* (1972) have shown that the automatic and manual techniques have similar precisions and accuracies.

In addition to being used for micronutrients, the AutoAnalyzer has also been employed for the determination of fluoride (Grasshoff, 1966; Chan and Riley, 1966), boron (Hulthe *et al.*, 1970; Charpiot, 1969) and hydrogen sulphide (Grasshoff and Chan, 1969, 1971). However, it seems likely that its sensitivity is too low for it to be applied to the determination of trace metals in sea water.

19.10.4.2. *Fluorimetry*

In fluorimetric analysis, the element to be determined is converted into a fluorescent compound by treating it with a fluorogenic reagent. The solution is irradiated with filtered, or preferably monochromatic, ultra-violet radiation having a wavelength corresponding to the absorption maximum of the complex. The fluorescence which this excites has its maximum intensity at a longer wavelength than that of the exciting radiation, usually in the visible region of the spectrum. The intensity of this fluorescence is proportional to the concentration of the fluorescing substance over a limited (low concentration) range and to the intensity of the radiation source. The fluorescent radiation which emerges at right angles to the exciting beam is passed through a monochromator or filter (in order to isolate the fluorescence maximum and eliminate any scattered ultra-violet radiation). Its intensity is then measured by means of a photomultiplier and associated circuitry. For most analytical purposes simple fluorimeters employing filters are adequate, but for studies of excitation and fluorescence spectra it is necessary to employ spectro-fluorimeters in which monochromators are used to isolate both the exciting and fluorescent radiation.

Fluorimetric methods in general have enjoyed much less popularity than absorptiometric ones, although they tend to be much more sensitive than the latter. This is partly because the lack of specific reagents necessitates a careful preliminary separation of the element to be determined and partly because certain substances, notably oxygen and nitrate, tend to quench the fluorescence.

Fluorimetry has been used only to a limited extent for the estimation of trace metals in sea water. Its most important application is to the rapid direct determination of aluminium using the fluorescence of the complexes which it forms with Pontachrome Blue Black R (Simons *et al.*, 1953; Sackett and Arrhenius, 1962) and Lumogallion (Nishikawa *et al.*, 1968; Shigematsu *et al.*, 1970). The technique is also of value for the determination of uranium (after preconcentration). In this instance, the uranium is fused under standard conditions with a flux containing sodium fluoride and the fluorescence of the resultant pellet is measured in a specially designed fluorimeter (see p. 399).

Minor uses of fluorimetry include the determination of beryllium (with morin, Ishibashi *et al.*, 1956) and gallium (with 8-hydroxyquinaldine, Ishibashi *et al.*, 1961). In addition to its application to inorganic problems, fluorimetry has also been employed extensively for the determination of chlorophyll in the sea (see p. 459).

19.10.4.3. *Emission spectroscopy*

In emission spectrographic analysis the sample (usually solid) is vaporized in an electric discharge (d.c. arc or, less frequently, an a.c. spark). The high temperature of the discharge excites the atoms in the sample to higher energy states. As they pass into the cooler part of the discharge they return to the ground state and emit their characteristic spectra. The radiation thus emitted is dispersed in a spectrograph and the spectrum is recorded photographically. Although the spectrographic technique permits multi-element analyses to be made with reasonable sensitivity, it suffers from a number of drawbacks which seriously limit its value in the analysis of sea water.

(i) The precision is poor because of the irreproducible nature of the excitation and photographic processes. Although the precision can be improved to some extent by the use of an internal standard technique, it is difficult to achieve a coefficient of variation of better than $\pm 5\%$, even when all analytical conditions have been rigidly standardized.

(ii) The composition of the sample exerts a considerable influence on the intensity of the spectral lines (matrix effect). For this reason it is essential that, even when an internal standard is used, the blanks and standards should have compositions similar to that of the standards so far as their major cations and anions are concerned.

(iii) The technique is too insensitive for it to be used for the direct determination of trace elements in sea water, and its imprecision makes it unsuitable for the estimation of the major elements.

Although spectrographic analysis of concentrates prepared by coprecipitation or co-crystallization was used in most of the early work on the occurrence of trace metals in sea water (see e.g. Wattenberg, 1943; Noddack and Noddack, 1940), the technique has now been superseded by more precise methods, such as atomic absorption spectrophotometry.

19.10.4.4. *Flame photometry*

Flame photometry is a specialized form of emission spectroscopy in which a solution of the sample is sprayed into a flame in order to excite the spectral lines of the desired element. In all, about 50 elements can be determined in this way, although the sensitivity is often too low for the technique to be used for trace element analysis. Because the flame is much more stable than an arc, and also because photoelectric methods of detection can be used instead of the photographic plate, flame photometry is inherently more precise than is arc spectroscopy. With care it is possible to achieve a coefficient of variation of $\pm 1\%$, and even $\pm 0.3\%$ can be attained (see e.g. Folsom, 1974). In order to obtain precisions of this order it is necessary to finely control not only the rate at which the sample is atomized, but also the flame conditions since the number of atoms in the excited state is an exponential function of the flame temperature. Although the flame spectra of the elements are considerably less complex than those obtained using arc excitation, interference may arise as the result of overlapping of the spectral lines of the element to be determined by those of other elements in the sample, or particularly by the band spectra of molecular species present in the flame. As with emission spectrographic analysis, the major anions and cations present in the sample may exert pronounced enhancing or depressing effects on the spectral emission of an element. This effect is not important if the trace element is preconcentrated before being submitted to flame photometry. However, it can cause difficulties in the direct analyses of sea water as it is frequently not feasible to prepare artificial standards having a similar composition to natural sea water. In such instances the method can be calibrated by examination of samples which have been spiked with known increments of the element to be determined (see Fig. 19.10). There is a further problem associated with the direct determination of trace elements in sea water by flame photometry; this is the irreproducibility which arises both from the presence of incandescent salt particles in the flame and from the build-up of salts in the burner as a result of the atomization of highly saline solutions. It is preferable, therefore, to preconcentrate the element if the highest precision is required. The monograph by Dean (1960) should be consulted for a discussion of the principles and applications of flame photometry.

In marine chemistry, the use of flame photometry has been practically entirely restricted to the determination of the alkali and alkaline earth metals. The technique has been used for the direct determination of both the major elements potassium, calcium, strontium (see Section 19.8.2) and the minor elements, lithium and rubidium (see Section 19.10.5). However, its precision is too poor for it to be used meaningfully in studies of the $K/Cl\%_0$ and $Ca/Cl\%_0$ ratios in sea water. For the remainder of these elements it is necessary to examine the undiluted sample and this reduces the precision significantly because of problems associated with atomization (see above). For this reason it is preferable to preconcentrate these metals before determining them by flame photometry, and preconcentration is essential in the determination of caesium and barium.

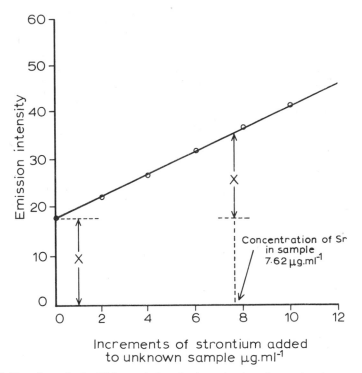

FIG. 19.10. Use of standard addition technique in determination of strontium in sea water by flame emission spectrophotometry (after Chow and Thompson, 1955).

19.10.4.5. *Atomic absorption spectrophotometry (AAS)*

The possibility of applying atomic absorption spectrophotometry to analytical problems was first appreciated about 20 years ago by Walsh (1955) who developed the first instrument for this purpose. Because of its sensitivity, specificity and comparative simplicity the technique has been widely adopted for the determination of trace elements in many different types of sample. Concurrently, a wide range of equipment for this purpose has been developed varying from simple instruments costing about £1,400 to elaborate double beam spectrophotometers costing upwards of £6,000. In the compass of this chapter it is only possible to outline the principles of the technique; for a more detailed account the reader is referred to the texts by Price (1972) and Slavin (1968). The use of AAS in the analysis of waters has been reviewed by Ediger (1973) and Burrell (1968).

Analytical atomic absorption is based on the measurement of the light absorbed by the unexcited atoms of the element at the wavelength of one of its resonance lines. In most atomic absorption methods a flame into which the sample is atomized effectively serves as an absorption cell. The heat of the flame serves to evaporate the solvent from the sample, to vaporize its constituents and to break down any molecular species. At the temperature of an air–acetylene (or similar) flame only a very small proportion of all atoms are raised to an excited state, and over 97% of them are in their ground states. The absorption arising from the transition of an element from its unexcited to its excited state is therefore effectively an absolute measure of the number of atoms of the absorbing element in the flame and hence of its concentration in the sample.

If the flame is regarded as a cell containing absorbing gas as in spectrophotometry, the absorption of light of the appropriate wavelength will be proportional to the concentration of the particular atoms in the flame and to the length of the light path in the flame. The integrated absorption can be related to the operating parameters by the expression (Willard et al., 1965)

$$\int K_v \, \partial v = (\pi e^2/mc) \, N_0 f.$$

Where K_v is the absorption coefficient for radiation of frequency v, m and e are respectively the mass and charge of an electron, c is the velocity of light, N_0 is the number of metal atoms per cm^3 in the ground state and f is the oscillator strength which is inversely proportioned to the lifetime of the excited state.

Atomic absorption spectrophotometry has two important advantages over flame emission methods. Firstly, it is very much less sensitive to variations in flame temperature, since the proportion of atoms in the ground state varies little with temperature. In contrast, in flame emission it is the proportion of atoms in a particular excited state which is important, and this varies exponentially with temperature. Secondly, it is easier to measure precisely the ratio of two luminous intensities (the intensity of a monochromatic source in the presence of and in the absence of absorbing atoms) than it is to measure emission intensities in absolute units.

In atomic absorption methods the absorption is usually measured at a wavelength corresponding with a resonance line; such absorption lines are inherently very sharp. The absorption intensity is a complex function of wavelength; it is controlled by spectral and flame parameters and is determined by the various factors causing the spectral line to broaden. Several factors can produce this broadening, an important one being the Doppler effect which arises from thermal motion of the atoms with respect to the point of observation; for temperatures of 1500–3000°K this will result, for most elements, in lines having half widths of 0·001–0·01 nm. Additional broadening results from collisions of the atoms both with one another and with other atoms and molecules present in the flame. In typical flames this effect gives half widths similar to those produced by the Doppler effect.

Instrumentation. Because of the important role which atomic absorption spectrophotometry plays in modern marine chemistry, the equipment used will be discussed in some detail. Basically it consists of four principal parts (see Fig. 19.11a), (i) a source of the resonance radiation of the element being determined, (ii) a means of atomizing and vaporizing the sample (usually a flame), (iii) a monochromator to separate the resonance line from any others in the flame (iv) a detector for the measurement of the extent to which the radiation of the source is absorbed by the element being determined.

(i) Radiation source. Since the absorption lines are very narrow it is not easy to measure the absorption using a continuous light source because of

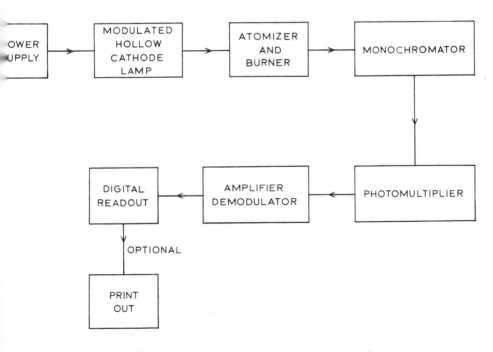

FIG. 19.11a. Block diagram of single beam atomic absorption spectrophotometer.

the background. The source used must therefore be essentially mono-chromatic. In practice, a hollow cathode discharge lamp is almost always used as it gives extremely sharp resonance lines. Such lamps consist of

L

glass envelopes containing an anode and a cylindrical cathode consisting of, or coated with, the metal in question;* if the resonance lines are in the ultra-violet the lamps are fitted with fused silica windows. The lamps are filled to a pressure of 1–2 Torr with an inert gas (usually neon). In operation, a potential in the range 600–1000 V is applied to the electrodes. This creates a discharge which fills the cathode. The inert gas ions which are formed by the discharge accelerate towards the cathode. When they collide with the hollow cathode they sputter the element of which it is composed into the discharge zone in which it is excited by the carrier gas ions. In order to stabilize the light the lamps are operated at a constant current which is usually kept quite low (ca. 4–10 mA) in order to keep the emission lines narrow and to improve the linearity of response.

(ii) In conventional atomic absorption spectrophotometry a flame into which the sample is atomized, is used as the absorption cell. The heat of the flame serves to evaporate the solvent and to reduce the metal to the atomic state. Introduction of the sample into the flame is usually achieved by using the support gas for the flame (usually air) to atomize the sample into a chamber in which large droplets are removed and the combustible gas (natural gas or acetylene) is admixed. The resultant mixture is then passed immediately to the burner. This is so designed as to give a long (ca. 10 cm) narrow flame in order to place a large number of the absorbing atoms in the light path. It is only necessary for the flame temperature to be sufficiently high to dissociate any molecular compounds of the analyte element into the free atoms. In most instances the required flame temperature can be achieved by using as the combustion mixture either air–natural gas or air–acetylene which is somewhat hotter. However, a few elements (e.g. Al, Ti, Mo) form very refractory oxides and for these a $N_2O-C_2H_2$ flame must be used. It is essential to have careful control over the flame and atomization conditions if reproducible results are to be obtained in atomic absorption spectrophotometry. This can be achieved by close regulation of the pressures and flow rates of the combustant gases and by careful design of the atomizer assembly.

Because of their simplicity, comparatively high sensitivity and reproducibility, flames are the most widely used atomic absorption "cells". However, they suffer from the disadvantages that the efficiency of nebulization is low and that the atoms are liberated into a very complex environment. A number of attempts have been made to overcome these drawbacks.†

---

* Although multi-element lamps are available, their performance tends to be inferior to that of single element ones.
† Kirkbright (1971) has reviewed the various devices available for the production of atomic vapour for atomic absorption analysis.

In one of these (see Kahn *et al.*, 1968) a known volume of the sample is evaporated in a small tantalum boat which is then inserted into an air–acetylene flame in the optical axis of the atomic absorption spectrophotometer. The sample is rapidly vaporized and the resulting pulse of absorption of the resonance radiation of the element is measured with a chart recorder. In another approach a carbon tube which can be heated electrically is used as the "cell" (see West, 1971). A few microlitres of the sample is placed within the tube using a syringe. A programmed progressively increasing potential is applied to the end of the tube, so as to successively evaporate the sample, ash any organic material and finally vaporize the residue. Again, the resultant absorption pulse is measured. Although it is usually possible to attain a much greater sensitivity with these techniques than with a flame,* the precision which can be achieved is considerably poorer ($\pm$ 5–10% compared with $\pm$1% with flame). In addition, problems caused by molecular absorption and matrix effects (Robinson and Sievin, 1973) are more severe.

Because of the volatility of elemental mercury it is possible to determine this element by atomic absorption spectrophotometry by a technique not involving heating the sample. In this "cold vapour" procedure the solution containing the mercuric ion is treated with stannous chloride. The elemental mercury thus produced is stripped from the sample with a current of gas and passed through a silica absorption cell. The element is determined specifically and with a very high sensitivity by measurement of the absorption of the 2537 Å resonance line caused by its vapour (see also Section 9.10.5.33).

(iii) The monochromator has two functions; to separate the resonance line from any others coming from the source, and to prevent the detector being overloaded with light from the flame and with luminescence from the inert gas discharge of the hollow cathode tube.

(iv) The detector, which consists of a photomultiplier and associated circuitry, is used to measure the intensity of the resonance line. In order to eliminate effects produced by emission from the flame it is usual to modulate the source electrically and to use a tuned amplifier in the detector circuit. The more sophisticated types of atomic absorption spectrophotometers operate on a double beam principle in order to increase the precision by compensating for variations in the intensity of the source and sensitivity of the detector (Fig. 19.11b). Most modern instruments are fitted with a range expansion facility which allows the sensitivity to be increased, in

* In favourable instances (e.g. for Cd and Mg) detection limits of a few picograms can be achieved using the carbon tube furnace. However, this is offset to some extent by the small volume of sample which can be used.

favourable instances, by a factor of 10 or even 20 times; this is of considerable value when determining low concentrations of trace metals.

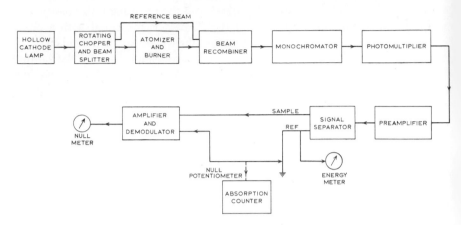

FIG. 19.11b. Block diagram of double beam, time-shared atomic absorption spectrophotometer.

Analytical applications of atomic absorption spectrometry. Since atomic absorption photometric methods are based only on atomic transitions from the ground state, they are free from many of the interferences which are experienced with emission procedures which depend on distribution of atoms over the various excited states. Thus, the influence of flame temperature will be small, provided that the temperature is sufficiently high to ensure reduction to the elemental state and to break down any molecular species, and so long as ionization is insignificant. Direct spectral interference of one element in the determination of another occurs very rarely.

In atomic absorption as in emission methods it is always essential that standards used for calibration should have very similar compositions to that of the solutions being examined. This is because there can be pronounced "matrix effects" which may arise in two ways:

(i) generally, absorption due to an element is decreased when high concentrations (> 0.1 M) of anions are present in the solution. This effect is important for those anions that form compounds with the element in question, which at the temperature of the flame, form refractory compounds and prevent it being reduced to the free atoms. Significant absorption suppression occurs with nitrate and sulphate ions, and the effect is particularly pronounced with phosphate ion, because of the formation of compounds such as $Ca_2P_2O_7$. The use of higher temperature flames reduces suppression of this type as they break down the refractory compounds. Interference can also be prevented by the use of a releasing agent which either combines preferentially with the interfering element (e.g. La^{3+} for PO_4^{3-}), or chelates the test element leaving it free to be reduced in the flame (e.g. EDTA for Ca in presence of PO_4^{3-}).

(ii) Changes in the concentrations of cations and anions or the presence of organic solvents or surfactants in the analyte will cause variations in its surface tension, viscosity and vapour pressure. These in turn will affect the rate at which the sample is drawn into the atomizer, the size of the droplets produced and the quantity of the aerosol which reaches the burner, and will thus influence the absorption.

The sensitivity of the atomic absorption spectrometric techniques varies considerably from element to element. Thus, using flame atomization sensitivities* of ca. 0·03 μg ml^{-1} can be achieved for Cd and Zn, whereas for barium the sensitivity is only about 1/100 of this. A several fold enhancement of the sensitivity can often be achieved if a solution of the element in an organic solvent (e.g. methyl *iso*-butyl ketone) is atomized. Calibration curves depart from linearity at quite low absorbances (usually less than 0·35 A). The causes of this deviation are complex and include pressure broadening, and factors connected with burner and source performance (see DeGalan and Samaey, 1969)

Application to marine chemistry. Although Fabricand *et al.* (1962) and Billings and Harriss (1965) have claimed that they were able to determine 1 μg l^{-1} levels of Cu, Zn, Fe and Mn directly in sea water by conventional AAS, other workers have not been able to achieve this high sensitivity. Even for strontium (8 mg l^{-1}), lithium (180 μg l^{-1}) and rubidium (120 μg l^{-1}) attempts at direct determination have given element/chlorinity ratios which do not agree well with those obtained by other methods and which show considerable scatter. In addition to effects produced on the rate and efficiency of atomization by the spraying of solutions of high ionic strength, a further source of error with such samples is the attenuation of the resonance radiation by scattering on salt particles in the flame.† This non-specific absorption becomes significant when spraying sea water having a salinity > 7‰. Compensation can be made for these effects with moderate efficiency by carrying out the calibration by a

---

* Defined as the concentration required to give an absorbance of 0·004.
† Effects arising from non-atomic attenuation of radiation (either by scattering on salt particles or through the presence of molecular species) can be corrected for by the use of a deuterium background corrector. This consists of a deuterium lamp which emits a continuum. The light from this is directed through the flame and the change in light absorbance which occurs when the sample is sprayed into the flame is taken to be the background absorption.

method of standard additions see above p. 300); however, this technique is more difficult to apply than it is with flame emission spectrometry because of the much shorter linear range of the calibration curve. Much of the random variation found during the direct analysis of sea water by AAS is caused by changes in the characteristics of the burner resulting from the deposition of salts along its slot.

In sea water analysis it is, therefore essential, if accurate results are required, to separate the analyte element from the major cations and anions before submitting it to AAS. Separation and preconcentration, usually by solvent extraction or ion exchange techniques, is of course essential in the determination of trace metals in sea water (see e.g. Joyner and Finley, 1966; Burrell, 1967; Brewer et al., 1969; Riley and Taylor, 1968). In such processes advantage can often be taken of the enhancement in sensitivity which results from spraying the element in an organic solvent, such as methyl iso-butyl ketone. Minor and trace metals which have been determined in sea water by conventional AAS following preconcentration include: Cd, Co, Cr, Cu, Fe, Li, Mn, Ni, Rb, Sc, Sr, and Zn (for details see Section 19.10.5).

As has been mentioned above (p. 305), some other atom sources are capable of giving much higher sensitivity in AAS than the flame. The possibility of applying these to the analysis of sea water has been investigated by several workers. However, considerable difficulties arise when these techniques are used directly with sea water, partly because of attenuation of the resonance radiation by scattering on salt particles and partly because of matrix effects which cause a considerable loss of sensitivity.* For these reasons the scope of these techniques is rather limited. Several attempts have been made to use the carbon tube furnace for the direct analysis of sea water. Thus, Segar and Gonzalez (1972) have found that, because of the interferences mentioned above, iron was the only trace element which could be determined directly. Ediger et al. (1974) have attempted to overcome this difficulty by volatilizing sodium selectively at 1000°C before atomizing the trace elements. With the normal type of carbon tube atomizer it was necessary to add ammonium nitrate to the sample to assist the volatilization; however, with a modified form of the apparatus in which the carbon tube was purged with inert gas, this was not necessary. Using the latter technique they were able to achieve a coefficient of variation of $\pm 3.4\%$ in the determination of 3 $\mu g\,l^{-1}$ of copper in sea water. The time for an analysis was $1\frac{1}{2}$ min per sample. The possibility of determining other elements in this way was also investigated, and detection

* Paus (1973) has recently suggested that this difficulty can be overcome by removing sodium from the samples by means of hydrated antimony pentoxide, before use of the carbon tube furnace atomizer.

limits of 1, 0·5 and 2 μg l^{-1} were found for Fe, Mn and As. These values are similar to the normal levels of the elements in ocean water, and it would, therefore, appear that the technique is of little value for analysis of unpolluted samples. However, it might be of use in the examination of contaminated estuarine waters. In view of this lack of sensitivity, it is preferable to precon-centrate the trace elements by solvent extraction (e.g. by means of the ammonium pyrrolidine dithiocarbamate-methyl *iso*-butyl ketone system) before determining them with the carbon tube atomizer. This technique has been adopted by Segar and Gonzalez (1972) for the determination of Ag, Co, Cu, Fe, Ni, Pb, V and Zn, (see also Segar, 1973), and by Paus (1973) for Fe, Cu, Mn and Zn. Similar difficulties arise when attempts are made to use the tantalum boat technique for the direct determination of trace metals in sea water and, in fact, zinc is the only element which has been estimated in this way (Burrell and Wood, 1969). However, Preston *et al.* (1972) have used this atomization method on a routine basis for the determination of Ag, Cd and Pb in concentrates prepared by solvent extraction.

19.10.4.6. *Atomic fluorescence spectrometry (AFS)*

In atomic fluorescence spectrometry, as in AAS, the unexcited atoms of the element to be determined are irradiated with one of its resonance fre-quencies. This results in their energies being raised from the ground state to the corresponding excited state. This energy can then be lost either by collision, or by the emission of a photon having the same frequency as the original one used for excitation. Measurement of the intensity of this emitted radiation is the basis of atomic fluorescence spectrometry. Quantitatively the measured intensity is the product of the absorption intensity and the fluorescence power yield of the atom Φ—the probability that it will emit a photon before it loses its energy in a quenching collision.

The apparatus used for AFS somewhat resembles that employed for AAS. The sample is atomized in a burner similar to that used in flame emission spectrometry. The flame is illuminated with the resonance frequency of the element in question using either a hollow cathode lamp or an RF excited electrodeless discharge tube (alternatively, it is possible to use an intense source of continuous radiation such as a xenon lamp). The fluorescent radiation from the element being determined is observed at right angles to the beam of exciting radiation by means of a monochromator and associated photomultiplier. In order to eliminate effects caused by molecular and other irrelevant emissions in the flame, the light source is modulated and the fluorescent radiation is observed using a tuned amplifier coupled to the photomultiplier.

An important advantage of AFS over AAS techniques is that the intensity of fluorescence, and hence the sensitivity depends directly on the intensity of the radiation used for excitation. In practice, there are limits to the increase in sensitivity which can be achieved by increasing the power of the source. The sensitivity of the method varies considerably from element to element, since the detection limit varies inversely with Φ. For a few elements, notably cadmium and mercury, sensitivities 100 times that attainable by AAS can be achieved; for others considerably better sensitivities can be obtained by flame emission or atomic absorption methods (see Table 19.7). As in AAS

TABLE 19.7

Comparative data on practical detection limits (μgml^{-1}) of some atomic spectroscopic methods

Element	Flame emission*	Atomic absorption conventional	carbon furnace†	Atomic fluorescence*
Al	0·005	0·1	0·00003	0·005
Ca	0·0001	0·002	—	0·0001
Cd	2	0·001	0·000001	2
Hg	40	0·05	—	40
Mg	0·005	0·0003	—	0·005
Pb	0·2	0·03	0·00006	0·2
Sn	0·3	0·02	—	0·3
Zn	50	0·002	0·0000006	50

*   Data from Dagnall and Sharp, 1973.
†   Data from Kahn, 1973.

the concentration range for linear response is somewhat limited. The observed intensity may fall below that expected from linear extrapolation through incomplete illumination of the flame cell and reduced rate of sample introduction at high concentrations, and also at low concentrations because of ionization of the atoms. In general, interferences are similar to those encountered in AAS as both methods depend on the populations of the unexcited analyte atoms in the flame. For a comparison of the two methods the paper by Barnett and Kahn (1972) should be consulted.

Despite their superior sensitivity for certain elements, e.g. cadmium and mercury*, atomic fluorescence techniques do not yet appear to have been employed for oceanographic purposes.

---

*   Mercury can be determined with high sensitivity by a flameless atomic fluorescence technique analogous to that used in AAS (see eg Thompson and Reynolds, 1971).

19.10.4.7. X-ray fluorescence spectrometry

In X-ray fluorescence analysis the sample is excited with an intense beam of unfiltered short-wave primary X-rays. Interaction between the beam and the elements in the sample causes the latter to emit their characteristic X-ray spectra. The fluorescent beam is collimated and directed onto the surface of a crystal mounted on a goniometer. The crystal serves as an analyzer. The resolved X-ray lines, reflected according to the Bragg condition, pass through a collimator and then to a detector in which the energy of the X-ray quanta is converted to electrical impulses. The primary collimator, the analyzer crystal and the second collimator are placed on the focal circle of the goniometer, in order to satisfy the Bragg equation as the goniometer is rotated to examine the X-ray fluorescence lines. Absorption of long wavelength X-rays by air and by the detector window limits the application of the X-ray fluorescence technique to elements with atomic numbers greater than that of sodium (12). However, with vacuum spectrometers it is possible to determine elements down to boron ($Z = 5$).

Although X-ray fluorescence spectroscopic methods can be used for the determination of major components of samples with a high precision, their sensitivity is too poor for them to be employed satisfactorily for the determination of trace components of samples at levels below ca. 20 p.p.m. In addition, there is a very marked matrix effect which makes it essential that standards should have the same physical and chemical form as the sample being examined. However, spectral interference of one element in the determination of another is very rare because of the simplicity of the X-ray spectra.

Multi-element X-ray techniques for the determination of trace metals in sea water have been described by Morris (1968) and by Armitage and Zeitlin (1971). In both instances the metals (Cu, Fe, Mn, Ni, Fe) were preconcentrated in order to raise them to sufficiently high concentrations and to eliminate matrix elements. Even so, it would seem that the methods are not sufficiently sensitive for the estimation of elements at concentrations below ca. 2 µg l^{-1} unless very large samples (> 20 litre) are taken.

19.10.4.8. Gas chromatography

The very high sensitivity (of the order of picograms) of the gas chromatographic electron capture detector for certain halogen compounds has led to efforts to use it for the determination of trace metals (see e.g. Tanikawa et al., 1970; Stephen, 1972; Ross and Siever, 1970). In these the metal to be determined is complexed as a volatile chelate with a fluorinated diketone. Preliminary attempts to apply this technique to the determination of aluminium in sea water have been made by Lee and Burrell (1973); however, they were unable to achieve quantitative results.

19.10.4.9. *Stable isotope dilution*

The availability of mass spectrophotometers sensitive to less than 10^{-9} g of isotopes has led to the development of precise and sensitive isotope dilution procedures for the determination of trace elements in sea water. In this technique a given volume of the sea water is spiked with a known amount of a concentrate of one of the least abundant isotopes of the element being determined. After equilibration of the spike, the element is separated from the water sample and the concentrate is applied to the filament of a mass spectrometer. The ratio of the intensities of the mass spectrum peak of the spiking isotope to that of a more abundant isotope is then measured. The weight of the element present in the water sample can then be calculated. The process can be conveniently illustrated by reference to work carried out by Chow and Goldberg (1962) on lithium in sea water. 'The samples (50 ml) were spiked with 14·9 μg of enriched lithium-6 (99·7 %at. ^{6}Li) as its sulphate. After concentration of the lithium, the atomic ratio of of ^{6}Li/^{7}Li (N) in the concentrate was determined mass spectrometrically. Then:

$$\mu g \text{ of Li in 50 ml of sample} = \frac{m(^6\mathrm{Li}_s - \mathrm{N}^7\mathrm{Li}_s)}{\mathrm{N}\,^7\mathrm{Li}_w - \,^6\mathrm{L}_w}$$

where $^6\mathrm{Li}_s$ and $^7\mathrm{Li}_s$ are the atomic percentages of these isotopes in the m μg of spike used, and $^6\mathrm{Li}_w$ and $^7\mathrm{Li}_w$ are the atomic percentages of the isotopes in common lithium (7·5 and 92·5 %at. respectively). A coefficient of variation of ca. ±3% was achieved in this work. It should be noted that one of the advantages of the isotope dilution procedure is that the recovery of the element in the concentration process need not be quantitative, nor need the element be separated in a pure state.

In addition to its application to the estimation of lithium, the stable isotope dilution technique has also been used in the analysis of sea water for a range of other metals, including strontium*, (p. 239), barium, (p. 342), rubidium, (p. 389), lead (p. 371), indium (p. 359), and uranium (p. 400). The technique is intrinsically one of the most precise available for the determination of trace metals in natural waters, and in this connection it is interesting to note that Brass and Turekian (1972) have achieved a coefficient of variation of ±0·2% in the determination of strontium in sea water. According to Wilson *et al.* (1962) uranium can be estimated not only more simply, but also with greater precision by isotope dilution than by either polarographic or fluorimetric methods. Chow (1968) has reviewed the application of isotope dilution

---

* Smales (1951) has used the radio-active nuclide strontium-89 in an isotope dilution procedure for the determination of strontium. Johannesson (1962) has described a radio-isotope dilution procedure for the estimation of perchlorate in sea water (see p. 387).

methods to sea water analysis; for a general review of the technique see the text by Toelgyessy *et al.* (1972).

19.10.4.10. *Electrochemical procedures*

It is only in the last five years, largely as the result of the development of more sophisticated instrumentation, that the possibility of using electrochemical procedures for the determination of trace metals in sea water has been exploited. Two principal types of technique are available for this purpose—polarography and anodic stripping voltammetry. Of these, the former, even in its more elaborate forms, is probably too insensitive to be used for the direct analysis of sea water. In contrast, the other has the necessary sensitivity for the direct determination of a few elements, but is limited in its scope and in many instances beset with difficulties arising from inter-element interferences. In addition to their normal analytical uses, both of these techniques have potentialities for the study of the speciation of trace elements in sea water (see e.g. Sections 20.5.4 and Table 20.12). Because of their complexity and increasing importance electroanalytical procedures will be discussed separately in Chapter 20.

NEUTRON ACTIVATION ANALYSIS*

By D. E. ROBERTSON, *Battelle Pacific Northwest Laboratory, Richland, Washington, U.S.A.*

19.10.4.11.1. *Introduction*

During the past two decades a remarkable analytical renaissance has evolved in the measurement of trace elements in the marine environment. Notwithstanding this, the various analytical methods at present in general use have produced such divergent data in interlaboratory comparisons that many of the published analyses must be regarded as highly suspect or meaningless.

The recent trace element intercalibration study organized by Brewer and Spencer (1970a) clearly illustrates the extreme difficulty of measuring elemental concentrations near, or below, the parts per billion concentration range in a complex mixture such as sea water. In this study analysts from 26 laboratories, working independently and using their preferred methods of analysis, measured the concentrations of various trace elements in two samples of deep Atlantic and Pacific Ocean Water. The lack of interlaboratory agreement revealed itself in a spread of data which often exceeded a factor of

* This paper is based on work performed under United States Atomic Energy Commission Contract AT(45-1)-1830.

ten. These discrepancies result from the fact that many of the measurements were made at concentrations close to the detection sensitivities of the various analytical methods.

Another severe problem which frequently plagues the trace element analysis of sea water is contamination of samples during the assay. It is, therefore, desirable to use a technique which is extremely sensitive, selective, reasonably precise, and which requires a minimum of sample handling and chemical manipulations.

For many elements neutron activation analysis (NAA) fulfils these important requirements and, in the hands of a competent analyst, it is a microchemical method well suited for the trace element analysis of sea water. Since the first application of neutron activation analysis to chemical oceanography for the determination of arsenic in sea water in 1952 by Smales and Pate, forty elements have been determined in sea water by NAA at many different laboratories during the succeeding two decades. These elements include Ag, As, Au, Ba, Ce, Co, Cr, Cs, Cu, Dy, Er, Eu, Fe, Gd, Hg, Ho, In, La, Lu, Mn, Nd, Ni, Pr, Rb, Re, Ru, Sb, Sc, Se, Sm, Sn, Sr, Ta, Tb, Tl, Tm, U, Y, Yb, and Zn.

With the increasing general availability of high flux nuclear research reactors to the scientific community, and the recent development of high resolution Ge(Li) gamma-ray spectrometry, NAA will continue to be an increasingly powerful analytical tool for determining trace element constituents in the marine environment. This chapter presents, in some detail, the state-of-the-art of neutron activation analysis as applied to the measurement of trace elements in sea water.

19.10.4.11.2. *Principles of neutron activation analysis.* In neutron activation analysis the elements to be determined in a sample are made radioactive by irradiation with neutrons, and the induced radioactive species are then identified and measured. The amount of a given neutron activation product that is formed during neutron irradiation is directly proportional to the amount of its parent isotope. Measurement of the radionuclide provides a measure of the total concentration of the parent element.

The basic equation for activation is

$$W = \frac{AM}{\sigma f \phi N(1 - e^{-\lambda t})},$$

where W = weight of element irradiated, in grams,

   A = induced activity at the end of irradiation (disintegrations sec^{-1}),

   M = atomic weight of that element,

   σ = activation cross section for the nuclear reaction concerned, (cm^2),

f = flux of neutrons used in the irradiation (neutrons cm^{-2} sec^{-1}),
ϕ = fractional abundance of the particular isotope of the element concerned,
N = Avogadro's number (6·02 × 10^{23} atoms mol^{-1}),
λ = the decay constant of the induced radionuclide (sec^{-1}),
t = irradiation time (sec)

The above equation shows that the detection sensitivity for a given element can be improved by increasing the neutron flux or the irradiation time, or both. The relative detection sensitivity for any element in the periodic table is a function of: (1) its activation cross section for the nuclear reaction concerned, (2) the decay constant of the induced radionuclide, (3) the atomic weight of the element and, (4) the fractional abundance of the parent isotope of the element concerned. Tables 19.10 and 19.12 show that for most elements NAA is an extremely sensitive method of analysis.

The concentration of an element in a sample can be calculated from the above equation if the neutron flux is monitored and if the counting efficiency of the radiation detector for the neutron activation product of interest is known. However, the easiest and most accurate method for quantifying the concentration of any element is to irradiate the sample along with a standard containing a known weight of the element being determined. Then, the weight of the element being determined is obtained from the following relationship

$$\frac{\text{Weight of element in unknown}}{\text{Weight of element in standard}} = \frac{\text{Activity of element in unknown}}{\text{Activity of element in standard}}$$

This method obviates the need for careful monitoring of the neutron flux and for calibration of the counting instrument, since the only measurements required are the relative counting rates of the neutron activation products of the elements of interest in the unknown and in the standard. When using this comparative method, the sample and standards must, of course, both be irradiated and counted under exactly the same conditions.

The great advantage which NAA possesses over other analytical methods is that once the radioactive isotopes have been produced, chemical manipulations of the samples can be performed without fear of contamination from additions of impurities from reagents, labware, etc. For this reason blank values are usually extremely small.

The neutron activation method of analysis does not distinguish between the various physicochemical forms of an element in sea water, but does provide a measure of its "total" concentration. However, if the different species of an element are separated prior to the neutron irradiation, NAA can provide a measure of each specific form.

In practice, the fundamental steps in a typical NAA are:
1. irradiate weighed quantities of the sample and standard in suitable containers for a sufficient time to induce adequate radioactivity in the element to be determined,
2. carry out instrumental NAA by counting the irradiated samples and standards with Ge(Li) and/or NaI(Tl) gamma-ray spectrometers at suitable periods after the irradiation so as to directly measure elements producing short, intermediate and long-lived neutron activation products,
3. compare the γ-spectral peak areas of samples and standards under identical counting conditions, making decay corrections, Compton corrections and background corrections when necessary,
4. if instrumental NAA does not provide the desired sensitivity, separate the activation product(s) of interest from the interfering radionuclides radiochemically and count the separated activity (activities) in the same geometries as the standards with gamma-ray or beta particle detectors,
5. confirm that the γ-activity measured arises solely from the radionuclide of interest by measuring the half-life and comparing it with that of the standard. If the radionuclide emits more than one gamma-ray per disintegration both the energies and relative intensities can be used for verification. Detailed discussions of the basic theory, fundamentals and applications of NAA, have been given by Guinn and Lukens (1967), Kruger (1971), Lenihan and Thompson (1969), Rakevic (1970), Ryan (1973) and Smales (1967).

19.10.4.11.3. *Neutron irradiation facilities and procedures.* The most generally applicable neutron source for trace level activation analysis is the high flux nuclear reactor. These reactors normally operate at power levels in the range of 10 to 1000 kW and produce thermal neutron fluxes of 10^{11} to 10^{13} n cm^{-2} s^{-1}. Reactors used for NAA are generally of the research type and are becoming increasingly available for use by the scientific community.

The neutron flux in such reactors varies spatially about the reactor core and inhomogeneities in the flux must be taken into account. Such inhomogeneities can be minimized by rotating the samples in the flux field in various types of specimen racks designed for that purpose. Typically, up to 50 samples and standards can be irradiated in one rotator simultaneously. Although the horizontal neutron flux gradients in each rotator are minimized by revolving the samples, there is a small vertical flux gradient which must be accounted for by placing a standard in each tier of samples if they are stacked vertically.

"Rabbit" irradiations are used for activating samples for periods of seconds to minutes, where the induced activity of interest is a short-lived isotope, i.e. one having a half-life in the range of seconds to minutes. Samples are activated and counted one at a time (usually purely instrumentally), with the reactor operating steadily and a standard sample of the element of interest is similarly activated and counted at intervals.

19.10.4.11.4. *Gamma-ray spectrometry and beta counting*

A. *Ge(Li) Diode detectors*

Lithium-drifted germanium detectors, with volumes of 20 to 140 cm^3, have revolutionized instrumental NAA because of their excellent energy resolution (about 2 to 3 keV full width at half maximum, above energies of about 200 keV). Even better resolution can be attained at energies below about 200 keV utilizing low energy photon detectors (LEP detectors). These detectors will allow increasing use to be made in NAA of radionuclides emitting low energy gamma-rays and X-rays.

Earlier γ-spectrometers using single NaI(Tl) crystals were capable of resolving only a few of the major constituents in complex mixtures of radionuclides. However, recently developed Ge(Li) detectors, can resolve the gamma-rays of nearly all the neutron activation products formed in sea water samples if their concentrations are sufficiently high to be detected. Gamma-ray energies can be determined to within ± 0.5 keV, and the energies of nearly all neutron activation products are now known to about ± 0.1 keV. A good example of the improved resolution of Ge(Li) detectors is shown in Fig. 19.15 which shows the gamma-ray spectra of neutron activated sea salts counted on both NaI(Tl) crystal and Ge(Li) diode detectors. The main disadvantage of the Ge(Li) detector is its relatively low efficiency, compared with that of NaI(Tl) crystals. However, the low efficiency of the Ge(Li) detectors is not a serious handicap for many elements since sufficiently high concentrations of their neutron activation products can normally be produced in high flux nuclear reactors.

The basic principles and applications of Ge(Li) diode detectors have been described in the literature (see e.g. Bertolini and Coché, 1968; Camp, 1967; Cooper, 1973; Dearnaley and Northrop, 1966; Heath, 1969; Keil and Bernt, 1972). Briefly, the hardware required for a basic Ge(Li) gamma-ray spectrometer is the Ge(Li) detector (including a cryostat), a preamplifier, an amplifier, and a 2000 or 4000 channel pulse height analyzer. Although Ge(Li) detector systems of greater versatility and sophistication have been constructed (e.g., Cooper and Perkins, 1972), ordinary Ge(Li) detector systems are adequate for the measurement of a large group of neutron activation products in marine samples.

In practice, the irradiated samples and standards are transferred from the irradiation containers into standard counting geometries, which may consist of polyethylene vials of various sizes or some other suitable containers. Care must be taken to ensure that the samples and standards are encapsulated in identical counting geometries. The samples and standards are positioned on or near to the Ge(Li) detector in a fixed geometry, since relatively small differences in the orientation of sample or standard can lead to serious errors. Counting is carried out for a sufficient length of time to achieve the necessary counting statistics. Differences in geometries in a set of samples can be minimized by counting the samples several centimetres away from the Ge(Li) detector, if sufficient activity is present. Samples should not be placed directly on the detector to increase counting efficiency without experimentally evaluating the geometry factors.

B. *NaI(Tl) detectors*

Thallium-activated sodium iodide crystal detector systems are most useful when sensitivity is the main concern, since NaI(Tl) detectors are much more efficient than Ge(Li) diode detectors. Sodium iodide crystals of various sizes and geometries are commercially available; the most commonly employed ones are well-type crystals or solid cylindrical crystals normally ranging in size from about 7 to 35 cm in diameter and about 7 to 25 cm thick. The background and Compton response of these crystals can be substantially lowered by anticoincidence shielding with large plastic phosphors or annular NaI(Tl) crystals (see e.g., Perkins and Robertson, 1965; Wogman et al., 1967). Two large anticoincidence shielded NaI(Tl) detectors can be combined to count as dual coincidence gamma-ray spectrometers, thereby greatly increasing the sensitivity and selectivity of NaI(Tl) gamma-ray spectrometry (Perkins and Robertson, 1965).

These large crystal NaI(Tl) systems are particularly useful for the measurement of traces of neutron activation products which have been radiochemically separated from an interfering matrix. Their high efficiencies permit very low levels of radioactivity to be measured in relatively short counting intervals.

The geometric considerations mentioned above for Ge(Li) detectors, are also applicable to NaI(Tl) detector systems. However, because NaI(Tl) detectors are normally much larger in size, these geometric factors are not as critical as they are for smaller Ge(Li) detectors.

C. *Beta counting*

The neutron activation products of some elements (e.g. P and Tl) decay primarily by beta emission. However, some other elements give rise to activation products which emit only low intensity gamma rays; for these elements, beta counting is the most sensitive means of detection.

Thin window, gas-flow proportional counters or Geiger–Muller counters are most frequently employed for beta counting. Again, care must be exercised to ensure that the samples are presented in the same geometries and matrices. This is even more critical for beta counting than for gamma spectrometry because β-particles are more easily absorbed or scattered by the sample matrix.

Very efficient radiochemical separations of the neutron activation products of interest from activated sea salts or preconcentrates from sea water are necessary for beta counting, since large amounts of other beta emitters (primarily ^{32}P and ^{35}S) are also produced. Even when the radionuclide of interest has been cleanly separated from other beta and gamma emitters, it is necessary to check its energy and half-life. Approximate beta energies are usually determined by measuring the beta particle absorption by a set of standard aluminum absorbers of varying thickness. Half-lives are determined by recounting the sample after appropriate time intervals. Only after these precautions have been exercised can it be certain that the beta activity is due solely to the radionuclide of interest. Because of the relatively unselective nature of beta counting, gamma-ray spectrometry is recommended whenever possible.

D. *Data handling*

Interpretation of gamma-ray spectra obtained from NaI(Tl) and Ge(Li) detectors typically involves (1) location of peaks in the spectra, (2) determination of peak energies, (3) measurement of peak areas and, (4) calculation of concentrations of the elements from which the gamma-ray emitters were formed. In NaI(Tl) spectrometry, if the radionuclide mixture is quite simple, the peak areas can be obtained by manual subtraction of background and Compton contributions. However, if the radionuclide mixture contains more than three or four gamma-emitting radionuclides, the spectrum stripping is most conveniently and accurately accomplished by using an appropriate computer program.

Because peak areas in Ge(Li) diode spectra are well resolved, determination of peak intensities by manual subtraction of the background and Compton contribution can be readily accomplished even for very complex mixtures of radionuclides. Normally, equal numbers of channels on each side of the peak area are selected to represent the contribution due to background and Compton interferences, and this contribution is manually subtracted from the peak area. However, if many routine analyses are required, it is less tedious and time consuming to process the data with a computer.

The development and application of computer programs for the analysis of gamma ray spectra is by no means an elementary task. For computer analysis to be feasible, spectrometer systems must be electronically very stable and precisely adjusted before counting. The spectra must, of course, be converted to a computer compatible form. Incorrect input of data confuses a computer, resulting in termination of the analysis or, even worse, in computation of the wrong answers. It is a good practice to check computer programs occasionally by desk calculations to ensure that they are operating as intended. Once these problems have been overcome, a computer analysis of the gamma-ray spectrum is capable of rapidly converting the spectra directly into concentrations of individual constituents. Baedecker (1971) has evaluated several computer methods for gamma-ray peak integration and has described a computer routine which processes spectral data, makes appropriate corrections for decay and provides a readout in concentration units. However, unless it is intended to analyse a large number of routine samples, it is often more trouble than it is worth to develop a computer program designed to quantify NAA. Detailed discussions of the use of computers for data reduction have been given by DeVoe and LaFleur (1969) and Salmon and Creevy (1970).

19.10.4.11.5. Scope of the technique

A. Accuracy and precision

The accuracy of neutron activation analysis can be affected by a number of systematic errors which lead to biased results. Some examples of the sources of such errors are: incorrectly prepared elemental standards, consistent unknown interference from another activation product, neutron self-shielding by large amounts of elements having high capture cross sections or systematic contamination. These errors are often difficult to recognize, and are best reduced by very careful design and preparation of the analytical procedure. Accuracy of a procedure is best evaluated by analysis of a standard reference material of known elemental composition. Unfortunately, no standard sea water samples are available for trace element analysis, but the multi-laboratory intercalibration exercise initiated by Brewer and Spencer (1970a) is a step in the right direction.

The precision of neutron activation analysis is affected mainly by random errors caused by such factors as poor counting statistics, instrumental fluctuations, faulty manipulation, random contamination and irreproducible chemical reactions. The cumulative effect of these types of errors can be characterized by the standard deviation, which can be determined by carrying out replicate analyses on a single sample.

The accuracy and precision required for trace element analyses of sea water varies greatly depending upon the element and the application of its measurement. For example, the precision of 2·4% at the 95% confidence level attained in the NAA of strontium in sea water (see Table 19.8) is about as good as as can be expected for trace element analyses of sea water by this technique. However, it barely attains the precision required for the detection of the very small (2–3%) variations in the Sr/Cl ratio encountered in the open ocean.

TABLE 19.8

Precision of determination of strontium in sea water by neutron activation with direct counting. Eight replicate analyses of a single sea water sample

Replicate sample	Sr concentration found (mg l^{-1})
1	7·73
2	7·81
3	7·75
4	7·78
5	7·92
6	7·79
7	7·67
8	7·95
Average concentration	7·80
% Standard deviation	1·2
% Standard deviation at 95% C.L.	2·4

Table 19.9 illustrates the typical precision which is attainable by instrumental NAA for the measurement of elemental concentrations in sea water near or below the µg l^{-1} range. Uranium, caesium and antimony concentrations in the open ocean appear to be almost directly proportional to the salinity, and these elements are rather homogeneously distributed in the oceans. Neutron activation analysis of these elements could not, with certainty, detect any oceanic variations in concentrations of less than 6 to 9%. However, the oceanic concentrations of elements, such as Co, Sc, Zn and many others, vary as much as an order of magnitude, and the precision attained by NAA is entirely adequate for most marine geochemical studies concerned with these elements.

To summarize, in very careful NAA of sea water it is generally possible to attain precisions of 2 to 10% standard deviation at the 95% confidence level, if the counting statistics are not a limiting factor. Where low counting statistics are the major source of error, the precision of a set of determinations is governed by the uncertainty associated with each individual counting rate measurement. This can be assessed using the equation

$$\sigma_n = \sqrt{\sigma_B^2 + \sigma_T^2}$$

TABLE 19.9

Precision of measurement of trace elements in a Pacific Ocean water sample by neutron activation and direct counting technique

Sample	\multicolumn{7}{c}{Concentrations in $\mu g\,l^{-1}$}
	U	Cs	Sb	Co	Sc	Zn	Fe
1	3·20	0·218	0·129	0·0083	0·00046	6·2	8·4
2	3·36	0·231	0·131	0·0098	0·00061	6·4	12·0
3	3·40	0·215	0·131	0·0085	0·00046	6·2	8·4
4	3·26	0·228	0·141	0·0089	0·00046	—	
5	3·49	0·230	0·131	0·0098	0·00050	—	
6	3·06	0·233	0·136	0·0084	0·00040	—	
Av. conc.	3·30	0·226	0·133	0·0089	0·00049	6·3	9·6
% Std. dev. at 95% confidence level	9·3%	6·6%	6·6%	15·5%	29·4%	3·9	43·3%

where σ_n is the standard deviation of the net count rate, σ_B is the square root of the background (including Compton interference) counts divided by the counting time, and σ_T is the square root of the total number of counts due to background and sample divided by the counting time. Thus, in NAA the measured concentration of an element will have an associated uncertainty that can be no better than the standard deviation of the net count rate of the activation product which was measured.

B. *Sensitivity*

The detection limit for a particular element by the NAA technique is dependent upon many factors. Numerous compilations of detection sensitivities computed for various sets of conditions have been published in the literature. A generally useful compilation of interference-free limits of detection is given in Table 19.10. It shows that NAA is an extremely sensitive method of analysis; indeed the detection limits of many of the elements in Table 19.10 could be improved tenfold by irradiating the samples for ten hours instead of one hour.

If the detection limits in Table 19.10 are compared with the typical concentrations of these elements present in the oceans (see Chapter 7), it is evident that NAA possesses adequate sensitivity to measure nearly all of these elements using only very small volumes of sea water.

TABLE 19.10

Interference-free limits of detection for 75 elements by neutron activation analysis using a 1 hour irradiation in a thermal neutron flux of $10^{13} \, n \, cm^{-2} \, sec^{-1}$
(after Guinn and Lukens, 1967)

Limit of detection (μg)	Elements
$1\text{--}3 \times 10^{-7}$	Dy
$4\text{--}9 \times 10^{-7}$	Eu
$1\text{--}3 \times 10^{-6}$	
$4\text{--}9 \times 10^{-6}$	Mn, In, Lu
$1\text{--}3 \times 10^{-5}$	Co, Rh, Ir
$4\text{--}9 \times 10^{-5}$	Br, Sm, Ho, Re, Au
$1\text{--}3 \times 10^{-4}$	Ar, V, Cu, Ga, As, Pd, Ag, I, Pr, W
$4\text{--}9 \times 10^{-4}$	Na, Ge, Sr, Nb, Sb, Cs, La, Er, Yb, U
$1\text{--}3 \times 10^{-3}$	Al, Cl, K, Sc, Se, Kr, Y, Ru, Gd, Tm, Hg
$4\text{--}9 \times 10^{-3}$	Si, Ni, Rb, Cd, Te, Ba, Tb, Hf, Ta, Os, Pt, Th
$1\text{--}3 \times 10^{-2}$	P, Ti, Zn, Mo, Sn, Xe, Ce, Nd
$4\text{--}9 \times 10^{-2}$	Mg, Ca, Tl, Bi
$1\text{--}3 \times 10^{-1}$	F, Cr, Zr
$4\text{--}9 \times 10^{-1}$	Ne
$1\text{--}3$	S, Pb
$4\text{--}9$	Fe

C. Errors and interferences

Apart from contamination, the most common errors in NAA result from irreproducible activation or counting geometries. These sources of error have been discussed in Sections 19.10.4.11.3 and 19.10.4.11.4. With care, these errors can be reduced to almost negligible proportions by irradiating and counting samples and standards in identical configurations. Neutron self shielding errors will be discussed in Section 19.10.4.11.6.A.3 and can be overcome by preparing the elemental standards in a matrix similar to that of the sample.

The absolute, or flux, method of NAA is subject to many more errors than is the comparative method using elemental standards for each element. Errors associated with half-life, decay schemes, counter calibrations, thermal neutron cross sections, or flux perturbations combine to reduce the accuracy of the absolute method of NAA. However, it is still feasible to achieve an accuracy and precision of 5 to 30 % if the factors are carefully controlled and evaluated.

Interferences from other radionuclides in neutron activated samples can be virtually eliminated by specific radiochemical separations. Frequently however, very complex mixtures of radionuclides can be resolved by counting with high resolution Ge(Li) gamma ray spectrometers. Rapid group chemical separations may be all that is necessary to achieve the desired radiochemical purity.

A more subtle form of interference can occur when the same activation product is produced from two different parent elements by different nuclear reactions. A good example of this type of interference could occur in the measurement of Al in sea water via the $^{27}Al\,(n,\,\gamma)^{28}Al$ thermal neutron reaction. If Al was preconcentrated by coprecipitation with $Fe(OH)_3$, Si and P would also be precipitated to some degree, and be present in the $Fe(OH)_3$ at concentrations orders of magnitude higher than the Al. If the $Fe(OH)_3$ sample is exposed to a neutron flux containing an appreciable fast neutron component, ^{28}Al could also be produced by $^{28}Si(n,\,p)^{28}Al$ and $^{31}P(n,\,\alpha)^{28}Al$ reactions. Thus, erroneously high Al results could be obtained. Usually, fast neutron cross sections are much smaller than thermal neutron cross sections, and for the majority of elements in sea water this type of interference is not a serious problem. This problem can be minimized by activating the samples in areas of the reactor characterized by well-thermalized neutron fluxes.

19.10.4.11.6 *Sea water analysis*

A. *Sample preparation*

1. *Materials.* Since the concentrations of most trace elements in sea water are near or below the parts-per-billion range, contamination of the samples during storage and analysis is a major problem. Therefore, sampling equipment, storage containers and reaction vessels should be made of materials which are relatively free from metallic impurities.

The preferred storage containers are made of high density polyethylene or polypropylene. Containers should be leached with high purity hydrochloric or nitric acid and rinsed with high purity water to remove traces of metals left on the surfaces during the manufacturing process. A good practice is to rinse the container with the actual sea water sample just prior to filling the bottle. Experience has shown that losses of trace elements from sea water by adsorption onto container surfaces can be eliminated or minimized by acidifying the sea water to a pH of 1 to 2 with high purity hydrochloric or nitric acid.

Reaction vessels and neutron irradiation containers should also be made of polyethylene or polypropylene since these materials are generally orders of magnitude purer with respect to trace elements than are borosilicate glass containers. An excellent discussion on the limitations of glass containers for the storage of ultrapure solutions is given by Adams (1972). When elevated temperatures (up to 200°C) are required for certain chemical manipulations, Teflon containers should preferably be used. These should also be acid leached and rinsed with high purity water before use.

Since chemical reagents often contain appreciable concentrations of various metal impurities, reagents of the highest purity should be used in the chemical manipulations. Water, and some solvents, usually need to be purified before use, and Smith (1972) has reviewed the state-of-the-art for preparing ultrapure water. Usually, reagent grade acids are surprisingly pure. Attempts at further purification by redistillation often result in an increase of various metallic impurity levels by leaching of the glass distillation apparatus. Isothermal distillation is a preferred method for purifying acids or other aqueous solvents which have high vapour pressures.

Airborne contamination, which is prevalent in most land-based or ship-board laboratories, can be minimized by working in positive pressure, laminar flow clean hoods (see Figure 19.12). Paulhamus (1972) has given a comprehensive review of airborne contamination problems and methods for their elimination.

More detailed discussions on the problems of contamination and methods for overcoming them are given by Robertson (1968b, 1972) and Thiers (1957a, b).

2. *Manipulations.* Sea water itself cannot be directly neutron-irradiated for an extended length of time in a high flux nuclear reactor using polyethylene or polypropylene containers. The gaseous radiolysis products (i.e., H_2, O_2 and other gases) which are generated from the sea water produce high pressures which may rupture the containers. Sea water has been neutron-irradiated in sealed quartz containers, but this is not to be recommended because the dangerously high pressures that are produced may shatter the quartz vials during handling.

The preferred method of handling sea water for instrumental NAA, or for post-irradiation radiochemical separations of the activated sea salt, is to evaporate weighed or pipetted aliquots of sea water to dryness. Either slow evaporation under infra-red heat lamps or freeze-drying can be employed.

The former method, which is the one preferred at the Battelle Laboratories involves accurately weighing approximately 35 ml aliquots of sea water in tared, cleaned 40 ml polyethylene beakers. The samples are placed under 4 l glass beakers and the water is evaporated slowly under several infra-red lamps. The entire operation is conducted in a positive pressure laminar flow clean hood to minimize contamination (see Figure 19.12). The dried sea salts, which form a crust at the bottom of the beaker, are then transferred into cleaned polyethylene or polypropylene irradiation vials.

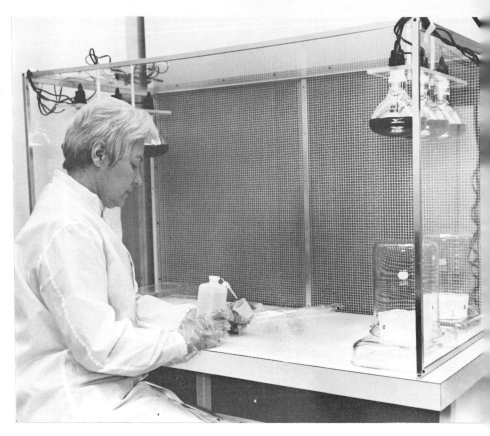

FIG. 19.12. Positive pressure laminar flow clean hood for minimizing air-borne contamination during manipulation of sea water samples.

The dry, encapsulated salts and appropriate elemental standards are then neutron-irradiated for periods of 7 to 24 hours in a thermal neutron flux of about 2–20×10^{12} n cm^{-2} s^{-1}. After the irradiation, the samples and standards are stored for one to two weeks in appropriate shielding to allow the high concentrations of interfering short-lived radionuclides (principally ^{24}Na, ^{38}Cl, ^{82}Br and ^{42}K) to decay to safe working levels. The concentrations and dose rates of the major short-lived interfering radionuclides in neutron activated sea salts are shown in Table 19.11. The initial high dose rates due to ^{24}Na and ^{38}Cl prevent any early post-irradiation manipulations of the samples to be performed, except in well-shielded facilities. It is more convenient to wait about one to three weeks to allow these interferences to decay to tolerable levels before working with the samples.

TABLE 19.11

Concentrations and gamma dose rates of major short-lived interfering radionuclides in neutron activated sea salts*

Element	mg of element per 1·2 g of sea salt	Activation product	Concentration of activation product at end of irrad. (mCi)	Dose rate (mR/h) 1 min. after irrad. at 1 cm	at 30 cm	Dose rate (mR/h) 2 weeks after irrad. at 1 cm	at 30 cm
Na	370	^{24}Na (15·4 hr)	370	$7·0 \times 10^6$	7800	1·9	0
Cl	670	^{38}Cl (37·3 m)	280	$1·0 \times 10^6$	1100	0	0
Br	2·3	^{82}Br (36 hr)	0·94	$1·4 \times 10^4$	15	22	0·024
K	13	^{42}K (12·5 hr)	1·3	1800	2	0	0
Mg	46	^{27}Mg (9·5 m)	0·83	4100	4·6	0	0
Ca	14	^{49}Ca (8·8 m)	0·091	1200	1·3	0	0

* For the instrumental NAA analysis described in the text: 1·2 g sea salts (35 ml sea water), are irradiated for 8 hours at a thermal neutron flux of 10^{13} n/cm^2/sec. The irradiated sea salts are stored in appropriate shielding for at least two weeks before handling and at least four weeks before counting. For purposes of comparison the maximum permissible occupational exposure to ionizing radiation is 75,000 mrem year^{-1} to the hands or extremities, and 5000 mrem year^{-1} to the total body. For gamma radiation, 1 mrem is nearly equal to 1 mR. Typically, an analyst may average about 2 hrs per week directly handling irradiated sea salts. Assuming that he works 50 weeks in a year, the accumulated gamma dose to the hands from handling 2-week-old irradiated sea salts would be about 2,400 mrem year^{-1} (using the dose rates given at 1 cm); the accumulated yearly dose to the total body would be about 2·4 mrem year^{-1} (using the dose rates for 30 cm). This exposure which originates mainly from ^{82}Br, is well below maximum permissible occupational exposure limits. The gamma dose could be reduced to negligible amounts by waiting for four weeks before handling the samples. The major activation products present four weeks after the neutron irradiation are the pure beta emitters ^{35}S and ^{32}P. The beta particles from these radionuclides are substantially absorbed by the sample containers, but some bremsstrahlung photons are produced which could contribute to the low energy gamma dose rate. The bremsstrahlung activity of the sample is a function of the neutron energy spectrum in the reactor, since ^{35}S and ^{32}P are produced mainly by fast neutron reactions on chlorine.

As shown in Table 19.11 the gamma dose rates of the samples two weeks after the irradiation will have fallen to completely safe working levels. The salts can then be quantitatively transferred into new polyethylene counting vials or some other appropriate counting containers, for instrumental analysis by direct gamma ray spectrometry. Alternatively, the salts can be dissolved and radiochemical separations of the elements of interest performed. The majority of the activity at this time is due to the beta emitters, ^{35}S and ^{32}P. If their activities are too high to permit direct instrumental gamma-ray spectrometry, the samples can be encapsulated in thin (1 mm) lead containers which attenuate the beta particles and much of their associated bremsstrahlung photon activity.

Simple pre-irradiation separations are frequently desirable for determining trace elements which cannot be measured by instrumental neutron activation analysis (especially those which form short-lived neutron activation products), since they greatly minimize the radiation interferences from ^{24}Na, ^{38}Cl, and ^{82}Br. The preconcentrates can be packaged in polyethylene irradiation vials and neutron activated with appropriate standards as described above. After the irradiation the samples are transferred to new counting containers and analyzed by gamma-ray spectrometry. Frequently, the pre-irradiation separation is all that is required to directly measure the activation product(s) of interest by Ge(Li) gamma-ray spectrometry. However, further purification may be necessary, and post-irradiation radiochemical separations of the preconcentrates can be employed to completely isolate the desired radionuclides.

Particulate matter suspended in sea water can contribute significant percentages of the total amount of several elements present. To determine the partitioning of the elements between the two phases, it is desirable to isolate the suspended particulate material by either centrifugation or membrane filtration. The preferred method is membrane filtration, with a filter having a pore size of about 0·3 to 0·5 μm. Both Millipore® and Nuclepore® filters have commonly been used for this purpose, and both filters contain roughly comparable traces of most impurities. It is recommended that as large a water sample as practical be filtered to obtain enough suspended matter for elemental analysis and to reduce the probability of contamination of the filtered water during the filtering process. If the filtered sea water is to be analyzed, the first several hundred ml of filtrate should be discarded because of contamination which may be leached from the filters and filtering apparatus. The filters should be rinsed several times with high purity water to remove all dissolved salts.

Several types of all-plastic filtering devices are commercially available (e.g. those made by Millipore Corporation and the General Electric Company) which are ideal for filtering sea water. The usual procedure for instrumental

NAA of suspended particulates is to collect the material on tared filters (Millipore,® Nuclepore,® etc.), desiccate to a constant weight and reweigh the filters in order to determine the particulate matter gravimetrically. Nuclepore® filters offer the most desirable taring characteristics compared to other types of membrane filters, because they absorb less moisture and are not appreciably reduced in weight by leaching during the filtration. Weighing membrane filters on an analytical balance may present some problems arising from the electrostatic charge which accumulates on the filters. This can be overcome by inserting a special α-emitting Po source (Watson Brothers, Northern California Division, Scientific Instruments, San Mateo, California) inside the balance case. The ionization of the air produced by the α-particles neutralizes the electrostatic charge on the membrane filter.

The weighed filters are then encapsulated in plastic irradiation vials and neutron activated, together with appropriate standards of the elements of interest. After the irradiation, the samples are counted on a Ge(Li) detector after optimum decay times. The suspended particulate matter in sea water is usually a combination of both geological and biological material, and up to 24 trace elements can be measured on the filters by instrumental NAA, depending upon the type and amount of material collected. If the activation products of interest cannot be measured instrumentally, the neutron-activated filters can be dissolved by acid digestion and radiochemical separations can be performed.

3. *Elemental standards.* In multielement instrumental NAA usually it is not practicable to use separate individual elemental standards. Composite standards consisting of compatible mixtures of elements are preferred. Standard solutions containing 5 to 10 elements at optimum concentrations can be conveniently prepared to provide good counting statistics and minimal interelement interferences. For example, at the Battelle Laboratories, a composite standard solution consisting of optimum concentrations of Rb, Cs, U, Sb, Co, Fe, and Zn is being used for instrumental NAA of sea salts. Thus, only a few standard capsules need to be included in a large set of samples for neutron activation. It is important that standards for each element being measured be included in each irradiation. The use of flux monitors for intercomparison of irradiations made at different times and reactor fuel configurations should be discouraged, since some activation products are produced to a large degree by resonance capture of fast neutrons. The production of such activation products depends not only on the neutron flux, but also the neutron energy spectrum.

Elemental standards should be encapsulated and irradiated in an identical fashion to the samples. Standards should be prepared in a matrix as similar to the sample matrix as possible in order to minimize problems such as neutron self-shielding, irreproducible activation geometries and interfering nuclear reactions. For example, when neutron activating large quantities of sea salts, the elemental standards should be added to an equivalent amount of NaCl encapsulated in the same geometry as the sample. However, if sample sizes of sea salts or preconcentrates from sea water are small, and the samples do not contain high concentrations of elements having large capture cross-sections (as is usually the case), exact matrix matching is not a critical requirement for the types of samples normally encountered in the trace element analysis of sea water.

B. *Instrumental analysis of neutron activated sea salts*

Because of the tremendous activity of ^{24}Na (half-life 15·4 h) which is produced during neutron irradiation, instrumental NAA of sea water is applicable only to those elements which have relatively long half-life activation products. Figure 19.13 illustrates the relative concentrations of various induced radio-nuclides in sea water following neutron activation and their change in concentration with time. It shows that none of the short-lived neutron activation products of interest could possibly be measured instrumentally, since the ^{24}Na, ^{38}Cl, ^{42}K and ^{82}Br are orders of magnitude more abundant, and completely mask the gamma ray contributions of the activation products of interest. Pre-irradiation separations of the elements of interest from the Na, Cl, K and Br are necessary to permit the measurement of these short-lived activation products.

Figure 19.14 shows that after about 26 days after removal from the reactor, the ^{24}Na has decayed to insignificant concentrations, and ^{82}Br (36 hours) is now the major gamma-emitting interfering radionuclide. Bromine-82 decays with the emission of a large number of high energy gamma rays which interfere with the measurement of the trace activation products of interest. Hence, it is necessary to store the samples for another one or two weeks to allow the ^{82}Br to decay to insiginificant levels. High concentrations of ^{35}S and ^{32}P remain in the samples, but they are pure beta emitters and can be tolerated in the gamma ray spectrometry. The ^{35}S and ^{32}P are both produced by the reactions of fast neutrons with stable chlorine-35, as well as by the usual thermal neutron (n, γ) reactions with sulphur and phosphorus.

Figure 19.15 shows a typical gamma ray spectrum of neutron activated sea salts counted on a Ge(Li) detector 54 days after the neutron irradiation. This figure also demonstrates the much greater resolution which can be obtained with Ge(Li) detectors compared with that obtainable using NaI(Tl) crystals. It is apparent that the eight trace elements, Sr, Rb, Cs, U, Sb, Zn, Fe and Co can be measured in sea salts by this technique.

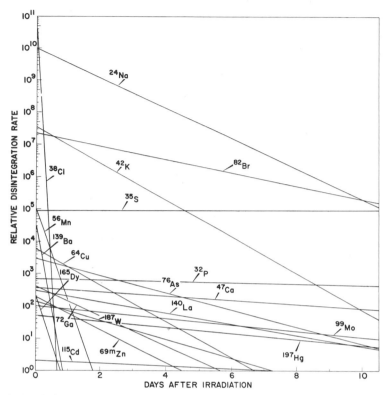

FIG. 19.13. Relative concentrations of induced radionuclides in sea water following an integrated thermal neutron exposure of 5×10^{16} n cm^{-2}. Emphasizing the problem of counting short-lived activation products in the presence of high concentrations of ^{24}Na, ^{82}Br, ^{42}K and ^{35}S.

Table 19.12 lists the trace elements in sea water, which can normally be measured by instrumental NAA, and compares the detection limits of this method with the actual concentrations of these elements observed in ocean water. This method is sufficiently sensitive to measure any of the concentrations of these elements observed in the oceans.

C. Preconcentration and post-irradiation separations

Since most trace elements exist in sea water at concentrations near or below the parts per billion range, rather large volumes of sample (0·1 to 1 l) are frequently desirable for analysis to give adequate counting statistics. Preconcentration of the desired elements is therefore necessary.

The simplest preconcentration procedure involves merely the evaporation to obtain dry salts. Since sea water contains approximately 3 to 3·5% of

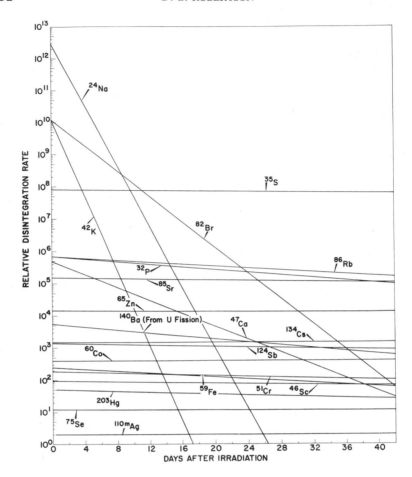

Fig. 19.14. Relative concentrations of induced radionuclides in sea water following an integrated thermal neutron exposure of 10^{18} n cm^{-2}. Emphasizing the problem of counting intermediate and long-lived activation products in the presence of high concentrations of ^{24}Na, ^{82}Br and ^{35}S.

dissolved salts, a concentration factor of about 97 can be achieved in this way. Sea salt samples ranging from about 0·03 to 5 g (equivalent to about 1 to 150 ml sea water) can be conveniently handled. Neutron activation analysis of dried sea salts is, for all practical purposes, limited to elements which form relatively long-lived (days to years) activation products, since the high concentrations of short-lived ^{24}Na, ^{38}Cl, ^{42}K and ^{82}Br prevent early access to the samples. In this case, post-irradiation radiochemical separations, performed several days or weeks after the neutron irradiation, are necessary to isolate the radionuclides of interest.

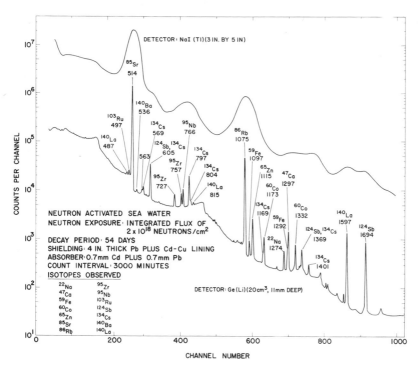

FIG. 19.15. γ-ray spectra of neutron activated sea water obtained with NaI (Tl) detector (upper) and Ge (Li) detector.

Group chemical post-irradiation separation schemes are frequently desirable for multielement analysis of sea water. Sequential group separations have been devised for the multielement NAA of biological and geological materials (Morrison and Potter, 1972; Morrison et al., 1969). These procedures could be slightly modified and applied to sea salts. Schutz and Turekian (1965b) have developed group radiochemical separation schemes, followed by isolation of individual radionuclides for multielement NAA of sea salts.

Radiochemical procedures to completely isolate the radionuclides of interest from nearly any type of matrix have been developed. As previously stated, post-irradiated separations have the advantage that once the radioactive species are produced, chemical manipulations can be performed using any amount or type of reagents without contaminating the samples with the elements being measured.

The most generally useful references on radiochemical separations for all

TABLE 19.12

*Estimated detection limits for instrumental neutron activation analysis of sea water**

Element	Typical oceanic concentration ($\mu g\,l^{-1}$)[†]	Detection limit ($\mu g\,l^{-1}$)
Sr	8000	100
Rb	120	5
Fe	1–20 (5)	1
Zn	0·5–10 (3)	0·2
U	3·3	0·1
Cs	0·3	0·003
Sb	0·2	0·005
Co	0·001–0·1 (0·03)	0·001
Sc	$1–20 \times 10^{-4}$	1×10^{-4}[‡]

* 35 ml sea water; 16 hour irradiation at 10^{13} n cm^{-2} sec^{-1}; 40 days decay; 1000 minute count on 20 cm^3 Ge(Li) diode detector.
† Estimated average concentrations in parentheses.
‡ Counted on a dual coincidence NaI(Tl) gamma ray spectrometer.

of the elements are those published by the U.S. Atomic Energy Commission under the auspices of the U.S. National Academy of Sciences, National Research Council. The radiochemistry of each element has been published as a monograph. This may be obtained for a nominal fee from the National Technical Information Service, U.S. Department of Commerce, Springfield, Virginia 22151 (e.g. see Roesmer, 1970). References to specific post-irradiation radiochemical separations used in neutron activation analysis of sea water have been compiled (see Section 19.10.5).

The other alternative for separating the elements of interest from the matrix interferences in sea water is to preconcentrate the elements by some chemical manipulation before the neutron irradiation. The most frequently employed methods are coprecipitation, cocrystallization, solvent extraction and ion exchange. The use of these techniques in trace element analysis has been adequately covered in the literature (see e.g. Goya and Lai, 1967; Joyner, et al., 1967; Minczewski, 1967; Mizuike, 1965; Pascual, 1966; Riley, 1965b; Riley and Taylor, 1968; Rottschafer, et al., 1972). Examples of their application to trace element analysis of seawater by NAA are given in Section 19.10.5.

The advantages of preconcentration are that very large sample volumes can be processed for analysis, and analysis can be commenced almost immediately after the neutron irradiation, if sufficient separation from the Na and Cl is achieved. Thus, elements which produce short-lived activation products can be measured.

19.10.5. METHODS FOR THE DETERMINATION OF INDIVIDUAL TRACE ELEMENTS
(excluding natural radioactive elements other than uranium)

Methods which have been used for the concentration and determination of both the dissolved and particulate forms of the individual trace metals in sea water are summarized below. No attempt has been made to review the analysis of fresh waters, although some of the methods which have been used for these might be applicable to sea water. As far as practicable an attempt has been made to cover the field critically. Much of the early work, which has been reviewed by Riley (1965b), has been omitted if it appeared to give results which differed widely from currently accepted values, or when outmoded or obviously unreliable techniques were used. Unfortunately, this unreliability is not only confined to the older investigations, and much modern work on the determination of trace elements in sea water is marred by failure to check the precision and accuracy of the analytical methods and to ensure that there is no interference from other elements. Although there have been great advances in analytical techniques and instrumentation in the past decade, many workers have failed to appreciate their limitations in sensitivity and specificity, and the matrix effects to which they are subject.

A few attempts have been made to carry out inter-laboratory intercalibrations of trace element analyses of sea water. In the largest of these, which was organized by workers at the Woods Hole Oceanographic Institution, three samples of sea water were distributed to 26 laboratories. Analyses of these samples were carried out by experienced analysts using methods of their own choice. The collected data from these studies (Brewer and Spencer, 1970a; see also Section 19.10.4.11) revealed a disquieting lack of consistency; the scatter for some elements being as great as an order of magnitude. Most of this variability probably arises from the use of inadequately tested methods and failure to prevent contamination and to determine reliable blanks. Much more reassuring results were obtained in a collaborative intercalibration experiment carried out by four laboratories as a preliminary to the Geosecs Expedition (Spencer et al., 1970d). Two or more laboratories reported data for Sr, Cs, Zn and U; in general the agreement was good. Bender et al. (1972) have given an account of an intercalibration test carried out for barium in waters from Geosecs I and III. Two sets of determinations made using mass spectrometric stable isotope dilution methods agreed to ca. $\pm 1\%$; results obtained using a combination of stable isotope dilution and neutron activation analysis agreed with them to $\pm 2\%$. Broecker et al. (1970) have presented results obtained by 4 groups of workers (see p. 435) for the profile of ^{226}Ra at the 1969 Geosecs calibration station in the Pacific Ocean. Most measurements agreed to $\pm 15\%$.

19.10.5.1. *Aluminium*

Aluminium occurs in sea water as both particulate and "dissolved" species. The former, which are those retained by a 0·45 μm filter, are very variable in concentration; they probably consist mainly of alumino-silicate minerals of detrital and authigenic origin. The dissolved species, which pass a 0·45 μm filter, are likely to include not only those in true solution, but also those in the colloidal size range. They are probably mainly hydroxy-complexes of the metal (Horn, 1967). The concentration of "dissolved" aluminium usually lies in range 1–10 μg l^{-1}.

Determination of dissolved aluminium. Riley (1965b) has reviewed the earlier work on the determination of aluminium in sea water. Most of it is now only of historical interest, because the results which were produced are 1–2 orders of magnitude higher than those currently accepted, perhaps as a result of failure to guard against interference from other elements.

Most recent work on dissolved aluminium in sea water has been carried out using fluorimetric methods, which, because of their sensitivity do not require a preconcentration stage. Simons et al. (1953) have made use of the orange-red fluorescent complex formed at pH 5·0 by reaction with Pontachrome Blue Black R (White and Lowe, 1937), and their procedure has been used by Sackett and Arrhenius (1962) in an extensive study of the distribution of aluminium in sea water. The effect of possible interfering elements on the method has been studied by Donaldson (1966) who found that although Fe^{3+}, Ga^{3+}, Co^{2+}, VO_3^-, Ti^{4+}, F^- and PO_4^{3-} can cause interference at concentrations appreciably greater than those in natural waters they will not interfere at their normal levels in sea water. The interference from unusually high concentrations of iron can be prevented by complexing it as the ferrous-bathophenanthroline complex. Lumogallion is a more sensitive reagent and has also been employed for the fluorimetric determination of aluminium in sea water down to 0·05 μg l^{-1} (Nishikawa et al., 1968; Shigematsu et al., 1970). Potential interfering elements are the same as those in the Pontachrome Blue Black R method. At its normal level in sea water only iron will interfere; however, its interference can be eliminated by complexing it with 1,10-phenanthroline.

Attempts have been made recently to apply atomic absorption spectrophotometry (Fishman, 1972) and gas liquid chromatography (Lee and Burrell, 1973) to the determination of dissolved aluminium in sea water. However, the methods lack the speed, simplicity and precision of the fluorimetric techniques.

Determination of particulate aluminium. Aluminium has been directly determined in the marine particulates retained by a $0.45\,\mu m$ membrane filter by X-ray fluorescence spectrometry (Cann and Winter, 1971). Other workers have dissolved the particulate matter on the filter by alkaline fusion or attack with hydrofluoric acid, and have determined aluminium either by atomic absorption spectrophotometry (Spencer and Sachs, 1969), colorimetrically (Joyner, 1964; Atkinson and Stefansson, 1969) or fluorimetrically (Sackett and Arrhenius, 1962; Feely et al., 1971; see also Donaldson, 1966).

19.10.5.2. *Antimony*. Both "dissolved" and particulate forms of antimony are known to occur in sea water. The former, which are probably hydroxy-species (perhaps $Sb(OH)_6^-$), appear to total ca. $0.25\,\mu g\,l^{-1}$. The particulate forms will include adsorbed species and Sb^{3+} present in the lattices of the suspended minerals.

Preconcentration. Antimony can be efficiently coprecipitated from sea water with hydrous manganese dioxide (Portmann and Riley, 1966a); ferric hydroxide is a much less effective carrier (Portmann, 1965) although it has apparently been used by Ryabinin and Romanov (1973). Gohda (1972) has concentrated antimony from sea water by precipitation with thionalide, oxine and ferric hydroxide. The element has also been concentrated from sea water by volatilization as stibine (SbH_3) (Braman et al., 1972).

Determination of dissolved antimony. Portmann and Riley (1966a) have determined antimony in sea water photometrically with Rhodamine B after concentrating it by coprecipitation with manganese dioxide and purifying it by solvent extraction of its iodide. The recovery of antimony, which was ca. 80%, was monitored by isotope dilution using [125]Sb.

The sensitivity of conventional atomic absorption or flame emission spectrophotometric techniques for antimony is only poor. However, high sensitivity for the element can be achieved using an electrical discharge on argon or helium to excite the emission spectrum. Braman et al. (1972) have described an elegant procedure based on this principle in which the degassed sample is treated with sodium borohydride, and the resultant stibine is swept with a current of helium into a discharge tube. The intensity of the 252·5 nm Sb emission is then measured with a sensitive spectrophotometer. A detection limit of ca. 1 ng of Sb was claimed.

Anodic stripping voltammetry has been used by Gilbert and Hume (1973) in a rapid method for the direct determination of antimony in sea water. Deposition of the element was carried out at $-0.5\,V$ from $4\,M$ hydrochloric acid medium. A precision of $\pm 18\%$ was achieved at an antimony level of $\pm 0.4\,\mu g\,l^{-1}$ (see Section 20.5.3).

On irradiation with thermal neutrons, antimony undergoes the nuclear reaction $^{121}Sb\ (n, \gamma)\ ^{122}Sb$ and $^{123}Sb\ (n, \gamma)\ ^{124}Sb$. With realistic irradiation times considerably greater sensitivity can be obtained in neutron activation analysis by counting the 2·8 day ^{122}Sb activity than by counting the longer lived (60 day) ^{124}Sb activity. However, in the analysis of sea water, the hazardously high ^{24}Na activity generated prevents the use of the shorter-lived radionuclide unless antimony is separated prior to counting (e.g. by volatilization as SbH_3 or coprecipitation (Ryabinin and Romanov, 1970, 1973; Ryabinin et al., 1971, 1972). This difficulty can be overcome by allowing the irradiated sea salts to cool for several weeks to allow short-lived activities to decay and then measuring the ^{124}Sb activity either by γ-spectrometry (Piper and Goles, 1969; Robertson et al., 1968; Spencer et al., 1970; Brewer et al., 1972) or by counting after a preliminary radiochemical separation (Schutz and Turekian, 1965; Gohda, 1972).

Determination of particulate antimony. The determination of antimony in marine particulates has been described by Piper and Goles (1969) and by Spencer et al. (1970, 1972). In each instance, the membrane filter bearing the particulates was submitted to neutron irradiation, and after prolonged cooling, ^{124}Sb activity was measured by γ-spectrometry.

19.10.5.3. *Arsenic*

The concentration of dissolved arsenic in the sea usually lies in the range $1-3\ \mu g\ l^{-1}$; the principal species probably being $HAsO_4^{2-}$ (see Table 3.10). However, Braman and Foreback (1973) have shown that a small proportion (4–7%) of the arsenic in certain inshore waters is in the 3+ oxidation state. The same workers have also demonstrated the presence of methylarsonic and dimethylarsinic acids ($< 1-4$ and $0·2-1\ ng\ l^{-1}$ respectively) in the same waters.

Determination of arsenic. Preconcentration. Arsenic (V) $(2\ \mu g\ l^{-1})$ can be nearly quantitatively ($> 99\%$) coprecipitated from sea water at pH 7 with ferric hydroxide (Portmann, 1965). This technique has been used for the analytical preconcentration of the element from sea water by several workers (e.g. Ishibashi et al., 1951, 1960b, Sugawara, 1955; Ryabinin and Romanov, 1972). Mercuric sulphide has also been used as a coprecipitant (Noddack and Noddack, 1940). Thionalide forms a highly insoluble complex with As^{3+}, and this reagent has been used by Portmann and Riley (1964) for the co-crystallization of arsenic from sea water in $97·5 \pm 1\%$ yields. Gohda (1972) has used a mixed precipitation with thionalide and ferric hydroxide for the same purpose. Preconcentration has also been carried out by the Gutzeit method in which arsenic is volatilized as arsine (see e.g. Rakestraw and Lutz, 1963; Young et al., 1959). However, this technique is undesirable because of the impossibility of evolving the arsenic quantitatively and also because of the difficulty of obtaining acceptable reagent blanks. Recently, Braman *et al.* (1972) have described an evolution technique in which arsine formed by the action of sodium borohydride at pH 4–9 is removed from the solution by a current of inert gas. According to Braman and Foreback (1973), only As^{3+} is reduced in this pH range, and before As^{5+} can be converted to arsine it must be reduced to the $3+$ oxidation state with sodium cyanoborohydride at pH 1–2. However, more recent work by Knudson and Christian (1974) has shown that As^{5+} can be reduced to arsine if the solution has an acidity of 0·2 M with respect to hydrochloric acid.

Determination of dissolved arsenic. Owing both to the lack of satisfactory analytical techniques, and the difficulty of obtaining reagents sufficiently free from this ubiquitous element, most of the determinations of arsenic in sea water published before 1952 are much too high and mainly of historical interest. The methods used in these investigations have been summarized by Portmann and Riley (1966) and will not be discussed here. It is is only since the advent of sensitive physico-chemical instrumental techniques that it has become possible to make reliable analyses for arsenic in sea water.

In an analogous fashion to phosphate, As^{5+} reacts in acid medium with molybate ion to give arsenomolybdic acid which can be reduced (e.g. by ascorbic acid) to give a molybdenum blue. This reaction is the basis of a sensitive photometric procedure for the determination of arsenic. It has been used to determine arsenic in sea water following concentration by coprecipitation (Sugawara, 1955) or co-crystallization (Portmann and Riley, 1964). Johnson (1971) and Johnson and Pilson (1972b) have recently described a different method for the direct determination of arsenate in sea water. This work is based on the fact that As^{5+}, but not As^{3+}, reacts with the reagent to produce molybdenum blue. Two aliquots of the sample are taken and one is treated with sulphite to reduce As^{5+} to As^{3+}. Both are then treated with the single solution phosphate reagent described by Murphy and Riley (1962). The difference between the absorbances of the two solutions is a measure of the amount of arsenate present in the sample, Johnson and Pilson 1972b) have extended this procedure by introducing an extra determination in which a further aliquot of the sample is treated with potassium iodate. This enables total arsenic $(As^{5+} + As^{3+})$ to be determined, and permits As^{3+} to be estimated by difference. These methods have little to recommend them because in most sea water samples phosphate–phosphorus, which gives ca. 3 times the absorbance as the same weight of arsenic, is present at concentration ten or more times that of the arsenic. The method therefore depends on the difference between two large absorbances.

The sensitivities of both flame emission and atomic absorption spectrophotometric procedures for arsenic are too low for these techniques to be of value in the analysis of sea water.* However, a high sensitivity can be achieved if a d.c. discharge in helium is used to excite the atomic spectrum of the element. This principle has been adopted in a method developed by Braman et al. (1972) for the determination of arsenic in sea water in which arsenic is volatilized from the sample as arsine (see above) and swept into the discharge tube with helium.

Natural arsenic consists of a single stable isotope—^{75}As. Irradiation of this nuclide with thermal neutrons leads to the production of ^{76}As which has a half life of 26·4 h and emits β-particles having an energy of 3·1 keV and multiple γ-rays. As a consequence of the high activation cross section of ^{75}As (5·4 barns), neutron activation analysis provides an attractive method for the determination of nanogram amounts of arsenic. The technique was first applied to the analysis of sea water by Smales and Pate (1952) who irradiated the water directly and then isolated the ^{76}As activity by radiochemical procedures for counting. The hazardously high induced activity of the irradiated sample (arising mainly from ^{24}Na) presents considerable shielding problems. This difficulty can be overcome by irradiating the arsenic after concentrating it from the sample. Thus, Ray and Johnson (1972) have described a procedure in which arsenic is preconcentrated by co-crystallization with thionalide, irradiated, and then determined by γ-spectrometry using the 0·559 MeV peak; a standard deviation of $\pm 0·15\,\mu g\,As\,l^{-1}$ was claimed. Similar neutron activation methods have also been used by Gohda (1972), Ryabinin et al. (1972) and Ryabinin and Romanov (1973).

Determination of organic forms of arsenic. Braman and Foreback (1973) have recently described a procedure for the determination of methylarsonic and dimethylarsinic acids in natural waters, including saline ones. In this, the buffered (pH 1–2) sample was treated with sodium borohydride which reduced methyl arsonic acid ($CH_3AsO(OH)_2$) to CH_3AsH_2, and dimethylarsinic acid ($(CH_3)_2AsO(OH)$) to $(CH_3)AsH$. These volatile compounds, along with AsH_3 formed by reduction of As^{3+}, were swept out of the reaction mixture with a stream of helium and condensed in a cold trap at $-196°C$. Subsequently the trap was allowed to warm to room temperature and, using helium, the compounds were swept out in the order of increasing boiling points into a discharge tube in which the arsenic emission lines were excited.

---

* Ediger et al. (1974) have recently shown that it is possible to determine arsenic in sea water directly with a sensitivity of $1\,\mu g\,l^{-1}$ if atomization is carried out by means of the carbon tube furnace (see p. 305).

Determination of particulate arsenic. For the determination of arsenic in marine particulates, Spencer *et al.* (1972) irradiated the pelletized membrane filter with thermal neutrons, and measured the induced ^{76}As activity by direct γ-spectrometry.

19.10.5.4. *Barium*

The concentration of dissolved barium in the sea appears to be variable, usually lying in the range 3–50 μg l^{-1}. The principal species present is probably Ba^{2+}, although a significant proportion will occur as BaSO$_4^0$. No information is available about the occurrence of the element in marine particulates.

Preconcentration. Although barium can be concentrated from sea water by coprecipitation with calcium oxalate, (Turekian and Johnson, 1966) the recovery is rather poor. Cation exchange techniques, which have been used by several workers (see e.g. Bowen, 1956; Andersen and Hume, 1968a; Wolgemuth and Broecker 1970; Bacon and Edmond, 1972), are probably the most effective means of preconcentrating the element. They also enable it to be efficiently separated from other alkaline earth metals.

Determination of dissolved barium. During the last decade there has been an upsurge in interest in the marine chemistry of barium, and this is reflected in the number of papers on its determination in sea water which have appeared recently. A wide range of physico-chemical techniques have been proposed for the analysis, and intercalibration studies have shown a reasonably satisfactory agreement between the results obtained by different methods (see e.g. Bender *et al.*, 1972).

Although attempts have been made (see e.g. Billings and Harris, 1965) to determine barium directly in sea water by flame emission or atomic absorption spectrophotometry this is infeasible, because of lack of sensitivity, inter-element interferences, and effects arising from the presence of salt particles in the flame. Before these techniques can be applied, barium must, therefore, be concentrated and separated from major ions, preferably by ion exchange. Both flame photometric (Andersen and Hume, 1968a; Andersen and Jademec, 1971) and emission spectrographic (Szabo and Joensuu, 1967) procedures have been used for the examination of concentrates prepared by ion exchange.

The principal relatively long-lived radionuclide produced when barium is irradiated with thermal neutrons is ^{131}Ba which decays by multiple γ-ray emission. Because of the low isotopic abundance (0·1 %) of the parent ^{130}Ba (activation cross section ca. 10 barns) and the relatively long half-life of ^{131}Ba (11·5 days), neutron activation analysis is comparatively insensitive for barium. Although the technique has been applied directly to freeze-dried sea salts (Bolter et al., 1964; Schutz and Turekian, 1965a), the high level of the ^{24}Na activity produced makes it desirable to concentrate the barium from the water sample before irradiation. This preliminary concentration has been carried out both by cation exchange (Bowen, 1956) and by coprecipitation with calcium oxalate (Turekian and Johnson, 1966). A coefficient of variation of ca. 11% has been obtained by Chan (cited in Turekian, 1971; see also Bender et al., 1972) using a modification of the method described by Turekian and Johnson (loc. cit.). In this, the yield of barium in the oxalate coprecipitation was determined using ^{140}Ba tracer. When the latter had decayed, the ignited oxalate precipitate was irradiated, and after radiochemical purification the ^{131}Ba activity was monitored by γ-spectrometry. The chemical yield of the purification stage was monitored by atomic absorption spectrophotometry.

Isotope dilution techniques have recently been used with considerable success for the determination of barium in sea water. They have the advantages of speed and simplicity and their precision (ca. 1%) is considerably superior to that of the other methods described above. In these procedures a known weight of the sample is spiked with the stable isotope ^{135}Ba. A small aliquot of either the spiked sample (Bernat et al., 1972), or preferably a barium concentrate prepared from it by ion exchange (Wolgemuth, 1970; Wolgemuth and Broecker, 1970; Bacon and Edmond, 1972) is evaporated on the filament of a mass spectrometer and the ^{135}Ba:^{138}Ba peak intensity ratio is measured. Bender and Snead (cited by Bender et al., 1972) have used a variant of the normal isotope dilution procedure in which a known weight of sea water is enriched with stable ^{130}Ba. Barium is concentrated from the sample by ion exchange, and its ^{130}Ba content is determined by neutron irradiation followed by γ-spectrometry of the resultant ^{131}Ba. A precision of 1–2% was claimed. Bender et al. (1972) have carried out intercalibration tests in which water samples collected at Geosecs Stations I and III were analysed by two mass spectrometric isotope dilution procedures and by the method of Bender and Snead (see above). A very satisfactory agreement (± 2%) was obtained between the various procedures.

Particulate barium. Spencer et al. (1972) have suggested that barium can be determined in marine particulates by neutron irradiation of the pelletized membrane filter followed by γ-spectrometry.

19.10.5.5. *Beryllium*

Dissolved beryllium occurs in sea water at a concentration of ca. $10 \, \text{ng} \, \text{l}^{-1}$, probably as hydroxy species such as $Be(OH)^+$. Nothing is known about the occurrence of beryllium in marine particulates.

Preconcentration. Beryllium has been concentrated from sea water by both coprecipitation with ferric or aluminium hydroxides (Ishibashi *et al.*, 1956) and adsorption on ferric hydroxide or hydrous manganese dioxide precipitated within the pores of a cation exchanger (Merrill *et al.*, 1960a, b).

Determination of dissolved beryllium. Beryllium has been determined in concentrates from sea water both by emission spectroscopy (Merrill *et al.*, 1960a, b) and by fluorimetry using morin (Ishibashi *et al.*, 1956).

19.10.5.6. *Bismuth*

The concentration of dissolved bismuth in sea water appears to be ca. 20 $ng\,l^{-1}$; the element is probably present as oxy and hydroxy species e.g. BiO^+, and $Bi(OH)_2^+$. Nothing is known of its occurrence in marine particulates.

Preconcentration. Bismuth can be coprecipitated from sea water with mercury(II) sulphide (Noddack and Noddack, 1940), ferric hydroxide and hydrous manganese dioxide (Portmann, 1965). The element can be quantitatively and very selectively adsorbed by strongly basic anion exchangers (e.g. Amberlite IR400 or Deacidite FF) from sea water which is $0.1\,M$ with respect to hydrochloric acid; desorption can be carried out with 1M nitric acid (Brooks, 1960; Portmann and Riley, 1966b).

Determination of dissolved bismuth. Bismuth has been determined in sea water using concentrates prepared by anion exchange both spectrographically (Brooks, 1960) and spectrophotometrically with dithizone (Portmann and Riley, 1966b). Gilbert and Hume (1973) and Florence (1972) have recently employed anodic stripping voltammetry in a rapid direct method for the semiquantitative determination of bismuth in sea water (see Section 20.5.3).

Boron – see Section 19.8.3.5.

Bromine – see Section 19.8.3.2.

19.10.5.1. *Cadmium*

The concentration of dissolved cadmium in sea water appears to be quite variable usually lying between 0.01 and $0.6\,\mu g\,l^{-1}$. The element probably occurs principally as chloro-complexes such as $CdCl_2^0$ and $CdCl^+$. Few data are available for the concentration of particulate forms of cadmium, but figures for British coastal waters (Preston *et al.*, 1972) suggest that its concentration may be about $\frac{1}{10} - \frac{1}{5}$ of that of the dissolved cadmium.

Preconcentration. It is necessary with almost all methods of analysis to concentrate cadmium from sea water before determining it. Although pre-concentration can be carried out by coprecipitation with copper(II) sulphide (Ishibashi *et al.*, 1962) or with strontium carbonate (Owa *et al.*, 1972) these techniques are too tedious for routine use. Rapid and selective separation of cadmium can be achieved by solvent extraction using the following extraction systems: dithizone – carbon tetrachloride (Mullin and Riley, 1956), am-monium pyrrolidine dithiocarbonate – methyl *iso*-butyl ketone (Preston *et al.*, 1972), sodium diethyl dithicarbonate – methyl *iso*-butyl ketone (Kuwata *et al.*, 1971). Solvent extraction methods are particularly valuable when the determination is to be completed by atomic absorption spectrophotometry.

When it is necessary to concentrate cadmium from several litres of sea water, ion exchange procedures are to be preferred to solvent extraction techniques. For this purpose either anion exchange or chelating ion exchange methods can be used. The former, which depends on the adsorption of a chloroion from a sea water medium 0·1 M with respect to hydrochloric acid, has been employed only by Brooks (1960). It is difficult to ensure quantitative uptake of cadmium in this method, and several workers have used less selective concentration procedures based on the chelating ion exchange resin Chelex-100 (Riley and Taylor, 1968; Abdullah *et al.*, 1971; Abdullah and Royle, 1972; Butterworth *et al.*, 1972; Windom and Smith, 1972). Muzzarelli and Marinelli (1972) have suggested that chitosan can be used as a chelating medium for the concentration of cadmium from sea water. Carritt (1953) has shown that cadmium can be taken up quantitatively from sea water by a column fitted with cellulose acetate saturated with a solution of dithizone in carbon tetrachloride.

Determination of dissolved cadmium. In early work cadmium was determined in concentrates from sea water by colorimetric techniques using dithizone (Mullin and Riley, 1956), or 1-(2-pyridylazo)-2-naphthol (Ishibashi *et al.*, 1962). Since these reagents are unspecific it was necessary to employ elaborate purification stages before the final photometric determination. For this reason colorimetric methods have now been superseded by more specific techniques which also permit the simultaneous determination of other elements.

Although the sensitivity of flame emission spectrophotometric procedures for cadmium is poor, a very high sensitivity can be achieved with atomic absorption spectrophotometry (AAS). Conventional AAS procedures have been used by Taylor and Riley (1968) and Windom and Smith (1972) for the determination of cadmium following preconcentration by chelating ion exchange. With 10 litre samples a sensitivity of ca. $0·005\,\mu g\,l^{-1}$ can be attained. Kuwata *et al.* (1971) have applied AAS to the examination of concentrates prepared by solvent extraction, but the limited concentration factor attainable in this way makes their method more suitable for the analysis of polluted waters. However, solvent extracts can be conveniently analysed by means of the tantalum boat technique (Preston et al., 1972; see also p. 305). Although the heated graphite atomizer has an attractively high sensitivity (ca. 10 pg Cd) it does not yet appear to have been applied to the determination of the element in sea water concentrates. Lund and Larsen (1974) have described a novel flameless atomic absorption spectrophotometric technique for the determination of cadmium. In this, the element is electrodeposited at constant potential onto a tungsten filament. After rinsing, this is then transferred to an optical cell in the light path of the spectrophotometer and cadmium is vaporized by heating the filament electrically to 1650°C.

The direct determination of cadmium in sea water at levels down to $0.03 \mu g \, l^{-1}$ can be achieved by anodic stripping voltammetry (Macchi, 1965; Whitnack and Sasselli, 1969; Smith and Redmond, 1971). The technique is simple and rapid, and is free from interference from other metals. Abdullah and Klimek (1974) have obtained a precision of $\pm 5\%$ at a level of $0.2 \mu g \, Cd \, l^{-1}$ using a pulsed anodic stripping method. Polarographic procedures are insufficiently sensitive to be applied directly to the determination of cadmium in sea water. However, after preconcentration, it has been determined in estuarine water by conventional polarography (Butterworth et al., 1972), and in sea water with a standard deviation of $0.03 \mu g \, l^{-1}$, by the more sensitive and selective pulse polarographic technique (Abdullah and Royle, 1972; Abdullah et al., 1971). Electrochemical procedures are discussed in detail in Chapter 20.

Determination of particulate cadmium. Preston et al., (1972) extracted cadmium from marine particulates with boiling 0·1 M hydrochloric acid containing 1% of 100 volume hydrogen peroxide. After concentration by evaporation cadmium was determined by atomic absorption spectrophotometry using the tantalum boat technique. A further atomic absorption technique with a sensitivity of 20 pg of Cd has been described by Belyaev and Gordeev (1972). In this, the atomization of the particulates was carried out by heating the filter in a hollow carbon electrode.

19.10.5.8. *Caesium*

Dissolved caesium occurs in sea water as the hydrated Cs^+ species at a concentration of ca. $0.4 \mu g \, l^{-1}$.

Preconcentration. Caesium has been concentrated from sea water by cation exchange techniques (Smales and Salmon, 1955; Yamagata, 1957; Riley and

Tongudai, 1966b). However, a much more specific separation can be achieved by adsorption onto ammonium 12-molybdophosphate (Feldman and Rains, 1964; Folsom *et al.*, 1964, 1970a) or by binding with crystalline copper, cobalt or zinc ferrocyanides as mixed caesium ferrocyanides e.g. $Cs_2Cu\,[Fe(CN)_6]$ (Ivanova, 1967; see also Folsom and Sreekumaran, 1970). The latter compounds have been used in cartridges to be towed behind ships to collect [137]Cs. Caesium has also been concentrated from sea water by extraction with a 0·025 M solution of sodium tetraphenylboron in methyl iso-butyl ketone (Folsom *et al.*, 1967, 1970b).

Determination of dissolved caesium. After preconcentration, caesium can be most satisfactorily determined by flame photometry using the near infra red 8521 Å emission line which is practically free from interference by other elements (see e.g. Folsom *et al.*, 1967, 1970a). Precision double beam differential flame spectrophotometers for the determination of caesium after solvent extraction as $CsB(C_6H_5)_4$ have been described by Spencer *et al.* (1970b) and by Folsom *et al.* (1974). In the instrument described by the latter workers a 4-port distributing valve was used so as to burn in succession concentrates from spiked and unspiked sea water as well as concentrates from spiked and unspiked reference sea water samples. To establish a zero reference signal, the Cs emission at 8521 Å was recorded alternately with that at 8525 Å. Burning of each sample concentrate was repeated 6 times and the signals were fed to a computer. A precision $\pm 0·3$ was claimed for the instrument and $\pm 0·5\%$ for the determination of caesium in sea water (see (see also Folsom, 1974). Riley and Tongudai (1966b) have employed D.C. arc excitation to determine caesium spectrographically in preconcentrates prepared by ion exchange; a standard deviation of $\pm 0·06\,\mu g\,l^{-1}$ was found.

Neutron activation techniques have a high sensitivity for caesium, and have been applied to the determination of the element in freeze-dried sea salts. After a cooling period of several weeks, the γ-activity of [134]Cs (half life 2·07 yr), arising from the reaction [133]Cs (n, γ) [134]Cs, is counted either directly by γ-spectrometry (Piper and Goles, 1969), or after radiochemical separation (Schutz and Turekian, 1965). The precision (± 10–15 %), is, however, inferior to that attainable by flame photometry.

19.10.5.9. *Cerium. See also lanthanides*

The analytical chemistry of cerium is distinguished from that of the other lanthanides by reactions associated with its 4+ oxidation state. For this reason it is convenient to discuss its determination in sea water separately. Dissolved cerium occurs in sea water at a concentration of 1–6 ng l^{-1}, probably as hydrated Ce^{3+} ion. No information is available about the occurrence of the element in marine particulates.

Preconcentration. Because of its very low concentration cerium must be concentrated before it can be determined in sea water. It can be efficiently ($> 98\%$) coprecipitated from sea water at pH 8–9 with ferric hydroxide, a technique which has been used by Goldberg *et al.* (1963), Høgdahl (1970), Carpenter and Grant (1967), Krishnamoorthy *et al.* (1970) and Shigematsu, *et al.* (1971). The cerium can be subsequently separated from iron by solvent extraction (Shigematsu *et al.*, 1971), anion exchange (Krishnamoorthy *et al.*, 1970), or cation exchange (Carpenter and Grant, 1967) techniques. Cerium has also been coprecipitated from sea water using hydrous manganese dioxide formed by the reduction of permanganate ion with hydrogen peroxide (Guegueniat, 1967).

Determination of dissolved cerium. Cerium after concentration from sea water, has been determined fluorimetrically (Shigematsu *et al.*, 1971), spectrophotometrically as Ce^{4+} (Carpenter and Grant, 1967; Krishnamoorthy *et al.*, 1970) and, as well as other lanthanides, by neutron activation analysis (see lanthanides below, and Goldberg *et al.*, 1963; Høgdahl *et al.*, 1968; Høgdahl, 1970).

19.10.5.10. *Chromium*

The concentration of dissolved chromium in sea water usually lies in the range $0\cdot2$–$2\,\mu g\,l^{-1}$. Although thermodynamics suggest that, in oxygenated sea waters, chromium should exist entirely as the CrO_4^{2-} ion (Fukai and Huynh-Ngoc, 1967) there is evidence that Cr(III) species, perhaps $Cr(H_2O)_4$-$(OH)_2^+$ or $Cr(OH)_3^0$, are significant, and may sometimes be dominant (Fukai and Vas, 1969; Elderfield, 1970). Perhaps as a result of biological activity the Cr(VI):Cr(III) ratio appears to be variable. On storage the relative amount of Cr(III) decreases with time (Fukai and Vas, 1969b).

Preconcentration. Coprecipitation with the hydroxides of iron (Chuecas and Riley, 1966; Fukai, 1967; Fukai and Vas, 1967; Fujinaga *et al.*, 1971) and aluminium (Loveridge *et al.*, 1960; Ishibashi and Shigematsu, 1950; Ishibashi, 1953; Fonselius, 1970) provides an efficient means for the analytical concentration of Cr(III) from sea water. Chromium(VI) is coprecipitated only to a slight extent, and this fact has been used as a basis for methods for differential determination of the two forms of chromium in sea water (see e.g. Fukai and Vas, 1967; Fukai, 1967). Black and Mitchell (1951) have cocrystallized chromium from sea water with tannin, thionalide and oxine.

Solvent extraction procedures have been used for the preconcentration of chromium by Chau et al. (1968); Morris (1968); Stanford (1971); Gilbert and Clay (1973). Chelating ion exchange resins only have a low efficiency for the uptake of Cr(III) from sea water (Riley and Taylor, 1968; Ueno, 1969) and do not adsorb CrO_4^{2-} significantly.

Determination of dissolved chromium. Spectrophotometric procedures using diphenylcarbazide have been extensively employed for the determination of chromium following preconcentration by coprecipitation (Ishibashi, 1953; Ishibashi and Shigematsu, 1950; Loveridge et al., 1960; Chuecas and Riley, 1966; Fukai and Vas, 1967; Fonselius, 1970). Iron and vanadium interfere, and must be removed before carrying out the determination. Chau et al. (1968) have determined chromium in sea water by atomic absorption spectrophotometry following solvent extraction with acetylacetone. Morris (1968) has attempted to apply X-ray fluorescence spectroscopy to the determination of chromium in concentrates prepared by extraction of its pyrrolidine dithicarbamate with methyl *iso*-butyl ketone. However, his method is too insensitive to be used with normal sea waters but it might prove of value for the analysis of polluted estuarine waters. Neutron activation techniques based on the nuclear reaction $^{50}Cr(n, \gamma)^{51}Cr$ have been applied to the determination of chromium in sea. Although Stanford (1971) prefers to concentrate the chromium by solvent extraction prior to irradiation and direct γ-spectrometry of the 0·328 MeV peak of ^{51}Cr, other workers irradiate the freeze dried salts, and after allowing to cool for several weeks, count the ^{51}Cr activity ($T^{\frac{1}{2}} = 28$ days) either directly by γ-spectrometry (Piper and Goles, 1969; Piper, 1971), or after radiochemical separation (Schutz and Turekian, 1965). Atomic absorption spectrophotometry using a $N_2O–C_2H_2$ flame has been used for the determination of chromium following solvent extraction (Gilbert and Clay, 1973).

Determination of particulate chromium. The determination of chromium in marine particulates by neutron irradiation followed by direct γ-spectrometry has been described by Piper and Goles (1969), Piper (1971) and Spencer et al. (1972).

19.10.5.11. *Cobalt*

It is likely that most early figures for the concentration of dissolved cobalt in sea water are considerably too high and that its normal concentration in oceanic waters does not exceed 0·1 $\mu g\, l^{-1}$. The principal dissolved cobalt species is probably hydrated Co^{2+}. The concentration of cobalt in marine particulate matter probably lies in the range 1–25 p.p.m. (Stoner, 1974).

Preconcentration. The very low concentration level of cobalt in ocean waters makes it essential to concentrate the element before analysis, except when neutron activation is to be employed. A considerable variety of separation techniques have been used for this purpose.

Although coprecipitation methods have been favoured by many workers because of the ease of processing large volumes of water, it is difficult, if not impossible, to obtain quantitative recoveries. However, this is of little consequence if carrier-free ^{58}Co is used as a tracer to measure the yield of the coprecipitation process. The complex precipitate of magnesium and calcium salts formed when sea water is made alkaline has been frequently used as a coprecipitant (Thompson and Laevastu, 1960; Joyner et al., 1967; Krishnamoorthy and Viswanathan, 1968; Krishnamoorthy, 1968). However, even under the optimum conditions, when the precipitate had been allowed to age for 7–21 days, Forster and Zeitlin (1966) were able to recover only $<96\%$ of the cobalt present. Ferric hydroxide, which has been used by Ishibashi (1953) for the preconcentration, is also not an efficient carrier of cobalt, and Burrell (1965) has found it necessary to carry out three precipitations in order to achieve a recovery of 92%. Viswanathan et al. (1965) and Doshi (1967) have attempted to carry down cobalt with the mixture of phosphates produced by the addition of disodium hydrogen phosphate; again the recovery was incomplete. A different approach has been adopted by Fukai (1968) who has recovered cobalt in 80–95% yield by adsorbing it onto solid manganese dioxide (see also Fukai et al., 1966). Harvey and Dutton (1973) concentrate cobalt with even greater efficiency by coprecipitating it with photochemically produced manganese dioxide. Co-crystallization is preferable to coprecipitation for the preconcentration of cobalt from sea water, and almost quantitative recoveries ($>99\%$) can be achieved by cocrystallization with 1-nitroso-2-napthol (Weiss and Reid, 1960).

Solvent extraction techniques have been widely adopted for the concentration of cobalt from sea water in the last decade. The systems which have been used for this purpose are shown in Table 19.13. Quantitative extractions can be readily achieved with most of these systems. However, it should be pointed out that, as a result of the limited concentration factor attainable (<200) and the insensitivity of the detection system, cobalt at its sea water concentration is normally undetectable by combined extraction–atomic absorption spectrophotometric procedures (see e.g. Brewer et al., 1969). Considerably better concentration factors can be attained by means of chelating ion exchange procedures, and these have been extensively adopted in recent years for the preconcentration of cobalt (see e.g. Callaghan et al., 1966; Lai and Goya, 1966; Joyner and Finley, 1966; Riley and Taylor, 1968; Ueno, 1969). The use of anion exchangers bearing complexing agents, such

TABLE 19.13

Systems for the solvent extraction of cobalt from sea water

Complexing agent	Solvent	Reference
2-nitroso-1-naphthol	toluene	Rozhanskaya (1964, 1965a, 1966, 1967)
		Emelyanov *et al* (1971)
2-nitroso-1-naphthol	chloroform	Oradovskii (1966)
2-nitroso-5-diethylaminophenol	dichloroethane	Motomizu (1973)
ammonium pyrrolidine dithiocarbamate	$CH_3COCH_2CH(CH_3)_2$	Brooks *et al.* (1967a)
		Brewer *et al.* (1969)
		Burrell (1967)
		Morris (1968)
ammonium pyrrolidine dithiocarbamate	$CH_3CO(CH_2)_2CH(CH_3)_2$ butyl alcohol	Segar (1971) Tsalev *et al.* (1972)
trifluoroacetylacetone	toluene + butyl alcohol	Lee and Burrell (1972)

as dithizone (Carritt, 1965) and dibromo-oxine (Topping, 1969), has also been suggested for the preconcentration of cobalt; however, they offer no advantage over the chelating exchangers.

Determination of dissolved cobalt. Aromatic compounds containing adjacent nitroso and hydroxy groups (e.g. 1-nitroso-2-naphthol) react almost specifically with cobalt yielding intensely coloured compounds. One such compound which has been widely used as a highly sensitive reagent for the photometric determination of cobalt is nitroso R salt (1-nitroso-2-naphthol-3,6-disulphonic acid). This reagent has been employed by several workers for the photometric determination of cobalt in sea water following preconcentration (Thompson and Laevastu, 1960; Weiss and Reid, 1960; Fukai *et al.*, 1966; Forster and Zeitlin, 1966; Fukai, 1968; Krishnamoorthy and Viswanathan, 1968). Although the cobalt complex formed by nitroso R salt is not extractable by organic solvents, those of some other nitroso compounds (c.g. 1-nitroso-2-naphthol) are, and have been employed in methods in which the cobalt derivative is concentrated by extraction prior to direct photometry. (Rhozanskaya, 1964). For one such method using 2-nitroso-5-dimethylaminophenol Motomizu (1973) has claimed a precision of $\pm 4\%$ at a cobalt level of 0·15 µg l^{-1}. Photometric catalytic procedures for the determination of cobalt in sea water concentrates have been described by Oradovskii (1966) and Emel'yanov *et al.* (1971).

Both atomic absorption spectrophotometric and X-ray fluorimetric techniques have only a comparatively poor sensitivity for cobalt. Although some workers (see e.g. Burrell, 1965; Brooks et al., 1967; Segar, 1971; Morris, 1968) have attempted to use these techniques for the estimation of cobalt in solvent extracts of sea water, it is unlikely that sufficient sensitivity can be attained for the determination of the element at the low level at which it occurs in ocean water (see Brewer, et al., 1969, Tsalev et al., 1972). Even with the tenfold greater concentration factor attainable by chelating ion exchange the sensitivity is barely adequate (Riley and Taylor, 1968). Cobalt, in the presence of dimethyl-glyoxime, gives a polarographic wave of great sensitivity. Both Abdullah and Royle (1972) and Harvey and Dutton (1973) have used this wave in pulse polarographic methods for the determination of cobalt in concentrates prepared by chelating ion exchange and coprecipitation respectively. A more extensive treatment of the determination of cobalt by electrochemical procedures is given in Chapter 20.

Irradiation of ^{59}Co ($\sigma = 20$; 100% of natural cobalt) with thermal neutrons leads to the production of ^{60}Co which decays with a half-life of 5·24 y with the emission of 1·17 and 1·33 MeV γ-radiation. This nuclear reaction has been used as the basis for neutron activation procedures for the determination of cobalt in sea water after preconcentration either by solvent extraction (Viswanathan et al., 1965; Doshi, 1967) or more frequently freeze-drying. In the latter instance counting has been carried out after cooling for several weeks either after radiochemical separation (Schutz and Turekien, 1965a, b) or directly by γ-spectrometry (Piper and Goles, 1969; Robertson, 1970; Spencer et al., 1970b; Piper, 1971; Brewer et al., 1972). In a typical method using direct γ-spectrometry Piper and Goles (1969) have achieved a coefficient of variation of $\pm 4\%$. It should be pointed out that there may be a sources of systematic error in Schutz and Turekian's (loc. cit.) procedure since the concentrations of cobalt in sea water which they found are more than an an order of magnitude higher than more recent figures by Robertson (1970).

Determination of particulate cobalt. Photometric methods based on its catalytic action on the oxidation of alizarin (Yatsimirski et al., 1970; Emel'yanov et al., 1971) or pyrocatechol-3,5-disulphonic acid (Blazhis et al., 1970) have been used for the determination of cobalt in acid digests of marine particulates. Where suitable facilities exist the estimation of particulate cobalt can be more readily carried out by submitting the pelletized membrane filter to instrumental neutron activation analysis (see e.g. Spencer et al., 1970a, 1972; Piper and Goles, 1969; Piper, 1971). Brooks et al. (1967a) have carried out the determination by dissolving the membrane filter in an acetone–hydrochloric acid mixture and spraying the solution into an atomic absorption spectrophotometer.

19.10.5.12. *Copper*

The concentration of copper in open ocean water probably never exceeds $4 \, \mu g \, l^{-1}$. However, considerably higher concentrations may be found in coastal and estuarine waters, particularly those subject to pollution. The principal inorganic species of copper present in the sea are probably hydrated Cu^{2+}, $CuCO_3^0$ and $CuOH^+$. Because of the stability of the complexes which it forms with many ligands, organic species of copper may also occur in significant amounts in sea water and, indeed, some evidence has been found for their existence (see e.g. Alexander and Corcoran, 1964, 1967; Slowey *et al.*, 1967; Williams, 1969; Foster and Morris, 1971).

Preconcentration. The concentration of copper in sea water is too low for the element to be determined directly, except by anodic stripping voltammetry (*vide infra*). Many workers have used solvent extraction techniques for the preconcentration of copper because they are rapid and the extracts can be examined directly by colorimetric and atomic absorption spectrophotometric methods. Extraction systems which have been used for this purpose are summarized in Table 19.14.

Copper is strongly adsorbed by chelating ion exchange resins and this process provides an efficient method for the preconcentration of copper from sea water (Riley and Taylor, 1968). Muzzarelli *et al.* (1970) and Muzzarelli and Rochetti (1974) have suggested that chitosan (see p. 286) can be used as a substitute for the more expensive chelating ion exchange resins. Kerfoot and Vaccaro (1973) have suggested the use of adsorptive carbon for the batchwise preconcentration of copper from sea water. Adsorbed copper is then removed by leaching with mineral acid; overall recoveries of copper of 70% were achieved. Among other media which have been recommended for the uptake of copper from sea water are anion exchangers and active carbon bearing complexing agents, such as dithizone, (see e.g. Carritt, 1965; Zharikov, 1970), and also dithizone coupled to carboxymethyl cellulose (Burrell, 1968).

Determination of dissolved copper. The earlier literature on the determination of copper in sea water has been reviewed by Chow and Robinson (1952). Most of the values recorded up to that time were much too high, either because of contamination or through the use of unsatisfactory analytical methods. Until comparatively recently the determination of copper was carried out mainly by colorimetric procedures, (Table 19.14) however, these are now being replaced by atomic absorption spectrophotometric techniques.

Dithizone is one of the most sensitive reagents available for the colorimetric determination of copper, but is unspecific. Over the pH range 2–9

TABLE 19.14

Solvent extraction systems for analytical preconcentration of copper from sea water

Complexing agent	Solvent	Method of determination	Ref.
dithizone	carbon tetrachloride	C	Morita (1948), Fonselius and Koroleff (1963)
		P	Tikhonov and Zhavoronkina (1960)
sodium diethyl-dithiocarbamate	xylene	C	Chow and Robinson
	amyl alcohol	C	(1952)
	carbon tetrachloride	C	Riley (1937)
		C	Atkins (1957); Molins (1957); Bougis (1962); Rhozanskaya (1965); Strickland and Parsons (1968); Williams (1969); Foster and Morris (1971)
		NAA	Slowey and Hood (1971)
	chloroform	C	Soares and Goncalves (1966)
	methyl *iso*-butyl ketone	AAS	Kuwata *et al.* (1971)
ammonium pyrrolidine dithiocarbamate	methyl *iso*-butyl ketone	AAS	Brooks *et al.* (1967 a, b); Brewer *et al.* (1969, 1972); Spencer and Brewer (1969); Orren, (1971); Preston *et al.* (1972); Burrell (1967); Paus (1973); Segar and Gonzalez (1972)
		XRF	Morris (1968)
	ethyl acetate	AAS	Magee and Rahman (1965)
8-hydroxy-quinoline	chloroform	OS	Brooks (1965)
		XRF	Armitage and Zeitlin (1971)
2,2'-biquinolyl	*n*-hexanol	C	Riley and Sinhaseni (1958); Bougis (1963); Loveridge *et al.* (1960)
quinoline-2-aldehyde-2-quinolylhydrazone	benzene	C	Abraham *et al* (1969)

copper can be extracted efficiently into carbon tetrachloride as its intensely red dithizonate. The complex is stable at very low pH values, and most interfering elements can be removed by back extraction with 0·01 N sulphuric acid. Only the dithizonates of silver and mercury are stable under these conditions; however, they are unlikely to interfere significantly in sea water analysis because their concentrations are much lower than that of copper. Photometric mixed colour dithizone procedures for sea water analysis have been described by Morita (1948) and Fonselius and Koroleff (1963).

Sodium diethyldithiocarbamate has been extensively used as a complexing agent in the extraction and subsequent photometric determination of copper in sea water. It reacts with copper(II) in slightly acidic or ammoniacal solution yielding a brown colloidal suspension of the copper complex. This complex can be extracted with various non-polar organic solvents to give a brown solution suitable for photometry; the solution is unstable and the colour fades in strong light. Several other heavy metals also form extractable complexes, but most of these are colourless and do not interfere. However, cobalt, nickel and iron form coloured complexes, but can be prevented from interfering by chelating them with EDTA at pH8; extra sodium diethyl-dithiocarbamate reagent must then be added to allow for competitive complexing of copper with the EDTA. The diethyldithiocarbamate extraction technique was applied to the spectrophotometric analysis of sea water by Chow and Robinson (1952) who carried out the extraction at pH8 with xylene. Equilibration with this solvent is slow, and Atkins (1957) suggested replacing it with carbon tetrachloride, with which equilibrium can be attained after only 2 minutes shaking, and which has the additional advantage of being denser than the aqueous phase. This technique has been adopted in modified form by several workers (see Table 19.14); for working details of the method (sensitivity $\pm 0·3 \, \mu g \, l^{-1}$) the reader should consult Strickland and Parsons (1968). (It should be noted that some interference may be experienced from iron in their method since no EDTA is added).

There are a number of colorimetric reagents which are completely specific for copper. Copper(I) reacts with both 2,2′-biquinolyl and 2,9-dimethyl-1, 10-phenanthroline* at pH 4–7 yielding strongly coloured complexes which can be extracted with polar solvents. Unfortunately, the molar absorbances of the copper complexes are only about half of that of copper(II) diethyl-dithiocarbamate. Riley and Sinhaseni (1958) have developed a method based on biquinolyl for the determination of copper (down to $1 \, \mu g \, l^{-1}$) in sea water, in which the copper is reduced to the 1 + oxidation state with hydroxy-

---

* 2,9-dimethyl-1,10,-phenanthroline has been used by Alexander and Corcoran (1967) as a reagent for the direct colorimetric determination of copper in sea water, but it is rather lacking in sensitivity for this purpose.

ammonium chloride and the Cu $(biquin)_2^+$ complex is extracted with hexanol for photometry. Loveridge et al. (1960) have examined their procedure using ^{64}Cu tracer and have found that the extraction is 98–100% complete even at a copper concentration of $2 \mu g \, l^{-1}$. Bougis (1963) has compared the biquinolyl method with one based on sodium diethyl dithiocarbamate (Bougis, 1962). He concluded that, although both gave similar results, the latter was preferable because of its simplicity. Because of its specificity the biquinolyl extraction procedure should be used for polluted or interstitial waters. A sensitivity 2–3 times that of diethyldithiocarbamate methods can be obtained by the use of quinoline-2-aldehyde-2-quinolylhydrazone in the extraction and photometric determination of copper. This reagent forms an intensely coloured compound with copper(II) ($\varepsilon = 5 \cdot 8 \times 10^4 \, l \, mol^{-1} \, cm^{-1}$) that can be extracted with benzene; in the presence of masking agents, the reaction is specific for copper. It has been used by Abraham et al. (1969) for the determination of 0–10 $\mu g \, Cu \, l^{-1}$ in sea water; a standard deviation of $0 \cdot 1 \mu g \, l^{-1}$ was found at a level of $2 \cdot 4 \mu g \, l^{-1}$. In connection with the leaching of copper from marine antifouling paints, Bowles and Nicks (1961) have investigated the use of oxalyldihydrazide and disodium ethyl bis-(5-tetrazolylazo) acetate for the direct photometric determination of 20–1000 μg $Cu \, l^{-1}$. They found that the former reagent gave greater sensitivity and was less susceptible to interference from other elements. The last of these reagents has been used in sea water analysis (Genovese and Magazzu, 1969), but its sensitivity makes it only suitable for the analysis of heavily polluted waters.

A high sensitivity and almost complete specificity for copper can be attained by atomic absorption spectrophotometry. The conventional technique is insufficiently sensitive to be applied directly to sea water. However, Ediger et al. (1974) have recently described a procedure for the direct AAS determination of copper in sea water in which the heated carbon tube atomizer is used, but its sensitivity ($0 \cdot 5 \mu g \, Cu \, l^{-1}$) is barely sufficient for the determination of the element at its normal levels in ocean water. It is thus essential to concentrate the element before determining it by the AAS technique, and both solvent extraction (usually with the ammonium pyrrolidine dithiocarbamate-methyl iso-butyl ketone system; see Table 19.14) and chelating ion exchange procedures (Riley and Taylor, 1968; Muzzarelli and Rochetti, 1974) have been widely adopted for this purpose. Although most workers have employed conventional atomization techniques for examination of the concentrate, some recent investigators have used the carbon tube furnace (Segar and Gonzalez, 1972; Paus, 1973; Ediger et al., 1974), but with some sacrifice of precision.

A number of electrochemical procedures have been used for the determination of copper in sea water. Polagraphic techniques are not sufficiently sensitive to be used for the direct determination of copper at its normal concentration in the oceans, but the more refined versions (e.g. a.c. polagraphy; Odier and Plichon, 1971) may be of value with enriched sea waters. However, both normal (Tikhonov and Zhavoronkina, 1960) and pulse polarography (Loveridge *et al.*, 1960; Abdullah and Royle, 1972) have been used for the examination of concentrates. The latter workers, who concentrated copper by chelating ion exchange, have claimed a coefficient of variation of 5% for their procedure. Anodic stripping voltammetric techniques offer considerable promise for the determination of copper in sea water, since they are rapid and can be applied without the need for preconcentration (see e.g. Ariel and Eisner, 1963; Ariel *et al.*, 1964; Whitnack and Sasselli, 1969). The adoption of the impregnated carbon electrode of Kemula in place of the hanging mercury drop used by earlier workers has led to a considerable improvement in the precision and convenience of the anodic stripping technique for the direct analysis of sea water (Smith and Redmond, 1971; Abdullah and Klimek, 1974). For a more extensive treatment of electrochemical methods for the determination of copper see Chapter 20.

Both neutron activation analysis (Slowey *et al.*, 1962; Hood and Slowey, 1964; Hood, 1966; Slowey and Hood, 1971; Gohda, 1972) and X-ray fluorimetry (Morris, 1968; Armitage and Zeitlin, 1971) have been used to a limited extent for the determination of copper in concentrates prepared by solvent extraction of sea water. However, these techniques would appear to have no advantages over the more precise and much simpler atomic absorption spectrometric methods.

Determination of total dissolved copper. The breakdown of organo-copper compounds before determination of total dissolved copper can be readily achieved by irradiation of the sample with intense ultra-violet radiation (Williams, 1969; Foster and Morris, 1971; Paus, 1973), alternatively, it can be brought about by oxidation with peroxydisulphate ion (Slowey and Hood, 1971; see also Muzzarelli and Rochetti, 1974).

Determination of particulate copper. Both atomic absorption spectrophotometric (Orren, 1971; Preston *et al.*, 1972) and colorimetric procedures (Yatsimirski *et al.*, 1971) have been used for the determination of copper in acid digests of marine particulate matter.

19.10.5.13. *Dysprosium.* See under Lanthanides

19.10.5.14. *Erbium.* See under Lanthanides

19.10.5.15. *Europium.* See under Lanthanides

Fluoride. See Section 19.8.3.4.

19.10.5.16. *Gadolinium.* See under Lanthanides

19.10.5.17. *Gallium*

The concentration of dissolved gallium in sea water appears to be about 30 ng l^{-1}; the dissolved species are probably hydroxy complexes—perhaps $Ga(OH)_4^-$. Nothing is known of the concentration of gallium in marine particulates.

Preconcentration. Gallium can be efficiently concentrated from sea water by coprecipitation with ferric hydroxide. (Burton *et al.*, 1959; Ishibashi *et al.*, 1961; Kraynov, 1961; Petrov and Pencheva, 1963). It has also been concentrated by extraction with *n*-hexanol using 2,2′,4′-trihydroxyazobenzene as a complexing agent (Liam-Ngog-Thu, 1967).

Determination of dissolved gallium. After coprecipitation, gallium has usually been separated from the iron carrier by extracting it from 6 M hydrochloric acid medium with ether; extraction of iron is prevented by reducing it to the 2+ state. Gallium is then determined photometrically after extracting it as the intensely pink coloured chloro-complex which it forms with Rhodamine B (Burton *et al.*, 1959; Ishibashi *et al.*, 1961; Petrov and Pencheva, 1963). A semiquantitative visual fluorimetric procedure based on extraction of the fluorescent complex formed with 2,2′4′-trihydroxyazobenzene has been described by Liam-Ngog-Thu (1967).

19.10.5.18. *Germanium*

Dissolved germanium probably occurs in sea water at a concentration of about 50 ng l^{-1} as $GeO(OH)_3^-$ and $GeO_2(OH)_2^{2-}$. No data appear to be available for the occurrence of the element in marine particulates.

Preconcentration. Germanium has been coprecipitated from sea water with ferric hydroxide (Wardani, 1958; Burton *et al.*, 1959); other metal hydroxide such as those of aluminium and chromium are much less efficient coprecipitants (Burton, 1957). The coprecipitation has also been carried out using antimony phenylfluoronate (Nazarenko and Shelikhina, 1967).

Determination of dissolved germanium. In all published work on the determination of germanium in sea water, phenylfluorone has been used for the photometric analysis of preconcentrates because of its selectivity and high sensitivity (Wardani, 1958; Burton *et al.*, 1959; Nazarenko and Shelikhina, 1967). Before carrying out the determination germanium must be separated from the iron used as coprecipitant and from other elements also coprecipitated. An almost specific separation can be achieved by extracting it from 9 M hydrochloric acid medium as its mainly covalent tetrachloride using carbon tetrachloride (Burton *et al.*, *loc. cit*; Nazarenko and Shelikhina, 1967).

19.10.5.19. *Gold*

The long history of the determination of gold in sea water has been reviewed by Putnam (1953) and more recently by Rosenbaum *et al.* (1969) and Jones (1970). With the exception of the very painstaking analyses by Haber (1928) most of the earlier figures appear to be one or more orders of magnitude too high, as a result of both contamination and the use of unsatisfactory analytical techniques. Recent work suggests that the concentration of dissolved gold is probably variable and lies in the range 1–50 ng l^{-1}. (See Table 7.16). Dissolved gold probably occurs principally as $AuCl_2^-$ and perhaps $AuClBr^-$ (Peshchevitskii *et al.*, 1965). No determinations of particulate gold appear to have been made.

Preconcentration. In the classical work by Haber (1928) gold was concentrated by coprecipitation with lead sulphide; coprecipitation with ferrous sulphide has also been used for the same purpose (Putnam, 1953). Weiss and Lai (1963) have concentrated gold in excellent yield from sea water by co-crystallizing it with 2-mercaptobenzimidazole. Gold can be very satisfactorily preconcentrated from sea water as its chloro-complex by the use of anion exchange resins (Brooks, 1960; Abdullaev *et al.*, 1968; Wood, 1971). However, it is so strongly adsorbed that it can only be recovered quantitatively by ashing the resin. Oka *et al.* (1964) have recovered gold by evaporating sea water to dryness, dissolving the dry salts in 3M hydrobromic acid and extracting with ethyl acetate.

Determination of dissolved gold. Although gold has been determined by fire assay (see e.g. Haber, 1928; Stark, 1943; Putnam, 1953) spectrographic (Brooks, 1960) colorimetric (Weiss and Lai, 1963) and atomic absorption spectrometric (Rosenbaum *et al.*, 1969) procedures, its very low concentration necessitates the use of samples of over 100 l. Neutron activation analysis is, at present, the only technique with sufficient sensitivity for the examination of conveniently sized samples (down to as little as 100 ml). Irradiation of ^{197}Au with thermal neutrons leads to the production of ^{198}Au which has a half-life of 2·7 days and decays by β and γ-emission. The reaction is analytically extremely sensitive since ^{197}Au comprises 100% of natural gold and has a high capture cross section for thermal neutrons (98 barns). When the neutron activation technique is applied to the analysis of sea water it is necessary to preconcentrate the gold before irradiation, because the half-life of the ^{198}Au is similar to that of short lived nuclides (mainly ^{24}Na) produced from major ions of the water. Neutron activation was first used for the determination of gold in concentrates from sea water by Hummel (1957) and has subsequently been adopted by numerous other workers (Oka *et al.*, 1964; Schutz and Ture-kian, 1965b; Cappadona, 1965; Abdullaev *et al.*, 1968; Wood, 1971).

19.10.5.20. *Hafnium*

No determinations of hafnium in sea water appear to have been carried out. Coprecipitation with ferric hydroxide followed by neutron activation analysis would probably be the most profitable line of approach to the problem.

19.10.5.21. *Holmium.* See lanthanides

19.10.5.22. *Indium*

There is considerable divergence between the two sets of data available for the concentration of indium in the sea. Chow and Snyder (1969) have obtained values of $4 \, \text{ng} \, l^{-1}$ for Pacific Ocean waters, whereas Matthews and Riley (1970b) have recorded figures of ca. $0.1 \, \text{ng} \, l^{-1}$ for Atlantic water (see also below). Barić and Branica (1969) have provided polarographic evidence that $In(OH)_2^+$ is the most probable principal species of the element in sea water.

Preconcentration. Indium can be quantitatively cocrystallized from sea water at pH 3·5–4·0 using 5,7-dibromo-8-hydroxyquinoline (Matthews, 1969). Although the element is taken up from acidified sea water by strongly basic anion exchangers, the retention is far from quantitative. Indeed, Matthews and Riley (1970a) have found that high recoveries (ca. 80%) can only be obtained from sea waters containing high concentrations of hydrochloric acid (optimum 6 M). However, they observed that indium could be concentrated from sea water at pH 9·2 in 98% yield using columns of the chelating ion exchanger Dowex A-1.

The possibility of determining indium directly in sea water by anodic stripping voltammetry has been discussed by Florence (1972) (see Section 20.5.3).

Determination of dissolved indium. Chow and Snyder (1969) have applied an isotope dilution technique to the determinatfon of indium in sea water. The water sample was spiked with a known weight of a concentrate of ^{113}In, and indium was separated from the water by anion exchange. The concentrate was applied to the filament of a mass spectrometer and the mass spectrum peak ratio of $^{113}In:^{115}In$ was measured and compared with that of natural indium (1:22.6). This technique is probably unsatisfactory for the determination of indium in sea water since its detection limit is ca. 2 ng l^{-1} and because ca. 40 ng l^{-1} of ^{115}In l^{-1} was added in the ^{113}In enriched spike used in the analysis.

The irradiation with thermal neutrons of ^{113}In (σ = ca. 50 barns) which constitutes 4·3% of natural indium gives rise to ^{114m}In (half-life = 49 days) which undergoes an internal transformation to ^{114}In ($t_{\frac{1}{2}}$ 72 s). When equilibrium has been attained, indium can be determined by counting the 1·98 MeV β-radiation of this daughter. Although considerably greater sensitivity could be achieved by counting the ^{116m}In originating from ^{115}In (isotopic abundance 95·7%, σ 145 barns), the half-life of this nuclide (54 min) is too short to permit the lengthy radiochemical separations necessary in the analysis of sea water. In the neutron activation scheme adopted by Matthews and Riley (1970a) an indium concentrate prepared by a combination of chelating and anion exchange techniques was irradiated with thermal neutrons. After decontamination of the activation product by solvent extraction techniques, the ^{114}In activity was counted. A sensitivity of 0·01 ng l^{-1} was claimed.

19.10.5.23. Iodine

The concentration of dissolved iodine in sea water is probably about 60 μg l^{-1}. Thermodynamics suggest that in fully oxygenated sea water the element should exist almost entirely as the iodate ion. However, most workers have found iodide to be present, often in major proportions. It is also possible that dissolved organic iodine compounds may occur in the sea although evidence for this is lacking. No data are avaiable on the existence of iodine in marine particulates; however, marine sediments are known to be rich in the element (up to 130 p.p.m.) (Shiskina and Pavlova, 1965). The extensive early literature on the determination of iodine in sea water has been reviewed by Reith (1930).

Preconcentration. Because of the very low solubility product of silver iodide (1·5 × 10^{-16}), iodide is effectively coprecipitated from sea water with silver chloride even if only ca. $\frac{1}{50}$ of the stoichiometrically necessary amount of silver ion is added. Iodate is not coprecipitated significantly. Coprecipitation with silver chloride has been used for the analytical concentration of iodide from sea water by Sugawara et al. (1955), Sugawara and Terada (1957) and Tsunogai (1971). The optimum conditions for carrying out the coprecipitation have been investigated by Matthews and Riley (1970c), who found that it was not possible to coprecipitate more than ca. 90% of the iodide by addition of silver nitrate. However, they were able to achieve a recovery of >99% by generating the silver ions homogeneously in the sea water by the gradual dissolution of the slightly soluble silver citrate.

Determination of iodine species in the sea. A considerable variety of techniques (mainly photometric) have been used for the determination of iodine species in sea water. In most instances they permit differentiation to be made between iodate and iodide.

Although procedures have been described (Reith, 1930; Yonehara, 1964) for the direct determination of iodide in sea water involving oxidation to the elemental state followed by extraction with carbon tetrachloride and photometry, it is unlikely that they have adequate sensitivity. A worthwhile increase in sensitivity can be achieved if the iodide ion is first oxidized to iodate (using permanganate, the excess of which is then reduced); when treated in acid medium with excess iodide this yields 6 atoms of iodine per original iodide ion

$$IO_3^- + 5I^- + 6H^+ \rightarrow 3I_2 + 3H_2O$$

This process has been used as the basis of both titrimetric (Skopintsev and Mikhailovskaya, 1933) and photometric (Schnepfe, 1972) methods for the estimation of total iodine in sea water. However, their precision is only poor. Iodate alone can be determined if the preliminary oxidation stage is omitted. Johannesson (1958) has claimed that he was able to determine iodate-iodide in sea water by liberating molecular iodine with iodide and acid and measuring the intense absorption at 365 nm arising from the tri-iodide ion formed from it ($I_2 + I^- \rightleftharpoons I_3^-$). His procedure has been studied by Truesdale and Spencer (1974) who have obtained a coefficient of variation of $\sim 2\%$ for their modification of it. Wong and Brewer (1974) have described a more precise procedure ($\sigma = \pm 0.2\,\mu g\,IO_3^- - I\,l^{-1}$) in which the liberated iodine is titrated with thiosulphate, the end point being determined photometrically by noting the disappearance of the absorption at 365 nm due to the I_3^- ion.

Because of the comparative insensitivity of direct methods for the determination of iodide in sea water Sugawara *et al.* (1955) have developed a procedure in which the iodide is concentrated by coprecipitation with silver chloride. The precipitate is digested with bromine water to oxidize the iodide to iodate. After an intermediate stage designed to remove any hypobromite, the iodate is allowed to react under acidic conditions with excess iodide ion and the iodine thus produced is determined photometrically as the starch-iodine chromogen. This procedure has been adapted and simplified by Tsunogai (1971a), for use in studies of the marine chemistry of iodine (see for example Tsunogai and Sase, 1969; Tsunogai and Henmi, 1971; Tsunogai, 1971b). Matthews and Riley (1970c) have carried out a careful study of the procedure of Sugawara *et al.* (1955). In addition to suggesting changes in the coprecipitation procedure (see above), they replaced the starch-iodine photometric procedure, which they considered to be a major source of error, by absorption spectrophotometric or photometric titration procedures based on the very intense absorption band of the I_3^- ion at 353 nm. The two methods which they developed gave overall iodine recoveries at a level of 40 μg I l^{-1} of 98–99% with coefficients of variation of $\pm 0.4\%$. Iodate is not coprecipitated significantly with silver chloride in the above procedures, and it is possible therefore, after removal of the precipitate to reduce the iodate to iodide which can then be determined in the same way as the original iodide.

Iodide exerts an almost specific catalytic effect on the reaction between cerium(IV) and arsenic(III). This catalytic action was first used as the basis of a sensitive method for the direct determination of iodide in sea water by Dubravčic (1955). In practice, the samples are diluted to the same chlorinity, an excess of arsenious acid solution is added followed by a solution of cerium (IV) sulphate. Under these conditions the reaction has first order kinetics and obeys the equation

$$k = \frac{2.303}{t_2 - t_1} \log \frac{[Ce^{4+}]_1}{[Ce^{4+}]_2} \text{ where}$$

$[Ce^{4+}]_1$ and $[Ce^{4+}]_2$ are the cerium(IV) concentrations at times t_1, and t_2 (Sandell and Kolthoff, 1934). Thus k, which is a function of the iodide concentration, can be computed if the cerium concentration is measured photometrically at intervals during the reaction, and compared with a calibration curve prepared by the method of standard additions. This procedure has been subsequently adopted by Barkley and Thompson (1960a, b), Voipio (1961) and Kappanna *et al.* (1962). Truesdale and Spencer (1974) have recently carried out a systematic investigation of the catalytic method, and have adapted it for automatic analysis. A further automated procedure has been developed by Revel (1969).

Two different types of electrochemical procedures have been used for the determination of iodate in sea water. In that described by Barkley and Thompson (1960a) the acidified sample is treated with excess iodide liberating 6 equivalents of iodine for each equivalent of iodate originally present. Excess thiosulphate is added and its excess is determined by amperometric titration with iodate. The method can be used to determine total iodine in the sample if iodide is first oxidized to iodate with bromine. A standard deviation of 2 μg IO$_3^-$–I l^{-1} was claimed. Truesdale and Spencer (1974) obtained little response at the end-point when they used the circuitry described by Barkley and Thompson (1960a), and cast doubt on whether, in fact, the end-point was an amperometric one, but rather considered it to be a potentiometric one. As a consequence, they redesigned the indicator circuit and were able to achieve a standard deviation of $\pm 0.4 \, \mu g \, IO_3^- - I \, l^{-1}$ with this modification (see also Section 20.6). Both Petek and Branica (1969) and Liss and Pointon (1973) have used pulse polarography for the direct determination of iodate in sea water. In the differential procedure used by the latter workers, which had a precision of $\pm 1.0 \, \mu g \, IO_3^- - I^-$, interference by zinc was masked with EDTA. Iodide was determined by difference after oxidizing it photochemically and subsequently determining total iodate polarographically. (For a further discussion of the use of electrochemical methods for the determination of iodine species see Sections 20.4.1 and 20.4.5; also Herring and Liss, 1974).

Storage of samples. Truesdale and Spencer (1974) have found that filtered (Millipore or glass fibre filters) or unfiltered, samples could be stored in the dark in glass bottles at 4 to $-20°C$ for several months without change of either the iodate or total iodine concentration. Losses of iodate-iodine occur if polyethylene containers are used for storage.

19.10.5.24. *Ionium.* See Section 19.12.2.8

19.10.5.25. *Iridium.* No determinations of iridium in sea water appear to have been carried out.

19.10.5.26. *Iron*

Iron occurs in sea water in both "dissolved" and particulate forms. The "dissolved" iron almost certainly contains a considerable fraction of colloidal species of the element since the concentration of analytically determinable iron decreases as the water is filtered through progressively finer membrane filters. When the water is filtered through the conventional 0.45 μm membrane filters the concentration of "dissolved" iron usually lies in the range 1–10 $\mu g \, l^{-1}$ (for a review of data on "dissolved" iron concentrations published before 1954 see Lewis and Goldberg (1954); for subsequent data see Head (1971a)). The species of "dissolved" iron present in sea water are probably mainly hydrolysed ones, e.g. $Fe(OH)_2^+$, but there is some possibility that soluble organic iron compounds may also be present. In oxygenated sea water soluble iron is present only in the $3+$ oxidation state, but iron(II) species probably predominate in euxinic waters. The concentration of particulate iron is very variable, being highest in inshore waters, especially those near the mouths of rivers. The particulate forms of the element are probably mainly silicates (e.g. clays) or ferric hydroxide flocs, but same may be associated with organic detritus or living organisms which may contain adsorbed iron and stable iron complexes (e.g. haemin haemoglobin).

Not all of these forms of iron will be available to marine organisms, and a determination of total iron in either the water, or the particulate fraction will lead to a considerable overestimation of the iron that can be utilized by phytoplankton. Strickland and Parsons (1960) have recognised this, and have attempted to measure biologically available iron by treating the filtered water sample and the particulate matter with hydrochloric acid before determining iron colorimetrically (see also, Strickland and Austin, 1959; Williams and Chan, 1966). However, it should be remembered that this approach is quite arbitrary and will not necessaily measure the iron available as a micro-nutrient.

Preconcentration. Techniques involving the extraction of iron complexes have been widely used for the preconcentration of the element from sea water (see Table 19.15). When extractable coloured complexes are formed the analysis can be completed photometrically (see below).

Iron can also be satisfactorily concentrated by chelating ion exchange on Chelex-100 (Abdullah and Royle, 1972) and on chitosan (Muzzarelli and Tubertini, 1969; Muzzarelli *et al.*, 1970).

Determination of dissolved iron. Until recently the routine determination of iron in sea water has been almost always carried out colorimetrically. For this purpose a number of specific reagents are available which react with iron(II) yielding stable intensely coloured complexes. The reaction is normally carried out in a buffered medium (pH 4–5) in the presence of hydroxy-ammonium chloride which serves to reduce iron to the $2+$ oxidation state.

2,2′-Dipyridyl and the closely related tripyridyl were first used for the determination of soluble iron in sea water by Cooper (1937a, b), and the former has been subsequently adopted for the estimation of particulate iron by several workers (Goldberg, 1952; Lewis and Goldberg, 1954; Strickland and Parsons, 1968). Simons *et al.* (1953), Armstrong (1957) and Dobizhanskaya and Tshenina (1959) have used the cheaper and slightly more sensitive 1,10-phenanthroline for the determination of dissolved iron in sea water. The sensitivity can be almost doubled if 4,7-diphenyl-1,10-phenanthroline sulphonic acid is used (Riley and Williams, 1959) but the cost is very much greater. No interference is experienced with any of these reagents from other ions at least ten times their normal concentrations in sea water.

The complexes formed by iron (II) with dipyridyl and phenanthroline cannot be extracted with organic solvents. In contrast the intensely coloured complex formed with 4,7-diphenyl phenanthroline (bathophenanthroline)

TABLE 19.15

Systems which have been used for extraction of iron from sea water

Complexing agent	Extractant	Method of determination	Reference
4,7-diphenyl-1,10-phenanthroline	iso-amyl alcohol[1]	P	Lewis and Goldberg (1954); Strickland and Austin (1959); Strickland and Parsons (1968); Williams and Chan (1966); Head (1971)
2,4,6-tripyridyl-s-triazine	nitrobenzene	P	Collins and Diehl (1960)
trifluoroacetyl acetone	toluene[2]		Lee and Burrell (1972)
ammonium pyrrolidine	chloroform	AAS	Preston et al. (1972)
dithiocarbamate	methyl iso-butyl ketone	XRF	Morris (1968)
		AAS	Brooks et al. (1967a) Brewer et al. (1969, 1972)
sodium diethyl dithiocarbamate	methyl iso-butyl ketone[3]	AAS	Joyner and Finley (1966)
8-hydroxyquinoline	chloroform	XRF	Armitage and Zeitlin (1971)
	methyl iso-butyl ketone	AAS	Orren (1971)

P   Direct photometry of extract.
AAS Atomic absorption spectrophotometry.
XRF X-ray fluorescence.
[1] It is probably preferable to use n-hexanol as it is less soluble and gives more complete extraction of the complex (Diehl and Smith, 1960).
[2] 85% extraction with a volume ratio of aqueous to organic phase of 40.
[3] Extraction efficiency only 33%.

can be efficiently extracted by polar solvents e.g. *iso*-amyl alcohol (Smith *et al.*, 1952). This reagent has been employed for the extraction and photometric determination of "biologically active" soluble iron and total soluble iron in sea water by Strickland and Parsons (1960) and Lewis and Goldberg (1954) respectively (see also Williams and Chan, 1966). Practical details for use of bathophenanthroline in the determination of soluble iron have been published by Strickland and Parsons (1968).

A number of other chromogenic reagents have been used for the photometric determination of "dissolved" iron in sea water. Although methods using thiocyanate have been employed for this purpose (see Lewis and Goldberg, 1954) they lack sufficient sensitivity and tend to have a poor precision (see Sandell, 1959). Nitroso-R salt has been employed by Bernhard (1971) for the automatic determination of total reactive iron following oxidative breakdown of organic complexes.

Lack of sensitivity and matrix effects prevent the direct application of conventional atomic absorption spectrophotometry to the determination of dissolved iron in sea water.* However, this technique has been extensively employed for the examination of concentrates prepared by chelating ion exchange or solvent extraction (see e.g. Brooks *et al.*, 1967; Brewer *et al.*, 1969, 1972; Orren, 1971; Preston *et al.*, 1972; Paus, 1973). It has of course, the advantage that several elements can be determined in the concentrate by the same method.

A number of other techniques have been employed for the determination of dissolved iron. γ-spectrometry following neutron irradiation of the freeze dried salts has been used for the estimation of total soluble iron by Robertson, *et al.* (1968), Piper and Goles (1969), Piper (1971) and Spencer *et al.* (1970b). X-ray fluorescence methods for the determination of soluble iron following preconcentration have been described by Morris (1968) and Armitage and Zeitlin (1971).

Decomposition of dissolved organic iron compounds. Most methods for the determination of total dissolved iron in sea water necessitate a preliminary decomposition of the organo-iron complexes. This has been achieved by Lewis and Goldberg (1954) by fuming the sample almost to dryness with perchloric acid. Although the method has been widely accepted (see e.g. Strickland and Parsons, 1968), many careful workers have experienced considerable difficulties with it owing to contamination (see Burton and Head, 1970). Even in a specially constructed laboratory with filtered air supply Alexander and Corcoran (1964) were able to obtain a coefficient of variation no lower than ca. 10%. As a consequence, it is probably desirable to use less drastic techniques in which it is easier to control contamination, even at risk of less complete breakdown of organo-iron complexes. Thus, Armstrong (1957) and Williams and Chan (1966) brought about the decomposition by autoclaving the acidified (pH 0·15) sample. However, Bernhard (1971) preferred to use an oxidative method involving autoclaving in the presence of peroxydisulphate ion. Burton and Head (1970), following the procedure developed by Tetlow and Wilson (1964) for the determination of total iron in boiler feed waters, autoclaved sea water samples with thioglycollic acid. However, although they were able to obtain almost quantitative recoveries of inorganic forms of iron, they found that certain organic species of iron (e.g. haemin) were only partly broken down.

* Segar and Gonzalez (1972) have found that it is feasible to determine total dissolved iron directly in sea water by AAS if the sample is vaporized by means of a heated graphite atomizer (see p. 305).

Determination of particulate iron. Particulate iron is distributed very inhomogeneously in sea water especially near to the coast. Cooper (1948) has pointed out that if significant results are to be obtained, the sampling programme must be designed in such a way that the results can be treated statistically. Studies of this type have been made by Lewis and Goldberg (1954). Schaefer and Bishop (1958), Menzel and Spaeth (1962) and Betzer and Pilson (1970, 1971).

Not all the particulate forms of iron are likely to be available for use by phytoplankton, and a number of biological workers have attempted to determine that fraction which is available by leaching the particulate matter with mold reagents. Thus, Strickland and Austin (1959) leached what they term "reactive particulate iron" from the particulate fraction using dilute hydrochloric acid (for practical details of the method see Strickland and Parsons, 1968). Burton and Head (1970) obtained a coefficient of variation of ca. 16% with this method, a value similar to that found by Williams and Chan (1966) using the same procedure. Ryther *et al.* (1967) who employed a similar technique suggested the use of potassium peroxydisulphate to break down organic complexes. However, it should be pointed out that there is no evidence that this method does determine only that iron which is available to phytoplankton.

A variety of other digestion techniques have been employed by other workers for the dissolution of particulate iron. These range from dissolution under comparatively mild conditions, for example extraction with the solution of hydrochloric acid in acetone favoured by Brooks *et al.* (1967a), or autoclaving with 0.1 N hydrochloric acid (Armstrong, 1957), to much more rigorous treatments involving fuming with mineral acids such as nitric acid (with (Paeda *et al.*, 1968) or without (Betzer and Pilson, 1970, 1971) sulphuric acid) or more frequently perchloric acid (Lewis and Goldberg, 1954; Williams and Chan, 1966; Burton and Head, 1970; Emel'yanov *et al.*, 1971). These more stringent digestion techniques, which have usually been used in conjunction with colorimetric or atomic absorption spectrophotometric methods of determination, provide a reasonable approximation to the total particulate iron. However, it is probable that the small amounts of iron associated with refractory silicates will not be quantitatively determined in this way, but will be recovered only after hydrofluoric acid digestion or sodium carbonate fusion (Atkinson and Stefansson, 1969).

In addition to the above wet chemical methods for the determination of total particulate iron a number of instrumental techniques have also been employed. Cann and Winter (1971) have claimed a coefficient of variation of $\pm 6\%$ for a non-destructive procedure in which the membrane filter was examined by X-ray fluorescence. The determination has also been carried

N

out by neutron irradition of the pelletized membrane filter followed, after cooling, by high resolution γ-spectrometry of the induced ^{58}Fe activity (Piper and Goles, 1969; Spencer et al., 1970a, 1972; Piper, 1971).

19.10.5.27. Lanthanides (including scandium and yttrium)

The dissolved lanthanides in sea water appear to be conservative; they have the following approximate concentrations (in $ng l^{-1}$): La 3; Ce 1; Pr 0·6; Nd 3; Sm 0·5; Eu 0·13; Gd 0·7; Tb 0·1; Dy 0·9; Ho 0·22; Er 0·87; Tm 0·2; Yb 0·8; Lu 0·2; Sc 5; Y 13. In each instance the principal dissolved species is believed to be the simple hydrated $3+$ ion of the element. Very little is known about the occurrence of these elements in marine particulate phases.

Preconcentration. The lanthanides can be efficiently coprecipitated with ferric hydroxide and this technique has been used for their preconcentration by most workers in this field (Wattenberg, 1943; Goldberg et al., 1963; Spirn, 1965; Høgdahl et al., 1968; Shigematsu et al., 1967; Hayes et al., 1966; Hayes 1969; Nagatsuka et al., 1971). Data by Høgdahl (1965) show that at pH 8·5 the lanthanide elements can be coprecipitated from sea water in this way with an efficiency of 92–98%. The lanthanides have also been concentrated from sea water by coprecipitation with calcium oxalate (Shigematsu et al., 1967) and by co-crystallization with methyl violet (Balashov and Khitrov, 1961).

Determination of dissolved lanthanides. Attempts have been made to determine lanthanides in concentrates from sea water both spectrographically (Wattenberg, 1943) and colorimetrically (Balashov and Khitrov, 1961). However, the results which were obtained are so much greater (100–1000×) than the currently accepted concentrations of these elements that they are only of historical interest. Because of their very high sensitivity and specificity neutron activation procedures have been used in all recent work (Goldberg et al., 1963; Hood and Slowey, 1964; Spirn, 1965; Høgdahl et al., 1968; Hayes et al., 1966; Shigematsu, 1967 (Eu and La only); Hayes, 1969; Nagatsuka et al., 1971). When uranium is irradiated with neutrons it undergoes nuclear fission yielding among other radionuclides those of several of the lanthanides. Uranium is coprecipitated with ferric hydroxide along with the lanthanides, and it is essential therefore that it should be separated before the neutron activation is performed. Although, Spirn (1965) has attempted to remove uranium by reprecipitating the ferric hydroxide with ammonium carbonate, this method is inefficient, and it is preferable to separate the element by ion exchange. In the procedure employed by Goldberg et al. (1963) both uranium and iron are adsorbed from concentrated hydrochloric acid medium using an anion exchanger. The lanthanides, which were not adsorbed, were again coprecipitated with ferric hydroxide ready for irradiation. Following irradiation with thermal neutrons, the precipitate was dissolved in hydrochloric acid and 10 mg amounts of the individual lanthanides were added. The lanthanides were then separated from one another by cation exchange using gradient elution with complexing agents, such as lactic or α-hydroxy-butyric acids. The separated individual lanthanides were converted to oxalates for gravimetric determination of chemical yield and β-counting of the induced lanthanide activities. According to Nagatsuka et al. (1971) the analysis can be considerably simplified if only those lanthanides which give rise to γ-emitting nuclides are to be determined. In this instance, after irradiation of the ferric hydroxide precipitate, the lanthanides are separated as a group by coprecipitation with lanthanum oxalate and then determined by γ-spectrometry.

Determination of particulate lanthanides. A neutron activation procedure for the determination of lanthanides in marine particulates has been described by Spirn (1965), and Spencer et al. (1970, 1972), who have used instrumental neutron activation for the estimation of lanthanum and scandium.

19.10.5.28. *Lanthanum.* See Lanthanides.

19.10.5.29. *Lead*

Recent reported values for the concentration of dissolved lead in open ocean waters lie in the range 0.02–$0.1\ \mu g\,l^{-1}$ (Tatsumoto and Patterson, 1963). Although considerably higher concentrations have been reported for coastal waters, it now seems likely these are erroneous and that the concentration rarely exceeds $80\ ng\,l^{-1}$ even close to industrialised areas (see Brewer et al. 1974). The principal dissolved lead species are probably $PbOH^+$ $PbCO_3^0$ with lesser proportions of $PbCl^+$ and $PbCl_2^0$ (Section 3.4.7). Lead is probably associated with both the inorganic and organic phases of marine particulates; however, little information is available about its concentration. Preston et al. (1972) have found that on average particulate lead comprises ca. 50–80% of the total lead in inshore water from the Irish Sea, but, in view of the findings of Brewer et al. (1974) even this may be doubtful.

Preconcentration. With the exception of anodic stripping voltammetry, all techniques for the determination of lead in sea water necessitate preconcentration of the element. Several different processes have been used for this purpose.

Both coprecipitation (with CuS (Boury, 1938); HgS (Noddack and Noddack, 1940); $Mg(OH)_2$ (Joyner et al. 1967)) and cocrystallization (with tannin, thionalide and oxine, Black and Mitchell, 1951) have been used for the preconcentration of lead from sea water. Recently, Krishnaswami *et al.* (1972), during a study of ^{210}Pb, have collected lead from sea water *in situ* by means of acrylic fibre loaded with ferric hydroxide.

Solvent extraction techniques have been employed for the concentration of lead from sea water by a number of workers. The extraction systems which have been used for this purpose include dithizone-chloroform (Tatsumoto and Patterson, 1963; Chow, 1958; Brewer *et al.*, 1974; see also Chow and McKinney, 1958), diethylammonium dithiocarbamate-chloroform (Loveridge *et al.* 1960) and ammonium pyrrolidine dithiocarbamate-methyl *iso*-butyl keton (Brooks *et al.* 1967a; Preston *et al.* 1972). Lead can also be readily concentrated by chelating ion exchange techniques employing Chelex-100 (Riley and Taylor, 1968; Abdullah and Royle, 1972), or the cheaper, but less satisfactory, chitosan (Muzzarelli and Marinelli, 1972). Lead has also been concentrated from sea water by uptake on active carbon modified with complexing agents such as dithizone and 8-hydroxyquinoline (Zharikov, 1970).

Determination of dissolved lead. The difficulty of determining the low concentrations of lead in sea water has been emphasized by a recent intercalibration study reported by Brewer *et al.* (1974). In this, four sets of water samples were analysed in 8 laboratories using flameless atomic absorption spectrophotometry (AAS), anodic stripping voltammetry (ASV) or pulse polarography (PP). Mass spectrometric isotope dilution analysis (MSID) was used as a referee method. The study revealed a very disquieting lack of agreement between the values obtained and the referee value, as well as among themselves; in general, results were considerably greater than the standard value. This lack of concordance was particularly evident for the water having the lowest lead content (14 ± 3 ng l^{-1}), for which concentrations of 1300, 50, 60, 600, 55, 180 and 120 ng l^{-1} were reported. Indeed, for all analyses of this water (except that carried out by MSID) the lead in the reagent blank exceeded that in the sample. The principal conclusions reached in this study were: (i) that in the procedures, other than isotope dilution, the ratio of analysed sample lead to procedural blank lead must be increased, by increasing the volume of sample used 10 fold; (ii) that it is necessary to carry out the analyses in a special clean laboratory kept pressurized with air which has been filtered to remove particles > 1 μm in diameter; if this cannot be arranged, work should be carried out under a positive pressure clean hood (see e.g. Fig. 19.12). (iii) that special care must be taken in purifying reagents (see Chow and McKinney (1958) and Tatsumoto and Patterson (1963) for methods of purification of reagents); (iv) that only polytetrafluorethylene or fused silica apparatus should be used in the analysis, and even this only after prolonged treatment with concentrated mineral acid. In view of the findings of the intercalibration study, further investigations are to be carried out. In these, special attention will be paid to the above points in order to try to produce reliable AAS and ASV procedures suitable for routine analysis of sea waters.

From the above discussion it will be realized that most of the determinations of lead in sea water carried out prior to the advent of the isotope dilution procedure are only of historical value (see Richards, 1957; Riley, 1965b). Thus, colorimetric procedures based on the use of dithizone which have been described by Boury (1938) and Loveridge et al. (1960), are insufficiently sensitive, although they might be of use with badly polluted estuarine waters.

At present, the most accurate methods available for the determination of lead in sea water are the time-consuming mass spectrometric isotope dilution procedures developed by Chow (1958) and Tatsumoto and Patterson (1963). In these, for the estimation of total lead, the sample (1 or 2 litres) is acidified to 0·1 N with respect to hydrochloric acid. A known amount of lead-208 is added and the solution is heated to 55°C for 20 hours to equilibrate the lead in the spike with that in the sample. After neutralization to pH 7·5 with ammonia the lead is extracted with a solution of dithizone in chloroform. The dithizone layer is back-extracted with dilute acid and lead is then re-extracted with dithizone, citrate and cyanide being added as complexing agents for other metals. Lead is again back-extracted with acid and a weighed amount of ^{206}Pb is added for measurement of yield. Finally, the solution is evaporated to dryness and the residue is transferred to the rhenium filament of the mass spectrometer with which the ^{208}Pb: ^{206}Pb and ^{208}Pb: ^{207}Pb ratios are then measured. The chemical yield of the process is assessed from the former measurement in order to evaluate the blank. The latter ratio measurement is used to assess the concentration of lead in the water sample.

Conventional atomic absorption spectrophotometric procedures, although specific, are insufficiently sensitive for the determination of lead even in 1000-fold concentrates of sea water. However, attempts have been made to apply both the tantalum boat technique (Preston et al. 1972) and more satisfactorily the heated carbon tube furnace (Segar and Gonzalez, 1972; Brewer et al. 1974) for atomization of concentrates prepared by solvent extraction. Difficulties associated with contamination have limited the sensitivity and precision attainable with these techniques (Brewer et al., loc. cit), but it is possible that they may be overcome by increasing the volume of sample used.

Two electrochemical procedures have been applied to the determination of lead in sea water (see Chap. 20). Both conventional (Butterworth, 1972) and pulse (Abdullah and Royle, 1972; Brewer et al., 1974) polarography have been used for the examination of concentrates. The only technique which can be used for the direct determination of lead in sea water is anodic stripping voltammetry. It was first applied to the problem by Whitnack and Sasselli (1969) who used a hanging mercury drop electrode. However, the concentrations of the element which they found were several orders of magnitude higher than those now accepted for sea water. More recently, Smith and Redmond (1971) have made a detailed study of the determination of trace metals by this technique and have concluded that other elements are unlikely to interfere in the estimation of lead. Very high results were obtained by the two workers who used it in the intercalibration study described by Brewer *et al.* (1974) (see above). The cause of this error appeared to be contamination and it was reported that further work was to be carried out using larger samples and taking elaborate precautions to protect against contamination.

Determination of particulate lead. Brooks *et al.* (1967a) digested the membrane filter bearing the particulate matter with 6M hydrochloric acid. After addition of acetone to dissolve the filter the extract was examined by conventional AAS. Preston *et al.* (1972) have described a procedure for the determination of lead in marine particulates in which the membrane filter bearing the particulates was boiled with dilute hydrochloric acid. Lead was subsequently determined in the digest by atomic absorption spectrophotometry using the tantalum boat technique. Belyaev and Gordeev (1972) determined particulate lead by atomic absorption spectrophotometry after vaporizing the membrane filter containing particulates in a hollow carbon electrode.

19.10.5.30. *Lithium*

Lithium appears to be a conservative element in sea water with a concentration of ca. $170 \, \mu g \, l^{-1}$ at a salinity of $35\%_0$. The principal dissolved species of the element is the hydrated Li^+ ion. No data appears to be available for the concentration of lithium in marine particulates.

Preconcentration. Most of the early methods for preconcentration of lithium employed time-consuming and inefficient precipitation or extraction processes (for reviews see Riley, 1965b and Morozov, 1968; see also Riva, 1967; Shigematsu *et al.*, 1969). These methods have been completely outmoded by cation exchange techniques which are rapid and simple and permit lithium to be separated quantitatively from the major cations of sea water (including ca. 80,000 its own weight of sodium) This technique was first applied to sea water analysis by Chow and Goldberg (1962) and has been subsequently employed by several other workers (Riley and Tongudai, 1964; Uesugi and Murakami, 1966; Murakami and Katsuya, 1967).

It is based on the fact that lithium has a very low affinity for strongly acidic cation exchangers and is very easily displaced from them by dilute mineral acids.

Determination of dissolved lithium. Although isotope dilution (Chow and Goldberg, 1962), colorimetric (Uesugi and Murakami, 1966) and chelatometric (Riva, 1967) methods have been used for the determination of lithium after preconcentration, most investigators have used optical spectroscopic techniques because of their simplicity and specificity. Both flame spectrophotometry (Riley and Tongudai, 1964; Morozov, 1968, 1969) and atomic absorption spectrophotometry (Fabricand *et al.*, 1966; Angino and Billings, 1966; Burrell, 1967; Friedman *et al.*, 1968; Shigematsu *et al.*, 1969) have been employed for this purpose, and there appears to be little to choose between them in terms of sensitivity and precision. When these techniques are applied directly to sea walter, errors associated with the efficiency of atomization of the sample and the presence of incandescent particles in the flame limit the precision to ca. $\pm 10\%$. If lithium is first preconcentrated, by e.g. ion exchange, a precision of $\pm 2\%$ can be easily achieved.

19.10.5.31. *Lutetium.* See lanthanides

19.10.5.32. *Manganese*

The concentration of dissolved manganese in the open ocean appears to lie in the range $0.2–4\,\mu g\,l^{-1}$, but somewhat higher concentrations may be found in inshore waters. The principal species of dissolved manganese are probably hydrated Mn^{2+} and perhaps $MnCl^+$. There has been no confirmation of the claim by Hood *et al.* (1962) that a substantial part of the dissolved manganese in sea water exists as organic complexes. Appreciable, but very variable amounts of manganese are associated with marine particulates.

Preconcentration. Coprecipitation with magnesium (Koroleff, 1947; Loveridge *et al.*, 1960; Burrell, 1965; Mokievskaya, 1965; Joyner and Finley, 1966), iron(III) (Rona *et al.*, 1960) and aluminium (Ishibashi *et al.*, 1960b) hydroxides has been used for preconcentration of manganese from sea water. Loveridge *et al.* (1960) have reported that the element at a level of $2\,\mu g\,l^{-1}$ could be recovered with an efficiency of $\sim 98\%$ by coprecipitation with magnesium hydroxide.

Particularly because of their convenience in subsequent atomic absorption spectrophotometry solvent extraction techniques have been extensively applied to the preconcentration of manganese. Extraction systems which have been used for this purpose are listed in Table 19.16. It should be noted that because of the relatively low stability constants of the manganese complexes produced it is usually necessary to employ a great excess of the complexing ligand in order to ensure quantitative extraction.

Manganese can also be preconcentrated efficiently from sea water by chelating ion exchange using Chelex-100 (Riley and Taylor, 1968b).

Determination of dissolved manganese. Several different techniques have been applied to the determination of manganese in sea water usually after pre-concentration.

Colorimetry. Because the permanganate ion has only a relatively low molar absorbance and reacts with chloride ion, it is not feasible to determine manganese photometrically directly in sea water by oxidation to perman-ganate. However, it has been determined in this way after preconcentration (Koroleff, 1947; Ishibashi *et al.*, 1960b; Loveridge *et al.*, 1960; Mokievskaya, 1965), but with rather poor precision. Although the sensitivity of the photo-metric formaldoxine method $(10 \,\mu g \, l^{-1})$ is not sufficiently high for the direct determination of manganese in normal sea waters (but see Genovese and Magazzu (1969)), the relatively high concentrations of manganese in anoxic waters can be determined in this way. Brewer and Spencer (1970), who employed a mixed reagent containing hydroxylamine hydrochloride, formaldoxime and ammonia, reported a coefficient of variation of $\pm 3.5\%$ at a level of $300 \,\mu g \, Mn \, l^{-1}$.

A number of direct photometric methods have been described for the direct determination of manganese in sea water. These are based on the fact that manganese catalyzes the oxidation of the leuco bases of certain *di-* and *tri-*phenyl methane dyes by the periodate ion, leading to the production of several hundreds of molecules of the intensely coloured parent dye per atom of manganese. Leuco compounds which have been used include those of tetra base (Harvey, 1949), malachite green (Yuen, 1958; Strickland and Parsons, 1968) and crystal violet (Kessick *et al.*, 1972). The reaction is reasonably specific for manganese and no interference is likely to be ex-perienced from other ions at their normal concentrations in sea water. The sensitivity of the technique is high (ca. $0.1 \,\mu g \, Mn \, l^{-1}$), and according to Kessick *et al.*, the sensitivity may be increased a further four-fold by extracting the reaction product into a small volume of a mixture of iso-butyl alcohol and benzene. However, the precision is poor, and it is difficult to achieve a coefficient of variation of better than $\pm 15\%$.

Atomic absorption spectrophotometry. Manganese cannot be determined directly in sea water by atomic absorption spectrophotometry owing to both lack of sensitivity and interference from the major ions. However, the technique has been widely used for the determination of the element in multielement schemes using concentrates prepared by solvent extraction (for references see Table 19.16) or chelating ion exchange (Riley and Taylor, 1968), and it probably provides the most satisfactory method at present available for the determination of manganese in sea water.

TABLE 19.16

Extraction systems for preconcentration of manganese

Complexing reagent	Solvent	Method of determination	Reference
sodium diethyl dithiocarbamate	chloroform	NAA	Rona et al. (1962); Slowey and Hood (1971)
		P	Tikhonov and Shalimov (1965)
Ammonium pyrrolidine dithiocarbamate*	methyl iso-butyl ketone	AAS	Joyner and Finley (1966) Burrell (1967) Brewer et al. (1969); Spencer and Brewer (1969); Preston et al. (1972); Paus (1973)
		XRF	Morris (1968)
8-hydroxyquinoline	chloroform	XRF	Armitage and Zeitlin (1971)
1-nitroso-2-naphthol	chloroform	C	Loveridge et al. (1960)

C   = colorimetry, AAS = atomic absorption spectrophotometry,
NAA = neutron activation analysis, P = polarography,
XRF = X-ray fluorimetry.
* The manganese-APDC complex is comparatively unstable and AAS measurements should be made within 30 min of carrying out the extraction.

Other methods. Neutron activation (Rona et al., 1962; Slowey and Hood, 1971), polarographic (Tikhonov and Shalimov, 1965) and X-ray fluorimetric (Morris, 1968; Armitage and Zeitlin, 1971) procedures have also been employed for the determination of manganese in concentrates prepared by solvent extraction. However, they appear to have no advantages in speed or precision over the atomic absorption method.

Determination of total dissolved manganese. Breakdown of complexed manganese has been effected by boiling or autoclaving the acidified (pH

1·0–1·5) sample (Mokievskaya, 1965; Tikhonov and Shalimov, 1965) or by treating it with potassium persulphate (Slowey and Hood, 1971).

Determination of particulate manganese. The estimation of particulate manganese has been carried out by colorimetry (Yatsimirski *et al.*, 1971), by atomic absorption spectrophotometry (Belyaev, 1972; Preston *et al.*, 1972) and by neutron activation followed by γ-spectrometry of the induced ^{56}Mn activity (Piper, 1971; Spencer *et al.*, 1970).

19.10.5.33. *Mercury*

Dissolved inorganic mercury in sea water probably occurs mainly as $HgCl_4^{2-}$ (see Section 3.2.2 Fig. 3.4). Its concentration in open ocean water usually appears to lie in the range $10–50 \, ng \, l^{-1}$, but considerably high concentrations may be present in inshore waters and especially in the waters of polluted estuaries. Because of its strong tendency to be adsorbed, the mercury content of the suspended particulate matter is frequently appreciable, and in estuarine waters subject to pollution by industrial effluents or domestic sewage the concentration in the particulate matter may reach 10 p.p.m. In such waters the concentration of particulate mercury may exceed that of dissolved mercury by a factor of ten.

Organo-mercury compounds (e.g. $HgCH_3^+$ and $Hg(CH_3)_2$) are known to be present in marine organisms and in many inshore sediments, particularly anoxic ones. Although there is indirect evidence for their presence in coastal waters (Fitzgerald and Lyons, 1973), they have not, as yet, been detected in ocean water.

Filtration. Because of the strong association of mercury with suspended marine inorganic and organic matter it is usually necessary to remove the particulate matter before determining dissolved mercury. Filtration of sea water through many types of filter can lead to significant (10–20%) loss of dissolved mercury by adsorption. Losses are least on glass fibre filters (e.g. 5% on Whatman GF/F), and can be reduced to negligible proportions by coating the glass fibre filter with a thin film of silicone oil (D. Gardner personal communication). Alternatively, particulate matter can be removed by centrifugation without significant loss of dissolved mercury. If the filtered sample is to be stored before analysis it should be rendered 0·1 N with respect to sulphuric acid and kept in a silicone coated glass bottle. Acidification should not be carried out prior to filtration as this may cause desorption of mercury from the particulate matter (Carr and Wilkniss, 1973).

Determination of dissolved inorganic mercury. Recent concern with environmental pollution with mercury has led to a spate of new methods for the determination of the element in sea water.

Preconcentration. The very low concentration of dissolved mercury in sea water makes it necessary to preconcentrate prior to analysis. In the earliest work on the occurrence of mercury in the sea the element was coprecipitated with copper sulphide (Stock and Cucuel, 1934) and the same technique has been used both by Weiss and Crozier (1972) and Williams and Weiss (1973) as a preliminary to neutron activation analysis. Voyce and Zeitlin (1974) have described a procedure in which mercury is separated from sea water by colloid flotation; however the recovery is not quantitative (80–90 %). Extraction from acidic medium (pH < 1) using a solution of dithizone in chloroform, or carbon tetrachloride provides a quantitative and selective method for the preconcentration of mercury, and has been employed by several workers (Hosohara *et al.*, 1961; Hosohara, 1961; Gardner and Riley, 1973); alkyl mercury ions are also extracted under these conditions (Kiwan and Fouda, 1968). Back extraction of mercury from the dithizone extract can be achieved with an acidified bromide or nitrite solution, or with 6 M hydrochloric acid (Chau and Saitoh, 1970, 1971).

Since dissolved inorganic mercury exists in sea water as the $HgCl_4^{2-}$ ion, it can be concentrated from sea water (usually acidified) by ion exchange using the chloride form of a strongly basic anion exchanger (Burton and Leatherland, 1971). Aqueous solutions of mineral acids are not efficient for the removal of the adsorbed mercury from the anion exchanger. However, it can be satisfactorily eluted with solutions of perchloric acid in non-aqueous media, such as ethyl acetate (Burton and Leatherland, *loc. cit.*) or acetone (P. D. Jones personal communication). Mercury is also adsorbed efficiently from sea water by chelating ion exchange resins (e.g. Chelex-100), but it is impossible to elute it quantitatively with solutions of mineral acids in either aqueous or non-aqueous media.

Brandenberger and Bader (1967a, b; 1968) have observed that traces of mercury can be collected from acidified media by amalgamation on a copper spiral. The copper spiral was subsequently heated electrically and the evolved mercury vapour was determined directly by atomic absorption spectrophotometry. This technique has been subsequently adopted for the determination of mercury (100–4000 ng l^{-1}) in natural (fresh) waters by Hinkle and Learned (1969) and by Fishman (1970) who employed silver spirals instead of ones made of copper. Smith *et al.* (1971) have used a similar method for the determination of 3–500 ng Hg l^{-1} in estuarine waters. Tests in the author's Laboratory (P. D. Jones private communication) have shown that mercury spikes can be recovered from 1 litre samples of stripped sea water with an efficiency of $\sim 90\%$ at a level of $100\,\mathrm{ng}\,\mathrm{l}^{-1}$ by gently shaking for 22 hours with a spiral made from $0.4\,\mathrm{m}$ of $0.15\,\mathrm{mm}$ silver wire. In this time uptake of mercury in the form of methyl mercury was less than 10%. Very much lower recoveries ($<25\%$) are obtained using copper spirals of the same dimensions.

The appreciable volatility of elemental mercury has been utilized by many workers for the preconcentration of the element from natural waters (April and Hume, 1970; Braman, 1971; Kopp *et al.*, 1972; Fitzgerald and Lyons, 1973; Topping and Pirie, 1972; Carr *et al.*, 1972; Cranston and Buckley, 1972). Inorganic forms of mercury have usually been reduced to the elemental state by means of a stannous salt. However, sodium borohydride (Braman, 1971) and Ti^{3+} (April and Hume, 1970) have also been used for the same purpose. The liberated mercury is then sparged from the sample by means of a current of air or nitrogen. The effects of variations in the operational parameters in this technique have been studied by Kopp *et al.* (1972). The mercury can be stripped from the gas by means of acidic permanganate solution (Topping and Pirie, 1972), with metallic gold (Carr *et al.*, 1972) or with a packed cold trap (Fitzgerald and Lyons, 1973) and then determined by atomic absorption spectrophotometry. Alternatively, it can be determined directly in the gas by atomic absorption or emission spectrophotometry (see below). Because of their rather poor sensitvity these direct methods are not suitable for the examination of ocean waters, but they may be of value in the analysis of estuarine waters.

Determination of dissolved mercury. A number of different techniques have been employed for the determination of mercury after its preconcentration from sea water. Although the results obtained by Stock and Cucuel (1934) by micrometry of the bead of elemental mercury are probably mainly of historical interest, they agree well with recent determinations. Photometric procedures depending on the extraction of the intensely coloured complex formed between mercury and dithizone have been employed by Hosohara (1961), Hosohara *et al.* (1961) and Burton and Leatherland (1971). The extraction is carried out from 0.1–$1\,\mathrm{N}$ acid medium preferably in the presence of EDTA in order to increase the selectivity of the reagent. However, even under these conditions certain other elements, notably silver and copper, are also extracted. The mercury must therefore be back extracted selectively (with, for example, $5\,\mathrm{N}$ hydrochloric acid or potassium bromide) before being re-extracted with dithizone and determined photometrically. Burton and Leatherland (1971) have claimed a standard deviation of better than $\pm 1\,\mathrm{ng}\,\mathrm{Hg}\,\mathrm{l}^{-1}$ (at a level of ca. $15\,\mathrm{ng}\,\mathrm{l}^{-1}$) for their method in which mercury is preconcentrated from 20–40 litre sea water samples prior to photometric determination.

The sensitivity of both flame emission spectrophotometry and conventional atomic absorption spectrophotometry are too low for these methods to be of value in the determination of mercury in preconcentrates from sea water. Mercury vapour absorbs very strongly at 253·7 nm—the wavelength of its resonance line—and this absorption has been used as a basis for the determination of traces of mercury in air for many years (Woodson, 1939). Poluektov *et al.* (1964) were the first workers to apply this principle to the determination of mercury in aqueous solutions. The acidified samples were treated with tin (II) in order to reduce inorganic mercury compounds to the element. Air was blown through the sample and passed through an absorption cell fitted with silica windows. The absorbance of the air was then measured with an atomic absorption spectrophotometer at 253·7 nm to give a measure of the mercury present in the sample. This technique, which is usually referred to as the *flameless* or *cold vapour* atomic absorption method, has been used for the determination of mercury in a wide range of different types of sample. It is rapid and simple to use, and has the advantages of very high sensitivity (under favourable conditions a detection limit of 0·5 ng can be attained) and almost complete freedom from interference by other elements. Even this high sensitivity could be increased approximately sixty-fold if it were possible to use the mercury resonance line at 184·9 nm (Dagnall *et al.*, 1972). However, interference from volatile organic compounds is likely to be severe at this wavelength. In addition, difficulty is likely to be encountered because of absorption arising from molecular oxygen. Even when measurement is made at 253.7 nm serious interference can be caused by volatile organic compounds which exhibit absorption at this wavelength. If only traces of such compounds are present in the sample, as with ocean waters, correction for their presence can be made using a deuterium arc background corrector (p. 307). Alternatively, correction can be made by circulating the carrier gas, after measurement of the absorbance, through a palladium chloride tube to absorb mercury and then remeasuring (Windham, 1972) or by aerating the sample before reducing the mercury to the elemental state.

The cold vapour method has been extensively applied to the determination of mercury in sewage, effluents and polluted waters. However, it is only in the last 2–3 years that it has been extended to the determination of the very much lower concentration levels of mercury existing in the sea. A few workers (e.g. Carr *et al.*, 1972; Cranston and Buckley, 1972; Fitzgerald *et al.*, 1974) have examined water samples directly by the cold vapour method, effectively using the stripping gas to preconcentrate the mercury. However, the sensitivity of such methods is inadequate for the analysis of ocean waters (thus, the detection limit of the procedure of Carr *et al.* (*loc. cit.*) is 15 ng l^{-1}). In addition, problems might arise if these techniques are used with polluted waters if volatile U.V.-absorbing compounds are present. Because of these difficulties, most investigators have preferred to preconcentrate mercury from sea water before applying the cold vapour method. In this way detection limits of ca. $2 \, \mathrm{ng} \, l^{-1}$ can be readily attained with 2–4 litre samples. Preconcentration of mercury has been carried out by dithizone extraction (Gardner and Riley, 1973; Chester et al., 1973), by reduction to the elemental state, stripping with air and trapping in acidic permanganate (Topping and Pirie, 1972; Harsányl et al., 1973), by trapping in a cold trap (Fitzgerald and Lyons, 1973), or on gold (Olafsson, 1974), by anion exchange (Burton and Leatherland, 1971) and by amalgamation on a silver wire which is subsequently heated electrically to expel the mercury (Smith et al., 1971). The last process is particularly attractive because of its comparative freedom from reagent blanks and its inherently greater sensitivity.

Conventional arc and flame spectral methods have a sensitivity too low for them to be used for the determination of mercury in sea water without recourse to inconveniently large samples. However, it is possible to attain up to 1000-fold greater sensitivity by the use of more elaborate excitation techniques. Braman (1971) has described a method in which the mercury in sea water samples was reduced to the elemental state and allowed to diffuse through a membrane into helium carrier gas. The carrier gas is passed into a quartz discharge tube in which the 253·7 nm mercury emission line is excited with a d.c. potential. A detection limit of $4 \, \mathrm{ng} \, l^{-1}$ is claimed; however, no figures for precision are given. Plasma emission spectrophotometry has been used for the determination of mercury in sea water medium by April and Hume (1970). However, as the working range of their method is 1 to $1000 \, \mu\mathrm{g} \, l^{-1}$ it is likely to be of little value even for the analysis of highly polluted estuarine waters. A very high sensitivity for mercury can be obtained by microwave excited emission spectrophotometry (Bache and Lisk, 1971), but this technique does not appear to have yet been employed in sea water analysis.

On irradiation with thermal neutrons mercury gives rise to the following radionuclides ^{197m}Hg, ^{197}Hg, ^{199m}Hg, ^{203}Hg, ^{205}Hg. Of these, ^{199m}Hg and ^{205}Hg are too short-lived to be of value for analytical purposes. For a given irradiation period ^{203}Hg ($t_{\frac{1}{2}}$, 46·9 days) will give a much lower activity than that of the ^{197m}Hg ($t_{\frac{1}{2}}$, 24 h) + ^{197}Hg ($t_{\frac{1}{2}}$, 65 h) arising jointly from ^{196}Hg (abundance 0·15%, $\sigma = 3100 \pm 100$). Weiss and Crozier (1972) have described a neutron activation procedure for the determination of mercury in sea water (see also Weiss et al., 1972 and Williams and Weiss, 1973). In this, mercury is concentrated by coprecipitation with copper sulphide, the precipitate is irradiated with thermal neutrons and, following a radiochemical separation, the combined $^{197m+197}$Hg activity is counted. The detection limit was found to be ± 4 ng. Replicate analyses using 0·5 litre aliquots gave a coefficient of variation of 11·1% at a mercury level of 83 ngl^{-1}.

Determination of organic mercury compounds. It is well known that organic mercury compounds (such as CH_3Hg^+, and $(CH_3)_2Hg$) are present in the tissues of aquatic animals, and it has been demonstrated that they can be synthesized by micro-organisms in both marine and fresh-water sediments. These facts suggest that such compounds may be present in both fresh and saline waters. Organo-mercury compounds are not determined by most of the methods in current use for the determination of dissolved inorganic mercury. However, they can be fairly readily oxidized to Hg^{2+}. If, after the oxidation has frequently been carried out with permanganate in acid medium inorganic mercury concentration should give an approximate value for the organic mercury level. In the analysis of effluents and waste waters, the oxidation has frequently been carried with permanganate in acid medium (see e.g. Omang, 1971). However, there is evidence that oxidation of many organomercury compounds under these conditions is incomplete (Goulden and Afghan, 1970; Kopp *et al.*, 1972)*. Kopp *et al.* (*loc. cit*) prefer to use peroxydisulphate ($S_2O_8^{2-}$) in conjunction with permanganate, although even this treatment has been found to give lower results than fuming with nitric and sulphuric acids (Nelson *et al.*, 1972). Ultra-violet irradiation (Armstrong *et al.*, 1966) has been used by Goulden and Afghan (1970) for the decomposition of organo-mercury compounds, and this technique has the advantage that reagent blanks are practically eliminated.

Little work has been published on the determination of total mercury in saline waters; indeed there is considerable doubt whether either CH_3Hg^+ or $(CH_3)_2Hg$ are stable in normal oxygenated sea water. Smith *et al.* (1971) have used permanganate oxidation at room temperature for the oxidation of organic mercury compounds in water from the tidal reaches of the River Thames. Organo-mercury compounds have been estimated in coastal and river waters from the vicinity of Long Island after photochemical decomposi-- tion (Fitzgerald and Lyons, 1973). Fitzgerald *et al.* (1974) determine "total" mercury in ocean waters after making the sample ca. 4N and ca. 1·5N with respect to sulphuric and nitric acids respectively followed by digestion at 60°C for 1 hour. In none of this work was evidence provided for the completeness of the breakdown of organo-mercury compounds. No attempt appears to have been made yet to identify the organo-mercury compounds present in the sea. Characterization might be expected to be achieved by solvent extraction followed by gas–liquid chromatography using an electron capture detector.

* Problems of the breakdown of organomercury compounds in waters, effluents etc. have been discussed in Environmental Protection Agency (1972).

19.10.5.34. *Molybdenum*

The concentration of dissolved molybdenum in sea water appears to lie in the range $10–12\,\mu g\,l^{-1}$. The principal dissolved species is probably MoO_4^{2-}. Molybdenum concentrations of 1·5–5·4 p.p.m. have been reported for marine particulates from coastal waters off Japan (corresponding to $0·01–0·04\,\mu g\,l^{-1}$) (Okabe and Toyota, 1967).

Preconcentration. No methods having sufficient sensitivity for the direct determination of molybdenum in sea water are available at present, and it is therefore always necessary to preconcentrate the element before analyses. Until recently, this has been usually achieved by coprecipitation or co-crystallization. Ferric hydroxide has been widely used as a coprecipitating agent (Ernst and Hoermann, 1936; Ishibashi, 1953; Ishibashi *et al.*, 1958; Kim and Zeitlin, 1969). Kim and Zeitlin (1971a, b) have used air flotation in the presence of a surface active agent to collect the ferric hydroxide precipitate, and claim to have achieved a 95·3 ± 2·6% recovery of molybdenum in this way. A drawback to the use of ferric hydroxide as a collecting agent is that careful pH control is necessary since the recovery falls off rapidly at pH values above and below the optimum at pH = 4·0 (Chan and Riley, 1966c). In contrast, hydrous manganese dioxide is free from this disadvantage and carries molybdenum quantitatively over the pH range 1·5–5·5 (Chan and Riley, 1966c); procedures using it as a carrier have been described by Sugawara *et al.* (1959), Sugawara and Okabe (1960, 1966), Bachman and Goldman (1964), Chan and Riley (1966c) and Head and Burton (1970). Other coprecipitants which have been used for the collection of molybdenum include mercuric sulphide (Noddack and Noddack, 1940), magnesium hydroxide at pH 10 (Kim and Zeitlin, 1969) and thorium hydroxide at pH 6, with which Kim and Zeitlin (1970) achieved a molybdenum recovery of > 99%. Co-crystallization techniques based on the use of tannin + thionalide + oxine in the presence of iron (III) have been employed by Black and Mitchell (1951), Young *et al.* (1959) and Kim and Zeitlin (1968). Weiss and Lai (1961) have co-crystallized molybdenum by means of α-benzoin oxime; the chemical yield being monitored radiochemically with ^{99}Mo.

Several workers have concentrated molybdenum from sea water by means of ion exchange or column chromatographic procedures. The adsorptive phases which have been used for this purpose include the thiocyanate form of diethylaminoethyl cellulose (Kawabuchi and Kuroda, 1969), Chelex-100 (Riley and Taylor, 1968a), p-aminobenzyl cellulose, diethylamino cellulose and chitosan (Muzzarelli and Roechetti, 1973). Preconcentration of molybdenum has also been carried out by solvent extraction (Zhavoronkina, 1960; Brooks, 1965).

Determination of dissolved molybdenum. The determination of molybdenum in concentrates from sea water using spectrographic techniques (see e.g. Noddack and Noddack, 1940; Black and Mitchell, 1951; Young et al., 1959; Zhavoronkina, 1960; Brooks, 1965), which are at best semiquantitative has, in recent years, been almost entirely superseded by the use of colorimetric methods. Two chromogenic reagents have been used for this purpose. Ishibashi (1953) pioneered application of a method in which an acidic solution of the concentrate is treated with stannous chloride and thiocyanate ion. The intensely red coloured molybdenum (V)-thiocyanate complex which is produced is extracted into butyl acetate for spectrophotometry. With minor modifications this technique has been adopted by several other workers (Sugawara et al., 1959; Sugawara and Okabe, 1960, 1966; Weiss and Lai, 1961; Bachman and Goldman, 1964; Kim and Zeitlin, 1968, 1970). The reaction involved is kinetically and mechanistically complex, and critical control of the several variables concerned (e.g. pH, thiocyanate concentration and reaction time) is necessary if reproducible results are to be achieved. In addition, it is somewhat lacking in sensitivity (see e.g. Sandell, 1959). Molybdenum (VI) with toluene-3,4-dithiol(dithiol) yields an intense blue-green complex which can be extracted with polar solvents (e.g. butyl acetate) for photometry (Sandell, 1959). This reaction, which is quite selective for molybdenum (VI) with toluene-3,4-dithiol (dithiol) yields an intense blue-green Difficulties which arose in this technique from interference by iron and copper have been overcome by modifications introduced by Chan and Riley (1966c) and Kawabuchi and Kuroda (1969). Head and Burton (1970) have obtained a coefficient of variation of $\pm 1.7\%$ for the method of Chan and Riley.

Although conventional atomic absorption spectrophotometry has only a poor sensitivity for molybdenum, this can be increased by at least an order of magnitude if a heated graphite atomizer is used. Muzzarelli and Roechetti (1973) have claimed a coefficient of variation of 1.3% for a method in which the latter method of atomization was employed for the examination of concentrates prepared by adsorption on *p*-aminobenzyl cellulose.

Determination of particulate molybdenum. A colorimetric technique for the determination of molybdenum in marine particulate matter has been described by Okabe and Toyota (1967).

19.10.5.35. *Neodymium.* See lanthanides.

19.10.5.36. *Neptunium.* See Section 19.13.

19.10.5.37. *Nickel*

The concentration of dissolved nickel in ocean water appears to lie in the range 0.5–$3\,\mu g\,l^{-1}$. However, considerably higher levels may occur in coastal waters. The principal species of dissolved nickel appears to be hydrated Ni^{2+}, but significant amounts of $NiCO_3^0$ may also perhaps be present.

Preconcentration. With most analytical techniques it is necessary to concentrate nickel before it can be determined in sea water. Until recently the preconcentration has usually been carried out by coprecipitation; carriers which have been used for this purpose include ferric hydroxide (Ishibashi, 1953; Burrell, 1965) and the mixed precipitate of magnesium and calcium compounds produced by the addition of sodium carbonate (Laevastu and Thompson, 1956; Forster and Zeitlin, 1966a, b). These time-consuming and somewhat inefficient procedures have now been largely superseded by techniques based on solvent extraction (see Table 19.17) or chelating ion exchange (Riley and Taylor, 1968; Abdullah and Royle, 1972) which are rapid and quantitative.

Determination of dissolved nickel. Most of the early determinations of nickel in sea water were carried out by spectrographic examination of concentrates prepared by coprecipitation (Ernst and Hoermann, 1936; Noddack and

TABLE 19.17

Solvent extraction systems for analytical preconcentration of nickel from sea water

Complexing agent	Solvent	Method of determination	Reference
Ammonium pyrrolidine dithiocarbamate	methyl *iso*-butyl ketone	AAS	Brooks *et al.* (1967a, b); Brewer *et al.* (1969) Spencer and Brewer (1969); Preston *et al.* (1972)
		XRF	Morris (1968)
Sodium diethyldithiocarbamate	chloroform	S	Zhavoronkina (1960)
		P	Tikhonov and Shalimov (1965)
Dimethylglyoxime	chloroform	C	Kentner *et al.* (1969)
		AAS	Rampon and Cuvelier (1972)
8-hydroxyquinoline	chloroform	XRF	Armitage and Zeitlin (1971)
pyridine-2-aldehyde-2-quinolyl hydrazone	benzene	C	Afghan and Ryan (1968)

AAS = atomic absorption spectrophotometry; C = colorimetry; P = polarography; XRF = X-ray fluorimetry; S = emission spectrography.

Noddack, 1940; Black and Mitchell, 1951; Zhavoronkina, 1960). This technique is, at best, semi-quantitative and the results obtained with it are probably mainly of historical interest. Because of its relatively high sensitivity and selectivity, the colorimetric dimethylglyoxime method (Sandell, 1959) has been used by several workers for the determination of nickel following preconcentration, usually by coprecipitation (Ishibashi, 1953; Laevastu and Thompson, 1956; Forster and Zeitlin, 1966b; Kentner et al., 1969). The latter workers have claimed a coefficient of variation of ca. 7% at the 0·5 μg Ni l⁻¹ level for their method. Research in the last decade has led to the discovery of a number of selective chromogenic reagents for nickel having sensitivities 2–3 times that of dimethylglyoxime. Quinozaline-2,3-dithiol has been employed by Forster and Zeitlin (1966a) for the analysis of pre-concentrates from sea water. Afghan and Ryan (1968) have described a technique for the direct determination of nickel in sea water in which the element is extracted for photometry as its intensely coloured complex with pyridine-2-aldehyde-2-quinolyl hydrazone ($\varepsilon = 6\cdot7 \times 10^4$ at 515 nm). No interference was reported from Cd, Co, Fe, Hg and Mn in 100-fold excess. A coefficient of variation of 2·5% was achieved at a level of 2·3 μg Ni l⁻¹.

In recent years, colorimetric methods for the determination of nickel in sea water have been largely superseded by atomic absorption spectrophotometric procedures. Although the latter still require the preconcentration of the nickel, they are considerably faster, and several elements can be determined in the same extract. Both solvent extraction (see Table 19.17) and chelating ion exchange (Riley and Taylor, 1968) have been used for this preliminary concentration. The application of X-ray fluorescence methods to the determination of nickel after concentration from sea water has been described by Morris (1968) and Armitage and Zeitlin (1971). However, such methods appear to have no advantage over atomic absorption spectrophotometry, and require much more expensive and elaborate equipment.

Two electrochemical techniques have been applied to the determination of nickel in sea water. Both normal (Tikhonov and Shalimov, 1965) and pulse (Abdullah and Royle, 1972) polarography have been used with a preliminary concentration stage. Smith and Redmond (1971) have investigated the possibility of determining the element in sea water by anodic stripping voltammetry. However, they found that this was not practicable owing to the formation of Ni–Hg compounds. In addition, the peak from nickel overlaps that of zinc. (See Chapter 20 for a more extensive discussion of the use of electrochemical methods for the determination of nickel in sea water.)

On irradiation with thermal neutrons nickel-64 (isotopic abundance 1·16%; $\sigma = 1\cdot6$ barns) undergoes the reaction $^{64}Ni(n, \gamma)^{65}Ni$. The use of this reaction in the direct determination of nickel in sea water is difficult owing to the short half life (2·56 h) of the nickel-65 produced. However, neutron activation of freeze-dried sea salts has been used for this purpose by Schutz and Turekian (1965a, b).

Determination of particulate nickel. Both colorimetry (Yatsimirski *et al.*, 1970) and atomic absorption spectrophotometry (Brooks *et al.*, 1967a; Preston *et al.*, 1972) have been used for the determination of nickel following acid leaching of marine particulates.

19.10.5.38. *Niobium*

The concentration of niobium in sea water has been reported to be < 0.05 $\mu g \, l^{-1}$. (Carlisle and Hummerstone, 1958.) It is probable that, in conformity with what is known of its chemistry, the element is not present in a dissolved form as it is reported to be completely removed by filtration.

Preconcentration. Niobium has been preconcentrated by coprecipitation with ferric hydroxide (Carlisle and Hummsterstone, 1958).

Determination of niobium. For the determination of niobium in sea water, Carlisle and Hummerstone have described a method in which, following coprecipitation with ferric hydroxide and dissolution of the precipitate in hydrofluoric acid, niobium was separated by paper chromatography. Methyl ethyl ketone containing 15% v/v of 40% hydrofluoric acid was used for development. The niobium spots were rendered visible by means of tannic acid and compared with those formed by known weights of the element. The chemical yield of the process, which ranged from > 50–80%, was determined radiochemically using ^{95}Nb.

19.10.5.39. *Osmium*

The osmium concentration of sea water has been reported to be $1 \, ng \, l^{-1}$ (Sharma and Parekh, 1968); the speciation of the element has not been established.

Preconcentration. No methods for the preconcentration of osmium from sea water have yet been described.

Determination of dissolved osmium. Sharma *et al.* (1966) have described an indirect method for the determination of osmium in sea water. This is based on the strong catalytic effect which osmium exerts on the reaction between Ce^{4+} and As^{3+}. The concentration of osmium present is assessed from a measurement of the change in the rate of the reaction. The reaction is also catalyzed by the iodide ion (which can be determined in the same way, see p. 362). This interfering ion is removed by ion exchange using a strongly acidic cation (sic) exchanger (see also Sharma and Parekh, 1967).

19.10.5.40. *Perchlorate*

Despite claims by Baas Becking *et al.* (1958), the presence of the perchlorate ion in sea water does not seem to have been established. No perchlorate could be detected by Greenhalgh and Riley (1961a) using a method having a sensitivity of 30 µg ClO_4^- l^{-1} in which the perchlorate ion was coprecipitated with $(C_6H_5)_4AsReO_4$ and determined by infra-red spectrophotometry of the ClO_4^- band. Johannesson (1962) has described an isotope dilution technique for the determination of perchlorate in natural waters. In this, the sample is spiked with a known weight of radioactive $K^{36}ClO_4$ and cations are removed by ion exchange. Potassium chloride is then added, the solution is evaporated and the resultant potassium perchlorate is weighed and its activity is counted with a Geiger counter.

19.10.5.41. *Platinum*

Platinum does not yet appear to have been determined in sea water. Work in the author's laboratory (K. Whitelaw, unpublished) suggests that its concentration is probably <0.5 ng l^{-1}.

19.10.5.42. *Plutonium.* See Section 19.13.

19.10.5.43. *Polonium.* See Section 19.12.2.3.

19.10.5.44. *Praseodymium.* See Lanthanides.

19.10.5.45. *Protactinium.* See Section 19.12.2.4.

19.10.5.46. *Radium.* See Sections 19.12.2.5 and 19.12.2.6.

19.10.5.47. *Radon.* See Section 19.12.2.7.

19.10.5.48. *Rare earth elements.* See Lanthanides.

19.10.5.49. *Rhenium*

The concentration of rhenium in sea water is ca. 10 ng l^{-1}. The element is probably present as the very stable perrhenate ion (Scadden, 1969).

Preconcentration. Rhenium has been concentrated from sea water by both anion exchange (Matthews and Riley, 1970c) and solvent extraction with *cyclo*-hexanone after making it 1M with respect to hydrochloric acid (Scadden 1968, 1969).

Determination of dissolved rhenium. Natural rhenium consists of 37·1% of ^{185}Re ($\sigma = 104$ barns) and 62·9% of ^{187}Re ($\sigma = 75$ barns). On irradiation with thermal neutrons these nuclides give rise to ^{186}Re ($t_{\frac{1}{2}}$, 89 h) and ^{188}Re ($t_{\frac{1}{2}}$, 16·7 h) respectively. Because of the high abundances and neutron capture cross sections of the parent nuclides neutron activation analysis is an extremely sensitive analytical technique for the element. This method has been used for the determination of rhenium in sea water following its preconcentration (Scadden, 1968, 1969; Matthews and Riley, 1970c; Olafsson and Riley, 1972). The latter workers achieved a sensitivity of 0·3 ng Re l^{-1}.

19.10.5.50. *Rhodium*
Rhodium does not yet appear to have been detected in sea water.

19.10.5.51. *Rubidium*
Recent data suggests that rubidium probably behaves conservatively and that its concentration is probably ca. 120 µg l^{-1}. The element probably occurs as the hydrated Rb$^+$ ion. Little is known about the association of the element with marine particulates.

Preconcentration. Rubidium has been preconcentrated from sea water by coprecipitation with potassium cobaltinitrite (Kovaleva and Bukser, 1940; Ishibashi and Hara, 1955, 1959) and, more recently with potassium tetra-phenylboron (Bolter *et al.*, 1964; Murakami and Uesugi, 1967) or $K_2AgBi(NO_2)_6$ (Bolter *et al.*, 1964). A more effective separation can be achieved by cation exchange techniques (Riley and Tongudai, 1966b; Murakami and Katsuya, 1967).

Determination of dissolved rubidium. Most of the data on the occurrence of rubidium in sea water published before 1955 (see Morozov (1968) and Riley and Tongudai (1966b) for summaries of earlier work) is so much at variance with currently accepted values that they can only be considered of historical interest. Accurate analysis for rubidium became possible only in the mid-1950's as a result of the development of precise physico-chemical techniques. Since then three main approaches to the problem have been used.

Irradiation of rubidium-85 (isotopic abundance 72·2%; $\sigma = 0·8$ barns) with thermal neutrons leads to the production of rubidium-86. This radionuclide, which has a half-life of 18·7 days, decays with the emission 1·78 MeV β-particles and 1·08 MeV γ-rays. This nuclear reaction provides a sensitive means for the determination of rubidium, and was first applied to the direct determination of the element in sea water by Smales and Salmon (1955). Subsequently, the neutron activation technique has been adopted by a number of other workers, who, after irradiation of the freeze dried salts counted the induced ^{86}Rb activity either after a radio-chemical separation (Bolter et al., 1964; Schutz and Turekian, 1965b), or after a suitable cooling period by direct γ-spectrometry (Piper and Goles, 1969; Spencer et al., 1970b). Piper and Goles (loc. cit.) have claimed a coefficient of variation of 4% for their technique.

A number of different optical spectrometric techniques have been employed for the determination of rubidium in sea water. Emission spectrography has been used both with dried sea salts (Borovik-Romanova, 1944) and with concentrates (Riley and Tongudai, 1966b; Schoenfeld and Held, 1969); however, it is difficult to attain satisfactory precision because of the irreproducible nature of arc excitation. Better precision can be attained by flame emission (FES) and atomic absorption (AAS) spectrophotometry, and these techniques have been used for the direct determination of rubidium in sea water by several workers (FES: Smith et al., 1965; Morozov, 1968, 1969; AAS: Fabricand et al., 1966; Friedman et al., 1968). Difficulties associated with the atomization of sea water and the fact that the sensitive resonance lines of rubidium (7800 and 7948 Å) are in a spectral region in which photomultipliers have a rather poor sensitivity, limit the precision attainable in such procedures to ca. 5% at best. For these reasons it is probably preferable to preconcentrate the rubidium before applying these techniques (see e.g. Murakami and Uesugi, 1967).

The stable isotope dilution technique was first applied to the determination of rubidium in sea water by Smales and Webster (1957, 1958), who confirmed the results which had been obtained previously by neutron activation analysis (Smales and Salmon, 1955). A similar procedure which has been used more recently by Smith et al. (1965) gave a coefficient of variation of $\pm 5\%$. The results obtained with it agreed, on average, to within $\pm 4\%$ of those obtained by flame spectrophotometry.

Determination of particulate rubidium. According to Piper and Goles (1969), neutron activation analysis can be used for the determination of rubidium in marine particulates.

19.10.5.52. Ruthenium

Dixon et al. (1966a) have reported that the concentration of dissolved ruthenium in sea water is variable in the range $0.15-1.4\,\mathrm{ng\,l^{-1}}$ (av. $0.74\,\mathrm{ng\,l^{-1}}$). A concentration of $0.04-0.09\,\mathrm{ng\,l^{-1}}$ (av. $0.07\,\mathrm{ng\,l^{-1}}$) was recorded for particulate forms of the element.

Preconcentration. Dixon et al. (1966a, b) have concentrated ruthenium from sea water (750 litres) by means of a double coprecipitation with ferric hydroxide. The carrier iron was removed by extraction of its chloro-complexes with di-*iso*-propyl ether, and ruthenium was purified by distillation of its tetroxide.

Determination of dissolved ruthenium. Of the currently available methods only neutron activation analysis has sufficient sensitivity for the determination of ruthenium after preconcentration from reasonable volumes of sea water. This method has been used by both Rose (1968) and Dixon et al. (1966a, b). In the procedure used by the latter workers the $^{97}\mathrm{Ru}$ ($t_{\frac{1}{2}}$, 2.9 days) produced by irradiation of $^{96}\mathrm{Ru}$ (isotopic abundance 5.5%) is purified by distillation as $\mathrm{RuO_4}$ and its $0.216\,\mathrm{MeV}$ γ-activity is measured with a γ-spectrometer; an overall precision of $\pm 12\%$ and a sensitivity of $0.08\,\mathrm{\mu g}$ of ruthenium was claimed.

Determination of particulate ruthenium. Neutron activation analysis has been applied to the determination of ruthenium in marine particulates by Dixon et al. (1966a).

19.10.5.53. *Samarium.* See Lanthanides.

19.10.5.54. *Scandium.* See also Lanthanides

The concentration of dissolved scandium in sea water has been reported to lie in the range $0.1-1.4\,\mathrm{ng\,l^{-1}}$; the element probably exists in the form of hydroxy species, such as $\mathrm{Sc(OH)_3}$. Little is known about the association of the element with marine particulates. Because of its chemical similarity to the lanthanides it has been determined in many of the investigations concerned with the distribution of these elements in the sea (*vide supra*). However, because of the extremely high sensitivity of neutron activation analysis resulting from the high capture cross section ($\sigma = 12$ barns) of $^{45}\mathrm{Sc}$ and the fact that element is mono-isotopic, it is possible to determine it by direct γ-spectrometry of the $^{46}\mathrm{Sc}$ ($t_{\frac{1}{2}}$, 84 days) produced when sea salts are irradiated (Robertson et al., 1968; Piper and Goles, 1969; Spencer et al., 1970b; Brewer et al., 1972). Chau and Wong (1968) have attempted to determine dissolved scandium in sea water by atomic absorption spectrophotometry following preconcentration by coprecipitation with ferric hydroxide. However, the concentration which they reported (ca. $10\,\mathrm{ng\,l^{-1}}$) for water from the China Sea is about an order of magnitude greater than that found by others, suggesting that there is a systematic error in their procedure.

Determination of particulate scandium. The determination of scandium in marine particulate matter by instrumental neutron activation analysis has been described by Piper and Goles (1969), Spencer *et al.* (1970b) and Brewer *et al.* (1972).

19.10.5.55. *Selenium*

There is considerable divergence between the published values for the concentration of dissolved selenium in sea water; the most likely value appears to be $\sim 0.1\,\mu g\,l^{-1}$. The element is probably present as SeO_3^{2-}. No information appears to be available about the concentration of the element in marine particulates.

Preconcentration. Although a variety of metal hydroxides are capable of coprecipitating selenium (IV) only ferric hydroxide does so efficiently (see Chan and Riley, 1965). This coprecipitant has been used frequently for preconcentration of selenium from sea water (Goldschmidt and Strock, 1935; Ishibashi, 1953; Chan and Riley, 1965). For maximum recoveries the pH should lie in the range 4–6 (Chan and Riley, 1965; Ishibashi *et al.*, 1967). Under these conditions selenium (VI) is carried only to an extent of ca. 2.5%. Selenium has also been concentrated from sea water by converting it to 5-nitropiazoselenol, through the action of 4-nitro-1,2-diaminobenzene, and then extracting with toluene (Shimoishi, 1973).

Determination of dissolved selenium. Several different physico-chemical techniques have been applied to the determination of selenium in sea water, almost always after preconcentration. In earlier work, colorimetric methods were used; these involved either reduction to the elemental state (Goldschmidt and Strock, 1935) or reaction with the chromogenic reagents thiourea (Ishibashi, 1953) and diaminobenzidine (Chan and Riley, 1965). However, their sensitivity is barely adequate unless unreasonably large samples are taken. A considerable improvement in sensitivity can be achieved by reaction of the selenium with 1,2-diamino aromatic compounds to give piazoselenol derivatives. After solvent extraction, these can be determined by either electron capture gas chromatography (Shimoishi, 1973; see above), or spectro-fluorimetry (K. Whitelaw, private communication).

A neutron activation procedure for the determination of selenium in sea water has been described by Schutz and Turekian (1965a). In this, the freeze dried salts are irradiated with thermal neutrons. After a few weeks cooling, selenium is separated radiochemically and the activity of ^{75}Se ($t_{\frac{1}{2}}$, 121d), arising from ^{74}Se (isotopic abundance 0·87%, $\sigma = 50$ barns), is counted.

19.10.5.56. Silver

The concentration of dissolved silver in ocean water probably lies in the range 0·01–0·08 µg l^{-1}. Considerably higher concentrations (up to at least 0·5 µg l^{-1}) may occur in coastal waters. The principal dissolved silver species appears to be $AgCl_2^-$. Little information is available about the occurrence of the element in oceanic particulate matter, but it has been found that in some coastal waters, more than half the total silver content is associated with the particulate phases (Preston et al., 1972).

Preconcentration. In early work, silver was concentrated by coprecipitation with the sulphides of mercury (Noddack and Noddack, 1940) or lead (Haber, 1928), and this mode of separation has been used even in recent times (Fukai and Huynh-Ngoc, 1971). Lai and Weiss (1962) have claimed that silver can be recovered from sea water in 92% yield by co-crystallization at pH 3·5–4·0 with thionalide (see also Robertson, 1971). Concentration of silver has also been carried out using copper dithizonate-chloroform (Soyer, 1963) and ammonium pyrrolidine dithiocarbamate (Preston et al., 1972). Both Davankov et al. (1962) and Chao et al. (1969) have attempted to adsorb silver as its chloro anion using an anion exchanger. However, Kawabuchi and Riley (1973) have shown that it is difficult to ensure quantitative uptake by the resin even if the sample is acidified to 0·5 M with respect to hydrochloric acid. They were able to recover the element completely if the sea water was first rendered 0·1 M and 0·05 M in hydrochloric acid and ammonium thiocyanate respectively. Elution was performed with 0·4 M thiourea. Au(III), Cu(II), Pd(II) and Pt(VI) were also concentrated with efficiencies of $> 50\%$. It is impracticable to preconcentrate silver using chelating ion exchange procedures, since although the element is quantitatively adsorbed it is not possible to elute it quantitatively.

Determination of dissolved silver. Only the neutron activation technique has sufficient sensitivity to determine silver in sea water without preliminary chemical preconcentration. Although much greater sensitivity could be theoretically attained in this technique by using the ^{107}Ag(n, γ) ^{108}Ag reaction, the very short half life (2·3 min) of the ^{108}Ag nuclide makes this infeasible. Most workers have therefore preferred to make use of the induced activity of ^{110m}Ag ($t_{\frac{1}{2}}$, 253 days) arising from ^{109}Ag (isotopic abundance 48·6%, $\sigma = 84$ barns) via the reaction ^{109}Ag$(n, \gamma)^{110}$Ag. This reaction was first used for the direct examination of freeze dried sea-salts by Schutz and Turekian (1965a, b). In order to increase the sensitivity other workers have preferred to irradiate concentrates prepared from the sea water (Robertson, 1971; Kawabuchi and Riley, 1973).

Both dithizone (Soyer, 1963; Fukai and Huynh-Ngoc, 1971) and p-dimethylaminobenzylidene rhodanine (Lai and Weiss, 1962) have been used as chromogenic reagents for the determination of silver following preconcentration from sea water. In addition, Lai and Weiss (loc. cit.) have determined silver in concentrates by making use of the catalytic action of silver on the reaction between Mn^{2+} and persulphate ion which leads to the production of permanganate ion which is estimated photometrically.

Although conventional atomic absorption spectrophotometry is too insensitive to be used for the direct determination of silver in normal sea waters, it has been employed in antifouling studies for the examination of silver-enriched waters (Carr, 1969). Similarly, the normal flame atomization technique is insufficiently sensitive for examination of concentrates prepared by direct solvent extraction; however, the much more sensitive tantalum boat (Preston et al., 1972) and carbon tube furnace (Segar and Gonzalez, 1972) techniques have been used for this purpose.

Determination of particulate silver. Preston et al. (1972) have described a method for the determination of particulate silver in which the filter is leached with hydrochloric acid. Silver is determined in the digest by atomic absorption spectrophotometry using the tantalum boat technique. Belyaev and Gordeev (1972) has developed an atomic absorption technique for the determination of silver (sensitivity 30 pg) in which the filter bearing the particulate material is vaporized electrically on a hollow carbon electrode. Spencer et al. (1972) determined silver in marine particulate matter by irradiating the pelletized filter with thermal neutrons, and after prolonged cooling measuring the ^{110m}Ag activity by γ-spectrometry.

19.10.5.57. Tantalum

According to Hamaguchi et al. (1963) the concentration of tantalum in sea water is $0.02 \mu g l^{-1}$. Nothing is known of the form in which it is present.

Preconcentration and determination. The only investigation of the occurrence of tantalum in sea water which has been made so far is that by Hamaguchi et al. (1963). In this, the element was coprecipitated from 10 l samples using ferric hydroxide and the washed precipitate was irradiated for 12 days in a thermal neutron flux of $3 \times 10^{11} n cm^{-2} s^{-1}$. After cooling, tantalum was separated by solvent extraction from an HF/H_2SO_4 medium and the ^{182m}Ta ($t_{\frac{1}{2}}$, 115 days) activity was determined with a proportional counter. It should be noted that, as it stands, this technique is barely sensitive enough (absolute sensitivity 0·5 μg Ta) for the determination of tantalum in sea water. However, the sensitivity could obviously be improved by using larger volumes of sample and a higher neutron flux.

19.10.5.58. *Tellurium*

Tellurium does not yet appear to have been determined in sea water, perhaps because of a dearth of sensitive methods for its determination.

19.10.5.59. *Terbium*. See Lanthanides.

19.10.5.60. *Thallium*

The concentration of dissolved thallium in sea water is ca. $10 \, \text{ng} \, l^{-1}$. Nothing is known about the speciation of the element, or its association with particulate matter.

Preconcentration and determination of thallium. Matthews and Riley (1969), following a suggestion by Brooks (1960), have developed an anion exchange procedure for the concentration of thallium from sea water. This is based on the fact when thallium is oxidized (with bromine) in an acidic chloride medium a stable chloro-anion of Tl(III) is produced which is very strongly adsorbed by anion exchangers. Elution of the exchanger with $0·5 \, \text{M}$ nitric acid will remove all other elements (except gold) which form adsorbable chloro-anions. Thallium is then eluted with sulphur dioxide water and recovered by evaporation.

To determine thallium in the concentrate, Matthews and Riley (1969) irradiated it with thermal neutrons. After cooling, thallium was separated by a multistage radiochemical process, and the induced ^{204}Tl ($t_{\frac{1}{2}}$, 4·1 years) activity was measured by β-counting. The method gave a coefficient of variation of $\pm 5\%$ at a level of $19 \, \text{ng} \, \text{Tl} \, l^{-1}$.

Florence (1972) has assessed the possibility of using anodic stripping voltammetry for the direct determination of thallium in sea water (see Section 20.5.3).

19.10.5.61. *Thiosulphate*

Although no detectable amounts of thiosulphate ion are present in normal sea water, the ion does occur in the transition zone above anoxic waters where it is produced as an intermediate in the oxidation of hydrogen sulphide ion (see Chapter 16). Grasshoff (1970c) has described a titrimetric technique having a sensitivity of $32 \, \mu$ equiv $S_2O_3^{2-} \, l^{-1}$ for its determination in such waters. Immediately after collection, the sample is treated with cadmium acetate solution taking the precautions that are used in the Winkler oxygen method. After standing for a few hours, the precipitated cadmium sulphide is filtered off, thus removing sulphide which interferes in the determination. Thiosulphate is then determined iodometrically in the filtrate.

Thorium. See Section 19.12.2.8

19.10.5.62. *Thulium.* See Lanthanides

19.10.5.63. *Tin*

There is a considerable divergence between published values for the concentration of tin in sea water. Most values lie in the range $0.6–3 \, \mu g \, l^{-1}$ (see e.g. Noddack and Noddack, 1940; Black and Mitchell, 1951; Shimizu and Ogata, 1963; Hamaguchi *et al.*, 1964; Kodama and Tsubota, 1971). However, much lower concentrations ($0.01–0.03 \, \mu g \, l^{-1}$) have been found by Smith and Burton (1972) who took particular care to avoid contamination of samples with this ubiquitous metal, both during collection and analysis. In view of the abundance and geochemical behaviour of tin it seems probable that more reliance is to be placed on these lower values. Dissolved tin probably exists in sea water as hydroxy species, e.g. $SnO(OH)_3^-$.

Preconcentration. Tin has been concentrated from sea water by coprecipitation with mercuric sulphide (Noddack and Noddack, 1940) and ferric hydroxide (Shimuzu and Ogata, 1963). Smith and Burton (1972) adsorbed tin from the sample at pH 9.6 ± 0.1 by pumping it through a column packed with high purity silica wool; elution was performed using dilute sulphuric acid. Recoveries of tin were found to lie in the range 67–82%. Kodama and Tsubota (1971) have described a concentration procedure in which tin is taken up as a chloro complex from 2м-hydrochloric acid medium by means of an anion exchange resin.

Determination of dissolved tin. No methods are available having sufficient sensitivity for the direct determination of tin in sea water,* and it is always necessary to use a preconcentration stage in the analysis. Emission spectrographic techniques which were used in the earliest investigations (Noddack and Noddack, 1940, Black and Mitchell, 1951) are only marginally sensitive enough, and the results obtained with them are probably only of historical interest. Considerably greater sensitivity can be obtained by colorimetric methods employing pyrocatechol violet (Kodama and Tsubota, 1971) and

* Smith and Redmond (1971) have studied the possibility of determining tin directly in sea water by anodic stripping voltammetry, but have concluded that it is not feasible.

phenylfluorone (Shimizu and Ogata, 1963; Smith and Burton, 1972). Hama-guchi *et al.* (1964) have used a neutron activation method for the estimation of tin in a concentrate prepared by coprecipitation.

19.10.5.64. *Titanium*

There is considerable uncertainty about the concentration of dissolved titanium in sea water. In view of what is known about the geochemical reactivity of the element it seems likely that the published values, which are of the order of $\mu g \, l^{-1}$, (Black and Mitchell, 1951; Griel and Robinson, 1952) are too high, perhaps because of contamination. Nothing is known about the speciation of the element in the sea.

Preconcentration. Titanium has been preconcentrated from sea water by coprecipitation with aluminium hydroxide (Griel and Robinson, 1952) and with the precipitate obtained by treating iron (III) with tannin, 8-hydroxy-quinoline and thionalide (Black and Mitchell, 1951).

Determination of dissolved titanium. Titanium has been determined, after preconcentration, by emission spectroscopy (Black and Mitchell, 1951) and also by colorimetry using thymol as a chromogenic reagent (Griel and Robinson, 1952). The results obtained by the latter workers are not convincing because the concentration of titanium found was less than the standard deviation of the method.

Determination of particulate titanium. Cann and Winter (1971) have described a non-destructive procedure with a precision of $\pm 6\%$ for the determination of particulate titanium. In this, the membrane filter bearing the particulate matter was examined by a thin film X-ray fluorescence technique.

19.10.5.65. *Tungsten*

Tungsten probably occurs in sea water as the WO_4^{2-} ion at a concentration of $\sim 0.10 \, \mu g \, W \, l^{-1}$. Nothing appears to have been published on the association of tungsten with marine particulate material.

Preconcentration. The coprecipitation of tungsten by ferric hydroxide has been studied by Ishibashi (1953) and Ishibashi *et al.* (1954; 1960c) who showed that coprecipitation is complete in the pH range 4–7.5; above and below these limits the efficiency with which the element is carried declines markedly. Hydroxides of other metals are less effective as carriers, thus Chan and Riley (1967) obtained recoveries of only ca. 70% with the hydroxides of aluminium and titanium under the optimum conditions of pH. Hydrous manganese dioxide is a more selective coprecipititant for tungsten, and recoveries of 97–98% can be achieved in the pH range 2·0–4·9 (Chan and Riley, 1967). Neither thionalide nor 5,7-dibromo-8-hydroxyquinoline are effective co-crystallizing agents for tungsten.

Determination of dissolved tungsten. Ishibashi *et al.* (1954) have described a photometric procedure for the determination of tungsten following co-precipitation with ferric hydroxide. In this, the precipitate was fused with fusion mixture and tungsten and molybdenum were leached from the fused residue with water. Most of the molybdenum was removed by precipitation as sulphide in the presence of citric acid. Tungsten was determined photo-metrically with dithiol (toluene-3,4-dithiol); molybdenum also reacts with the reagent and correction must be made for residual traces of it. Chan and Riley (1967) have found that losses of tungsten of ca. 20% occur in this process through failure to leach it completely after the fusion and because it coprecipi-tates with molybdenum sulphide. In order to surmount these difficulties Chan and Riley (*loc. cit.*) coprecipitated tungsten with hydrous manganese dioxide, and after removal of manganese by ion exchange, separated molybdenum by extracting its dithiol complex using citrate as a complexing agent for tungsten. After destruction of the citrate, tungsten was determined by extraction and photometry of its dithiol complex. Recoveries of tungsten were found to be 94–95%.

19.10.5.66. *Uranium.* (See also Table 18.2)

Uranium will be discussed in the present section, since its concentration in sea water, unlike those of other naturally occurring radioactive elements, is sufficiently high (ca. $3\cdot4\,\mu g\,l^{-1}$)* for it to be determined by chemical techniques. Dissolved uranium is probably present in sea water as carbonato complexes, perhaps $UO_2(CO_3)_3^{4-}$ (Takai and Yamabe, 1971).

Preconcentration. The preconcentration of uranium from sea water has been reviewed by Yamabe and Nobuhara (1969). Procedures which have been used analytically include coprecipitation, ion exchange, solvent extraction and foam separation. The first of these techniques is the one which has been employed most extensively. Ogata and Kakihana (1969) and Ogata (1968) have investigated the coprecipitation of uranium from sea water by 34 insoluble compounds over the pH range 3–9; of these, the hydroxides of Fe (III), Al and Ti (IV) were the most effective, and carried the element with an efficiency exceeding 85%. Following its initial use by Hernegger and

* Although uranium appears to be a conservative element in the oceans, concentrations of up to $17\,\mu g\,l^{-1}$ have been reported for coastal waters of the Gulf of Mexico (Sackett and Cook, 1969).

Karlick (1935), coprecipitation with ferric hydroxide has been used by many workers for the preconcentration of uranium (Føyn et al., 1939; Koczy, 1949, 1950; Nakanashi, 1952; Rona and Urry, 1952; Ishibashi et al., 1961a; Miyake et al., 1964; Sarma and Krishnamoorthy, 1968; Bhat et al., 1969). Kim and Zeitlin (1971c) have employed foam separation in place of filtration or centrifugation to separate the ferric hydroxide carrier from sea water, but were able to achieve a uranium recovery of only ca. 82%; a similar process using Th as carrier has been proposed by Leung et al. (1972). Ogata and Inoue (1970) and Ogata et al. (1970) have suggested the use of preformed titanium hydroxide for the adsorption of uranium from sea water.

In addition to metal hydroxides, phosphates have also been frequently used for the coprecipitation of uranium from sea water. Aluminium phosphate is particularly satisfactory for this purpose and has been selected by several workers (Wilson et al., 1960; Milner et al., 1961; Torii, 1964; Torii and Murata, 1964; Viswanathan et al., 1965; Blanchard, 1965; Koide and Goldberg, 1965; Veeh, 1968; Hashimoto, 1971). The mixed precipitate of calcium and magnesium phosphate obtained by treatment of the sample with NaH_2PO_4 has also been found to be an efficient carrier for the element (Yamabe and Takai, 1970).

Uranium can be efficiently preconcentrated by co-crystallization. Carriers which have been used for this purpose include methyl violet thiocyanate (Kuznetsov and Akimova, 1958) and 1-nitroso-2-naphthol (Weiss et al., 1961).

Preconcentration of uranium has been carried out using a number of solvent extraction systems including 8-hydroxyquinoline-chloroform (Wilson et al., 1960; Milner et al., 1961), dibutylphosphoric acid-kerosene (Streeton, 1953) or CCl_4 (Stewart and Bentley, 1954), bis-(2-ethylhexyl)phosphoric acid-CCl_4 (Rona et al., 1956; Milner et al., 1961). Extraction with most of these systems tends to be incomplete, necessitating the use of multiple extractions, or the use of a radio-isotope dilution technique employing [237]U to determine the percentage extraction.

Several different ion exchange processes have been employed for the analytical preconcentration of uranium. Although Davankov et al. (1962) and Korkisch et al. (1956) have attempted to adsorb it onto an anion exchanger, perhaps as sulphato complexes, other workers prefer to take up its organic complexes. Thus, Hazan et al. (1965) adsorbed the element onto Dowex-1 from a medium containing 2-methoxy-ethanol and hydrochloric acid. A number of other trace elements are also adsorbed, but many of them can be eluted with a mixture containing 80% methoxyethanol and 20% 3N hydrochloric acid prior to eluting uranium with aqueous 1N hydrochloric acid. Miyake et al. (1966, 1970b) recovered uranium in 99·7% yield by absorbing its complex with cyclo-hexane diamine tetraacetic acid on the chelating ion exchange resin Dowex-A1 and then eluting it with 5 M hydrochloric acid.

Determination of dissolved uranium. In most methods for the determination of uranium in sea water, interference from other elements and/or lack of sensitivity make it essential to preconcentrate the element. In many instances it is also necessary to carry out a further separation stage in order, for example to remove the carrier used for coprecipitation. This separation can be readily achieved by (i) extracting the uranium as its nitrate using ethyl acetate (see e.g. Sandell, 1959), (ii) ion exchange methods (Korkisch, 1969), or (iii) carbonate precipitation methods in which the uranium remains in solution and the diverse ions are precipitated.

Fluorimetric determination. Uranium can be determined with high sensitivity ($<10^{-10}$ g) by fluorimetry. This technique depends on the fact that if uranium is fused with sodium fluoride (with, or without, sodium carbonate as flux) the resultant pellet emits a yellowish fluorescence when irradiated with ultra violet radiation. Complete specifity for uranium can be achieved if the fusion is carried out with a flux having $>90\%$ sodium fluoride and if the fluorescence is excited with 365·0 nm quanta and measured at 554·6 nm. Although visual or photographic techniques were formerly used for the measurement of the fluorescence, these have been entirely superseded by photoelectric methods, and a number of specialized fluorimeters have been developed for this purpose (see e.g. Kennedy, 1950; Fletcher *et al.*, 1950). The fluorescence of the pellets is influenced by many factors; the composition of the flux is of prime importance as are also the temperature and time of fusion and the cooling cycle. Because uptake of atmospheric water vapour enhances the fluorescence, the time elapsing before measurement is also important. It is thus apparent that if reproducible results are to be achieved it is necessary to control closely these many variables (see Centanni *et al.*, 1956). Since many elements decrease or quench the fluorescence of uranium it is necessary to pre-separate the element before carrying out the fusion.

The fluorimetric technique was first applied to the determination of uranium in sea water by Hernegger and Karlick (1935) and has subsequently been adopted by many other workers (Føyn *et al.*, 1939; Koczy, 1949, 1950; Nakanishi, 1952; Rona and Urry, 1952; Smith and Grimaldi, 1954; Grimaldi *et al.*, 1954; Wilson *et al.*, 1960; Sugimura *et al.*, 1964; Torii, 1964; Torii and Murata, 1964; Hazan *et al.*, 1965; Viswanathan *et al.*, 1965, 1968; Ogata and Inoue, 1970; Leung *et al.*, 1972). It should be noted that most of the values which have been obtained using this technique are appreciably (1·5–10 times) lower than that currently accepted for the concentration of uranium in sea water. This discrepancy probably results from a combination of failure to recover uranium completely from sea water and quenching effects arising from incomplete separation of interfering ions. In a very careful investigation the fluorimetric method (in which correction was made for the incompleteness of the recovery) Wilson *et al.* (1960) found that the results agreed very well with those obtained by isotope dilution and polarographic techniques.* However, principally because of the rather irreproducible nature of the fusion process, the precision ($\pm 5\%$) was inferior to that attainable by the other two procedures.

Colorimetric determination. Although there are a number of very sensitive chromogenic reagents for uranium all of them are subject to serious interference from other elements, particularly iron. When such methods are applied to the analysis of sea water it is, therefore, essential that the preconcentration stage should be followed by one in which uranium is separated from such interfering elements. Chromogenic reagents which have been used for the determination of uranium in sea water include dibenzoylmethane (Tatsumoto and Goldberg, 1959; Miyake *et al.*, 1964), Arsenazo III (Miyake *et al.*, 1970a; Davankov *et al.*, 1962) and Rhodamine B (Kim and Zeitlin, 1971c).

Determination by isotope dilution. Because of the very favourable ratio of $^{238}U:^{235}U$ (1:0·0071) in natural uranium, the element can be readily determined by isotope dilution methods and both Rona *et al.* (1956) and Wilson *et al.* (1960) have applied the technique to the analysis of sea water. In both of their procedures the water sample was spiked with a uranium-235 concentrate. After equilibration, uranium was concentrated by solvent extraction and the $^{235}U:^{238}U$ ratio was determined by mass spectrometry. Wilson *et al.* (1960), who carried out an intercomparison with two other methods, considered the the isotope dilution procedure was to be recommended on the grounds of both speed and precision (coefficient of variation $\pm 5\%$).

Polarographic determination. Both conventional (Korkisch *et al.*, 1956; Ishibashi *et al.*, 1961a) and pulse (Milner *et al.*, 1961) polarographic procedures have been used for the determination of uranium in sea water following preconcentration by coprecipitation or solvent extraction. Milner *et al.* (1961) have pointed out that in order to obtain accurate results it is necessary to carry out extensive chemical separations prior to polarography, and to use ^{237}U as a tracer in order to correct for incomplete recovery of uranium.

* In a recent study of uranium content of Mediterranean water by this fluorimetric technique using ^{237}U for the determination of recovery Franchini (1973) has obtained results similar to those obtained by isotope dilution.

When these precautions were taken Wilson *et al.* (1960) were able to achieve a coefficient of variation of 1·4% with the method developed by Milner *et al.* (1961). (For further discussion of the polarographic determination of uranium see Section 20.4.2.)

Determination by α-spectrometry. All three naturally occurring nuclides of uranium decay with the emission of α-particles. Provided that the element is first separated from its daughter products, many of which are also α-emitters, α-counting can be used as a sensitive means for its determination. Since the α-particles from the various nuclides differ considerably in energy ($^{238}U = 4·19$ MeV; $^{235}U = 4·39$ and $4·21$ MeV; $^{234}U = 4·77$ and $4·22$ MeV) there is the possibility of determining isotope abundance ratios by α-spectrometry. Methods depending on α-spectrometry have been used for both the determination of uranium in sea water (Koide and Goldberg, 1965; Sarma and Krishnamoorthy, 1968; Veeh, 1968; Bhat, *et al.*, 1969), and the estimation of the $^{234}U:^{238}U$ ratio (Blanchard, 1965; Koide and Goldberg, 1965; Sarma and Krishnamoorthy, 1968; Bhat *et al.*, 1969; Sackett and Cook, 1969; Miyake *et al.*, 1966, 1970; Somayajulu and Goldberg, 1966). In a typical procedure the sample is spiked with ^{232}U (an artificially produced 5·3 MeV α-emitter of half life 77 years) for the determination of the chemical yield, and uranium is coprecipitated with ferric hydroxide. The precipitate is dissolved in concentrated hydrochloric acid, the uranium is purified by anion exchange procedures and finally electroplated onto a silver disc for α-spectrometry.

Determination by techniques involving neutron irradiation. The two principal uranium isotopes differ considerably in their behaviour on irradiation with thermal neutrons. Uranium-238 (isotopic abundance 99·3%, $\sigma = 2·7$ barns) undergoes the reaction $^{238}U (n, \gamma)^{239}U$. The ^{239}U ($t_{\frac{1}{2}}$, 23·5 min) decays by β-emission to ^{239}Np ($t_{\frac{1}{2}}$, 2·35 days) which in its turn decays, again by β-emission, to ^{239}Pu ($t_{\frac{1}{2}}$, 2·44 yr). This process has been utilized for the determination of uranium in sea water. Robertson *et al.* (1968); Spencer *et al.* (1970) and Brewer *et al.* (1972) have employed direct γ-spectrometry of the irradiated freeze dried salts. However, Bilal *et al.* (1971) preferred to isolate the neptunium activity by counter current ion migration prior to γ-spectrometry for the 0·278 MeV radiation of the ^{239}Np.

In contrast to ^{238}U, uranium-235 (isotopic abundance = 0·71%) when irradiated with neutrons undergoes nuclear fission. The fission fragments produced are mainly radionuclides having atomic masses lying in the range 76–157. This process was first applied to the analysis of sea water by Stewart and Bentley (1954) who irradiated a uranium concentrate in a heavy water reactor and examined the product in a fission fragment counter. Although their results showed a very good precision, they are considerably lower (2·36–1·64 av. 2·49 µg U l^{-1}) than the currently accepted values. Recently, two methods depending on the counting of the tracks of the induced fission products have been adopted for the examination of sea water. In the procedure described by Bertine et al. (1970) the freeze dried sample ($\sim$ 50 mg) was mixed with cellulose and made into flat pellets. These and analogous standards, were wrapped in 10 µm thick polycarbonate film and irradiated with neutrons. Subsequently, the polycarbonate films were removed and, after treatment with alkali, the numbers of holes in them were measured with a digital discharge counter. The coefficient of variation of the method appears to be ca. $\pm 7\%$, and sea water was found to contain 3·31 $\pm$ 0·51 µg U l^{-1}. A considerably better precision ($\pm 3\%$) has been claimed by Hashimoto (1971) who sealed a solution of a concentrate of uranium along with pieces of muscovite in a silica ampoule. The tube, along with similar standards, was irradiated for 15 minutes in a flux of fast neutrons. Subsequently, the muscovite was recovered and, after etching with hydrofluoric acid, a photomicrograph was taken from which the fission track density could be assessed and compared with those of standards.

Determination of particulate uranium. Spencer et al. (1972) have described an instrumental neutron activation procedure for the determination of uranium in marine particulates.

19.10.5.67. *Vanadium*

The concentration of vanadium in sea water, according to most recent work, lies in the range 1–3 µg l^{-1}. It seems probable that the principal dissolved species of the element is $H_2VO_4^-$. Little information is available about the association of vanadium with marine particulate matter.

Preconcentration. Coprecipitation with ferric hydroxide has been frequently used for the concentration of vanadium from sea water (Ernst and Hoermann, 1936; Ishibashi, 1953; Sugawara et al., 1953; Naito and Sugawara, 1957; Ishibashi et al., 1951, 1962a; Sugawara et al., 1963; Mulikovskaya, 1964; Chan and Riley, 1966d). The coprecipitation has been shown to be quantitative over the pH range 3·5–8 (Ishibashi et al., 1964; Chan and Riley, 1966d; Kuwamoto, 1960). Chan and Riley loc. cit. have found that quantitative coprecipitation can also be achieved with chromium (III) hydroxide (at pH 6·5), hydrous manganese dioxide (at pH 3·2–5·3) and the mixed precipitate obtained by addition of Fe^{3+}, tannin, thionalide and 8-hydroxyquinoline (see Black and Mitchell, 1951). Ishibashi et al. (1951) have used iron (III) cupferrate as a carrier for vanadium. Only two solvent extraction systems appear to have been employed for the pre-concentration of vanadium. Zhavoronkina (1960) used a sodium diethyldithiocarbamate-carbon tetra-chloride extraction system for concentrating the element for spectrographic examination. Nishimura *et al.* (1973) reacted the vanadium with 4-(2-pyridyl) azo-resorcinol and extracted the resultant complex with chloroform after converting it to an ion pair with tetradecyl-dimethyl-benzyl ammonium chloride. Extraction of other ions was prevented by complexing them with a complexone and cyanide ion. Both anion exchange and chelating exchange methods have been used for the pre-concentration of vanadium. Kirijima and Kuroda (1972) adsorbed the element as a thiocyanato complex on the thiocyanate form of Dowex-1 and subsequently eluted it with hydrochloric acid; organic forms of vanadium are also said to be retained. Riley and Taylor (1968a) have used the chelating resin Chelex-100 for concentrating vanadium from sea water at pH5; elution was carried out with 2M ammonia solution.

Determination of dissolved vanadium. In the earliest work on the occurrence of vanadium in sea water the element was determined spectrographically following preconcentration (Ernst and Hoermann, 1936; Black and Mitchell, 1951; see also Zhavoronkina, 1960). This technique is, at best, semi-quantitative, and in most recent investigations colorimetric procedures have been used.

Two methods are available for the direct colorimetric determination of vanadium in sea water. Okabe and Moringa (1968, 1969) have developed a procedure based on the catalytic effect which the element exerts on the oxidation of phenylhydrazine-*p*-sulphonic acid by chlorate ion. The *p*-diazobenzene sulphonic acid which is produced is coupled with 1-naphthyl-amine yielding a pink azo dye which is estimated photometrically (see also Tanaka and Awata, 1967). Nishimura *et al.* (1973) have described a technique in which the vanadium is determined photometrically after extraction of the ion pair which is formed between the vanadium-4-(2-pyridyl)-azo-resorcinol complex and a quaternary ammonium ion. Formation of extractable com-plexes of other elements is inhibited by the use of masking agents; however, the method has a salt error, but this is constant over the salinity range 30–40‰. A coefficient of variation of $\pm 3\%$ was claimed at the $3 \,\mu\text{g V l}^{-1}$ level. Other photometric methods for the determination of vanadium necessi-tate a preconcentration stage usually involving coprecipitation with ferric hydroxide. Since iron interferes strongly in most colorimetric procedures it must be removed before carrying out the determination. This can be readily achieved by a cation exchange technique involving elution of vanadium with hydrogen peroxide (Chan and Riley, 1966d). A number of colorimetric methods have been used for the determination.

The technique based on the formation of a phosphovanadotungstic heteropoly acid, which was first employed for the purpose (Ishibashi *et al.*, 1951, 1964) is insufficiently sensitive. 8-Hydroxyquinoline has been used as the chromogenic reagent by Naito and Sugawara (1957), Sugawara *et al.* (1953) and Mulikovskaya (1964). However, it suffers from lack of specificity and this necessitates careful purification of the vanadium before this reagent can be used. Almost complete specificity and fairly high sensitivity for vanadium can be obtained using diaminobenzidine as the chromogenic reagent (Cheng, 1961). A method using this compound has been adopted by Chan and Riley (1966d). This gave a coefficient of variation of $\pm 2 \cdot 8\%$ at a vanadium level of $1 \cdot 8 \ \mu g \, l^{-1}$. Kirijima and Kuroda (1972) use 4-(2-pyridyl-azo-resorcinol for the photometric determination of vanadium following preconcentration by ion exchange.

A few attempts have been made to apply electrochemical methods to the determination of vanadium in sea water. Ishibashi *et al.* (1962a) estimated the element polarographically after preconcentrating it by coprecipitation. Smith and Redmond (1971) have investigated the possibility of determining vanadium directly in sea water by anodic stripping voltammetry, but concluded it was not feasible because of interference by other elements, particularly copper.

19.10.5.68. *Ytterbium*. See Lanthanides.

19.10.5.69. *Yttrium*

Yttrium closely resembles the lanthanides in its chemical properties and for this reason its determination is discussed with that of that group of elements.

19.10.5.70. *Zinc*

The concentration of dissolved zinc in ocean waters appears to lie in the range $0 \cdot 5 - 4 \ \mu g \, l^{-1}$. However, considerably higher concentrations (up to $100 \ \mu g \, l^{-1}$, or even more) may be found in inshore waters, particularly those adjacent to industrialized or highly populated areas. The speciation of dissolved zinc is very sensitive to pH changes; at pH $8 \cdot 2$ the principal species are $ZnOH^+$, hydrated Zn^{2+} and $ZnCO_3^0$ (see Section 3.4.7). Little information is available about the zinc levels in oceanic particulate matter; however, in inshore waters up to 50% of the total zinc may be associated with particulates (see e.g. Preston *et al.*, 1972). Methods of determination of zinc in sea water have been reviewed by Branica (1970) and Rozhanskaya (1970a).

Preconcentration. With most of the currently available methods it is necessary to concentrate zinc before carrying out the determination. Although coprecipitation (Noddack and Noddack, 1940: Rona *et al.*, 1962; Burrell, 1967) and cocrystallization (Black and Mitchell, 1951; Young *et al.*, 1959; Topping, 1969) have been used for the preconcentration of zinc, most recent investigations have used solvent extraction or ion exchange techniques as they are simpler and can often be combined directly with the method of determination. Solvent extraction systems which have been used for this purpose are shown in Table 19.18. Although zinc can be taken up from acidic sea water medium

TABLE 19.18

Solvent extraction systems for the concentration of zinc from sea water

Complexing agent	Solvent	Method of Determination	Reference
Dithizone	carbon tetrachloride	P	Morita (1950, 1953); Chipman *et al.* (1958); Fonselius and Koroleff (1963); Fonselius (1970); Rozhanskaya (1970a, b); Fukai *et al.* (1973)
Sodium diethyldithiocarbamate	carbon tetrachloride	NAA	Rona *et al.* (1962); Slowey *et al.* (1962); Hood and Slowey (1964); Hood (1966); Slowey and Hood (1971)
Ammonium pyrrolidine-dithiocarbamate	methyl *iso*-butyl ketone	AAS	Burrell (1967); Brewer *et al.* (1969), (1972); Orren (1971); Preston *et al.* (1972); Segar and Gonzalez (1972); Paus (1973)
8-hydroxyquinoline*	chloroform	ES	Brooks (1965)
Trifluoroacetyl acetone	toluene–*iso*-butylamine		Lee and Burrell (1972)

P = photometric; NAA = neutron activation analysis; AAS = atomic absorption spectrophotometry; ES = Emission spectroscopy; * – Counter current liquid–liquid extraction.

by anion exchangers as a chloro-anion, this method of concentration does not seem to have been applied in any published methods. Chelating ion exchange provides an efficient means for preconcentration of the element from sea water; exchangers which have been used for this purpose include Chelex-100 (Riley and Taylor, 1968; Butterworth *et al.*, 1972; Abdullah and Royle, 1972) and chitosan (Muzzarelli and Marinelli, 1972). Attempts have been made to concentrate zinc by means of complexing agents adsorbed onto anion exchangers; thus, (Carritt (1965) has examined the uptake of zinc by dithizone adsorbed onto an anion exchanger. Although Topping (1969) found that zinc was quantitatively retained by a column of anion exchanger impregnated with 5,7-dibromo-8-hydroxyquinoline, he was able to elute only about 75% of it with 0·2 N sulphuric acid.

Determination of dissolved zinc. Several different physico-chemical techniques have been employed for the determination of zinc in sea water, usually following preconcentration.

Dithizone and its analogues are the only satisfactory chromogenic reagents known for the colorimetric determination of zinc. Dithizone reacts with zinc in faintly acidic to weakly basic medium yielding an intense purple coloured complex which can be extracted with chloroform or carbon tetrachloride. A number of other metals react in the same way, but their interferences can be prevented, with some loss of sensitivity, by carrying out the extraction at pH 4·0–5·5, using thiosulphate as a masking agent. Dithizone has been used extensively for the determination of zinc in sea water. Some workers (e.g. Fonselius and Koroleff, 1963; Fonselius, 1970) simply extracted the sample with a dilute solution of dithizone in carbon tetrachloride (50 mg l^{-1}) and, after back extraction of the excess reagent with 0·02 N ammonia, determined the zinc dithizonate photometrically. Correction was applied for copper which is also extracted. Fukai *et al.* (1971) prevented interference from this element by removing it by a preliminary extraction at pH 2·0. Other investigators (e.g. Morita, 1950, 1953; Chipman *et al.*, 1958) have preferred to concentrate the element by means of a preliminary extraction with dithizone in CCl$_4$. It was then back extracted with dilute acid (thereby separating it from copper) and finally re-extracted with more dilute dithizone solution at pH 5–6 in the presence of thiosulphate as a masking agent. The zinc dithizonate was finally determined photometrically by the mixed colour method.

Optical emission spectroscopic techniques using either flame or arc excitation are insufficiently sensitive for the determination of zinc in sea water. However, atomic absorption spectrophotometry provides a sensitive and specific means for its determination. Despite claims by Fabricand *et al.* (1962), zinc cannot be determined directly in sea water using conventional atomization procedures, but according to Burrell and Wood (1969), the tantalum boat volatilization technique can be employed for this purpose; its precision is, however, only poor ($\pm 8\%$). Conventional atomic absorption spectrophotometry following preconcentration, is probably the most convenient and precise method at present available for the determination of zinc in sea water, and has been adopted by many recent workers (see e.g. Burrell, 1966, 1967; Riley and Taylor, 1968a; Brewer *et al.*, 1969; Orren, 1971; Preston *et al.*, 1972). Segar and Gonzalez (1972) and Paus (1973) have employed the carbon tube atomizer for the determination of zinc in concentrates prepared by solvent extraction.

Electrochemical techniques have been extensively employed for the determination of zinc in sea water (see Branica, 1970). The sensitivity of conventional polarography is too low for the direct determination of the element, but it has been used for the examination of concentrates prepared by chelating ion exchange (Butterworth *et al.*, 1972). Petek and Branica (1969) have claimed that they can carry out the determination directly by means of pulse polarography; Abdullah and Royle (1972), however, preferred to apply this technique only after a preliminary concentration stage. Several workers have investigated the possibility of using anodic stripping voltammetry for the direct determination of zinc in sea water (Macchi, 1965; Guiseppe, 1965; Whitnack and Sasselli, 1969; Bernhard, 1971; Smith and Redmond, 1971; Zirino and Healy, 1971, 1972). However, the application of this technique has proved difficult because of interference from copper which forms an intermetallic compound with zinc and so it inhibits its removal during the stripping stage. Zirino and Healy (1970) have used anodic stripping voltammetry at various pH values in an attempt to study the speciation of zinc in sea water, and Lieberman and Zirino (1974) have made an unsuccessful attempt to employ the technique to the continuous monitoring of zinc in sea water. Chapter 20 should be consulted for a more complete discussion of the application of electrochemical methods to the determination of zinc in sea water.

When zinc is irradiated with thermal neutrons the only relatively long-lived induced activity arises from the γ-emitter ^{65}Zn ($t_{\frac{1}{2}}$, 245 days) which is produced from ^{64}Zn (isotopic abundance 48·9%; $\sigma = 0.44$ barns). This reaction has been used as the basis for neutron activation procedures involving both γ-spectrometry of irradiated sea salts (Robertson *et al.*, 1968; Spencer *et al.*, 1970; Piper and Goles, 1969; Piper, 1971; Brewer *et al.*, 1972) and irradiation of preconcentrates followed by radiochemical separation and counting of the ^{65}Zn activity (Rona *et al.*, 1962; Slowey *et al.*, 1962; Hood and Slowey, 1964; Hood, 1966; Slowey and Hood, 1971).

Determination of total dissolved zinc. The breakdown of organo-zinc complexes before determination of total dissolved zinc can be carried out most satisfactorily by exposure of the sample to intense ultra-violet radiation (see e.g. Williams, 1969); alternatively, it can be achieved by oxidation with peroxydisulphate ion (Burrell, 1967; Slowey and Hood, 1971).

Determination of particulate zinc. The zinc content of marine particulate matter has been determined by atomic absorption spectrophotometry following acid digestion (Orren, 1971; Preston *et al.*, 1972) and by instrumental neutron activation analysis (Piper, 1971; Spencer *et al.*, 1972).

19.10.5.71. *Zirconium*

The concentration of dissolved zirconium in sea water has been reported to be 0.01–$0.04\,\mu g\,l^{-1}$ (Shigematsu *et al.*, 1964). The element probably exists in the form of hydroxy species, such as $Zr(OH)_4^0$.

Preconcentration. Zirconium has been preconcentrated from sea water by coprecipitation with both aluminium (Shigematsu *et al.*, 1964) and iron (III) (Hedges and Hood, 1966a) hydroxides. It has apparently also been separated by leaching dried sea salts with EDTA solution (Sastry *et al.*, 1969) and by ion exchange (Hedges and Hood, 1966b).

Determination of dissolved zirconium. Following preconcentration and separation from the carrier and other elements (by, for example, anion exchange or extraction with thenoyl trifluoroacetone (Shigematsu *et al.*, 1964)), zirconium has been determined colorimetrically as the Zr-alizarin sulphonic acid lake (Hedges and Hood, 1966a, b; Sastry *et al.*, 1969). Shigematsu *et al.* (1964) have carried out the determination fluorometrically using the rather unspecific reagent morin (3, 5, 7, 2′, 4′ flavone). However, Hedges and Hood (1966a) were unable to duplicate the results of the latter workers.

19.11. DETERMINATION OF MICRONUTRIENT ELEMENTS

Certain trace elements are essential for the healthy growth of phytoplankton. The most important of these *micronutrient elements* are nitrogen and phosphorus, the availability of which may exert a controlling influence on the fertility of sea water for phytoplankton. In addition to these elements, silicon and certain trace metals may be essential for algal growth, and there is some evidence that shortages of certain of the latter (e.g. iron, manganese and perhaps copper) may, under some circumstances, limit growth in the sea. Because of the importance of these micronutrient elements in the marine ecosystem their determinations merit more detailed consideration than do those of the other trace elements.

19.11.1. NITROGEN

Most species of phytoplankton are able to utilize only fixed forms of nitrogen (although in the absence of the latter many of the blue green algae are able to fix molecular nitrogen). The most important forms of inorganic fixed nitrogen in the sea are ammonia* and the nitrite and nitrate ions. The concentrations of these are considerably lower than those in fresh water and usually lie in the ranges <0.15–3 µg-at. NH_3-$N l^{-1}$, <0.01–3 µg-at. NO_2^--N l^{-1} and <0.1–35 µg-at. NO_3^- $N l^{-1}$ in oxygenated waters. Some short-lived inorganic nitrogen species, such as hydroxylamine and the hyponitrite ion, may also be present in the sea, but their presence has not so far been detected†. In addition to inorganic nitrogen species, sea water contains a wide range of organic nitrogen compounds both dissolved and particulate. Although these nitrogen compounds have concentrations which rarely total more than 15 µg-at. $N l^{-1}$ and 1.5 µg at. $N l^{-1}$ respectively, they play a very important role in the nutrition of heterotrophic organisms. Their determinations, and that of total organic nitrogen, are discussed in Section 19.14.

19.11.1.1. *Ammonia*

Because of its much lower concentration, and the interference from alkaline earth metals, ammonia is much more difficult to determine in sea water than in fresh water. Both of these difficulties can be overcome by separating the ammonia by distillation. However, the distillation process is tedious and there is a serious risk of the formation of ammonia by the breakdown of organic nitrogen compounds unless the pH during the distillation is less than 10.0. Suitable pH conditions for the distillation can be produced by the use of metaborate (Riley, 1953), borate (Konnov, 1965), or carbonate (Jenkins, 1967) buffers. Jenkins (*loc. cit.*) has cautioned against the employment of phosphate buffers (American Public Health Association, 1965) because of the risk of high reagent blanks.

Photometric methods. Of the various photometric procedures available for the determination of ammonia few are satisfactory for its direct estimation in sea water, although many of them can be used if a preliminary distillation is carried out. Because of its lack of sensitivity and reproducibility, and also the necessity for prevention of interference from the alkali earth metals, the classical Nessler method, still widely used for fresh water analysis, is only of historical interest for the examination of sea water. Four principal colorimetric methods which have been proposed for the direct determination of ammonia in sea water will be discussed below. Of these, the indophenol blue procedure,

---

* Although calculations show that ca. 90% of the ammonia in sea water exists in the form of the ammonium ion following convention, the sum of these two species will be referred to as ammonia (see Whitfield, 1974).

† Hydroxylamine has, however, been detected in the anoxic waters of the Baltic (Sen Gupta, 1970).

although it is somewhat lacking in sensitivity, is undoubtedly the most satisfactory, because of its reproducibility, low blank, and freedom from interference by organic nitrogen compounds.

Indophenol blue procedure. The indophenol blue reaction was first discovered over a century ago by Berthelot (1859). In this reaction monochloramine (NH_2Cl), produced by reaction of ammonia with the hypochlorite ion, is allowed to react in alkaline solution with phenol. This leads to the production, *via* an intermediate quinone chlorimide (I), of the intensely coloured indophenol blue ion (II) (Tetlow and Wilson, 1964). The reactions involved are complex, and much effort has been expended in evaluating the optimum

I   red   II   +

indophenol blue

conditions of pH, reagent concentration, and reaction time for them. In order to minimize the production of dichloramine and the decomposition of urea to ammonia the formation of monochloramine should be carried out between pH 8 and 11. Generation of the indophenol blue colour requires a pH of >9.6. The final reaction is slow, but can be promoted either by heating, or by catalysis with manganese (II) (see e.g. Riley, 1953), acetone (Crowther and Large, 1956), or more effectively sodium nitroprusside (see Harwood and Kuhn, 1970) or potassium ferrocyanide (Liddicoat *et al.*, 1974). There is an extensive literature on the application of the indophenol blue reaction to the analysis of natural waters. For reasons of space only the more important aspects of its use for the determination of ammonia in sea water can be discussed here.

The principal difficulty in the application of the indophenol blue method to sea water lies in the necessity for using a high pH for the development of the colour. This leads to the precipitation of calcium and magnesium compounds, the turbidity of which interferes in the spectrophotometry. This interference has been overcome in three ways:

(i) by allowing the precipitate to settle and decanting the supernatant liquid (Sagi, 1966; Emmet, 1969; Koroleff, 1970; Matsunaga and Nishimura, 1971), there is however, some loss in sensitivity,

(ii) by addition of a complexing agent before making alkaline. Compounds which have been used for this purpose include *cyclo*-hexyl-*trans*-1,2 diaminotetra-acetic acid (Roskam and de Langen, 1964); EDTA (Sagi,

1969; Manabe, 1969); citrate (Solorzano, 1969; Head, 1971; Slawyk and MacIsaac, 1972; Liddicoat *et al.*, 1974),

(iii) by treating the monochloramine at pH 3·4 with phenol and extracting the intermediate quinone monochlorimide with hexanol. Indophenol blue is then generated in the hexanol layer by addition of sodium hydroxide (Newell and dal Pont, 1964; Newell, 1967).

The second of these three procedures is probably the most satisfactory for routine use.

Other variants of the indophenol blue procedure include (i) the replacement of phenol by other phenolic compounds such as *m*-cresol (Yamaguchi *et al.*, 1969), thymol (Roskam and de Langen, 1964; Matsunaga and Nishimura, 1971) and salicyclic acid (Benesch and Mangelsdorf, 1972) (ii) substitution of other oxidizing agents for the rather unstable sodium hypochlorite used in the formation of monochloramine, e.g. chloramine T (Newell and Dal Pont, 1964; Newell, 1967; Matsunaga and Nishimura, 1971), sodium dichloro-iso-cyanurate ($ClNCON(Cl)CON=CONa$) (Benesch and Mangelsdorf, 1972*; Grasshoff and Johannsen, 1972; Liddicoat *et al.*, 1974). The latter reagent has the advantage of being a stable solid, and the production of hypochlorite when it hydrolyses is both rapid and quantitative.

Although Solorzano's (1969) technique has been widely adopted because of its simplicity, some workers (e.g. McCarthy and Kamykowski, 1972; Liddicoat *et al.*, 1974) have found it to lack reproducibility and to suffer from high and erratic blanks. The latter group of workers traced the difficulties with blanks to the sodium nitroprusside used as catalyst and substituted potassium ferrocyanide in its place. They considered that the instability of commercial hypochlorite solution also contributed to the irreproducibility and recommended that sodium dichloro-*iso*-cyanurate (see above) be used instead. Even with these modifications the response of the method varied from day to day; this variation was attributed to the effect of light on the reaction, the colour being more intense when the light was strong. For this reason they recommended that the reactants should be irradiated with a long-wave U.V. lamp (maximum energy at 365 nm) during colour development. Using their recommended conditions they were able to achieve standard deviations of 0·02 and 0·04 µg–at. NH_3–$N l^{-1}$ with samples containing 1 and 4 µg–at. $N l^{-1}$ respectively. Beer's Law was obeyed up to 20 µg–at. $N l^{-1}$. The indophenol blue technique is subject to negligible interference from amino acids, urea and nucleic acids at their normal concentrations in sea water. For samples containing very low levels of ammonia it may be possible to enhance the sentivity by extraction of the indophenol blue into a small

* Grasshoff and Johannsen (1974) have pointed out that due to magnesium interference there is a very large salt error in the procedure described by Benesch and Mangelsdorf.

volume of hexyl alcohol, as suggested by Matsunaga and Nishimura (1971) (see also Fujinuma *et al.*, 1971) or, alternatively, by acidifying and extracting the red unionized compound and finally back extracting into a small volume of sodium hydroxide solution.

Automated indophenol blue procedures designed for the AutoAnalyser system have been described by several authors (Grasshoff, 1970; Head, 1971; Grasshoff and Johannsen 1972; Benesch and Mangelsdorf, 1972; Slawyk and MacIsaac, 1972). There appears to be little to choose between them in terms of sensitivity and precision. The last named workers have carried out a comparison of their method with an automated rubazoic acid technique (see p. 413).

Methods involving oxidation to nitrite. Richards and Kletsch (1964) have described a very sensitive method for the determination of ammonia in sea water. This is based on the fact that in highly alkaline media the hypochlorite ion oxidizes ammonia to nitrite ion, the reaction being catalyzed by bromide ion. The nitrite can be readily estimated by the method of Bendschneider and Robinson (1952) (see p. 415), after first reducing the excess hypochlorite with sodium arsenite and acidifying. An overall yield of nitrite of ca. 73 % was found. Erratic results have been obtained by a number of workers using this technique, and several modifications, including an automated method, have been suggested (Strickland and Parsons, 1968; Føyn, 1968). An extensive investigation of the method has been carried out by Truesdale (1971) who found that a major source of error was variable decomposition of nitrite following the acidification stage. These variations could be minimized by buffering to pH 1·6 with a citrate solution containing sulphanilamide and carrying out the coupling reaction exactly 1 min later. It was also observed that the time for the oxidation of ammonia could be reduced from about 3·5 h to 17 min by addition of 0·25 g of bromide per litre. Truesdale claimed a coefficient of variation of ca. 2 % at an ammonia level of $1·5\,\mu g$–at. $N\,l^{-1}$ for his procedure. Although urea at its normal levels in sea water causes little interference, a considerable proportion (up to 70 %) of amino-acid nitrogen is also measured. This method should, therefore, not be used where a measure of ammonia *per se* is required, but it may be of value in productivity studies as the amino-acids are probably useful sources of nitrogen for many algae.*

A novel approach to the oxidation of ammonia to nitrite has been described by Yoshida (1967a) who used nitrifying bacteria for this purpose. A sensitivity of $1·4\,\mu g$–at. $N\,l^{-1}$ was claimed for the method, but as might be expected the precision is rather poor.

* Matsunaga and Nishimura (1974) have suggested that the oxidation can be reduced to 1 min from the 17 min used in Truesdale's procedure. Under these conditions interference from urea and amino acids was found to be negligible.

Methods depending on hypobromite oxidation. Several workers have oxidized ammonia in sea water with an excess of hypobromite. Residual hypobromite has been determined colorimetrically, either after addition of iodide and starch (Matsudaria and Tsuda, 1957; Grasshoff, 1968), or by the bleaching of azo-dyes, such as Bordeaux Red (Gillbricht, 1961). These methods have little to recommend them for the direct determination of ammonia since hypobromite will also be reduced by many of the organic compounds present in sea water, such as urea and amino-acids (see e.g. Nehring, 1968). However, they may, because of their simplicity, be of value if ammonia is first separated by distillation (see e.g. Konnov, 1965).

Rubazoic acid method. In 1953 Kruse and Mellon developed a highly sensitive technique for the determination of ammonia. This was based on its reaction at pH 3·7 with chloramine T and a mixture of pyridine, pyrazolone and *bis*-pyrazolone to produce a pink-violet coloured compound. The latter could be extracted with carbon tetrachloride as a yellow product, the absorbance of which was proportional to the concentration of ammonia. This procedure has been adopted for the analysis of sea water by Atkins (1957) and Strickland and Austin (1959); for practical details the reader should consult Strickland and Parsons (1968). Although the method is sensitive and specific for ammonia, it has the drawbacks of being time-consuming and necessitating the use of pyridine—an obnoxious and highly toxic compound. An extensive study of the reaction has been made by Procházková (1964) who showed that the yellow coloured product is rubazoic acid (III)

He also demonstrated that the presence of pyridine was not essential. In the method which he subsequently developed, the formation of rubazoic acid was carried out in two stages. In the first, chloramine T was allowed to react at pH 6·5 with ammonia. The solution was then buffered to pH 10 with sodium carbonate and bis-pyrazolone and pyrazolone were then added. After formation of rubazoic acid was complete the solution was acidified and the free acid was extracted with trichloroethylene for photometry. Procháková's procedure has been modified for manual analysis of sea water by Johnston (1966). Johnston's method, in its turn has been automated by MacIsaac and Olund (1971). The latter procedure has been compared with a simpler, but less specific, automated indophenol blue procedure by Slawyk and MacIsaac (1972) who found a very close correlation ($R = 0.96$) between the results obtained by the two methods for water from an area of upwelling.

Electrometric methods. Although ion specific electrodes are available for the estimation of ammonia they are of no value for use with sea water owing to both their poor sensitivity, and the interference caused by the major cations (see Section 20.3.1.1 and Table 20.5).

Avoidance of contamination. Great care must always be taken during the analysis to avoid contamination with ammonia or particles of ammonium salts. It is desirable to carry out determinations in a separate laboratory in which ammonia and ammonium salts are never used. Water to be used in the analysis and for the preparation of reagents should be purified by passage through a mixed bed ion exchange column.

19.11.1.2. *Hydroxylamine*

Hydroxylamine is unstable in oxygenated sea water at pH8 and probably for this reason has not so far been detected in such waters. However, it has been detected occasionally in anoxic waters, in which it is more stable (see e.g. Sen Gupta, 1970). Strickland *et al.* (1967) have described a procedure having a sensitivity of ca. 0.01 µg–at. NH_2OH–N l^{-1} for the analysis of sea waters. Hydroxylamine is oxidized to nitrite in acidic medium by means of iodine. After reduction of the excess iodine by means of arsenite, the nitrite is determined photometrically by Bendschneider and Robinson's (1952) procedure (see p. 415). Allowance must be made for pre-existing nitrite. Hydrogen sulphide interferes, but can be stripped from the acified sample by bubbling with nitrogen.

19.11.1.3 *Nitrite*

All the methods available for the determination of nitrite in sea water are based on the classical photometric Griess–Ilosvay technique. In this, the nitrite is used to diazotize an aromatic primary amine. The resultant diazonium ion is coupled with another aromatic amine to form a pink azo dye, the absorbance of which is proportional to the amount of nitrite in the sample. Only a few combinations of amines give azo dyes having sufficiently high absorbances for the determination of nitrite at the low concentration at which it is present in sea water.* In most of the earlier work, which has been reviewed by Riley

* Celardin *et al.* (1974) have shown recently that a very much higher sensitivity ($\varepsilon = 89,000$) can be achieved by carrying out the diazotization with 4-amino-acetophenone and the coupling with N-phenyl-1-naphthylamine. The colour intensity is, however, very sensitive to pH changes.

(1965b), sulphanilic acid was diazotized and coupled with 1-naphthylamine. Barnes and Folkard (1951) have compared several versions of this method and have found that described by Rider and Mellon (1946) to be the best in both sensitivity and precision. Although a few workers (e.g. Fernandez and Okuda, 1967) still favour this combination of reagents, most workers now avoid the danger of working with potentially carcinogenic 1-naphthylamine by employing the method developed by Shinn (1941), and modified for sea water analysis by Bendschneider and Robinson (1952). In this, the diazotization and coupling reactions are performed with sulphanilamide and N-(1-naphthyl)-ethylenediamine respectively. Since both reactions proceed rapidly at pH 3–4, it is not necessary to buffer the solution to pH 5–6 before carrying out the coupling reaction as is the case with the Griess method. The method, which has a sensitivity of ca. 0.02 μg–at. NO_2–$N\,l^{-1}$, obeys Beers Law over a wide range of nitrite concentrations. It is devoid of salt error and is subject to very few interferences. For practical details of both manual and automated versions of the method, the monograph by Strickland and Parsons (1968) should be consulted. The preparation of stable standard nitrite solutions for the calibration of Bendschneider and Robinson's (1952) method has been studied by Sugawara (1969); autoclaved (110°C) standards containing 0.1–20 μg–at. $N\,l^{-1}$ prepared from trebly recrystallized sodium nitrite were found to be stable for at least 150 days.

The concentration of nitrite in sea water is frequently below the level at which it can be precisely determined by the above methods. Several attempts have been made to increase the sensitivity of the Bendschneider and Robinson (1952) procedure. These have involved carrying out the reaction with a large volume of sample and concentrating the azo dye into a small volume for photometry. The concentration has been carried out by ion exchange (Wada and Hattori, 1971) and also by solvent extraction with both polar (Zeller, 1955) and non-polar solvents (Matsunaga et al., 1972; Cescon and Macchi, 1970, 1972). With the latter class of solvents it is necessary to back-extract the azo compound with dilute hydrochloric acid before carrying out photometry.

19.11.1.4. Nitrate

Despite the large amount of analytical development work which has been carried out, there is still no completely satisfactory method for the determination of nitrate in sea water. Owing to lack of sensitivity, and interference from chloride or alkali earth ions, most of the methods used for the determination of nitrate in fresh waters are inapplicable to sea water. In most of the earlier work, nitrate was reduced by means of Devarda's alloy to ammonia which, after distillation, was determined photometrically with Nessler's reagent. These procedures are time-consuming, and are open to the objection that the drastic experimental conditions cause many organic nitrogen compounds to break down to ammonia (Harvey, 1926).

Although the nitrate ion has an intense absorption band in the far ultra-violet at 202 nm ($\varepsilon = 9900$), the strong absorption in this spectral region due to bromide ion and dissolved organic compounds prevents it being used for the determination of nitrate in sea water. However, according to Merten and Massart (1971) it is possible to determine nitrate concentrations in the parts per million range in sea water by spectrophotometry of the acidified sample at 210 nm. Measurement of the absorbance of the sample after treatment with Raney nickel, which reduces nitrate to ammonia, enables correction to be made for interfering absorbances. When sea water containing traces of nitrate is treated with an equal volume of concentrated sulphuric acid it develops a strong absorption band at 230 nm, probably arising from the formation of nitrosyl chloride. This process has been adopted by Armstrong (1963) for the determination of up to 200 µg-at. $NO_3^- - N\,1^{-1}$ in sea water. Allowance for absorption arising from organic compounds was made by measuring the absorbance of a similarly treated sample to which hydrazine sulphate had been added to reduce the nitrosyl chloride. Because of the hazards involved, particularly on shipboard, in mixing equal volumes of sea water and concentrated sulphuric acid, Armstrong's method has been little used, although it has been adopted by Leroy (1967).

A number of direct methods for the photometric determination of nitrate in sea water have been described in the literature. These depend on its ability, in a strong sulphuric acid medium, to oxidize certain readily oxidizable organic compounds to intensely coloured products. Compounds which have been used for this purpose include diphenylamine, diphenylbenzidine, strychnidine and brucine. Riley (1965b) has reviewed these methods and has pointed out that although they are sensitive, their precision is poor. The main cause of error appears to be variations in the dispersal of the heat generated on adding the solution of the reagent in concentrated sulphuric acid, as this affects the colour intensity. It is thus important to try to standardize the methods of mixing and subsequent cooling (see e.g. Marvin, 1955; Dal Pont et al., 1963; Ostrowski and Nowak, 1964). In addition, the colour intensity produced is critically dependent upon the final sulphuric acid concentration. In the analysis of estuarine waters Jenkins and Medsker (1964) were able to reduce the variability caused by heating by first treating the sample with the necessary amount of sulphuric acid. After cooling, brucine solution was added and the solution was heated to 100°C for 20 min. to produce the coloured complex (see also Jenkins, 1967). This method suffers from the drawbacks that the calibration curve is non-linear below 7 µg–at. $NO_3 - N\,1^{-1}$

and that the slope of the calibration curve differs considerably from one batch of brucine to another. Kahn and Brezenski (1967a) and Petriconi and Papee (1971) have proposed modifications to the technique of Jenkins and Medsker in order to overcome these difficulties. Kahn and Brezenski (1967b) have developed an automatic brucine procedure for which they claim a precision double that of their manual method.

If nitrate could be reduced quantitatively to nitrite, the determination of the latter by the Griess–Ilosvay method, or one of its modifications, would provide an almost specific method for the estimation of nitrate. The main difficulty in developing a method based on this principle is that of finding a reducing agent which will give a nearly quantitative yield of nitrite, without tending to reduce the latter to lower oxidation states of nitrogen or to reduce the nitrate to alternative products. Procedures have been described employing both homogeneous and heterogeneous reducing conditions.

In the homogeneous reduction procedure developed for the analysis of sea water by Mullin and Riley (1955a) nitrate was reduced with hydrazine in a medium buffered to pH 9·6 with sodium phenoxide, in the presence of catalytic amounts of copper. In sea water the yield of nitrite was ca. 70% This method has been reinvestigated by Strickland and Parsons (1960) and more recently by Koroleff (1968) who have each introduced several modifications. An automated version of this technique has been described by Henriksen (1965a).

The hydrazine reduction procedure suffers from a number of disadvantages:

(i) the yield of nitrite depends on the internal surface area of the flask used, and is also influenced by the degree of agitation of the solution. The latter factor, which is of importance if measurements are made at sea, has been attributed by Strickland and Parsons (1968) to the presence of air above the solution,

(ii) the reduction process is slow and requires about 24 hrs for completion,

(iii) a number of substances inhibit the reduction; these include, particulate matter and hydrogen sulphide (see Williams and Coote, 1962; Shiro and Zeitlin, 1962). Kahn and Brezenski (1967a) have noted interference when they applied the method to estuarine waters; they attribute this to the action of pollutants which destroyed the hydrazine.

In view of these various drawbacks, and the fact that the method is liable to give erratic results, the hydrazine reduction technique is now little used for sea water analysis. However, it finds extensive application to the estimation of nitrate in fresh waters in which it is possible to carry out the reduction at pH values > 12 at which it proceeds more satisfactorily.

Heterogeneous reduction at a metal surface provides an alternative approach to the quantitative reduction of nitrate to nitrite in sea water. Several workers have attempted to carry out the reduction by treating the sample with granulated or powdered zinc under a variety of conditions of and temperatures (see Riley, 1965b for earlier references, and Chow and Johnson, 1962; Matsunaga and Nishimura, 1969). However, it is difficult to achieve reproducible results with this reducing agent, as nitrite is not the only product of the reduction of nitrate. This is particularly the case in acid solution for which the standard redox potential for the reaction

$$NO_3^- + 2H^+ + 2e^- \rightleftarrows NO_2^- + H_2O \qquad (E_0 = 0.94 \text{ V})$$

is almost identical with that of the competing reaction

$$NO_3^- + 4H^+ + 3e^- \rightleftarrows NO + 2H_2O \qquad (E_0 = 0.97 \text{ V})$$

which is very strongly influenced by pH.

Grasshoff (1964) has pointed out that the redox equilibrium

$$NO_3^- + H_2O + 2e^- \rightleftarrows NO_2^- + 2OH^-$$

which occurs in neutral or alkaline medium has $E_0 = 0.015$ V. He suggested that cadmium or cadmium amalgam ($E_0 = -0.40$ V) would be much more suitable for bringing about this reaction than zinc which has a considerably higher standard potential ($E_0 = 1.22$ V). He claims that nitrate can be reduced quantitatively to nitrite, in both sea and fresh water medium, by passage through a reactor column packed with coarsely powdered amalgamated cadmium. Ammonium chloride is added to the sample in order to buffer the solution and to complex cadmium which has been oxidized during the reduction of nitrate.

$$NO_3^- + Cd(s) + 2NH_4^+ + 2NH_3 \rightarrow NO_2^- + Cd(NH_3)_4^{2+} + H_2O$$

A similar method of analysis was developed independently by Morris and Riley (1963); this differed mainly in the omission of the ammonium chloride.

A number of modifications of these techniques have been subsequently proposed. Thus Wood et al. (1967) have replaced the amalgamated cadmium by a copper/cadmium couple and have used EDTA as a complexing agent in place of ammonium chloride.* Koroleff (1968) has reported that this modifica-gave consistent results which agreed well with those obtained by his hydrazine reduction technique (loc. cit.). He did, however, stress the necessity for a preliminary activation of the cadmium reductor by washing it with sea water containing ca. 50 µg–at. NO_3–N l^{-1}. Although Sugawara (1969) has found the method of Wood et al. (1968) to be satisfactory, Strickland and Parsons (1968)

* Olsen (1968) found that reduction with amalgamated cadmium was too slow and has prepared a more active reductant by treating the Cd/Hg couple with copper sulphate.

have experienced difficulties with it and have found that it is advisable to revert to ammonium chloride as activator. Lambert and Du Bois (1971) prefer to carry out the reduction by shaking the buffered sample with cadmium filings in the presence of copper sulphate and ammonium chloride instead of passing the sample through a coloumn packed with cadmium.

The technique developed by Strickland and Parsons (1968) is probably the best nitrate method currently available. However, as with other methods using cadmium, successful operation depends very much upon the care devoted to the preparation of the cadmium reductant in a sufficiently active state (see Nehring, 1968; Isaeva and Bogoyavlensky, 1968). Some batches of cadmium give lower yields of nitrite than do others which suggests that the purity of the metal used may be of importance. With repeated use the cadmium columns gradually lose their activity and must be reactivated; the rate at which this occurs often differs from one column to another. This drawback is eliminated in automatic methods of analysis (Brewer and Riley, 1965; Armstrong et al., 1967; Strickland and Parsons 1968; Watanuki, 1969) since the standards and samples all pass through the same reductor column. Hager et al. (1972) have found a small but significant difference between the results obtained by the manual method of Wood et al (1968) and the automated technique of Armstrong et al. (1968), but they were unable to ascertain its cause.

Yoshida (1967b) has suggested a novel method for the determination of low concentrations of nitrate (0.06–0.6 µg–at. NO_3–N l^{-1}) in sea water in which nitrate is reduced to nitrite by means of a nitrate-reducing bacterium.

19.11.2. PHOSPHORUS

19.11.2.1. *Fractionation of phosphorus species*

Phosphorus occurs in sea water in several chemical forms, both organic and inorganic. It is usually almost impossible to make a clear-cut differentiation between them analytically, and all that can normally be done is to categorize them arbitrarily. Schemes for fractionating them have been discussed by Jenkins (1967), Olsen (1967), Piton and Vorturiez (1968) and Strickland and Parsons (1968). The general problem of the determination of phosphorus in natural waters has been extensively reviewed by Armstrong (1965), Burton (1973a) and Chamberlain and Shapiro (1973).

Filtration through a membrane filter having an average pore size of ca. 0.5 µm is frequently used to distinguish between the particulate and dissolved forms of phosphorus in sea water. The selection of filters having this pore size is purely arbitrary, and represents a compromise between speed of filtration and the fineness of particle which can be retained (see Section 19.4).

The phosphorus contained in the filtrate which is termed the "dissolved phosphorus" will comprise not only that in true solution but also that in any colloidal or very fine particulate material. It is not known, in fact, whether phosphorus does occur in the latter forms in sea water in significant amounts (see e.g. Breger, *et al.*, 1973). However, Rigler (1964) has adduced evidence for the existence of these forms in a sample of lake water from his observation that nearly half of the "dissolved" organic phosphorus which passes a $0.45 \mu m$ membrane filter is retained by a $0.1 \mu m$ one. It should be borne in mind that it is not practicable to carry out particle sizing using these filters, and that the porosity of a filter will vary during the filtration of samples containing even small amounts of particulate matter as the pores become progressively blocked (see Section 19.4).

It is customary to divide the dissolved phosphorus into two fractions. That which reacts rapidly with an acidic molybdate reagent under conditions sufficiently mild to prevent hydrolysis of condensed phosphates or labile organic' phosphates is often terms "reactive" phosphorus (Strickland and Parsons, 1968). This fraction probably corresponds closely to the inorganic orthophosphate (see e.g. Chamberlain and Shapiro, 1970). The remainder of the dissolved phosphorus is usually referred to as the dissolved combined phosphorus. Little is known about the compounds comprising this fraction, but it is surmised that they are principally-derived esters of orthophosphoric and polyphosphoric acids, e.g. sugar esters, phospholipids, adenosine triphosphate etc. In addition, aminophosphonic acids (which contain the very stable C–P bonds) may also be present as they have been detected in phytoplankton and other marine organisms. In samples from coastal areas this fraction may also include inorganic condensed phosphates (e.g. sodium hexametaphosphate) which, owing to their use as detergents and in water treatment, are found in increasing amounts in effluents. Dissolved combined phosphorus cannot be determined directly, but is estimated by difference after breaking down the organic phosphorus compounds to orthophosphate and deducting the pre-existing reactive phosphate. If present, condensed phosphates can be determined by difference after hydrolysing them to orthophosphate under mild conditions. The concentrations of free orthophosphate and dissolved organic phosphorus in the sea are highly variable and range betwen 0–$4 \mu g$–at. $P\, l^{-1}$ and 0–$0.5 \mu g$–at. $P\, l^{-1}$ respectively.

The particulate fraction may contain both organic and inorganic forms of phosphorus. Although specific methods are available for the determination of several organic species, e.g. adenosine triphosphate (Holm–Hansen and Booth, 1966), it is more usual to estimate the total phosphorus (excluding that associated with resistant minerals) by carrying out an oxidative mineralization and measuring the amount of orthophosphate produced. Probably only a proportion of the phosphorus determined in this way can be utilized by marine organisms. However, although the need for a method for the determination of this ecologically important phosphorus fraction is generally recognized (see e.g. Pomeroy *et al.*, 1965) no satisfactory techniques is yet available.

Studies of the phosphorus species in the sea are mainly carried out for ecological purposes, and for these it is only those forms of the element which can be assimilated by marine organisms which are important. When it is remembered that both the particulate and dissolved organic phosphorus fractions in the sea are highly variable mixtures of many compounds, some of which are biologically active at very low concentrations, the crude nature of the above fractionation scheme will be apparent. Jenkins (1968) has pointed out that present interpretative ability may not justify the determination of more than the total phosphorus (including particulate), the total dissolved phosphorus and the reactive phosphate. Even this scheme is time-consuming, and frequently only the reactive phosphate and, more rarely, the total dissolved phosphorus are determined.

19.11.2.2. *Determination of orthophosphate*

Because the determination of all forms of combined forms of phosphorus in sea water nearly always involves conversion to orthophosphate, the determination of the latter is of fundamental importance. All the methods currently used for this purpose are based on the formation of 12-molybdophosphoric acid and its subsequent reduction to a phosphomolybdenum blue complex, the absorbance of which is measured. Although several reducing agents are capable of bringing about this reduction, only a few of them produce a molybdenum blue complex having sufficiently high absorbance to permit them to be used for the estimation of phosphate at its normal levels in sea water. Of these, the most frequently used are stannous chloride and ascorbic acid.

Chalmers (1959) has shown that 12-phosphomolybdic acid exists in two isomeric forms. The unstable β-form is produced when the deca- or dodeca-molybdate ion is the principal molybdate species in the solution. This isomer undergoes spontaneous transformation to the α-form, but can be stabilized by the addition of polar organic solvents (for the mechanisms of the isomerization, see Chalmers and Sinclair, 1965). The α-isomer is also produced directly from phosphate when the reacting species is the octamolybdate ion. It is essential that the acidity and molybdate concentration in the solution are carefully controlled since these, together with the $H^+ : Mo$ ratio, control the species of molybdate present. Jones and Spencer (1963) have shown that if the acid concentration is less than ca. 0·2 N the molybdate ion itself undergoes some reduction to molybdenum blue in the subsequent reduction stage. At this acidity, and above, the β-form of the 12-phospho-molybdic acid is the principal product. Although its formation can be carried out at acidities up to at least 1·0 N, this is not desirable in practice because the increasing acidity decreases the rate of reduction to molybdenum blue and increases the risk of hydrolysing labile organic phosphorus compounds. Probably the conditions recommended by Harvey (1955) represent a reason-able compromise (acidity 0·28 M, ammonium molybdate tetrahydrate 500 mg l^{-1}).

In most early work on the determination of phosphorus in sea water reduction of the 12-phosphomolybdic acid has been carried out using stannous chloride* (for a review of these procedures see Burton and Riley. 1956). These procedures suffer from a number of serious disadvantages, including the instability of the colour, the dependence of the rate of reduction on tempera-ture and salinity, and the considerable salt error (Riley 1965b). Many of these drawbacks can be overcome if the 12-phosphomolybdic acid is extracted and the reduction carried out in the organic phase (Henriksen, 1965b; Jenkins, 1968; Pakalns and McAllister, 1972).

The use of ascorbic acid as a reductant in the determination of ortho-phosphate in sea water was first suggested by Greenfield and Kalber (1955). Their technique was subsequently modified by Murphy and Riley (1958) to produce a single solution method which was free from most of the difficulties associated with the use of stannous chloride. However, because colour development was slow there was some risk of hydrolysis of labile phosphate esters. In a later modification of their procedure Murphy and Riley (1962) overcame this drawback by incorporating antimony (Sb^{3+}) into the reagent. This single solution reagent reacts rapidly with phosphate ion to give a very stable purple-blue complex which contains antimony and phosphorus in a 1:1 atomic ratio. Crouch and Malmstadt (1967a, b) have shown that phosphomolybdic acid is an intermediate in the reaction and that ascorbic acid acts stoichiometrically as a 2-electron reductant.

The method of Murphy and Riley (1962) has been widely adopted for the determination of dissolved orthophosphate in both fresh water and sea water, and a number of modifications of the reagent exist (Table 19.19). It is impor-tant that the final acidity in the solution should be sufficiently high to prevent reduction of molybdate itself occurring, but low enough to minimize any risk of hydrolysis of very labile organophosphorus compounds. Probably the conditions recommended by Strickland and Parsons (1968) strike the best balance between these two factors. If it is necessary to analyse waters

---

* The optimum conditions for the stannous chloride method have been reinvestigated recently by Sugawara (1969).

TABLE 19.19

Final concentrations of reagents in phosphate determination using ascorbic acid in the presence of antimony (after Burton, 1973a)

Authors	Sulphuric acid N	Ammonium molybdate tetrahydrate $mg\,l^{-1}$	Ascorbic acid $mg\,l^{-1}$	antimony $mg\,l^{-1}$
Murphy and Riley (1962)	0·40	960	845	8·0
Stephens (1963)	0·23	550	980	2·3
Edwards et al. (1965)	0·11	270	240	18·7
Vogler (1965a)	0·20	480	950	4·8
Strickland and Parsons (1968)	0·23	550	980	4·6
Harwood et al. (1969)	0·40	960	4000	40·0

highly polluted with phosphate the range of linear response can be extended by increasing the concentration of antimony in the reagent (Harwood et al., 1969; Edwards et al., 1965).

Calibration. Sugawara (1969) has investigated the preparation of stable standard phosphorus solutions. He has found that autoclaved (110°C) working standards in 30·5‰ NaCl solution prepared using potassium dihydrogen phosphate were stable indefinitely if stored in sealed glass ampoules.

Interferences. The major cations and anions of sea water have very little effect on the molybdenum blue method if ascorbic acid is used as reductant and Murphy and Riley (1962) have reported a salt error of $<1\%$ at a salinity of 36‰. The method is also relatively free from interference from other elements including many of those capable of forming heteropoly acids with molybdate ion. Thus, no intereference is caused by 10 mg silicate–silicon l^{-1}. Arsenic (V) does, however, react and yields a corresponding heteropoly blue complex which has a similar molar absorbance to that of the complex produced by phosphorus. Interference from arsenic, which, according to Jones (1966) varies from day to day, can be prevented if it is reduced to the non-reactive $3+$ oxidation state with metabisulphite before addition of the combined reagent (Pett, 1933; Harvey, 1948; Jones and Spencer, 1963; Johnson, 1971). Alternatively, the interference can be overcome by increasing the sulphuric acid concentration in the reagent from 2·5 to 3·6 N (Portmann, 1965). Condensed phosphates only react very slowly with the reagent (see e.g. Solorzano and Strickland, 1968).

Since the determination of phosphate is carried out under moderately acid conditions (pH <0·7), there is some risk that naturally occurring labile organic phosphates may undergo hydrolysis, and thus cause interference. However, tests carried out using sugar phosphates and other labile phosphorus compounds have given little evidence that significant hydrolysis occurs (Murphy and Riley, 1962; Vogler, 1965; Strickland and Solorzano, 1966). Thus, the last named authors found that, of the range of compounds which they tested, only adenosine monophosphate reacted appreciably within 1 hr. Independent work in the author's laboratories has confirmed these findings, with regard to sea water, but has shown that a small amount of hydrolysis (<5%) of the most labile organic phosphates can occur in non-saline media. Chamberlain and Shapiro (1969) have suggested that possible interference by hydrolysis could be minimized by using a short period of colour development and then increasing the acidity to prevent further formation of molybdenum blue.

Automatic procedures for the determination of orthophosphate. Several automated molybdenum blue methods based on the AutoAnalyzer system have been described for the determination of orthophosphate in sea water (Grasshoff 1967, 1970b; Chan and Riley, 1966b; Armstrong *et al.*, 1967; Strickland and Parsons, 1968; Atlas *et al.*, 1971; Coote *et al.*, 1971; Goulden and Brooksbank, 1974). Ascorbic acid is used as reductant in most of them; however, stannous chloride and hydrazine sulphate have also been used. Slow plating out of the molybdenum blue complex on the walls of the Auto-Analyzer causes some difficulty if ascorbic acid is used in the presence of antimony, but no plating occurs if antimony is omitted. Hager *et al.* (1972) have carried out comparative tests between manual and automatic phosphate methods and have found that they have similar precisions (coefficient of variation ±1%), and that there were no significant differences between the results obtained by the two techniques.

Determination of low concentrations of orthophosphate. When the phosphate concentration in sea water lies below 0·2 μg-at.l^{-1} it is not practicable to determine it directly by the molybdenum blue method. Under such circumstances it is possible to obtain a ten-fold increase in sensitivity by extracting the molybdenum blue complex into a small volume of a polar organic solvent. Stephens (1963) has used *iso*-butanol for this purpose; however, its high solubility presents difficulties (Fossato, 1968), and *iso*-butyl acetate (Cescon and Scarazzatto, 1973) or *n*-hexanol (J. Murphy, private communication) are to be preferred as extractants. Pakalns and McAllister (1972) have described a technique for the determination of low concentrations of phosphate in sea water in which the yellow phosphomolybdic acid is extracted with *iso*-butyl acetate, and then reduced to molybdenum blue with a mixed reagent containing ascorbic acid and stannous chloride. This method, which has a precision of ± 0.005 µg–at. P at a level of 0.2 µg–at. $P l^{-1}$, may be of particular value in determining phosphate in oligotrophic waters because of its freedom from interference by arsenic.

Comparative studies of methods for the determination of orthophosphate. Comprehensive inter-comparisons of several methods for the determination of orthophosphate in sea water have been made by Jones and Spencer (1963) and Jones (1966), using water samples collected from several oceanic regions and from coastal regions throughout the year. They found that results obtained with the stannous chloride method of Harvey (1948) were usually higher than those given by the ascorbic acid procedures of Murphy and Riley (1958, 1962). The discrepancies with the latter (in which antimony is used) were generally $<10\%$, except at low phosphate levels. Although they believed that these differences might, in part, reflect the greater sensitivity of the stannous chloride method to interference by arsenate, they were unable to draw any conclusions about which of the methods was the most accurate.

A number of collaborative inter-laboratory studies of the determination of phosphate in sea water have been carried out (see e.g. Jones and Folkard, 1968; Koroleff, 1965; Rochford, 1963, 1964; Palmork, 1969; Bogoyavlenskii, 1965a; Spencer, 1973). Burton (1973a) has derived the coefficients of variation given below for results from an intercalibration trial carried out at Copenhagen in 1966 (Jones and Folkard, 1968). In this, six water samples derived mainly from the Baltic, but including one from the Thames Estuary, were analysed in triplicate by seven or eight analysts using the ascorbic acid–antimony method. The standard deviations attained by the individual analysts ranged from 0.008–0.038 µg–at. $P l^{-1}$.

Average value (µg–at. $P l^{-1}$)	0.026	0.86	1.51	1.92	3.25	5.19
Coefficient of variation (%)	100	11	3.8	2.9	3.8	3.0

Considerably better precision (standard deviation $= 0.8$–1.0%) was achieved in calibration work carried out on samples of ocean water during the preliminary stages of the Geosecs Expedition (Spencer, 1973). A more detailed assessment of inter-laboratory precision in the analysis of oceanic samples should be possible when the results of the recent extensive SCOR-UNESCO intercomparison experiments become available.

Although many of the methods used for the determination of dissolved orthophosphate give similar results with sea water, there is some possibility that other phosphorus species may also be determined. Rigler (1966, 1968) has suggested that, with lake waters at least, molybdenum blue methods estimate more than just dissolved orthophosphate. His conclusions were based on the inability of anion exchangers to remove all the reactive "orth-phosphate" (Jones and Spencer (1963) have observed a similar phenomenon for sea water), and on the fact that a bio-assay method using ^{32}P gave much lower results, particularly at low phosphorus concentrations than did the molybdenum blue procedure. He attributed these discrepancies to hydrolysis of labile organic phosphate during the photometric determination. The problem has been discussed by Chamberlain and Shapiro (1970) who concluded that neither a hidden blank (see p. 293), nor hydrolysis of organic phosphorus esters will seriously bias the results. They have pointed out that results for lake waters obtained by an algal bio-assay technique agreed well with those found by their arsenate-insensitive molybdenum blue method (Chamberlain and Shapiro, 1969), and suggested that at least part of the discrepancies noted by Rigler could have been caused by interference of arsenate. They were of the opinion that if steps are taken to exclude this interference the molybdenum blue method will determine inorganic phosphate, but not necessarily only free orthophosphate ion; thus, the analysis would include any colloidal inorganic phosphate which was not removed by filtration (Olsen, 1967). However, the agreement between results obtained by bioassay and chemical techniques suggests that the latter will provide a reasonable measure of the amount of phosphorus which can be utilized by organisms. This is probably more important from an ecological stand-point than a knowledge of the free orthophosphate concentration. Although these deductions were made for fresh waters, it is probable that they apply with equal force to sea water.

19.11.2.3. *Determination of dissolved combined phosphorus*

Until recently, it was possible to determine only the free orthophosphate and the total orthophosphate remaining after an oxidative digestion under acidic conditions. The difference between these two quantities was generally considered to be organic phosphorus. However, inorganic condensed phosphates, which are now known to be present occasionally (see Section 12.3.1.2) will also be included in this estimate.

Several workers have used acid hydrolysis in the determination of total phosphorus. Thus, Harvey (1948) carried out the hydrolysis of samples under pressure at 130°C after rendering them 0·5 N with respect to sulphuric acid. However, it is possible that acid hydrolysis, even at the much higher acidity recommended by Kabanova (1961), will not completely decompose all the organic phosphates likely to be present in sea water, let alone compounds containing the much more stable C–P linkages. Thus, Vogler (1966), using conditions similar to those employed by Harvey, found recoveries ranging from 5% for phosphorocholine to 100% for glucose-1-phosphate for the twelve compounds which he examined. In order to ensure complete decomposition of the very wide range of organic phosphorus compounds present in sea water it is therefore obviously necessary to use an oxidative process. In early methods the oxidation was brought about by drastic means, usually involving fuming with sulphuric or perchloric acids (see Riley 1965b; Armstrong, 1965). Although most of the naturally occurring phosphorus compounds can be oxidized efficiently in this way (see e.g. Burton and Riley, 1956), the procedures are time-consuming and it is difficult to avoid contamination. For these reasons these methods are now little used in the determination of dissolved organic phosphorus; however they do still find application in the determination of particulate phosphorus (see e.g. Strickland and Parsons, 1968). A simpler and more satisfactory oxidation procedure has been developed by Menzel and Corwin (1965) who suggested autoclaving the sample with potassium persulphate. This process has been found to give excellent recoveries with a broad spectrum of organic phosphorus compounds including some having carbon-phosphorus linkages. e.g. phenylphosphonic acid (Gales et al., 1966; Jenkins, 1967, 1968). Hydrogen peroxide has also been used for carrying out the oxidation in fresh water medium (Harwood et al., 1969), but has been found by Jenkins (1967) to be less efficient than persulphate for estuarine waters.

Armstrong et al. (1966) and Armstrong and Tibbitts (1968) have demonstrated that many organic phosphorus compounds, including some with C–P bonds, undergo photo-oxidation in sea water medium when they are irradiated with short wavelength ultra-violet radiation. This observation provides the basis for an elegant and extremely satisfactory method for the determination of total dissolved phosphorus in sea water (for practical details see Strickland and Parsons, 1968). In it, the sample is treated with a small quantity of hydrogen peroxide and irradiated with a 0·8–1·0 kW medium pressure mercury vapour lamp. After 1 hour the orthophosphate ion produced is determined photometrically. Since no significant hydrolysis of condensed phosphates occurs if the temperature is not allowed to rise above 60°, it is possible to determine them separately by determining the phosphate liberated on acid hydrolysis of an aliquot of the irradiated sample (Solórzano and Strickland, 1968; Armstrong and Tibbitts, 1968). There is at present no means of differentiating between inorganic and organic condensed phosphates, but it might be possible to achieve separation of these species by gel filtration (see e.g. Yoza, 1973).

19.11.3. SILICON

Silicon occurs in sea water as both particulate and dissolved species. The former are principally silicate minerals, which may be of either detrital or authigenic origin, and siliceous organisms and their remains. The concentration of particular silicon in the seas varies widely, and is high in neritic waters and near the mouths of large rivers. The concentrations in the ocean are comparatively low, and probably average about 50 µg–at. 1^{-1}.

Dissolved silicon is commonly regarded as a micronutrient element as it is essential for the healthy growth of a number of classes of lowly plants and animals (e.g. diatoms and radiolaria) which have skeletons of opaline silica ($SiO_2 x H_2O$). In the sea, it ranges in concentration from ca. 1–150 µg–at. Si 1^{-1}, and exists almost entirely as ortho-silicic acid, which at pH 8·2 is ca. 5% ionized. It seems unlikely that polymeric forms of silicic acid are present as they are known to be rapidly depolymerized in sea water (Burton, *et al.*, 1970), but experimental data about this are contradictory. Only silicic acid and its dimer are estimated in most of the methods used for the determination of dissolved silicon; however, these are probably the only forms of the element which can be used by siliceous organisms. This fraction has been terms "reactive silicate" by Strickland and Parsons (1968).

Two basic techniques are in use for the determination of dissolved silicon in sea water. Both are based on the formation of a hetero-poly acid by treatment of the sample with an acidic molybdate solution; in one of these procedures, the absorbance of the yellow silicomolybdic acid itself is measured. In the other, which is considerably more sensitive, spectrophotometry is carried out on the molybdenum blue complex which is formed by reduction of the heteropoly acid.

The development of methods for the determination of dissolved silicon is made difficult by the existence of two forms of silicomolybdic acid. Strickland (1952b) has designated these the α- and β-forms and has shown that they are isomeric and differ only in their degree of hydration. The proportions in which these species are formed depends on the concentrations of the various molybdate species in the solution; these, in turn, are controlled by the acid/molybdate ratio. According to Strickland (1952b) the β-isomer arises from the $Mo_4O_{13}^{2-}$ ion. Its formation is favoured by a low pH and an H^+ : Mo molar ratio of 3:5 (the maximum yield of the β-form is obtained when the solution is 0·08 N and 0·3% with respect to hydrochloric acid and ammonium molybdate respectively). The β-isomer is unstable and slowly transforms irreversibly into the less intensely coloured α-form. The latter is formed as the principal product when the pH of the solution lies within the range 3·7–4·0; there is some uncertainty about which is the dominant molybdate species under these conditions, but it may be the octamolybdate ion (see e.g. Liss and Spencer, 1969; Chalmers and Sinclair, 1965).

Methods for the determination of silicon based on photometry of the yellow silicomolybdic acid have been reviewed by Mullin and Riley (1955b).

Although a few workers still favour them (see e.g. Grasshoff, 1964a; Lopez-Benito, 1970),* they have largely fallen into disuse as they suffer from several important drawbacks.

(i) Under the reaction conditions chosen by most workers β-silicomolybdic acid is the principal product (the actual ratio of α- to β-isomers obtained depends on the relative concentrations of acid and molybdate). The β-heteropoly acid changes rapidly to the α-isomer, with the result that the colour, after reaching a maximum, fades fairly quickly. In an attempt to overcome this difficulty some workers have preferred to carry out photometry on the α-isomer itself. Thus, Anderson (1958) heated the β-compound to 100°C to bring about the transformation to the α-form. Grasshoff (1964a) recommended that the formation of the heteropoly acid should be carried out at a pH value of 3·7–4·0 (chloroacetic acid buffer) so as to produce the α-isomer. However, Liss and Spencer (1969) found that Grasshoff's technique tended to give high results and cast doubts on whether α-silicomolybdic acid is, in fact, the principal reaction product. †

(ii) The sensitivity is insufficient for the determination of silicon at concentrations below 3 μg–at. l^{-1}. However, it can be increased several fold by extracting the silicomolybdic acid with ethyl acetate and measuring the absorbance at 335 nm, the absorption maximum of β-silicomolybdic acid (Schink, 1965).

(iii) The calibration curve is non-linear at high silicon concentrations.

(iv) There is an appreciable salt error, the magnitude of which has been evaluated by Saeki (1950), Bien (1958) and Kato and Kitano (1966).

Most workers now favour methods in which β-silicomolybdic acid is reduced to an intensely coloured molybdenum blue complex, having its absorption maximum at ca. 815 nm (reduction of the α-isomer gives a similar complex). It is necessary to control the acidity of the solution carefully during the reduction process in order to avoid the formation of molybdenum blue, either from the excess molybdate of the reagent, or from heteropoly acids of other elements. If high precision is to be attained it is essential, because of the instability of β-silicomolybdic acid, to control the time elapsing before addition of the reducing agent (Fanning and Pilson (1973)). Several reducing agents have been employed and all give similar sensitivities.

---

* The optimum conditions for the photometric determination of silicon in sea water by this technique have been reinvestigated by Sugawara (1969). However, he concluded that methods involving reduction to molybdenum blue were more reliable, particularly at low silicon concentrations.

† Conditions for the formation of molybdosilicic acids have been recently investigated by Truesdale and Smith (1975).

Stannous chloride (Armstrong, 1951; Ostrowski and Czerwinska, 1959) gives a rapid reduction, but the resultant molybdenum blue is unstable and its colour fades quite rapidly. Mullin and Riley (1955b) who investigated the use of a number of organic reducing agents, including hydroquinone, and 1-amino-2-naphthol-4-sulphonic acid (see Wilson, 1965), recommended p-methylamino phenol sulphate (Metol) because of the stability of the colour which it produces; this reagent has been adopted by other workers (e.g. Strickland and Parsons, 1968; Novoselov and Simolin, 1965; Lopez-Benito, 1970). In the procedure developed by Mullin and Riley loc. cit. interference from phosphate is prevented by the addition of oxalic acid to decompose phosphomolybdic acid.* The method is subject to a salt error which varies linearly with salinity, and amounts to a reduction in absorbance of ca. 10% at a salinity of 34‰. The determination can be carried out in glass vessels provided that they have been allowed to stand for several hours filled with a mixture of equal volumes of concentrated nitric and sulphuric acids and then well washed. However, polyethylene bottles are also satisfactory and require little pre-treatment (Folkard and Jones, 1968). For practical details of this determination the monograph by Strickland and Parsons (1968) should be consulted.

The Metol reduction procedure has been adapted to the automatic determination of silicate in sea water by several workers (Brewer and Riley, 1966; Grasshoff, 1966; Strickland and Parsons, 1968; Lopez-Benito, 1970). Wilson (1965) and Armstrong et al., (1967) have devised similar methods using 1-amino-2 naphthol-4-sulphonic acid and stannous chloride as reductants respectively. Brewer and Riley (1967) showed that there was a high degree of correlation between the results obtained for ocean water with their automatic method and those found by the manual technique of Mullin and Riley (1955b). An analogous study by Hager et al. (1972) has given similar results.

Standard silicate solutions. Standard silicate solutions are best prepared by fusing pure silica with a five-fold excess of sodium hydroxide, dissolving the fused cake in water and adjusting to $\sim$pH 5·0 before dilution to volume. The resultant standard is stable indefinitely if kept in a polyethylene bottle. The stability of the standard is reduced if its pH is much less than 5·0. Sugawara (1969) has studied the problems of storing the standard silicate solution. According to Fanning and Pilson (1973) the use of sodium silicofluoride (Strickland and Parsons, 1968) is not to be recommended for the preparation

---

* According to Chalmers and Sinclair (1966) mannitol is preferable to oxalic acid for the prevention of phosphate interference.

of standard silicate solutions as it gives less colour in the molybdenum blue procedure than is obtained with the equivalent amount of silica.

Other methods for the determination of reactive silicate. Few alternatives to the above photometric methods are available for the determination of reactive silicate in sea water. A direct emission spectrographic procedure which has a sensitivity of 10 µg–at. Si l^{-1} has been described by Lotrian and Johannin-Gilles (1969). A promising approach which does not appear to have been exploited yet is the amplification technique. When silicomolybdic is formed, and solvent extracted away from the excess molybdate it is possible to determine the molybdenum which it contains by a sensitive colorimetric or atomic absorption spectrophotometric technique, and thus take advantage of the very favourable atomic ratio of Mo:Si (12:1) in the heteropoly acid.

Comparative studies of reactive silicate procedures. A number of collaborative intercomparisons of the determination of reactive silicate have been carried out (see Palmork, 1968, 1969; Bogoyavlenskii, 1965c; Stefansson and Olafsson, 1970; Spencer, 1973). Although Palmork (*loc. cit.*) has reported divergences of up to 25% between silicate determinations carried out by different techniques; Liss and Spencer (1969) found that the results obtained with the stannous chloride technique of Armstrong (1951) agreed well with those given by the method of Strickland and Parsons (1968) in which Metol is used as the reducing agent. Coefficients of variation of ca. 0·6% were obtained in preliminary intercalibration work for the Geosecs Expedition (Spencer, 1973).

Determination of total dissolved silicon. Only monomeric and dimeric forms of silicic acid are measured by the above techniques. If it is desired to determine total "dissolved" silicon in water it is necessary to heat the sample with sodium hydroxide in a PTFE beaker. This treatment converts colloidal silica and polysilicic acids into orthosilicic acid which can then be determined (see e.g. Morrison and Wilson, 1967). However, it should be pointed out that it is improbable that these polymeric forms of silica are normally present in significant concentrations in sea water (Burton, *et al.*, 1970; Lisitsyn *et al.*, 1965).

19.12. DETERMINATION OF THE RADIOELEMENTS, URANIUM-238, URANIUM-235 AND THORIUM AND THEIR DECAY PRODUCTS

19.12.1. INTRODUCTION

Studies of the marine distribution of the naturally occurring actinide elements and their decay products have proliferated in the last few years. Such work has

P

been concerned not only with the geochemistry of these nuclides, but also with their use as tracers for the study of oceanic mixing processes. Of these elements, only uranium is present in sea water at a sufficiently high concentration for it to be determined by chemical methods (see Section 19.10.5.56). The other nuclides are determined by radioactive counting techniques. Since the activities are extremely low it is necessary to concentrate the nuclide concerned from a large volume of water (up to 1000 l). and to separate it from other activities. When β- or γ-activity is to be counted it is essential that the background count is kept to a minimum by the use of heavy shielding and anticoincidence circuitry.

The determination of the actinides and their radioactive decay products in sea water is usually carried out according to the following sequence of operations.

(1) Equilibration of the sample with a known amount of the stable form of the element to be determined, or if the element has no stable nuclides, with one of its synthetically produced nuclides (which decays with the emission of either a different type of radiation or particles of different energy, e.g. the use of the γ-emitter ^{234}Th in the determination of α-active nuclides of thorium). This serves to provide a means of determining the overall yield of the concentration and separation process, and in addition, if weighable amounts are added, it acts as a carrier.

(2) Concentration of the nuclide by procedures similar to those used for preconcentration of trace elements (see Section 19.10.2), particularly coprecipitation with ferric hydroxide (Rose, 1959).*

(3) Separation of the nuclide in a state of radiochemical purity by use of appropriate techniques, such as solvent extraction, ion exchange, co-precipitation, scavenging, etc.

(4) Conversion of the nuclide into a form suitable for counting and determination of chemical yield. If the stable element is being used as carrier, this may involve precipitation of an insoluble compound which can be filtered, weighed and counted. In other instances the desired radio-nuclide along with its radioactive carrier is electrodeposited onto a silver or stainless steel disc.

(5) Determination of total recovery using the tracer. When the stable element is used as tracer this is usually done gravimetrically, but alternatively, may be achieved by suitable physicochemical methods, such as spectrophotometry. If the tracer itself is a radionuclide the recovery is assessed

---

* Krishnaswami et al. (1972) have suggested the use of acrilan fibre loaded with ferric hydroxide for the in situ preconcentration of radio-elements from very large volumes of sea water, particularly for the estimation of ^{228}Ra:^{226}Ra and thorium isotope ratios and perhaps ^{210}Pb.

by use of counting techniques which will differentiate between its activity and that of the nuclide being determined.

(6) Counting of the desired activity. Since uranium-238, uranium-235 and thorium-232 and many of their decay products are α-emitters this is frequently achieved by α-spectrometry.

The choice of techniques to be used for concentration and purification depends largely on the radionuclide concerned. Although rather out-dated, the series of publications of the U.S. Atomic Energy Agency on the radiochemistry of the elements still provides a useful guide to separation methods. Procedures which have been used for thorium and both its decay products and those of uranium-238 and uranium-235 are reviewed below. (The determination of uranium itself and the estimation of the $^{234}U:^{238}U$ ratio are discussed in Section 19.10.5.56).

19.12.2. DETERMINATION OF RADIONUCLIDES

19.12.2.2. *Ionium.* See thorium-230.

19.12.2.2. *Lead-210*

The decay of lead-210 ($t_{\frac{1}{2}}$, 22 years) proceeds by the following sequence $^{210}Pb \xrightarrow{\beta^-} {}^{210}Bi$ ($t_{\frac{1}{2}}$, 5·0 days) $\xrightarrow{\beta^-} {}^{210}Po$ ($t_{\frac{1}{2}}$, 138·4 days) $\xrightarrow{\alpha} {}^{206}Pb$ (stable). Several approaches have been used for its determination in sea water. In the earliest of these (Rama *et al.*, 1961), iron(III) and lanthanum hydroxides were used as coprecipitants and the ^{210}Bi daughter was separated by ion exchange and determined by β-counting. Tsunogai and Nozaki (1971, 1973), Shannon and Orren (1970) and Shannon *et al.* (1970) preconcentrated lead-210 together with its decay products. After removal of ^{210}Po by plating the concentrate was allowed to stand for at least 3 months to re-establish equilibrium between the lead-210 and its granddaughter ^{210}Po. The latter was then plated out and determined by α-counting.

19.12.2.3. *Polonium-210*

Shannon and Orren (1970) determined polonium in sea water by extracting it as its pyrrolidine dithiocarbamate complex at pH2. The residue remaining after evaporation of the extract was dissolved in hydrochloric acid. Polonium was plated out on a silver disc and determined by α-counting. Tsunogai and Nozaki (1971, 1973) carried out the preconcentration of polonium, using bismuth as carrier, coprecipitated it by addition of a mixture of sodium and ammonium carbonates. After dissolution of the precipitate polonium was plated out and determined by counting its 5·298 MeV α-activity in a 2π gas flow counter.

19.12.2.4. *Potassium- 40*

Although potassium-40 accounts for more than 90% of the total radio-activity in sea water very little attention has been paid to it. Riel *et al.* (1965) have described an *in situ* method for its determination.

19.12.2.4. *Protactinium- 231*

Methods for the determination of protactinium-231 ($t_{\frac{1}{2}}$, $3\cdot43 \times 10^4$ years) in sea water have been described by Sackett (1960), Moore and Sackett (1964), Kuznetsov *et al.* (1960a, b) and Imai and Sakanoue (1973). Basically all these methods are very similar. Concentration is often effected by coprecipitation with ferric hydroxide; sometimes barium sulphate is also used. The protactinium is subjected to a separation process, usually concluding with an extraction stage in which it is extracted from hydrofluoric acid medium using methyl *iso*-butyl ketone. The protactinium is back extracted with 8M HCl–$0\cdot1$–$0\cdot5$ M HF, and recovered either by evaporation or electrodeposition. Determination is carried out by α-counting or α-spectrometry. Protactinium-234 ($t_{\frac{1}{2}}$, $27\cdot4$ days; which decays by β- and γ-emission) can be used as a tracer for yield measurement (see e.g. Kuznetsov, 1966a; Imai and Sakanoue, 1973).

19.12.2.5. *Radium- 226*

Two different approaches have been made to the problem of determining radium-226 in sea water. In the first of these, which has been used by Sugimura and Tsubota (1963), Miyake *et al.* (1964) and Nomura (1971), the element is preconcentrated by coprecipitation and, after purification is determined by α-spectrometry. In the most widely used approach, the radium is allowed to come to equilibrium with its gaseous daughter radon-222 ($t_{\frac{1}{2}}$, $3\cdot83$ days) and the α-activity of this is then counted. In most earlier work (see e.g. Føyn *et al.*, 1939; Rona, 1943; Rona and Urry, 1952; Pettersson, 1955; Koczy and Szabo, 1965; Szabo, 1967a, 1971), radium was preconcentrated by coprecipitation with barium sulphate and, after standing for 1–2 weeks for attainment of radioactive equilibrium, the radon generated was swept into an ionization chamber. However, recent emphasis has been mainly on simpler and more direct procedures (see for example, Drozhzhin *et al.*, 1965; Lazarev *et al.*, 1965; Broecker, 1965; Broecker *et al.*, 1967, 1970; Ku *et al.*, 1970; Broecker and Peng, 1971, 1971a; Chung and Craig, 1972). In these, radon is first stripped from the sample with a current of helium or nitrogen. The sample is then allowed to stand for 1–2 weeks for the radium to come to equilibrium with its daughter radon. At the end of this time the radon is stripped from the sample and transferred to a trap cooled with liquid nitrogen* by circulating

* Alternatively the radon can be adsorbed onto activated charcoal at $-40°C$; however, desorption requires a temperature of 350–400°C (Drozhzhin *et al.*, 1965).

helium through it. Subsequently the trap is allowed to warm up and the radon is swept into the cell of an α-scintillation counter or α-spectrometer. The main source of error in this technique is the radon discharged into the sample from the walls of the container during the period while equilibration is taking place. The amount of radon entering in this way, even under favourable circumstances (i.e. with bottles of flint glass or polyethylene), may amount to ca. 30% of that generated from the radium in the sample (Broecker et al., 1970).

Broecker et al. (1970) have reported the results of an intercalibration exercise on the determination of radium in sea water which was carried out as a preliminary to the Geosecs Project. In this, samples from a vertical profile at a station in the N. Pacific were analysed by four workers. Three used the direct radon stripping method, and the other employed a highly elaborate technique involving a preliminary coprecipitation of the radium with barium sulphate (to obviate errors associated with the bleeding of radon which are troublesome in direct methods). Results from all four methods agreed to better than $\pm 15\%$.

19.12.2.6. Radium-228

Radium-228 ($t_{\frac{1}{2}}$, 6·7 years), the daughter of thorium-232 ($t_{\frac{1}{2}}$, $1·39 \times 10^{10}$ years) is much more difficult to determine than its isotope radium-222. Its estimation was first described by Moore (1969a, b). The total radium was concentrated by coprecipitation with barium sulphate. The barium sulphate was converted to barium chloride which was washed with hydrochloride acid containing 5% of diethyl ether to remove Ca and Sr. At this stage the chemical yield was assessed by allowing the concentrate to stand and counting the ^{222}Rn generated from the ^{226}Ra originating from the sample. It was not possible to measure ^{228}Ra by counting its β-activity, because the activities of the daughters generated from the relatively large amount of ^{226}Ra present in the sample swamped the β-activity arising from the ^{228}Ra and its daughter ^{228}Ac. Instead, after purging with nitrogen to remove ^{222}Rn, actinium-228 ($t_{\frac{1}{2}}$, 6·13 hours) was extracted as its chelate with thenoyltrifluoroacetone into benzene. The extract was evaporated on a planchet and the ^{228}Ac activity was counted using a heavily shielded gas-flow β-ray counter with anti-coincidence circuitry. Further work on ^{228}Ra in sea water has been reported by Sakanoue et al. (1970), Kaufman (1969), Imai and Sakanoue (1973), Krishnaswami et al. (1972) and Trier et al. (1972).

19.12.2.7. Radon-222

Although the occurrence of radon-222 in sea water was first reported almost 40 years ago (Evans, 1935; Evans et al., 1938), it is only within the last decade that realization of the potential use of the element in studies of gas exchange

(Matthieu, 1969), vertical eddy diffusion (Broecker *et al.*, 1967) and bottom water processes (Chung and Craig, 1972) has stimulated work on its oceanic distribution.

In practically all the methods which have been described for the determination of radon in sea water the gas is stripped from the water with a current of nitrogen, or preferably, helium. A few workers transfer the gas directly to the counter (Evans, 1935; Evans *et al.*, 1938; Iwasaki and Ishimori, 1951). However, most of them purify the radon first by condensing it in a cold trap (Jacobi, 1949; Broecker, 1965; Broecker *et al.*, 1967; 1968; 1970; Broecker and Kaufman, 1970; Broecker and Peng, 1971; Schink *et al.*, 1970; Chung and Craig, 1972). Counting of the purified radon is usually carried out using an α-scintillation counter or α-spectrometer. Although these direct methods for the determination of radon in sea water have obvious advantages, attempts have also been made to determine it using its short-lived decay products. For example, Kimura *et al.* (1955) separated and β-counted its ^{214}Pb granddaughter ($t_{\frac{1}{2}}$, 26·8 min) and Kerr *et al.* (1962) have made use of its great granddaughter ^{214}Bi ($t_{\frac{1}{2}}$, 19·7 min). Such indirect methods have, however little to recommend them.

19.12.2.8. *Thorium*

Although a number of photometric (see e.g. Ishibashi and Higashi, 1954; Lazarev *et al.*, 1965) and even spectrographic (Ishibashi and Azuma, 1949) methods for the determination of thorium in sea water have been described, the results obtained with them are so much higher than those obtained by radiometric methods that they can be discounted.

Unsuccessful attempts to apply radioactive counting techniques to the problem were made by both Føyn *et al.* (1939), who counted the thoron (^{220}Rn) emitted from a preconcentrate, and by Koczy *et al.* (1957) who examined α-tracks in nuclear emulsions exposed to a concentrate from the sea water. It was not until the late 1950's that α-counting techniques had been refined sufficiently for the extremely low levels of thorium present in the sea to be established (Sackett *et al.*, 1958). Subsequently, the development of reliable α-spectrometers has widened the scope of the analysis to include not only the determination of thorium, but also measurement of the ^{230}Th:^{232}Th and ^{228}Th:^{232}Th ratios (Moore and Sackett, 1964; Kuznetsov *et al.*, 1966c; Somayajulu and Goldberg, 1966; Miyake *et al.*, 1970b; Nomura, 1971; Krishnaswami *et al.*, 1972; Imai and Sakanoue, 1973; for further references see Table 18.4). In such procedures, typified by the method described by Imai and Sakanoue (*loc. cit.*), the sample (1000 l) is spiked with ^{234}Th (a β- and γ-emitter with half life of 24·1 days) which serves as a tracer for estimation of chemical yield. Thorium, together with U Pa and Pu, are coprecipitated with ferric hydroxide at pH 10. The precipitate is dissolved in hydrochloric acid and iron is extracted from 8 M HCl–0.1 M HF medium using di-*iso*-butyl ketone. Aluminium is then added to mask the fluoride ion, and Pa is removed by extraction with di-*iso*-butyl ketone. The aqueous phase is evaporated to dryness and the residue is taken up in 8 M HNO_3 and Pu and Th are extracted with tri-*n*-octylamine. Thorium is back-extracted with conc. hydrochloric acid and purified by cation exchange, before electrodeposition and α-spectrometry for ^{228}Th, ^{230}Th and ^{232}Th.

Instrumental neutron activation analysis has been used by Spencer *et al.* (1972) for the determination of thorium-232 in marine particulate matter. Other determinations of particulate thorium have been carried out by Miyake *et al.* (1970b) and Kuznetsov *et al.* (1964).

Uranium. See Section 19.10.5.56

19.13. Determination of Radionuclides other than those of the Uranium and Thorium Series

by J. W. R. Dutton and N. T. Mitchell

Fisheries Radiobiology Laboratory, Lowestoft, Suffolk, England

19.13.1. introduction

Naturally-occurring radioactivity in the sea has either a terrigenous or cosmic origin. The most abundant (in terms of radioactivity) are 3 radionuclides from the first group—potassium-40 (670 dpm* l^{-1}), rubidium-87 (64 dpm l^{-1}), and uranium-238 (about 2 dpm l^{-1}), with which are associated its daughters, the $4n + 2$ series. Other terrigenous radionuclides identified and measured in sea water include thorium-232 and its daughters (the $4n$ series) see Section 19.12 and uranium-235 (and the $4n + 3$ series) (see Sections 19.10.5.56 and 19.12). The distribution of these radionuclides is relatively uniform, unlike those of cosmic origin whose concentrations might be expected to be higher in surface water: these latter include ^{3}H, ^{7}Be, ^{14}C and ^{32}Si. Radionuclides from both these groups have been used to study mixing and removal processes, although not all have been extensively determined in sea water. Sediments, suspended matter and biological materials are often studied as tracers for chemical and physical processes occurring in the sea.

The testing of nuclear weapons has resulted in the presence in the hydrosphere of measurable quantities of many more radionuclides (see also Chapter 18). The method of production varies: radioactive fissile material will already be present in the weapon, and further activity is produced during the explosion. This is caused by fission of the plutonium or uranium, by neutron activation of the fissile material and of the weapon hardware, and by neutron activation

* dpm = disintegrations per minute.

of the environment at the time of detonation. The pattern of dispersion varies with the type of explosion: direct mixing may occur between local, high-level contamination, and the sea, especially for sea-bursts, but these effects are generally confined to a relatively small area and the largest part of the radio-activity from uncontained explosions enters the stratosphere, from which some will return to the oceans as fallout following meteorological mixing processes. These uncontained tests were carried out on a relatively large scale between 1945 and 1958, and again during 1961 and 1962, but since that time the frequency of testing has decreased considerably. As a result, the rate of deposition of radioactivity has also decreased, but concentrations in the surface waters, after increasing between 1957 and 1966, have remained substantially constant.

The increasing use of nuclear reactors to produce electricity and the subsequent processing of the irradiated fuels has resulted in an economic requirement for the discharge to sea of a very small proportion of the waste as liquid radioactive effluent. At present, the quantities discharged are only a small fraction of the annual deposition from fall-out, but they result (as does the nuclear weapon sea burst) in local distributions of relatively high activity. The types of radionuclide encountered in these circumstances are similar to those derived from nuclear weapon testing; variations will, however, be encountered mainly due to the longer irradiation times, and also on account of the different materials used in reactor systems. Other man-made sources of radioactivity exist which result, or have resulted, in measurable contamination of the marine environment; the use of nuclear reactors for submarine propulsion, heat sources for aerospace power generators, radioactive tracers for hydrological studies—all these have resulted in the deliberate or accidental release of measurable quantities of radionuclides.

As a result of the varied modes of introduction of radionuclides into the marine environment, a wide range of different concentrations and physical states may be encountered, and analytical techniques will differ accordingly. Volumes of sample required will vary from 1 to 10^3 litres, different initial treatment will be necessary, separation techniques will vary depending on the concentrations of interfering radionuclides and stable elements and the purity of reagents required will sometimes be very high indeed; different methods of measurement may be possible, and may, indeed, be necessary because of the presence of more than one radioisotope. These different analytical requirements have resulted in a wide variety of reported methods for the measurement of radioactivity in the sea.

19.13.2. SAMPLE STORAGE

Although some initial separation or concentration steps can undoubtedly be carried out on board ship, it is general practice to store the samples in plastic containers for subsequent analysis in the laboratory, and consideration must, therefore, be given to the stability of radionuclides during storage. Problems associated with storage are discussed in an International Atomic Energy Agency technical report (1970) and by Riley (1965b; see also Section 19.5), and it appears that for many radionuclides acidification (usually with HCl) to a pH of about 1–1·5 is generally recommended. It is usually assumed that the radioisotopes of strontium (^{90}Sr and ^{89}Sr), and caesium (^{137}Cs and ^{134}Cs) are stable during storage at the natural pH of sea water because isotopic equilibrium with the simple ionic forms of the stable elements (present at about 8 mg l^{-1} and 0·3 μg l^{-1}, respectively) will be readily achieved, but with other radionuclides—especially those which have no stable counterparts in sea water—additional precautions may need to be taken, such as the use of reducing agents to stabilize plutonium as Pu(III). It should also be noted that, because radioactivity in the marine environment comes from many different sources in a variety of chemical and physical states, it may be misleading to assume that, for instance, ruthenium-106 derived from fall-out has the same storage characteristics as that from discharge of liquid radioactive wastes from the reprocessing of irradiated fuel. The storage stability should, therefore, be checked by analysing samples from the environment under investigation, and it is important that details of precautions taken are included in any report, together with the wash-out procedures used to remove any absorbed material from the container prior to analysis.

19.13.3. DIRECT MEASUREMENT OF RADIOACTIVITY

The measurement of gross β activity could hardly be expected to result in useful data—the potassium-40 alone would result in a β activity of about 600 dpm l^{-1}—and any other radionuclides present would need to be calculated by difference from this natural level. The more specific properties of γ photon emission have enabled γ-spectrometric techniques to be used with some success to obtain direct information of the radioactive content of sea water, both by *in situ* measurement (Chesselet *et al.*, 1964), and by measuring the activity of a portion of the liquid or dried material, using either NaI(Tl) or Ge(Li) spectrometric systems. These measurements are, in general, restricted to higher concentrations (> 10 pCi l^{-1}), and it is usual to resort to separation and concentration steps for lower concentrations of activity. Direct measurements of radioactivity in biological materials and sediments have been extensively made by γ-spectrometric techniques, and have been used to trace water movements and to investigate transfer processes in the oceans.

19.13.4. RADIOCHEMICAL SEPARATIONS

If direct measurement of the radioactive content of sea water is impracticable, then recourse must be made to chemical separation and measurement of the activity, and the following general approach typical of radiochemical separations can be formulated.

(a) The first step is usually carrier addition and isotopic equilibration. The use of stable carriers generally assists the ease of handling of precipitates, etc., and enables the yield of the chemical process to be measured; it must therefore be ensured that the chemical (and physical) states of the radioactive isotope and the stable carrier are the same. It has already been noted that isotopic equilibration can be assumed to be naturally adequate in the case of radiostrontium and radiocaesium; equilibrium with the stable carrier can be ensured for other radioisotopes by the use of oxidation/reduction cycles, or by the formation of organic or inorganic ligand complexes. Examples of equilibration procedures are the oxidation of Ru(III)→Ru(VIII) (World Health Organization, 1966) in the determination of ^{106}Ru, the oxidation/ reduction cycle of iodine from $I^- \rightarrow IO_3^- \rightarrow I^0$ (Bhat et al., 1972) in the determination of ^{131}I, and the use of complex formation with oxalic acid and carrier zirconium for ^{95}Zr (Hampson, 1963). If it is proposed to shorten or omit altogether the equilibration step, then it is advisable to test the method both with, and without, the complete exchange cycle.

The use of radioactive tracers to measure yields is becoming more general; they enable yields to be determined for radionuclides with no stable isotopes, and examples are the use of γ-emitting ^{234}Th to measure the yield of α-emitting thorium isotopes (Sackett et al., 1958), and ^{236}Pu as the yield indicator for ^{239}Pu (and ^{238}Pu) (Pillai et al., 1964).

(b) Concentration of the radionuclide (with, or without, carrier addition and isotope equilibration) from the sea water is generally necessary to ensure separation from the salt water matrix and to permit subsequent chemical procedures to be carried out on a normal laboratory scale. Techniques used include coprecipitation, ion exchange (using organic or inorganic exchangers), adsorption (onto manganese dioxide (Yamagata and Iwashima, 1963) for example) and solvent extraction. These concentration procedures may not be specific, nor may they need to be, for many are useful as a means of freeing groups of radionuclides from much of the matrix material so that γ-spectrometric analysis can be applied to the separated material. An example of this approach is the use of alumina sorption beds by Robertson et al. (1968) who passed several thousand litres of sea water through such a filter; the radioactivity was then measured by γ-spectrometry. In the north-eastern Pacific, measurable quantities of ^{46}Sc, ^{51}Cr, ^{60}Co, ^{65}Zn, ^{54}Mn, ^{106}Ru, ^{110m}Ag, ^{144}Ce, ^{228}Th and ^{226}Ra have been collected from sea water in this way. A further example of a multi-nuclide scavenger is the use of manganese dioxide in the determination of radionuclides in sea water in the environs of the Cap de la Hague discharge (Guegueniat, 1967); in this instance, ^{95}Zr (and ^{95}Nb) and ^{106}Ru were determined. In both of these examples, the yields for the different radionuclides were measured by careful preliminary experiments.

(c) After the initial concentration step has been carried out, standard radiochemical procedures may be used to produce a separated source of the radionuclide with adequate yield and the necessary purity.

(d) The chemical yield of the separation process is carried out by determining the amount of stable carrier in the source presented to the detector of radioactivity; this is often measured by converting the final solution or extract to a precipitate of known stoichiometry which is then weighed prior to counting. Examples of suitable precipitates are caesium perchlorate or chloroplatinate (for ^{137}Cs), silver iodide (for ^{131}I), ruthenium metal (for ^{103}Ru and ^{106}Ru) and zirconium oxide (for ^{95}Zr). Other methods of measuring the yield by determining the final stable carrier concentration include UV spectrophotometry (e.g. of potassium perruthenate solution (World Health Organization, 1966), flame-emission photometry (e.g. of the concentrated Cs naturally present in sea water (Folsom and Sreekumaran, 1970)) and volumetric analysis (for strontium (Noshkin and Mott, 1967)).

The use of radioactive tracers to measure the yield of thorium and plutonium, where no stable element counterpart exists, has already been noted, but a further application of the technique is the addition of a radioactive tracer as a preferred method of yield determination even when the stable carrier is present. Strontium-85, for example, is now in common use to measure the yield of the ^{90}Sr separation from sea water. This radionuclide decays by electron capture, emitting 0·51 MeV γ photons: the yield of the procedure is measured by simple γ-counting.

(e) Measurement of radioactivity. Multi-nuclide separations must be followed by γ-spectrometry using either NaI(Tl) or Ge(Li) detection systems, and it is important to ensure that a correct interpretation is made of the resultant spectrum. The higher resolution of the latter detector enables a more certain identification of the γ-emitting nuclides to be made and should result in fewer systematic errors due to interference from γ-photon emissions of similar energy; on the other hand, the efficiency of the NaI(Tl) crystals is considerably greater, as is the ease with which large detectors can be made. Both systems are improved by the use of anticoincidence screening detectors, and some very sophisticated arrangements have been described by Wogman and others of the Battelle Northwest Laboratories (Wogman, 1970; Cooper *et al.*, 1968); the relative merits of the different solid-state detection systems have been evaluated by the same group (Cooper, 1969). For β-emitting radionuclides which do not emit γ-photons in significant quantity, β-counting of the separated radioactive fraction is necessary. Gas-filled counters operating in the proportional or Geiger region can be used, usually with anticoincidence shielding to reduce cosmic background, and the technique generally results in greater precision than does γ-spectrometry. The low energy β-emissions of tritium (0·018 MeV) and carbon-14 (0·159 MeV) can also be measured in this way, by incorporating the radionuclide, generally as a hydrocarbon, into the counting gas; carbon-14 can also be measured by liquid scintillation counting (Mathews *et al.*, 1973). A survey of systems for tritium concentration and counting has been made by Cameron (1967). The measurement of α-activity is usually effected by silicon surface-barrier detectors after chemical separation of the radionuclide. Because there are no stable element counterparts, it is usual to add another radioisotope as a tracer; if this also decays by α-emission, then it is necessary to use spectrometry to separate the different energies of emission and, because α particles are easily absorbed, special attention must be paid to the purity of the counting source.

The purity of the separated source (or the freedom from interference in the interpretation of activity from a γ-spectrum) must be carefully checked. The identity of the radionuclide can be confirmed thus: by half-life measurements (of the decay of the nuclide itself, or the in-growth of a daughter), by constancy of specific activity (determined by carrying out a purification step after counting and comparing the radioactive:stable isotope ratio with that of the original source), or by tests of spectral purity. These last checks are possible not only on α or γ spectra – where the relative abundance of several peaks can be measured and compared – but also on β-sources, whose purity when measured by a proportional or Geiger detector can be tested by counting with different absorber thicknesses between source and detector, or, for proportional or liquid scintillation counters, by the use of discriminators to measure the energy distribution of the β-spectrum.

The amount of activity in the reagents which contributes to the sample count may become significant when very low environmental levels are being measured, and it is important that the size and variability of this radiochemical blank be properly established. Deep-water strontium-90 data have been discussed by Volchok *et al.* (1971) who have stressed the extreme importance of the blank. Bowen *et al.* (1970) has discussed the radiochemical blank for strontium-90 determinations; his comments are also applicable to all other low-level radiochemical measurements.

It is clear that there is much value in participation in intercomparison exercises. Samples of sea water, sediment and biological material are available from the International Atomic Energy Agency through its Marine Laboratory at Monaco; concentrations of activity available vary from low fall-out levels to higher levels typical of the immediate environment of waste discharge, and some data from these exercises have already been published (Fukai *et al.*, 1973). The US Atomic Energy Commission has also organized intercomparisons for strontium-90 measurements in sea water; the results of some of these exercises have been published by Bowen *et al.* (1970).

19.13.5. DETERMINATION OF INDIVIDUAL RADIONUCLIDES

Summaries of methods used to determine radioactivity in sea water are presented in Table 19.20. The analysis of biological materials and sediments is generally similar to that of sea water, except that the initial treatment of the sample may involve the destruction of the organic fraction, leaching by acetic acid or strong nitric acid, or complete dissolution using hydrofluoric acid to volatilize silica. The initial treatment used and the radiochemical separations required will vary depending on the radionuclide composition of the sample and the objectives of the analyst; reviews of methods of separation and measurement can be found in the IAEA technical report series (e.g. IAEA (1970) and IAEA (in press)) in Nucleonics reviews published biennially in Analytical Chemistry, and in the older, but still very valuable, Radiochemistry series produced by the Sub-committee on Radiochemistry of the National Academy of Sciences—National Research Council. Nuclear Science Abstracts (issued bi-monthly) include methods for the determination of radionuclides in the marine environment. A list of 170 references to literature either reporting results on, or describing methods for, the determination of 26 radionuclides in the marine environment is given in the chapter "Oceanic Distributions of Radioactivity from Nuclear Explosions" by Volchok *et al.* in "Radioactivity in the Marine Environment" (1971).

19.14. DETERMINATION OF ORGANIC COMPONENTS

by P. J. LE B. WILLIAMS
Department of Oceanography, The University, Southampton, England

19.14.1. INTRODUCTION

The organic analysis of sea water is a subject area which has evolved very slowly. For some time it has been recognized that organic material, both in solution and suspension, has a profound effect on biological processes in the sea. Only recently has there been an awareness that the organic fractions, especially the dissolved part, may have an effect on the distribution of many inorganic species. In view of this, it is natural that much of the existing

TABLE 19.20

Determination of radionuclides other than those of the uranium and thorium series

Radionuclide	Method of separation	Method of determination	Reference
Tritium	Distillation. Electrolytic concentration. $H_2O \xrightarrow{Zn} H_2$	GM gas counting Proportional gas counting Proportional gas counting	Kaufman and Libby (1954) Brown and Grummitt (1956) Munnich and Roether (1967)
	Distillation. Electrolytic concentration. $H_2O \xrightarrow{Mg} H_2$	GM gas counting	Östlund and Werner (1962) Östlund (1962)
	Distillation. Electrolytic concentration. $H_2O \xrightarrow{Mg} H_2$; $2H_2 + C_2H_2 \xrightarrow{Pd} C_2H_6$ REVIEW	Proportional gas counting	Bainbridge *et al.* (1961) Cameron (1967)
Beryllium-7	Jute sponge impregnated with $Fe(OH)_3$; collection by towing. Iron dissolved and Be separated chemically	β^- counting	Lal *et al.* (1964)
	Sea water passed through filter beds followed by ^7Be adsorption on Al_2O_3	NaI (Tl) γ-spectrometry	Perkins *et al.* (1966)
Carbon-14[a]	CO_2 stripped from acid sample with N_2, collected in KOH; acidified; CO_2 collected by CaO and desorbed.	Proportional gas counting	Broecker *et al.* (1959, 1960)
	CO_2 stripped from acid sample with N_2, converted to $SrCO_3$: $SrCO_3 \rightarrow SrC_2 \rightarrow C_2H_2$	Proportional gas counting	Bien *et al.* (1965)
	CO_2 stripped from acid sample with N_2. Collected in KOH; CO_2 liberated $\rightarrow LiC_2 \xrightarrow{H_2O} C_2H_2 \xrightarrow{V} C_6H_6$	Liquid scintillation counting	Mathews *et al.* (1973)
Silicon-32	Jute sponge impregnated with $Fe(OH)_3$. Collection of SiO_2 by towing. Iron/SiO_2 dissolved and Si separated chemically. ^{32}P daughter separated	β-counting	Lal *et al.* (1964)

Phosphorus-32	Solvent extraction of molybdenum blue complex. Molybdate removed by extraction with tributyl phosphate. Aqueous phase evaporated to dryness	β-counting	Chakravarti et al. (1964)
Scandium-46	Sea water passed through filter beds followed by ^{46}Sc adsorption on Al_2O_3 filter	NaI (Tl) γ-spectrometry	Perkins et al. (1966); Robertson et al. (1968)
Chromium-51	Cr carrier (and others) added, precipitated with ammonia solution and filtered. Precipitate wet ashed with HNO_3 and H_2O_2 and dried	NaI (Tl) γ-spectrometry	Chakravarti et al. (1964)
	Cr carrier (and others) added, reduced to Cr^{III}, and precipitated with ammonia solution	NaI (Tl) γ-spectrometry	Cutshall et al. (1966)
	Sea water pumped through filter beds, followed by ^{51}Cr adsorption on Al_2O_3	NaI (Tl) γ-spectrometry	Perkins et al. (1966)
Manganese-54	Mn coprecipitated on $Fe(OH)_3$. Fe removed by extraction with methyl isobutyl ketone. Zn separation by ion exchange. Mn extracted into chloroform after complexing as permanganate with tetraphenyl arsonium chloride	NaI (Tl) γ-spectrometry	Shah and Rao (1972)
	(i) Coprecipitation on $Fe(OH)_3$ (ii) Oxidation, with carrier Mn, to MnO_2 (iii) Precipitation with H_2S	NaI (Tl) γ-spectrometry	Slowey et al. (1965)
Iron-55 and Iron-59	Coprecipitated on $Fe(OH)_3$ and $Mg(OH)_2$ at pH 10·0–10·2. Fe separated with thiourea and purified by precipitation with hexamethylenetetramine and Na_2CO_3. Reprecipitated with thiourea	Gas-flow proportional counting and NaI (Tl) γ-spectrometry	Kautsky and Schmitt (1962)
Cobalt-60	Adsorption on granular MnO_2	NaI (Tl) γ-spectrometry	Yamagata and Iwashima (1963)
	Co carrier added, pH adjusted to 11 with ammonia solution, sample saturated with H_2S. Filtered	NaI (Tl) γ-spectrometry	Chakravarti et al. (1964)

TABLE 19.20—*continued*

Radionuclide	Method of separation	Method of determination	Reference
	Sea water pumped through filter beds and ^{60}Co then adsorbed on Al_2O_3	NaI (Tl) γ-spectrometry	Perkins *et al.* (1966)
	Co carrier and hydroxylamine hydrochloride added to sea water. Co^{II} coprecipitated on $Fe(OH_3)Co$ separated by anion exchange as chloro complex	β^--counting	Bhat *et al.* (1972)
	REVIEW		Preston and Dutton (1970)
Nickel-63	Ni complexed with dimethylglyoxime, extracted into chloroform and back-extracted into the aqueous phase with dilute H_2SO_4	Liquid scintillation counting	Lai and Goya (1964)
Zinc-65	(i) Zn carrier added, precipitated with ammonia solution and filtered	NaI (Tl) γ-spectrometry	Chakravarti *et al.* (1964)
	(ii) Zn carrier added, pH adjusted to 11 with ammonia solution, and sample saturated with H_2S. Filtered		
	Sea water pumped through filter beds and ^{65}Zn then adsorbed on Al_2O_3	NaI (Tl) γ-spectrometry	Perkins *et al.* (1966)
	Zn carrier added and coprecipitated with $Fe(OH)_3$. Fe removed by extraction with methyl *isobutyl* ketone. Zn separated by ion exchange	NaI (Tl) γ-spectrometry	Shah and Rao (1972)
Strontium-89	Sr and Ca precipitated as oxalates. Sr/Ca separation by nitrate precipitation (correction for ^{90}Sr)	β-counting	Cigna (1963)

Element	Procedure	Measurement	References
Strontium-90[b)(c)]	Precipitation of alkaline earths as: (a) Carbonate with Na_2CO_3 $Na_2CO_3 + NH_4Cl$ $(NH_4)_2CO_3$ (b) Oxalate		Sugihara et al. (1959) Higano and Shiozaka (1960) Azhazha (1964) Sutton and Kelly (1968) Miyake et al. (1960) Rocco and Broecker (1963)
	Purification of strontium from calcium, etc, with: (a) fuming nitric acid (b) Ion exchange Final purification with ferric hydroxide and chromate scavenge		Rocco and Broecker (1963) Sugihara et al. (1959) Higano and Shiozaka (1960) Azhazha (1964) Noshkin and Mott (1967) All
	Milking of ^{90}Y daughter with: (a) Ferric hydroxide (b) Yttrium hydroxide Strontium recovery estimated by: (a) Gravimetry (b) Volumetry (c) ^{85}Sr counting Measurement of radioactivity	β-counting of ^{90}Y daughter	Sugihara et al. (1959) Higano and Shiozaka (1960) Azhazha (1964) Rocco and Broecker (1963) Miyake et al. 1(960) Noshkin and Mott (1967) Rocco and Broecker (1963) All
	REVIEW		Bowen (1970)
	Coprecipitated with Fe and La hydroxides, and purified by ion exchange	β-counting	Miyake and Sugiura (1955)
Zirconium-95	Zr carrier added and equilibrated with oxalic acid. Coprecipitated with	NaI (Tl) γ spectrometry	Hampson (1963)

TABLE 19.20—*continued*

Radionuclide	Method of separation	Method of determination	Reference
	zirconium phosphate. Rare earths scavenged with LaF_3, Ru scavenged with thioacetamide. Zr precipitated as hydroxide and ignited to oxide		
(with niobium-95)	Adsorption onto granular MnO_2	NaI (Tl) γ-spectrometry	Yamagata and Iwashima (1963)
(with niobium-95)	*In situ* measurement	NaI (Tl) γ-spectrometry	Chesselet *et al.* (1964)
(with niobium-95)	(i) Coprecipitated with $Fe(OH)_3$ (ii) Coprecipitated with mixed carrier sulphides	NaI (Tl) γ-spectrometry	Chakravarti *et al.* (1964)
	Sea water pumped through filter beds, and ^{95}Zr adsorbed onto Al_2O_3	NaI (Tl) γ-spectrometry	Perkins *et al.* (1966)
	Adsorption on MnO_2 produced in the sample by treating $KMnO_4$ with H_2O_2	NaI (Tl) γ-spectrometry	Guegueniat (1967)
	REVIEW		Chesselet (1970)
	REVIEW		Dutton (In press)
Niobium-95	Coprecipitated with Fe and La hydroxides, and separated by ion exchange using oxalic acid	β-counting	Miyake and Sugiura (1955)
	Coprecipitated with Zr phosphate and niobic acid. Rare earths removed on LaF_3 and ruthenium with thioacetamide. Zr removed as Ba ZrF_6. Nb extracted as cupferron complex and back extracted. Nb precipitated as niobic acid and ignited	NaI (Tl) γ-spectrometry	Hampson (1963)
Ruthenium-106[a]	Coprecipitated with Fe and La hydroxides. Fe removed by solvent extraction and Ru	β-counting[e]	Miyake and Sugiura (1955)

Nuclide	Procedure	Counting method	Reference
	separated by cation exchange		
	Adsorption onto granular MnO_2	NaI (Tl) γ-spectrometry	Yamagata and Iwashima (1963)
	(i) Coprecipitation with $Fe(OH)_3$	NaI (Tl) γ-spectrometry	Chakravarti et al. (1964)
	(ii) Carrier addition and sulphide precipitation		
	Coprecipitation with $Fe(OH)_3$. Fe removed by isopropyl ether extraction. Ru purified by distillation and solvent extractions as RuO_4. Ru back extracted and converted to lower oxides for counting. Chemical yield on aliquot of back extract by spectrophotometry	β-counting	Yamagata and Iwashima (1965)
	Carrier addition and equilibration. Ru separated by solvent extraction of RuO_4, back extracted and converted to lower Ru oxides. Chemical yield (after counting) determined by spectrophotometry	β-counting	World Health Organization (1966)
	Sea water pumped through filter beds, and ^{106}Ru adsorbed on Al_2O_3	NaI (Tl) γ-spectrometry	Perkins et al. (1966)
	Adsorption onto MnO_2 produced in the sample by reacting $KMnO_4$ with H_2O_2	NaI (Tl) γ-spectrometry	Guegueniat (1967)
	REVIEW		Chesselet (1970)
	REVIEW		Shiozaki et al. (In press)
Silver-110m[f]	Sea water passed through filter beds, and ^{110m}Ag then adsorbed on Al_2O_3	NaI (Tl) γ-spectrometry	Robertson et al. (1968)
Antimony-124	^{124}Sb and carrier coprecipitated with $Fe(OH)_3$	NaI (Tl) γ-spectrometry	Cutshall et al. (1966)
Antimony-125	Sea water filtered and Sb coprecipitated at pH 9·0–9·5 with $Fe(OH)_3$. Precipitate dissolved and Fe extracted with di-isopropyl ether. Aqueous phase evaporated to dryness	NaI (Tl) γ-spectrometry	Slowey et al. (1965)
Iodine-131	Iodide carrier added and sample treated	β-counting	Bhat et al. (1972)

TABLE 19.20—*continued*

Radionuclide	Method of separation	Method of determination	Reference
	with $NH_2OH.HCl$. I (and Br) removed on AgCl column. Chloride dissolved in ammonia solution. Residue dissolved in KCN solution and Br volatized with $H_2SO_4/K_2Cr_2O_7$. Iodine precipitated as AgI, purified by KCN solution and re-precipitated as AgI.		
Caesium-137	Initial decontamiration of sample by:		
	(a) Ferric hydroxide scavenging		Yamagata and Yamagata (1960)
	(b) Sulphide and carbonate precipitation		Sodd et al. (1960)
	Concentration by:		
	(a) $K_3Co(NO_2)_5$ coprecipitation		Sugihara et al. (1959)
			Folsom et al. (1960)
			Schmitt and Kautsky (1961)
			Malavicini and Vido (1961)
			Schroeder and Cherry (1962)
	(b) Co-crystallization with K-dipicrylamine complex		Yamagata and Yamagata (1960)
	(c) Ion exchange on Tl phosphotungstate		Caron and Sugihara (1962)
	(d) Precipitation with silicotungstate		Sodd et al. (1960)
	(e) Coprecipitation with nickel ferro-cyanide		Mohanrao and Folsom (1963)
			Folsom and Saruhashi (1963)
	(f) Adsorption onto ammonium molybdophosphate		Miyake et al. (1961)
			Folsom and Saruhashi (1963)
	(g) Adsorption onto potassium hexacyanoferrate		Boni (1966)[g]
			Shirasawa and Schuert (1968)
	(h) Adsorption onto copper or zirconium ferrocyanides		Folsom and Sreekumaran (1970)

	Procedure	Counting	Reference
Counting			
	(a) Directly on coprecipitate with $K_3Co(NO_2)_6$	NaI (Tl) γ-spectrometry	Folsom et al. (1960) Malavicini and Vido (1961) Yamagata and Yamagata (1960) Schmitt and Kautsky (1961)
	(b) As $CsPtCl_6$	β-counting	
	(c) As $Cs_3Bi_2I_9$	β-counting or NaI (Tl) γ-spectrometry	
	(d) As $CsClO_4$	β-counting	Sugihara et al. (1959) Sodd et al. (1960)
	(e) As silicotungstate	β-counting or NaI (Tl) γ-spectrometry	Schroeder and Cherry (1962) Mohanrao and Folsom (1963) Folsom and Saruhashi (1963)
	(f) Directly on coprecipitate with nickel ferrocyanide	NaI (Tl) γ-spectrometry	Miyake et al. (1961) Folsom and Saruhashi (1963) Folsom and Sreekumaran (1970)
	(g) Directly on coprecipitate with ammonium molybdophosphate	NaI (Tl) γ-spectrometry	Weiss and Lai (1960)
Barium-140	**REVIEW** Cocrystallized with potassium rhodizonate, together with other alkaline earths	NaI (Tl) γ-spectrometry	
	Seawater passed through filter beds, followed by ^{140}Ba adsorption on Al_2O_3	NaI (Tl) γ-spectrometry	Perkins et al. (1966)
Radio-lanthanides Cerium-144 and Promethium-147	Coprecipitated with carbonates, dissolved and coprecipitated with $Fe(OH)_3$:		Sugiharaha et al. (1959)
	Separated from U, Th (and daughters) and Fe by precipitation and ion exchange		
	Separation of Ce as $CeIO_4$	β-counting	
	Separation of ^{147}Pm with Sm and Nb carriers by ion exchange	β-counting	
Cerium-144	Ce coprecipitated with Fe or Ti as hydroxide and purified by solvent extraction using di-(2-ethyl hexyl) hydrogen phosphate in heptane	β-counting	Hampson (1964)
	Carrier added and precipitated as hydroxide at pH2. Purified by oxalate precipitation, oxidized to Ce(IV) and	β-counting	Shiozaki et al. (1964)

TABLE 19.20—continued

Radionuclide	Method of separation	Method of determination	Reference
	extracted with methyl isobutyl ketone. Ce back extracted and precipitated as oxalate		
Cerium-144	Sea water filtered and Ce coprecipitated with Fe(OH)$_3$ at pH 9·0–9·5. Precipitate redissolved and Fe extracted with di-isopropyl ether. Aqueous phase evaporated to dryness	NaI (Tl) γ-spectrometry	Slowey et al. (1965)
	Sea water passed through filter beds and ^{144}Ce then adsorbed on Al$_2$O$_3$	NaI (Tl) γ-spectrometry	Perkins et al. (1966)
	^{144}Ce (and carrier) precipitated as ferrocyanide complex, dissolved in HCl, and Ce purified by precipitation as hydroxide and then as oxalate	NaI (Tl) γ-spectrometry	Ivanova (1967)
	REVIEW		Høgdahl (1970)
	Lead-210 see Section 19.12.2.2		
	Polonium-210 see Section 19.12.2.3		
	Radon-222 see Section 19.2.2.7		
	Radium-226 see Section 19.12.2.5		
	Radium-228 see Section 19.12.2.6		
	Thorium-228 see Section 19.12.2.8		
	Uranium see Section 19.10.5.6 and Table 18.2		
	^{238}U; ^{234}U see Section 19.10.5.6 and Tables 18.2 and 18.3		
Plutonium (239Pu (with ^{240}Pu) and ^{238}Pu)	Pu reduced to Pu^{III} and coprecipitated with BiPO$_4$. Separated from Bi by coprecipitation with LaF$_3$, purified by anion exchange, and electrodeposited. ^{236}Pu used as tracer	α-spectrometry	Pillai et al. (1964)
	Separated by coprecipitation with	α-spectrometry	Miyake and Sugimura (1968)

hydroxides, purified by anion exchange. Electrodeposited	α-spectrometry	Wong (1971)
Pu reduced to Pu^{III} with $NaHSO_3$ and coprecipitated with $Fe(OH)_3$. Purified by anion exchange on nitrate and chloride columns and electrodeposited. ^{236}Pu used as tracer		
Coprecipitated with $Fe(OH)_3$, dissolved in 8M HNO_3. Separated from Fe, U and Th by solvent extraction with trioctylamine and Pu stripped from solvent by extraction with 8M HCl and 0·2M HF. Electrodeposited for counting	α-spectrometry	Sakanoue et al. (1971)

(a) The earlier literature on the determination of carbon-14 in sea water has been reviewed by de Vries (1959).
(b) The earlier literature on the determination of strontium-90 in sea water has been summarized by Riley (1965b).
(c) Strontium-90 purification cycles have been separated for ease of presentation.
(d) Chemical separations are of radioruthenium. Measurement of radioactivity involves the detection of β or γ emissions of the ^{106}Rh daughter (half life 30s).
(e) With absorbers to separate ^{103}Rh and ^{106}Rh.
(f) A review of methods for the determination of radiosilver in the marine environment has been recently presented: Dutton (In press).
(g) A modification of this method is used to measure ^{137}Cs and ^{134}Cs in sea water by NaI (Tl) γ-spectrometry (Dutton, 1970).

analytical work on organic material in sea water has been undertaken as part of biological programmes. The problem of the organic composition of sea water has not attracted the attention of the analyst to the same degree as has its inorganic composition, and this is probably responsible for its slow growth.

There is a pressing need for simple and reliable methods for the organic analysis of marine waters. Such methods are essential if the flow of material through the food chain is to be fully understood, and if the factors controlling the behaviour and growth of planktonic organisms are to be elucidated. Sensitive analytical methods for the organic analysis of seawater are vital for the determination of specific compounds (particularly synthetic ones which may present environmental hazards), and for the examination of their pathways and kinetics in the marine food chain.

19.14.2. FILTRATION (see also Section 19.4)

The organic analysis of sea water is almost invariably preceded by separation into particulate and dissolved fractions (these terms are to be preferred to suspended and soluble; see Olsen, 1968). The act of filtration introduces certain problems; (i) the physico-chemical one of the molecular interpretations of the terms particulate and dissolved, (ii) analytical ones such as the release or removal of dissolved organic material during filtration, and (iii) the size of the particles removed by various filter types and pore sizes.

A variety of filters have been used for marine work. In early analytical work conventional filter papers were used, but these have been largely superseded by glass fibre mat filters and cellulose ester membrane filters. The membrane filters can be manufactured with a narrow distribution of pore sizes; the glass fibre filters on the other hand are depth filters and have no exact pore size. Despite certain shortcomings both types of filters have been extensively used in aquatic chemistry. Two other types of filters have been introduced in recent years, polycarbonate filters (Nuclepore) made by perforating polycarbonate sheets with nuclear radiation and metal filters made by compressing small silver particles. Both types of filters are made with rated pore sizes.

The performance of various filters has been studied by Sheldon and Sutcliffe (1969) and Sheldon (1972). Using a Coulter Counter to examine the particles remaining in the filtrate, they concluded that none of these filters acted as perfect sieves with a sharp cut off size. Nuclepore and metal filters gave the best separation of particles in terms of their rated pore size. The membrane filters removed particles much smaller than their rated pore size, probably because of the tortuous shape of the channels within them. The glass fibre filters removed most particles having a size of 1 µm or more.

The release and uptake of material by filters is an important practical consideration in the selection of a filter and in the interpretation of analyses of the retained particulate material and of the filtered samples. There are few detailed systematic studies of this problem. Williams (1965) reported the release of microgram amounts of fatty acids from membrane filters, and it is known that there is a change in the micronutrient content of water after passage through certain filters (Marvin et al., 1972).

Quinn and Meyers (1971) reported the removal of heptadecanoic acid from sea water by various filters. Best recoveries of this compound were achieved with paper filters; glass fibre filters retained about half of the fatty acid and membrane filters about 85%. Menzel (1966) and Banoub and Williams (1972) have concluded that glass fibre filters sorb dissolved organic material during filtration; if no account is taken of this loss, small but significant errors in the determination of particulate organic material may arise.

When dealing with the gross analysis of a complex organic mixture such as exists in natural waters, the terms particulate and dissolved can only be regarded as operational ones as they have no fundamental meaning. Sharp (1973a) examined the size distribution of organic material in sea water and concluded, with some justification, that the present practice of reporting separate particulate and dissolved fraction analyses should be abandoned. He recommended that total organic material should be determined and suggested a re-evaluation of the analysis of particulate organic material.

19.14.3. PARTICULATE ORGANIC MATTER

Organic analyses of particulate material in sea water are made to gain an indication of plankton abundance as well as to obtain a measure of the distribution of particulate organic material in the course of chemical or biological studies. A complete discussion of the analysis of particulate organic material (POM) in the sea would include a detailed consideration of the biochemical analysis of plankton. As this subject is too extensive to review here, the present account will only consider the conventional elemental analysis of POM and those biochemical analyses that are used to gain an indication of biomass. At present, unfortunately, no comprehensive review of techniques for the biochemical analysis of plankton is available, but the subject has been covered in part in the reviews by Strickland (1965) and Giese (1967) and practical details of the methods have been described by Strickland and Parsons (1968).

19.14.3.1. *Elemental analysis*

Methods for the elemental analysis of POM typically involve the initial oxidation of the particulate material followed by the determination of the appropriate inorganic combustion product. Carbon, being one of the major elements in organic material, is commonly used as a measure of the total POM, although other methods such as ignition loss and the reduction of dichromate by the POM on heating at 100°C have been used. These latter two methods lack sensitivity for much oceanographic work; for inshore work the dichromate method has certain advantages. It is simple, comparatively rapid and does not suffer from interference from inorganic carbonate; it is probable that some of the organic material resists oxidation by dichromate under the conditions used, although the error is comparatively small – probably less than 10% (Wakeel and Riley, 1957; Copin-Montegut and Copin-Montegut, 1973).

Current procedures for the determination of particulate organic carbon (POC) involve a combustion at high temperature in oxygen. In most cases the sample is burnt in a combustion tube in a stream of oxygen at temperatures in excess of 600°C; the exception to this is the method described by Holm-Hansen et al. (1967) in which the sample is heated in a sealed glass ampoule at 500°C. (This particular method was originally described for the determination of the lipid content of extracts – but it has subsequently been extended to POC determinations, see Strickland and Parsons, 1968). The carbon dioxide produced during combustion is determined by infra red analysis (e.g. Menzel and Vaccaro, 1964; Wangersky and Gordon, 1965; Holm-Hansen et al., 1967; Banoub and Williams, 1972), by thermal conductivity in the gas phase, or by changes in electrical conductivity which result when it is absorbed in a base (Dal Pont and Newell, 1963). Commercial microanalytical C, H, N analysers have been used for the analysis of POM in the sea (see Strickland and Parsons, 1968, pp. 221–222).

Contamination during sampling, storage and analysis usually sets a lower limit to the sensitivity and precision of about $1 \mu g \, C \, l^{-1}$. Gordon (1969) compared two methods for measuring POC and discussed in detail the problems of sensitivity and some of the factors controlling it. Errors resulting from the decomposition of calcium carbonate during the combustion can be minimized by careful selection of the temperature of combustion (Telek and Marshall, 1974).

Both dry and wet combustion methods have been used in the analysis of particulate organic nitrogen (PON) in sea water. Most commercial instruments, e.g. the Coleman Nitrogen Analyser and the Perkin-Elmer CHN analyser use the classical Dumas combustion technique, the nitrogen gas produced being determined manometrically in the former instrument and by thermal conductivity in the latter. Various wet combustion procedures have been described which are based on the Kjeldahl method. In the earlier ones, e.g. that of Strickland and Parsons, 1965, a more or less conventional Kjeldahl digestion was employed, followed by distillation and determination of ammonia in the distillate. More recently, methods have been devised in which the ammonia is determined directly in the digest. Dal Pont and Newell (1963) and Holm-Hansen (1968) have determined ammonia with a photometric ninhydrin procedure, whereas Banoub (1972) used an indophenol blue method. When ammonia is determined directly in the digest, the selenium–copper catalyst is omitted from the sulphuric acid digestion mixture.

A summary of published methods for the analysis of sea water for particulate carbon and nitrogen is given in Table 19.21.

19.14.3.2. *Biochemical Analysis of Particulate Material*

Only methods used to estimate the biomass of the population or some part of it will be considered here. These are of two types: those used to estimate total planktonic biomass and those for plant biomass, the latter being based on chlorophyll determination. Krey *et al.* (1957) measured the so-called albuminoid nitrogen content of particulate material using the biuret method which determines alkali-soluble protein; this approach presumes that this type of protein is most probably only present in living material and that it is related to biomass. The method has not been widely used for the determination of biomass, probably because of its lack of sensitivity. Holm-Hansen and his co-workers (Holm-Hansen and Booth, 1966; Holm-Hansen *et al.*, 1968; Holm-Hansen, 1970) have examined the possibility of using other cell constituents as measures of biomass. They concluded that the polynucleotide, deoxyribonucleic acid (DNA), was a poor indicator of biomass because it appeared to be present mainly in detrital material. Adenosine triphosphate (ATP), on the other hand, seems to have more merit as a measure of biomass. The method for measuring the ATP content of particulate material (Holm-Hansen and Booth, 1966) is based on the luciferin–luciferase assay technique, and is both straightforward and very sensitive. Holm-Hansen and Booth reported a detection limit of better than 1×10^{-10} g of ATP; this is equivalent to a biomass of about 0·1 µg dry weight. The central problem with the ATP method is the conversion factor which relates the ATP content to biomass. Hamilton and Holm-Hansen (1967) and Holm-Hansen (1970) have examined the ATP content of a range of algae and bacteria; although the extreme range was almost 50-fold, most values were within a factor of five of the median value of 0·4 % of cell carbon content of the cell.

The analysis of particulate material for its chlorophyll content is used as a means of obtaining a measure of algal biomass; in addition the relative abundances of the various types of chlorophyll may give an indication of the

TABLE 19.21

Analysis of sea water for particulate carbon and nitrogen

Element	Combustion procedure	Detection procedure	Range	Sensitivity or Precision	Reference
Total organic matter	Wet oxidation with dichromate at 100°C	Titration	100–2000 µgC	60 µg as C	Strickland and Parsons (1968)
Carbon	Dry combustion at 800°C in a muffle tube	Collection of a constant volume of gas and measurement of CO_2 concentration with an I.R. analyser	5–500 µgC	10 µgC	Menzel and Vaccaro (1964)
				2 µgC(S.D.)	Banoub and Williams (1972)
Carbon	Dry combustion at 700°C in a muffle tube	Collection of CO_2 in alkali and measurement of conductivity change	50–700 µgC	25 µgC(S.D.)	Dal Pont and Newell (1963)
Carbon	Dry combustion in a glass ampoule at 500°C	Ampoule opened and the CO_2 flushed through an I.R. analyser and the CO_2 content determined by integration of pulse	10–200 µgC	2 µgC(S.D.)	Holm-Hansen et al. (1967)
Carbon	Dry combustion at 750–800°C in muffle tube	Measurement of CO_2 pulse produced during combustion with an I.R. analyser	not given	7·4 µgC(S.D.)	Wangersky and Gordon (1965)
Nitrogen	Dumas dry combustion	Automatic measurement of N_2 gas (Model 29 Coleman Nitrogen Analyser)	50–1000 µgN	10 µgN(S.D.)	Strickland and Parsons (1968)
Nitrogen	Kjeldahl combustion	NH_3 determination by ninhydrin method	0·5–50 µgN	0·24 µgN(S.D.)	Holm-Hansen (1968)
Nitrogen	Kjeldahl combustion	NH_3 determination by ninhydrin method	not given	not given	Dal Pont and Newell (1963)
Nitrogen	Kjeldahl combustion	NH_3 determination as indophenol blue	0–100 µgN	1 µgN(S.D.)	Banoub (1972)

S.D. = standard deviation.

gross floristic composition of the population. Chlorophyll analyses were originally restricted to extracts from cells collected on filters, but now with the development of highly sensitive fluorimeters it is possible to determine chlorophyll directly in water samples. It is thus now possible for the first time to monitor and profile sea water continuously for its algal population. The measurement of chlorophyll and its interpretation will be discussed in greater detail in Chapter 14 and consequently it will be considered here in outline only.

Practical details of the extraction and analysis of chlorophyll from particulate material are available in various texts (e.g. Strickland and Parsons, 1968; SCOR/UNESCO, 1966). The trichromatic analysis of the extract, introduced by Richards and Thompson (1952), involves the determination of the extinction of the chlorophyll extract at three wavelengths. From these values with the aid of three equations the absolute amount of the individual chlorophylls a, b and c can be calculated. The original equations proposed by Richards and Thompson have been modified by Parsons and Strickland (1963) and also by the SCOR/UNESCO (1966) working group. The determination of the chlorophyll content of extracts by measurement of extinction is usually satisfactory down to concentrations of about $0.1 \mu g l^{-1}$; below this it is preferable to determine chlorophyll by fluorescence techniques. Chlorophyll fluoresces with a high quantum yield, and with a suitable fluorimeter it is possible to extend the sensitivity of chlorophyll determinations down to $0.01 \mu g l^{-1}$. The present method for the analysis of chlorophyll in marine particulate material by fluorescence is based essentially on procedures described by Yentsch and Menzel (1963) and Holm-Hansen et al. (1965). The sample is excited at 450 nm, and the emission in the region of 660 nm is used to determine the chlorophyll content. As with the absorptiometric method, if measurements are made at three wavelengths in the 660 nm region, it is possible to resolve a mixture of the three chlorophylls a, b and c (see e.g. Loftus and Carpenter, 1971). An alternative approach to the measurement of the chlorophyll composition of an extract, is to separate the individual chlorophylls and allied pigments by a form of chromatography. One such procedure has been described by Riley and Wilson, 1965, this entails the separation of the extracted pigments on silica gel thin-layer plates; the individual pigments are identified by their R_f values; the spots are then removed and after elution the pigments are determined quantitatively by absorption spectrophotometry. Alternatively, the plate can be scanned with a densitometer (Garside and Riley, 1969). These particular methods permit the separation and determination of the principal chlorophylls, phaeophytin and several accessory pigments.

The phaeophytins are the principal degradation products of the chlorophylls (see Yentsch, 1967). Their chemical and spectral properties are sufficiently similar to those of the parent chlorophylls to lead to erroneous values for them unless a correction is made for their presence in the sample. At low pH values, chlorophylls lose magnesium and are degraded to their equivalent phaeopigments. This change is accompanied by an approximately 40% decrease in the absorption at 665 nm. This fall in extinction on acidification has been used to resolve chlorophyll-phaeophytin mixtures in extracts (Lorenzen, 1967; Strickland and Parsons, 1968). It is usual to restrict this analysis to the chlorophyll *a* and phaeophytin *a* series.

There is also a change in fluorescence yield when chlorophyll is degraded to its equivalent phaeopigment (Yentsch and Menzel, 1963). It would appear that the resolution of chlorophyll-phaeophytin mixtures by fluorescence measurements is more difficult than it is by measurement of absorption changes. The reason for this is that the fluorescence ratios of the three chlorophyll-phaeophytin pairs differ considerably (Loftus and Carpenter, 1971) and as a consequence, the resolution of mixtures becomes complex.

The direct *in situ* measurement of chlorophyll by fluorimetry was first described by Lorenzen (1966). With present instrumentation, the method has a sensitivity limit of better than $0 \cdot 1$ µg chlorophyll l^{-1}. The main drawback of the method is that the measured *in situ* fluorescence of a given amount of chlorophyll depends upon a variety of factors that include species composition, cell physiology and morphology and ambient light intensity (Kiefer, 1971; 1973). For this reason the interpretation of the results is not straightforward.

19.14.4. DISSOLVED ORGANIC MATTER

There are two fundamentally different types of analyses carried out on the dissolved organic fraction of sea water; the first entails the determination of the total amount of one of the principal elements (usually carbon, nitrogen or phosphorus), the second involves the determination of some specific compound or group of compounds. Optical methods for the estimation of dissolved organic material, in sea water based on the absorption of u.v. radiation, have been used by various workers (Kalle, 1966; Ogura and Hanya, 1967; Foster and Morris, 1971a). However, this approach has generally not been successful because of the difficulty in interpreting the results (see Section 12.3).

19.14.4.1. *Elemental analysis*

In common with the analysis of particulate organic material, the elemental analysis of the dissolved fraction involves an initial oxidation followed by the determination of the appropriate oxidation product. The major difficulty in the analysis of DOM centres around the oxidation step: with present instrumentation and analytical methods the determination of the oxidation products is usually a secondary problem, although the two steps in the analysis are inevitably interlinked to some degree.

There are two approaches to the analysis of dissolved organic carbon (DOC): (i) the oxidation of the organic material at high temperature after the water has been evaporated, (ii) chemical or photochemical oxidation of the organic material in solution. The dry combustion methods have the advantage that the oxidation may be accepted as being essentially complete, whereas with wet oxidation procedures this assumption cannot be made and it is necessary to establish the completeness of the oxidation. In practice, this turns out to be difficult, principally because of our ignorance of the nature of most of the organic material in sea water.

Dry combustion procedures for DOC have been mainly used by Soviet marine chemists. Their procedure has been modified in detail over the years, but basically consists of evaporating about 50 ml of an acidified sample to dryness and heating at 700°C in a combustion tube; the carbon dioxide produced is measured by absorbing it in barium hydroxide and titrating the excess base (Datsko and Datsko, 1950; Skopintsev, 1960). The method would appear to require a great deal of skill to avoid high blanks due to contamination. Gordon and Sutcliffe (1973) have recently published a modified procedure whereby the sea water samples are freeze dried, prior to combustion. The salt residues are then ignited in a CHN analyser, as a result it is possible to determine both DOC and DON, which is especially valuable. Van Hall et al. (1963) devised a different approach to the dry combustion of aqueous samples. Their method (which is intended for the analysis of waste water) entails injection of a small volume (e.g. 30 µl) of the water sample directly into the hot part of a combustion tube where evaporation and oxidation takes place. Their method, although simple and rapid, lacks the requisite sensitivity for sea water samples, owing mainly to the high blank, equivalent to $1-5 \, mg \, C \, l^{-1}$. A modified version of their procedure has been recently described by Sharp (1973b). The difference between these two methods lies in the technique used to introduce the sample into the combustion tube.

Various chemical oxidizing agents have been used to combust the dissolved organic material in sea water. Early procedures, which have been reviewed by Barnes (1959) and Riley (1965b), usually incorporated an acid dichromate oxidation; the method used by Duursma (1961) represents the most refined version of this type of procedure. Duursma's analytical set-up was elaborate, and only about three samples could be analysed per day. Wilson (1961) introduced an entirely new approach in which persulphuric acid was used as the oxidizing agent and the carbon dioxide produced was measured with a non-dispersive infra-red analyser. This method was refined by Menzel and Vaccaro (1964) and their modification which has been used extensively, permits about 30–50 samples to be analysed per day (see Table 12.2, for examples). In common with most wet oxidation procedures, there is some uncertainty about the completeness of the oxidation. When Menzel and Vaccaro introduced their procedure they reported essentially complete oxidation of a variety of organic compounds including chitodextrin, adenylic acid, cysteine and sulphanilamide. Williams (1969a) has presented data on photo-oxidation and persulphate oxidation of 20 samples of sea water, the agreement between the DOC content determined by the two methods being good. He argued convincingly that it is unlikely that oxidation by the two procedures, having such different chemistry, would be equally incomplete. It is, however, clear that dry combustion analysis of sea water gives significantly higher results than do wet oxidation procedures (e.g. compare Skopintsev *et al.*, 1966; Starikova, 1970; Sharp, 1973b and Gordon and Sutcliffe, 1973; with Duursma, 1961; Menzel and Ryther, 1968; Banoub and Williams, 1972; Table 12.2). This could arise from underestimation of the blank of the dry combustion procedure, or incomplete oxidation in the case of the wet oxidation procedures, or a combination of both. Gordon and Sutcliffe (1973) and Sharp (1973b) concluded that their dry combustion methods were more accurate than the persulphate wet combustion method. Gordon and Sutcliffe obtained results comparable to those of Skopintsev, but significantly higher than those obtained by Sharp. The results of these two recent papers are convincing except in one important aspect; in neither paper is the blank of the method dealt with thoroughly. The method of Gordon and Sutcliffe, which involves several handling steps with opportunities for contamination, would seem to be especially vulnerable in this respect, and it is to be noted that it appears to give higher results than Sharp's method. Thus, whereas one is inclined to accept the authors' conclusion that the wet combustion is substantially incomplete, final proof is, as yet, still awaited.

It is convenient to consider the analysis of sea water for dissolved organic nitrogen (DON) and dissolved organic phosphorus (DOP) at the same time. Wet oxidation procedures are normally used for these determinations. In contrast to the analysis of sea water for DOC, it is not a simple matter to remove the inorganic forms of the element prior to oxidation, and consequently the organic fraction is determined from the increase in concentration of an appropriate inorganic oxidation product(s). Most of the earlier procedures used an initial evaporation followed by oxidation with, for example, sulphuric acid for DON, or perchloric acid for DOP; these methods have been reviewed by Barnes (1959), Strickland and Parsons (1965) and Riley (1965b). They were generally tedious, exacting and often imprecise, and have been largely superseded by the photochemical oxidation procedure developed by Armstrong et al. (1966). This method involves irradiating the sample with a high wattage medium pressure mercury ultra-violet source. Armstrong et al. (1966) and subsequently Armstrong and Tibbitts (1968) examined the technique critically and have demonstrated that a wide range of compounds is oxidized essentially to completion. One great advantage of this method is that the only addition to the sample is a small quantity of hydrogen peroxide. The major analytical application of the method in sea water analysis has been the determination of dissolved organic nitrogen and phosphorus. But it has also been used for the determination of organically bound heavy metals, the preparation of sea water free from vitamins, ammonia and DOM to serve as blanks for various analytical methods. For determination of DOP it is only necessary to undertake orthophosphate analyses before and after oxidation. The estimation of DON is somewhat complicated by the fact that the end products of the photo-oxidation are nitrite and nitrate, and it is therefore usual to measure nitrate plus nitrite before and after oxidation. Further, ammonia is also oxidized to nitrite plus nitrate, and unless separate analyses are made of ammonia before oxidation, the "organic nitrogen" will include ammonia and DON. In contrast to its analytical simplicity, the detailed interpretation of organic phoshorus analyses is complex because of the various forms of phosphorus that can occur in natural water. This has been discussed by Olsen (1967), Strickland and Parsons (1968, p. 45) and Burton (1973). Tables 12.3 and 12.4 list results of DON and DOP analyses of sea water. The photo-oxidation method has also been used for the analysis of freshwater samples (Henricksen, 1970; Afghan et al., 1970; Manny et al., 1971). These studies revealed that the photo-chemical reaction is pH sensitive, the rates are fastest in the range pH6–9; with freshwater samples it may thus be necessary to adjust or control pH in some manner. Grasshoff (1966a) has adapted the photo-oxidation method for continuous automatic analysis of total phosphorus by irradiating sea water in a long narrow bore silica coil; Ehrhardt (1969) modified this method for the semi-continuous analysis of dissolved organic carbon. This appears to be the first successful application of the photo-oxidation technique to the analysis of DOC. Previous workers (see Armstrong and Tibbitts, 1968) had been unsuccessful, apparently because carbon dioxide is released from the quartz vessels. Before leaving the photo-oxidation procedures it should be pointed out that they are purely empirical, and that our understanding of the chemical and physical processes involved is quite inadequate.

19.14.4.2. *Detailed organic analysis of sea water*

Before discussing the analysis of sea water for various classes of dissolved organic materials it is worthwhile to consider first some general approaches to the analysis of compounds at the high dilutions at which they occur in the sea.

(a) *Analytical approaches.* Although most analytical methods are insufficiently sensitive for the determination of organic compounds at the very low concentrations encountered in the sea (normally $< 10^{-7}$ M), certain compounds can be determined by bioassay procedures down to concentrations as low as 10^{-12} M. Two types of bioassay procedures are used for sea water analysis. There is the conventional bioassay procedure, which will be referred to here as growth bioassay. It entails obtaining an organism whose growth rate is dependent upon the concentration of the compound to be measured. In an analysis, the growth of the assay organism is measured in the sterilized water sample, and from this the concentration of the compound in the sample is determined. For this type of bioassay procedure to be successful it is necessary that the compound is only a minor growth requirement of the organism. This fact, almost by definition, restricts the method to vitamins. Belser (1959, 1963) described growth bioassay procedures for a range of compounds (e.g. amino acids, purines and pyrimidines) using mutants of the marine bacterium *Serratia marinorubra*, and suggested they could be used for carbohydrates and other compounds. It would appear that he was over optimistic; Litchfield and Hood (1966) persevered with Belser's method but the greatest sensitivity they could achieve with non-vitamins was only about 500 μg l^{-1}. A summary of successful growth bioassay procedures is given in Table 19.22. Some of the earlier work has been omitted but may be found in Riley (1965b).

The other form of bioassay may be termed uptake bioassay. It was first described by Hobbie and Wright (1965) for the determination of glucose in freshwater. It differs from the growth bioassay method in that it may be applied to a major growth substrate such as glucose. It entails obtaining a suitable bioassay organism, adding it to the sterilized water sample along with varying amounts of the ^{14}C-labelled form of the compound to be analysed. The rate of uptake of the radio-isotope is then determined; from this and the predetermined kinetic parameters of the assay organism, the concentration of the compound in the sample may be calculated. This technique has been used for the determination of glucose in sea water by Vaccaro and Jannasch (1966) and Vaccaro *et al.* (1968).

Other methods for direct organic analysis of sea water include enzymatic methods for glucose (Hicks and Carey, 1968), and urea (McCarthy, 1970) and colorimetric methods for urea (Newell *et al.*, 1967; Emmet, 1969).

TABLE 19.22

Growth bioassay procedures for vitamins in sea water

Vitamin	Bioassay Organism	Sensitivity ($ng\ l^{-1}$)	Analogue Response	Reference(s)
B_{12}	*Ochromonas malhamensis*	not given	B_{12}^{III}; Pseudo vitamin B_{12}	Cowey (1956)
B_{12}	*Lactobacillus Leichmanii*	1·0	B_{12}^{II}; Pseudo vitamin B_{12}; Factor A	Cowey (1956)
B_{12}	*Euglena gracilis*	0·1	Pseudo vitamin B_{12}; Factor A	Daisley (1958)
B_{12}	*Thraustochytrium globosum*	1·0	B_{12}^{III}	Vishniac and Riley (1961)
B_{12}	*Cyclotella nana*	<0·1	B_{12}^{III}; Pseudo vitamin B_{12}; Factors A and B and others	Ryther and Guillard (1962); Gold (1964); Carlucci and Silbernagel (1966a)
Thiamine	unnamed phycomycete	20	none	Vishniac and Riley (1961)
Thiamine	*Monochrysis lutheri*	2	not given	Carlucci and Silbernagel (1966b)
Thiamine	*Cryptococcus albidus*	10	Thiazole	Natarajan and Dugdale (1966)
Biotin	*Amphidinium carteri*	0·2	—	Carlucci and Silbernagel (1967)
Biotin	*Serratia marinorubra*	not given	—	Natarajan (1968)

Recently, Newell's method for urea has been successfully automated (De Manche *et al.*, 1973). This appears to be the first automatic method for a single organic compound in sea water. These methods will be discussed in more detail in subsequent sections.

The analysis of sea water for most other organic compounds has involved an initial separation procedure. Several desalting procedures were examined in principle by Jeffrey and Hood (1958), but they have found little subsequent application. The separation procedures that are commonly used in the analysis of organic compounds in sea water may be considered under five major headings: solvent extraction, strategic sorption onto synthetic resins, non-strategic sorption onto various surfaces, co-precipitation and volatilization.

Solvent Extraction. Either the compound itself may be extracted from the sample, or alternatively if it is possible to prepare a suitable derivative, the compound may be extracted in this form. Solvent extraction of the parent compound has been restricted almost exclusively to lipids, hydrocarbons and their halogenated derivatives. The only exception appears to be vitamin B_{12}, which has been extracted from sea-water with phenol (Cowey, 1956). A summary of various solvent extraction procedures is given in Table 19.23.

The formation and extraction of lipophilic derivatives is an approach with many potential advantages and has been used with considerable success in the analysis of certain trace metals in sea water. It has, however, found restricted use in the organic analysis of sea water. No doubt the principal reason for this is that few suitable derivatives can be formed in aqueous solution. At least two procedures have been published, both of them are for amino acids, but apparently neither is quantitative. Palmork (1963a) formed 2:4-dinitrophenyl derivatives of amino acids, extracted them into diethyl ether and separated the individual amino acid derivatives by thin-layer chromatography (TLC). Chau and Riley (1966) reported poor recoveries with this procedure. The method has the additional disadvantage that the reagent 2:4 dinitrofluorobenzene is highly toxic. Litchfield and Prescott (1970) prepared 1-dimethylaminonaphthylene-5-sulphonic acid (dansyl) derivatives of amino acids, these were extracted into ether and separated by TLC. The derivatives may be located and quantified by their strong fluorescence. Again the method is not quantitative.

Strategic sorption onto synthetic resins. This approach has been used extensively for amino acids. Two procedures have been described. The first is the more or less conventional removal of amino acids with a cation exchange resin, followed by their selective elution with ammonia or piperidine. Unless

TABLE 19.23

Solvent extraction procedures used in the analysis of sea water

Class of compound	Solvent system	Reference
Fatty acids	Chloroform at acid pH	Williams (1965); Treguer et al. (1972); Stauffer and MacIntyre (1970)
Fatty acids	Ethyl acetate at acid pH	Slowey et al. (1962a)
Fatty acids	Petroleum ether at acid pH	Jeffrey (1966)
Fatty acids and hydrocarbons	Chloroform-methanol or Chloroform at acid pH	Quinn and Wade (1972); Duce et al. (1972)
Hydrocarbons	Hexane	Zsolnay (1971); Parker et al. (1972)
DDT and related compounds	Hexane	Cox (1971), Seba and Corcoran (1969)
Polychlorinated biphenyls	Hexane-ether	Harvey et al. (1973)
Vitamin B_{12}	Phenol	Cowey (1956)

the bulk of the inorganic cations is removed before the ion exchange step, an excessive amount of ion exchange resin is required. As a result, it is normal to precede the ion exchange with some partial desalting, either by evaporating to dryness and extracting the amino acids with alcohol (Degens *et al.*, 1964; Starikova and Korzhikova, 1969; Pocklington, 1971) or, alternatively, the sample may be partially evaporated and the salt crystals removed from time to time and washed with alcohol and, the resultant brine processed (Palmork 1963b; Chau and Riley, 1966; Riley and Segar, 1970). The second procedure is to remove the amino acids from the sample with a copper-loaded chelating resin. This method has the advantage that no preliminary desalting is needed. A small column of about 10 cm^3 bed volume will remove essentially all the amino acids from 1–2 l of sea water. The amino acids are subsequently eluted from the column with a small volume of ammonia (Siegel and Degens, 1966; Webb and Wood, 1966; Andrews and Williams, 1971; Clark *et al.*, 1972). The method is a great deal simpler than the ion exchange methods, but it suffers from two disadvantages: firstly the resin "bleeds" off glycine and as a result it is difficult to obtain a satisfactory analysis of glycine using this procedure. The glycine appears to be a decomposition product of the resin. Secondly, it appears that some batches of the chelating resin have given low recoveries of amino acids.

Non-strategic sorption. This approach includes a miscellany of procedures. Various materials have been used as sorbants: charcoal, alumina, nylon and polystyrene; the procedures are summarized in Table 19.24.

Co-precipitation. This is a technique which has been successful in the inorganic analysis of sea water but has had less successful application in marine organic chemistry. Co-precipitation with ferric hydroxide has been used to separate amino acids (Tatsumoto *et al.*, 1961; Park *et al.*, 1962) and fatty acids (Williams, 1961; Garrett, 1967) from sea water. Chau and Riley (1966) examined a range of metal hydroxides, but could find none suitable for the quantitative removal of amino acids. Williams and Zirino (1964) examined the removal of total DOC by various hydrated metal oxides, singly and in combination. They found that the ferric hydroxide precipitate was the most effective scavenger of organics; however, the maximum removal they achieved was only 63 %.

Volatilization. It is possible to separate volatile organic compounds from sea water by stripping them out with a current of gas. Thus, the lower aliphatic hydrocarbons and dimethyl sulphide have been displaced with helium from which they are subsequently removed using charcoal, alumina

TABLE 19.24

Non-strategic sorption procedures used in the organic analysis of seawater

Sorbant	Compound	Eluting solution	Reference(s)
Charcoal	Rhamnose, ribose, sucrose	Ethyl alcohol	Schaefer (1965)
	Glucose	Ethyl alcohol	Andrews and Williams (1971)
Aluminium oxide	Glycollic acid	6N HCl	Shah and Fogg (1973); Shah and Wright (1974)
Nylon	Gelbstoff	0·1 N NaOH	Sieburth and Jensen (1968)
Polystyrene (Amberlite XAD-1)	Fatty acids (*n*-heptanoic, *n*-heptadecanoic)	2N NH_4 OH	
	Vitamins B_2 and B_{12}	Ethyl alcohol	
	Surfactants	Ethyl alcohol	Riley and Taylor (1969)
	Insecticides (lindane, DDT, endrin, malathion)	Ethyl alcohol	
	Humic acid*	2N NH_4 OH	
	Polychlorinated biphenyls	Acetonitrile	Harvey et al. (1973)

* Mantoura and Riley (1975) have investigated the application of Amberlite XAD-2 for the uptake of humic substances (including Gelbstoff) from sea water.

or paraffin oil (Linnenbom and Swinnerton, 1970; Lovelock et al., 1972). An alternative procedure is to shake the sample in a closed container and analyse the head space gas. This procedure has been used for the analysis of sea water for some lower aldehydes and ketones (Corwin, 1970) as well as various halogenated methane derivatives (Lovelock et al., 1973).

The analysis of sea water for specific groups of compounds will now be briefly considered. A summary of methods with representative results is given in Table 19.25.

Carbohydrates. Sea water has been analysed for total carbohydrate content as well as for individual sugars. The methods for total carbohydrate rely upon the formation of furfural derivatives from carbohydrates in the presence of hot concentrated sulphuric acid. With this type of colorimetric procedure it is difficult to know whether or not only the group in question is reacting to produce the colour, and also whether the group is completely determined. Handa (1966b) examined a range of compounds, the only one he found to interfere was euglenol.

The concentrations of individual sugars in sea water have been determined by a variety of procedures. Glucose has been determined directly by an uptake bioassay procedure (Vaccaro and Jannasch, 1966; Vaccaro et al., 1968) as well as by an enzymatic method (Hicks and Carey, 1968). The latter procedure involves determining the end product of a series of enzyme reactions starting from glucose; the substrates, other than glucose, and the enzymes are added in excess. The reaction sequence involves the phosphorylation of glucose to glucose-6-phosphate and the subsequent dehydrogenation of glucose-6-phosphate with nicotinamide adenine dinucleotide phosphate; the reduced nucleotide which is produced is reacted with resazurin in the presence of the enzyme diaphorase and the fluorescence of the resultant compound is measured. Because of the specifity of enzyme reactions, this method should, therefore, be highly specific for glucose. When the enzyme and biossay methods were compared they gave good agreement down to 10 µg glucose 1^{-1}, below this concentration the precision of both methods deteriorated (Vaccaro et al., 1968). Other methods for individual sugars involve extracting the sugar from the sample in some manner. Degens et al. (1964) evaporated sea water to dryness, extracted the salt crystals with 80% ethyl alcohol, then separated and determined the sugars by paper chromatography; Andrews and Williams (1971) adsorbed sugars from sea water with charcoal, eluted them with ethyl alcohol and determined glucose with the enzyme glucose oxidase. This latter method will detect less than 1 µg glucose 1^{-1}. Tables 12.7 and 19.25 give a selection of the results of analyses of sea water for carbohydrates.

Amino acids. All methods for the determination of amino acids in sea water entail the initial separation of the compounds from the sample. A wide variety of separation methods have been used and are summarized in Table 19.25. Amino acids have been extracted into an organic solvent as either their 2:4 dinitrobenzene or dansyl derivatives (Palmork, 1963a; Litchfield and Prescott, 1970), co-precipitated with ferric hydroxide (Tatsumoto *et al.*, 1961; Park *et al.*, 1962), or extracted from evaporates with ion exchange resins (Degens *et al.*, 1964; Chau and Riley, 1966; Starikova and Korzhikova, 1969; Riley and Segar, 1970; Pocklington, 1971). In addition, they have been extracted with a copper-chelating resin (Siegel and Degens, 1966; Andrews and Williams, 1971; Clark *et al.*, 1972).

The extracted amino acids are separated chromatographically, usually by automatic amino acid analysis on ion exchange resins, although GLC, TLC and paper chromatographic techniques have also been used.

At present, the copper-chelating resin technique is probably the simplest quantitative method for separating amino acids from sea water, although it has certain shortcomings (see above). However, in principle the most promising type of technique is the formation and extraction of an organic derivative, since this usually involves much less manipulation.

Vitamins. Vitamins are invariably determined in sea water by growth bioassay procedures and methods have been described for vitamin B_{12}, thiamine, biotin and nicotinic acid. The principles of the technique have been discussed in Section 19.14.4.2. In early work, the growth of the bioassay organism was measured by counting cells or measuring the turbidity of the medium arising from the cells; Gold (1964) measured growth of photosynthetic assay organisms by determining the rate of $^{14}CO_2$ fixation. This approach has increased the sensitivity of the assay about 10-fold (cf. Carlucci and Silbernagel, 1966a, with Ryther and Guillard, 1962).

Lipophilic compounds. The lipophilic compounds – lipids, hydrocarbons and chlorinated hydrocarbons – are usually analysed by extraction into an immiscible solvent, followed by analysis by GLC. Some initial fractionation of the major types of lipophilic compounds may be achieved by means of the solvent system used for the extraction; a semi-polar solvent such as chloroform or chloroform-methanol will be most suitable for fatty acids and triglycerides, whereas non-polar solvents such as hexane are more appropriate for hydrocarbons and their halogenated derivatives. In some instances the extracts have been purified prior to GLC analysis on either silica gel columns or thin-layer plates. The higher fatty acids are determined by GLC as their methyl esters, whereas the lower fatty acids, hydrocarbons and chlorinated

TABLE 19.25

Organic analyses of seawater for specific compounds or class of compounds

Compound	Method	Sensitivity (μg l^{-1})	Typical Range of results	Reference
Total carbohydrates	Anthrone and n-ethyl carbazole		<700 μg l^{-1}	Lewis and Rakestraw (1955)
Total carbohydrates	Phenol-sulphuric acid		600–200 μg l^{-1}	Handa (1966a, b)
Total carbohydrates	Anthrone		800–200 μg l^{-1}	Walsh and Douglas (1966)
Galactose and glucose	Evaporation, extraction with alcohol, ion exchange desalting and paper chromatography		16–1·3 μg l^{-1}	Degens et al. (1964)
Glucose	Uptake bioassay	10	150–<10 μg l^{-1}	Vaccaro and Jannasch (1966); Vaccaro et al. (1968)
Glucose	Direct enzyme assay	3	200–<3 μg l^{-1}	Hicks and Carey (1968) Vaccaro et al. (1968)
Glucose	Sorption onto charcoal, elution and determination enzymatically with glucose oxidase	0·5	5·7–0·5 μg l^{-1}	Andrews and Williams (1971)
Amino acids	Co-precipitation with Fe(OH)$_3$, solution in HCl, desalting with ion exchange resin, separation by paper and liquid chromatography	—	—	Tatsumoto et al. (1961) Park et al. (1962)
Amino acids	Extraction of 2:4 di-nitrophenyl derivatives with ether, separation of individual amino acids by thin-layer chromatography	—	—	Palmork (1963a)
Amino acids	Evaporation of seawater to a brine,	<1	10–0·5 μg l^{-1}	Palmork (1963b), Chau

Substance			Method	References
	individual	individual	desalting on cation exchange resin, separation and determination by paper, thin-layer or liquid chromatography	and Riley (1966), Riley and Segar (1970), Bohling (1970, 1972)
Amino acids	$10-0.5\ \mu g l^{-1}$ individual	<1 individual	Evaporation of seawater to dryness, extraction with ethanol, desalting on cation exchange resin, separation and determination by paper, liquid or gas–liquid chromatography	Degens et al. (1964) Pocklington (1971)
Amino acids	$10-0.5\ \mu g l^{-1}$ individual	<1 individual	Extraction on copper loaded Chelex-100 resin, separation and analysis of individual amino acids by liquid and thin layer chromatography	Siegel and Degens (1966) Webb and Wood (1966) Andrews and Williams (1971), Clark et al. (1972)
Fatty acids	$10-0\ \mu g l^{-1}$ individual fatty acids	about 0.1 individual fatty acids	Extraction from acidified seawater with chloroform, chloroform methanol, ethylacetate or carbon disulphide; methylated, analysed by GLC	Williams (1961), Williams (1965), Slowey et al. (1962a), Jeffrey (1966), Stauffer and MacIntyre (1970), Duce et al. (1972), Quinn and Wade (1972)
Fatty acids	not given	not given	Co-precipitation with $Fe(OH)_3$, acidification and extraction with $CHCl_3$; methylation and analysis by GLC	Garrett (1967)
Fatty acids	$10-50\ \mu g l^{-1}$ total fatty acids	about 1 total	Extraction with chloroform, formation of copper soap, analysis of copper content of soap by atomic absorption	Treguer et al. (1972)
Methane and other lower hydrocarbons	$100-1\ \mu g l^{-1}$ individual hydrocarbons	about 0.01	Flushed off with helium, collected on alumina or charcoal, analysed by GLC	Linnenbom and Swinnerton (1970)

TABLE 19.25—*continued*

Compound	Method	Sensivity	Typical range of results	References
Dimethyl sulphide	Flushed off with helium, collected in paraffin oil, displaced by heating and analysed by GLC	not given	$12\ \mu g l^{-1}$	Lovelock et al. (1972)
Freon, methyl iodide and carbon tetrachloride	Headspace analysis, GLC	not given	CCl_3F $60\ pg l^{-1}$ CH_3I $750\ pg l^{-1}$ CCl_4 $400\ pg l^{-1}$	Lovelock et al. (1973)
Acetone, butyraldehyde	Head space analysis, GLC	not given	$50–0\ \mu g l^{-1}$ individual	Corwin (1970)
Hydrocarbons (especially *n*-alkanes)	Extracted with chloroform or hexane, purified on silica gel or silicic acid column or TLC plate, analysed by GLC or combustion	not given	$50–1\ \mu g l^{-1}$ total	Duce et al. (1972) Parker et al. (1972) Zsolnay (1971) Barbier et al. (1973)
DDT group	Extracted with hexane, analysed by GLC	not given	about $1\ ng l^{-1}$	Seba and Corcoran (1969) Cox (1971)
PCB group	Extracted with hexane, or hexane–ether or XAD; purified on silicic acid column and analysed by GLC	not given	$4\cdot2–0\cdot05\ \mu g l^{-1}$ $1–50\ ng l^{-1}$ $0\cdot4–4\cdot3\ ng l^{-1}$	Duce et al. (1972) Harvey et al. (1973) Harvey et al. (1973)
Urea	Reaction with diacetyl monoxime and semicarbazide, measurement of red derivative	$0\cdot1\ \mu g\text{-at}\ N.l^{-1}$	$3–0\cdot5\ \mu g\text{-at}\ N.l^{-1}$	Newell et al. (1967) De Manche et al. (1973) Rensen (1971)
Urea	Determination of NH_3 produced from urea after hydrolysis by urease	$0\cdot06\ \mu g\text{-at}\ N.l^{-1}$	$0\cdot3–0\cdot1\ \mu g\text{-at}\ N.l^{-1}$	McCarthy (1970)
Urea	Reaction with phenol and hypochlorite	$0\cdot2\ \mu g\text{-at}\ N.l^{-1}$	—	Emmett (1969)

Glycollic acid	Extracted with alumina, eluted with sulphuric acid, determined colorimetrically	not given	$20\text{--}40\ \mu\text{g}\,\text{l}^{-1}$	Shah and Wright (1974)
Vitamin B_{12}	Bioassay with *Cyclotella nana*	$<0{\cdot}1\ \text{ng}\,\text{l}^{-1}$	$5\text{--}0{\cdot}1\ \text{ng}\,\text{l}^{-1}$	Menzel and Spaeth (1962) Carlucci and Silbernagel (1966c) Carlucci (1970) Natarajan (1971)
Vitamin B_1 (thiamine)	Bioassay with *Monochrysis lutheri* Bioassay with *Cryptococcus albidus*	$2\ \text{ng}\,\text{l}^{-1}$ $10\ \text{ng}\,\text{l}^{-1}$	$8\ \text{ng}\,\text{l}^{-1}$ $100\text{--}{<}10\ \text{ng}\,\text{l}^{-1}$ $3\ \text{ng}\,\text{l}^{-1}$	Carlucci (1970) Ohwada and Taga (1972)
Biotin	Bioassay with *Amphidinium carteri* Bioassay with *Achromobacter*	$0{\cdot}2\ \text{ng}\,\text{l}^{-1}$	$2\ \text{ng}\,\text{l}^{-1}$ $2{\cdot}5\text{--}5\ \text{ng}\,\text{l}^{-1}$	Carlucci (1970) Ohwada (1972) Ohwada and Taga (1972)

hydrocarbons are sufficiently volatile to be analysed without the necessity of forming a derivative. The lower hydrocarbons and their derivatives have been successfully removed from sea water by flushing out with helium or by shaking the sample with air in a sealed container and analysing the gas in the headspace. Details of the various procedures which have been used are summarized in Table 19.25 and results are given in Table 12.6.

Treguer *et al.* (1972) used a novel approach to determine the fatty acid content of extracts from sea water; they formed copper soaps of the fatty acids, and these were extracted in to a chloroform-heptane mixture. Finally, the soap content was determined by measuring the copper concentration in the extract by atomic absorption spectrophotometry.

Urea. This is a compound of interest to the marine biologist for it is an alternative source of nitrogen for the phytoplankton. Three methods have been described for its determination in sea water. Two are colorimetric procedures: Newell *et al.* (1967) measured urea in sea water by reacting it with diacetyl monoxime; the product was further reacted with semi-carbazide to form a coloured semicarbazone. Emmet (1969) described a method for urea based on a modification of the phenol-hypochlorite ammonia determination. An automated procedure has now been described for Newell's method (De Manche *et al.*, 1973). A quite different approach to urea analysis was used by McCarthy (1970). His method entails adding the enzyme urease to the sample and measuring the ammonia produced from the hydrolysis of urea. The enzymatic method appears to be potentially more sensitive than are the colorimetric procedures; however the precision of the enzymatic method will fall if the sample also contains high concentrations of ammonia. Pineda (1973) has found good agreement between results for duplicate samples analysed by the methods of Newell *et al.* and of McCarthy.

19.14.5. CONCLUSION

In some respects considerable progress has been made in the organic analysis of sea water over the past decade. The dissolved organic carbon, nitrogen and phosphorus contents may now be measured on a semi-routine basis; similarly, simple and sensitive methods are available for the determination of particulate carbon and nitrogen. Fluorimeters are now available that make possible the *in situ* measurement of chlorophyll in ocean water. Bioassay methods have been perfected for the measurement of nanogram amounts per litre of the vitamins B_{12}, thiamine and biotin. Chemical methods are available for the measurement of amino acids and, to a lesser extent, of sugars in sea water. One of the most important developments has been the introduction of the photochemical oxidation technique as it has opened up a wide range of new possibilities in marine chemistry and biology.

Despite these undoubtedly real achievements an array of problems and deficiencies still exist. With probably the single exception of gas liquid chromatography, organic analysis of sea water has not seen the advances in instrumentation that have occurred in the inorganic field. For certain biological and microbiological studies there is a need to have a detailed knowledge of the concentration of sugars, amino acids and such like compounds in sea water and of their variation in space and time. The present analytical methods for these compounds are too exacting and time consuming for such work, and there is a pressing need to develop much simpler procedures. There has been a long standing need for a method to measure the total biomass of sea water. It is not certain at present if the ATP method satisfies this need. The analysis of particulate material for chlorophyll has now reached a high degree of refinement, but the full exploition of its *in situ* measurement by fluorimetry is held back by the difficulty of the exact interpretation of the data.

Finally, our present knowledge of the detailed molecular composition of the organic material in sea water is still very rudimentary. At present, reliable analyses are available for perhaps 50 or so compounds, which constitute less than 10% of the total mass of organic matter, see Table 12.12. It is probable that the present analytical approach to the study of the molecular composition of sea water will progress only slowly, and for some time yet will fall short of a complete organic description of sea water. Fundamentally different approaches are needed to the problem, although it is far from clear at the present which are likely to be the most fruitful lines of attack.

REFERENCES

Abdullaev, A. A., Gineev, E. S. Giakhov, V. A., Zhuk, L. I. and Zakhidov, A. S. (1968). *Izv. Akad. Nauk. Uzb. SSR Ser. Fiz.-Mat. Nauk.* **12**, 59.

Abdullah, M. I. and Klimek, R. (1974). in press.

Abdullah, M. I. and Royle, L. R. (1972). *Analyt. Chim. Acta*, **58**, 283.

Abdullah, M. I., Royle, L. G. and Morris, A. W. (1971). *Nature, Lond.* **235**, 158.

Abraham, J., Winpe, M. and Ryan, D. E. (1969). *Analyt. Chim. Acta*, **48**, 431.

Adams, P. B. (1972). *In* "Ultrapurity—Methods and Techniques" (M. Zief and R. Speights, eds.) p. 293. Marcel Dekker, New York.

Afghan, B. K. and Ryan, D. E. (1968). *Analyt. Chim. Acta*, **41**, 167.

Afghan, B. K., Goulden, P. D. and Ryan, J. F. (1970). In "Advances in Automated Analysis"), pp 291–297. Technicon International Congress, New York.

Alcock, G. P. and Coates, K. B. (1958). *Chem. & Ind. (Rev.)* 554.

Alexander, J. E. and Corcoran, E. F. (1964). *Bull. Mar. Sci., Gulf Caribb.* **14**, 594.

478   J. P. RILEY

Alexander, J. E. and Corcoran, E. F. (1967). *Limnol. Oceanogr.* **12**, 236.
Aleksandruk, V. M. and Stepanov, A. V. (1968). *Okeanologiya.* **8**, 746.
Alexander, J. E. (1964). Ph.D. Thesis, University of Miami.
Alexander, E. J. and Corcoran, E. F. (1967). *Limnol. Oceanogr.* **12**, 235.
Allan, J. E. (1961). *Spectrochim. Acta,* **17**, 467.
Alsterberg, G. (1925). *Biochem. Z.* **159**, 36.
American Public Health Association (1965). "Standard Methods for the examination of water and waste water," 12th ed. American Public Health Association, New York.
Andersen, A. T. and Føyn, L. (1969). *In* "Chemical Oceanography" (R. Lange, ed.) Universitets Forlaget Oslo.
Andersen, L. H. (1958). *Acta. Chem. Scand.* **14**, 495.
Andersen, N. R. and Hume, D. N. (1968a). *Analyt. Chim. Acta.* **40**, 207.
Andersen, N. R. and Jadamec, J. R. (1971). Symposium on Investigation of the Resources of the Caribbean Sea and adjacent regions 1968. Preprints.
Andersen, N. R. and Hume, D. N. (1968b). *In* "Trace Inorganics in Water". (R. A. Baker, *ed.*) American Chemical Society, Washington, D.C.
Andersen, N. R. and Hume, D. N. (1968c). *Adv. Chem. Ser. No. 73*, 296.
Andersen, N. R. and Jadamec, J. R. (1971). *Symp. Invest. Resources Caribb. Sea Adjacent Regions Pap.* 1968, 15.
Andersen, N. R., Gassaway, J. D. and Maloney, W. E. (1970). *Limnol. Oceanogr.* **15**, 467.
Andrews, P. and Williams, P. J. leB. (1971). *J. mar. biol. Ass. U.K.,* **51**, 111.
Anfält, T. and Jagner, D. (1970). *Analyt. Chim. Acta,* **50**, 23.
Anfält, T. and Jagner, D. (1971). *Analyt. Chim. Acta,* **53**, 13.
Anfält, T. and Jagner, D. (1971a). *Analyt. Chim. Acta,* **57**, 177.
Anfält, T. and Jagner, D. (1973). *Analyt. Chem.* **45**, 2412.
Angino, E. E. and Billings (1967). *Geochim. Cosmochim. Acta,* **30**, 153.
Angino, E. E., Billings, G. K. and Andersen, N. (1966). *Chem. Geol.* **1**, 145.
Anselm, C. D. and Robinson, R. T. (1951). *J. mar. Res.* **10**, 203.
April, R. W. and Hume, D. N. (1970). *Science,* **170**, 849.
Ariel, M. and Eisner, U. (1963). *J. Electroanalyt. Chem.* **5**, 362.
Ariel, M., Eisner, U. and Gottesfield, S. (1964). *J. Electroanalyt. Chem.* **7**, 307.
Armitage, B. and Zeitlin, H. (1971). *Honolulu Inst. Geophys. Contrib.* 245.
Armstrong, F. A. J. (1951). *J. mar. biol. Ass. U.K.* **30**, 149.
Armstrong, F. A. J. (1957). *J. mar. biol. Ass. U.K.* **36**, 509.
Armstrong, F. A. J. (1963). *Analyt. Chem.* **35**, 1292.
Armstrong, F. A. J. (1965). *Oceanogr. Mar. Biol. Ann. Rev.* **3**, 79.
Armstrong, F. A. J. and Tibbitts, S. (1968). *J. mar. biol. Ass. U.K.* **48**, 143.
Armstrong, F. A. J., Williams, P. M. and Strickland, J. D. H. (1966). *Nature,* **211**, 481.
Armstrong, F. A. J., Stearns, C. R. and Strickland, J. D. H. (1967). *Deep-Sea Res.* **14**, 381.
Arnon, D. I. (1958). "Perspectives in marine biology", p. 351. University of California Press, Berkeley.
Atkins, W. R. G. (1957). *J. Cons. int. Explor. Mer.* **22**, 271.
Atkinson, L. P. (1972). *Analyt. Chem.* **44**, 885.
Atkinson, L. P. and Richards, F. A. (1967). *Deep-Sea Res,* **14**, 673.
Atkinson, L. P. and Stefansson, U. (1969). *Geochim. Cosmochim. Acta* **33**, 1449.

Atlas, E. L., Hager, S. W., Gordon, L. I. and Park, P. K. (1971). "Practical Manual for use of the Technicon Auto Analyzer in sea water nutrient analysis," Dept. of Oceanogr., Oregon State Univ. Rept. 71–22, 49 pp.

Atwood, D. K., Froehlich, P. N. and Kinard, W. F. (1973). *Limnol. Oceanogr.* **18**, 771.

Azhazha, E. G. (1964). *In*: "Radioactive Contamination of the Sea" (Baranov, V. I. and Khitnov, L. M. eds.), *Akad. Nauk. USSR*—Oceanographic Commission (1964).

Baas Becking, L. G. M., Haldane, A. D. and Izzard, D. (1958). *Nature, Lond.* **182**, 645.

Bache, C. A. and Lisk, D. G. (1971). *Analyt. Chem.* **43**, 951.

Bachmann, R. W. and Goldman, C. R. (1964). *Limnol. Oceanogr.* **9**, 143.

Bacon, M. P. and Edmond, J. M. (1972). *Earth Planet. Sci. Lett.* **16**, 66.

Baier, R. E. (1972). *J. Geophys. Res.* **77**, 5062.

Bainbridge, A. E., Sandoval, P. and Suess, H. E. (1961). *Science,* **134**, 552.

Balashov, Y. A. and Khitrov, L. N. (1961). *Geokhimiya*, 796; *Geochemistry Int.* 877.

Banoub, M. W. (1972). *Intern. J. Environ. Anal. Chem.*, **2**, 107.

Banoub, M. W. and Williams, P. J. leB. (1972). *Deep-Sea Res.*, **19**, 433.

Barbier, M., Joly, D., Saliot, A. and Tourres, D. (1973). *Deep-Sea Res.*, **20**, 305.

Barić, A. and Branica, M. (1969). *Limnol. Oceanogr.* **14**, 796.

Bark, L. S. and Rixon, A. (1970). *Analyst.* **95**, 786.

Barkley, R. A. and Thompson, T. G. (1960a). *Analyt. Chem.* **32**, 154.

Barkley, R. A. and Thompson, T. G. (1960b). *Deep-Sea Res.* **7**, 24

Barnes, H. (1959). "Apparatus and Methods of Oceanography, Part 1: Chemical," 341 pp. Allen and Unwin, London.

Barnes, H. and Folkard, A. R. (1951). *Analyst.* **76**, 599.

Barnes, R. O. (1973). *Deep-Sea Res.* **20**, 1125.

Barnes, W. J. and Parker, C. A. (1960). *Analyst,* **85**, 828.

Barnett, W. B. and Kahn, H. L. (1972). *Analyt. Chem.* **44**, 935.

Basargin, N. N. and Nogina, A. A. (1969). *Tr. Kom. Anal. Khim.* (USSR), **17**, 331.

Bather, J. M. and Riley, J. P. (1954). *J. Cons. int. Explor. Mer.* **20**, 145.

Bauman, A. and Tagliatti, S. (1964). *Archiv. Hig. Rada. Toksikol.* **15**, 399.

Bauman, E. W. (1968). *Analyt. chim. Acta.* **42**, 127.

Bauman, R. P. (1962). "Absorption Spectroscopy." Wiley, New York.

Belcher, R., Leonard, M. A. and West, T. S. (1959). *Talanta*, **2**, 92.

Belser, W. L. (1959). *Proc. Nat. Acad. Sci., Wash.* **45**, 1533.

Belser, W. L. (1963). *In* "The Sea" (M. N. Hill, Ed.), Vol. II, p. 220. Interscience, New York.

Belyaev, Y. I. and Ovsyanyi, E. I. (1968). *Oceanologiya*, **8**, 734 and 920.

Belyaev, Y. I. and Gordeev, V. V. (1972). *Okeanologiya*, **12**, 905.

Bender, M., Snead, T. and Chan, L. H. (1972). *Earth Planet Sci. Lett.* **16**, 81.

Bendschneider, K. and Robinson, R. J. (1952). *J. mar. Res.* **11**, 87.

Benesch, R. and Mangelsdorf, P. (1972). *Helg. Wiss. Meeresunters,* **23**, 365.

Benson, B. B. and Parker, P. D. M. (1961). *Deep-Sea Res.,* **7**, 244, 254.

Berge, H. and Brügmann, L. (1970). *Beitr. Meeresk.* **27**, 5.

Berge, H. and Brügmann, L. (1971). *Beitr. Meeresk.* **28**, 19.

Berge, H. and Brügmann, L. (1972). *Beitr. Meeresk.* **29**, 115.

Bernat, M., Church, T. and Allegre, C. J. (1972). *Earth Planet Sci. Lett.* **16**, 75.

Bernhard, M. (1955). *Publ. Staz. zool. Napoli,* **29**, 80.

Bernhard, M. (1968). *Conseil Nationale Energie Nucleaire-EURATOM Ann. Rept. 1967* (RT/B10 (68) 60).

Bernhard, M. (1971). *Conseil Nationale Energie Nucleaire Report No*. RT/B10 (71) 4.

Berthelot, M. P. (1859). *Rép. Chim. Appl.*, 284.

Bertine, K. Chan, L. H. and Turekian, K. K. (1970). *Geochim. Cosmochim. Acta*, 34, 641.

Bertolini, G. and Coché, A. (1968). "Semiconductor Detectors". Wiley-Interscience, New York.

Betzer, P. R. and Pilson, M. E. Q. (1970). *J. Mar. Res.* 28, 251.

Betzer, P. R. and Pilson, M. E. Q. (1971). *Deep-Sea Res.* 18, 753.

Bewers, J. M. (1971). *Deep-Sea Res.* 18, 237.

Bhat, I. S., Kamath, P. R. and Ganguly, A. K. (1972). Bhaba Atomic Research Centre, Bombay, India. Rept. B.A.R.C., 644.

Bhat, S. G., Krishnaswamy, S., Lal, D., Rama and Moore, W. S. (1969). *Earth Planet. Sçi. Letters*, 5, 483.

Bhavnagary, H. M. and Krishnaswamy, N. (1967). *Indian J. Tech.* 5, 170.

Biechler, D. G. (1965). *Analyt. Chem.* 37, 1056.

Bien, G. S. (1958). *Analyt. Chem.* 30, 1525.

Bien, G. S., Rakestraw, N. W. and Suess, H. E. (1960). *Tellus*, 12, 436.

Bien, G. S., Rakestraw, N. W. and Suess, H. E. (1965). *Limnol. Oceanogr.* 10, Suppl., R25.

Bieri, B. H. (1965). *J. mar. Res.* 23, 33.

Bieri, R. H., Koide, M. and Goldberg, E. D. (1966). *J. geophys. Res.* 17, 5243.

Bieri, R. H., Koide, M. and Goldberg, E. D. (1968). *Earth Planet. Sci. Letters*, 4, 329.

Bilal, B. A., Braetter, P., Muehlig, B., Roesick, U. and Zimen, K. E. (1971). *Radiochim. Acta*, 16, 191.

Billings, G. K. and Harris, R. C. (1965). *Texas J. Sci.* 17, 129.

Billings, G. K., Bricker, O. P., Mackenzie, F. T. and Brooks, A. L. (1969). *Earth Planet. Sci. Lett.* 6, 231.

Black, W. A. P. and Mitchell, R. L. (1951). *J. mar. biol. Ass. U.K.* 30, 575.

Blanchard, R. L. (1965). *J. Geophys. Res.* 70, 4055.

Blasius, R. and Brozio, B. (1967). *Chelates Analyt. Chem.* 1, 49.

Blazhis, I. K., Junevicius, R., Paeda, R. and Valiukevicus, C. (1970). *Liet. TSR Aukstuju Mokuklu Mokslo Darb. Chem. Chem. Technol.* 11, 43.

Bodman, R. H., Slabaugh, L. V. and Bowen, V. T. (1961). *J. Mar. Res.* 19, 141.

Bogoyavlenskii, A. N. (1965a). *Tr. Inst. Okeanol. Akad. Nauk SSSR*, 79, 34.

Bogoyavlenskii, A. N. (1965b). *Tr. Inst. Okeanol. Akad. Nauk SSSR*, 79, 49.

Bogoyavlcnskii, A. N. (1965c). *Tr. Inst. Okeanol. Akad. Nauk SSSR*, 79, 60.

Bohling, H. (1970). *Mar. Biol.*, 6, 213.

Bohling, H. (1972). *Mar. Biol.*, 16, 281.

Bojanowski, R. and Ostrowski, S. (1968). *Acta Geophys. Polonica.* 16, 351.

Bolter, E., Turekian, K. K. and Schutz, D. F. (1964). *Geochim. Cosmochim. Acta*, 28, 1459.

Boni, A. L. (1966). *Analyt. Chem.* 38, 89.

Borovik-Romanova, T. F. (1944). *Dokl. Acad. Nauk. SSSR*, 42, 221.

Bougis, P. (1962). *Cah. Biol. Mar.* 3, 317.

Bougis, P. (1963). *J. Cons. int. Explor. Mer.* 28, 171.

Boury, M. (1938). *Rev. Trav. Off. Pêches. Marit.* 11, 157.

Bowen, H. J. M. (1956). *J. Mar. biol. Ass. U.K.* 35, 451.

Bowen, V. T., Strohal, P., Saiki, M., Ancellin, J., Merten, D. and Ganguly, A. K. 1970). *In* "Reference methods for marine radioactivity studies" (Y. Nishiwaki and R. Fukai, eds) International atomic agency, Vienna.

Bowles, P. F. and Nicks, P. F. (1961). *Analyst.* **86**, 483.
Bradbury, J. H. and Hambly, A. N. (1952). *Australian J. Sci. Res. Ser.* A. **5**, 541.
Braman, R. S. (1971). *Analyt. Chem.* **43**, 1462.
Braman, R. S. and Foreback, G. C. (1973). *Science,* **182**, 1249.
Braman, R. S., Justen, J. L. and Foreback, C. C. (1972). *Analyt. Chem.* **44**, 2195.
Brandenberger, H. and Bader, H. (1967a). *Helv. Chim. Acta,* **50**, 1409, 1414.
Branderberger, H. and Bader, H. (1967b). *At. Absorption Newslett.* **6**, 101.
Brandenberger, H. and Bader, H. (1968). *At. Absorption Newslett.* **7**, 53.
Branica, M. (1970). *In* "Reference methods for marine radioactivity studies," *Tech. Rept. Ser. Internat. Atomic Agency* **No. 118**, 243.
Brass, G. and Turekian, K. K. (1972). *Earth Planet. Sci. Lett.* **16**, 117.
Breger, I. A., Zubovic, P. and Chandler, J. C. (1973). *U.S. Geol. Survey Prof. Paper* **800C**. C263.
Brewer, P. G. and Riley, J. P. (1965). *Deep-Sea Res.* **12**, 765.
Brewer, P. G. and Riley, J. P. (1966). *Analyt. Chim. Acta,* **35**, 514.
Brewer, P. G. and Riley, J. P. (1967). *Deep-Sea Res.* **14**, 219.
Brewer, P. G. and Spencer, D. W. (1970). *Limnol Oceanogr.* **15**, 107.
Brewer, P. G. and Spencer, D. W. (1970a). "Trace element intercalibration study", *Woods Hole Oceanographic Institution Report No. 70–62.* 63 pp.
Brewer, P., Frew, N., Cutshall, N., Wagner, J. J., Duce, R. A., Walsh, P. R., Hoffman, G. L., Dutton, J. W. R., Fitzgerald, W. F., Hunt, C. D., Girvin, D. C., Clem, R. G., Patterson, C., Settle, D., Glover, B., Presley, B. J., Trefry, J., Windom, H. and Smith, R. (1974). *Mar. Chem.* **2**, 69.
Brewer, P. G., Spencer, D. W. and Smith, C. L. (1969). *Amer. Soc. Testing Materials, Publ. No.* **443**, 70.
Brewer, P. G., Spencer, D. W., and Wilkniss, P. E. (1970). *Deep-Sea Res.* **17**, 1.
Brewer, P. G., Spencer, D. W. and Robertson, D. E. (1972). *Earth Planet. Sci. Lett.* **16**, 111.
Broecker, W. S. (1965). *In* "Symposium on diffusion in oceans and freshwaters, Lamont Geological Observatory, 1964", Palisades, N.Y., 116.
Broecker, W. S. and Kaufman, A. (1970). *J. Geophys. Res.,* **75**, 7679.
Broecker, W. S. and Peng, T. H. (1971). *Earth Planet. Sci. Lett.* **11**, 99.
Broecker, W. S. and Peng, T. H. (1971a). *Earth Planet. Sci. Lett.* **11**, 95.
Broecker, W. S., Cromwell, J. and Li, Y. H. (1968). *Earth Planet. Sci. Lett.* **5**, 101.
Broecker, W. S., Gerrard, R., Ewing, M. and Heezen, B. C. (1960). *J. geophys. Res.,* **65**, 2903.
Broecker, W. S., Kaufman, A., Ku, T. L., Chung, Y. C. and Craig, H. (1970). *J. Geophys. Res.* **75**, 7682.
Broecker, W. S., Li, Y. H. and Cromwell, J. (1967). *Science,* **158**, 1307.
Broecker, W. S., Tucek, C. S. and Olson, E. A. (1959). *Int. J. appl. Radiat. and Isotopes,* **7**, 1.
Broenkow, W. W. and Cline, J. D. (1969). *Limnol. Oceanogr.* **14**, 450.
Brooks, R. R. (1960). *Analyst,* **85**, 745.
Brooks, R. R. (1965). *Geochim. Cosmochim. Acta* **29**, 1369.
Brooks, R. R., Presley, B. J. and Kaplan, I. R. (1967a). *Talanta,* **14**, 809.
Brooks, R. R., Presley, B. J. and Kaplan, I. R. (1967b). *Analyt. Chim. Acta,* **38**, 321.
Brooks, R. R., Presley, B. J. and Kaplan, I. R. (1968). *Geochim. Cosmochim. Acta,* **32**, 397.
Brown, R. W. and Grummitt, W. E. (1956). *Canad. J. Chem.* **34**, 220.

Buch, K. (1933). *J. Cons. int. Explor. Mer.* **8**, 309.

Burger, K. (1973). "Organic reagents in metal analysis." Pergamon Press, Oxford.

Burrell, D. C. (1965). *Atomic Absn. Newslett.* **4**, 309.

Burrell, D. C. (1967). *Analyt. Chim. Acta,* **38**, 447.

Burrell, D. C. (1968). *At. Absn. Newslett.* **7**, 66.

Burrell, D. C. and Wood, G. C. (1969). *Analyt. Chim. Acta,* **48**, 45.

Burton, J. D. (1957). Ph.D. Thesis, University of Liverpool.

Burton, J. D. (1973). *Water Res.* **7**, 291.

Burton, J. D. and Head, P. C. (1970). *Limnol. Oceanogr.* **15**, 164.

Burton, J. D. and Leatherland, T. M. (1971). *Nature,* **231**, 440.

Burton, J. D. and Riley, J. P. (1956). *Mikrochim. Acta.* 1350.

Burton, J. D., Culkin, F. and Riley, J. P. (1959). *Geochim. Cosmochim. Acta,* **16**, 151.

Burton, J. D., Leatherland, T. M. and Liss, P. S. (1970). *Limnol. Oceanogr.* **15**, 473.

Butterworth, J., Lester, P. and Nickless, G. (1972). *Mar. Pollut. Bull.* **3**, 72.

Buyanov, N. I. and Anisimov, Y. A. (1966). *Mater. Rybokhoz. Issled. Ser. Basseina,* No. 7, 183.

Byrne, R. H. and Kester, D. R. (1974). *J. mar. Res.* **32**, 119.

Callahan, C. M., Pascual, J. N. and Lai, M. G. (1966). *U.S. Clearinghouse Fed. Sci. Tech. Inf. AD647661.*

Cameron, J. F. (1967). *In* "Symp. on Radioactive Dating and Low-level Counting", p. 543. International Atomic Energy Agency (Vienna). (Symposium held at Monaco 1967).

Camp, D. C. (1967). "Applications and Optimization of the Lithium Drifted Germanium Detector System". UCRL—50156, Lawrence Radiation Laboratory, Livermore, California.

Cann, J. R. and Winter, C. K. (1971). *Mar. Geol.* **11**, M33.

Cappadona, C. (1965). *Atti Accad. Sci. Lettere Arti Palermo Pt 1* **24**, 71.

Carey, A. G., and Paul, R. R. (1968). *Limnol. Oceanogr.* **13**, 545.

Carlisle, D. B. and Hummerstone, L. G. (1958). *Nature, Lond.* **181**, 1002.

Carlucci, A. F. (1970). *Bull. Scripps Inst. Ocean.* **17**, 23.

Carlucci, A. F., and Silbernagel, S. B. (1966a). *Can. J. Microbiol.* **12**, 175.

Carlucci, A. F. and Silbernagel, S. B. (1966b). *Can. J. Microbiol.* **12**, 1079.

Carlucci, A. F. and Silbernagel, S. B. (1966c). *Limnol. Oceanogr.* **11**, 642.

Carlucci, A. F., and Silbernagel, S. B. (1967). *Can. J. Microbiol.* **13**, 979.

Caron, H. I. and Sugihara, T. T. (1962). *Analyt. Chem.,* **34**, 1082.

Carpenter, J. H. (1957). *Limnol. Oceanogr.* **2**, 271.

Carpenter, J. H. (1965a). *Limnol Oceanogr.* **10**, 135.

Carpenter, J. H. (1965b). *Limnol Oceanogr.* **10**, 141.

Carpenter, J. H. (1966). *Limnol Oceanogr.* **11**, 264.

Carpenter, J. H. and Grant, V. E. (1967). *J. Mar. Res.* **25**, 228.

Carpenter, J. H. and Manella, M. E. (1973). *J. Geophys. Res.* **78**, 3621.

Carr, R. A. (1969). *At. Absn Newsletter,* **8**, 69.

Carr, R. A. (1970). *Limnol. Oceanogr.* **15**, 318.

Carr, R. A. and Wilkniss, P. E. (1973). *Environ. Sci. Tech.* **7**, 62.

Carr, R. A., Hoover, J. B. and Wilkniss, P. E. (1972). *Deep-Sea Res.* **19**, 747.

Carritt, D. E. (1953). *Analyt. Chem.* **25**, 1927.

Carritt, D. E. (1962). *J. Geophys. Res.* **67**, 3548.

Carritt, D. E. (1963). *In* "The Sea" (M. N. Hill, ed.), Vol. II, p. 120, Interscience, New York.

Carritt, D. E. (1965). *Rhode Island Graduate School of Oceanography Publ.* No 3, 203.
Carritt, D. E. and Carpenter, J. H. (1966). *Deep-Sea Res.* 24, 286.
Cescon, B. and Macchi, G. (1970). *Analyt. Chem.,* 42, 1809.
Cescon, B. and Macchi, G. (1972). *Comm. Ital Oceanogr. Racc Dat. Oceanog.* Ser A No 61, 22.
Cescon, B. S. and Scarazzatto, P. G. (1973). *Limnol Oceanogr.* 18, 499.
Celardin, F., Marcantonatos, M. and Monnier, D. (1974). *Analyt. Chim. Acta,* 68, 61.
Centanni, F. A., Ross, A. M. and DeSesa, M. A. (1956). *Analyt. Chem.* 28, 1651.
Chakravarti, D., Lewis, G. B., Palumbo, R. F. and Seymour, A. H. (1964). *Nature Lond.* 203, 571.
Chalmers, R. A. (1959). Proc 3rd Conf. Analyt. Chem. Prague.
Chalmers, R. A. and Sinclair, A. G. (1965). *Analyt. Chim. Acta,* 33, 384.
Chalmers, R. A. and Sinclair, A. G. (1966). *Analyt. Chim. Acta,* 34, 412.
Chamberlain, W. and Shapiro, J. (1969). *Limnol. Oceanogr.* 14, 921.
Chamberlain, W. and Shapiro (1970) *U.S. Nat. Tech. Inform. Serv. PB Rept.* No 200824, 22pp
Chamberlain, W. and Shapiro, J. (1973). *In* "Environmental Phosphorus Handbook" (E. J. Griffith, A. Beeton, J. M. Spencer, and D. T. Mitchell eds), Wiley, New York.
Chan, K. M. and Riley, J. P. (1966a). *Analyt. Chim. Acta,* 35, 365.
Chan, K. M. and Riley, J. P. (1966b). *Deep-Sea Res.,* 13, 467.
Chan, K. M. and Riley, J. P. (1966c). *Analyt. Chim. Acta,* 36, 220.
Chan, K. M. and Riley, J. P. (1966d). *Analyt. Chim. Acta,* 34, 337.
Chan, K. M. and Riley, J. P. (1967). *Analyt. Chim. Acta,* 39, 103.
Chao, T. T., Jenne, E. A. and Heppting, L. M. (1968). *U.S. Geol. Surv. Prof. Paper,* 600D, 16.
Chao, T. T., Fishman, M. J. and Ball, J. W. (1969). *Analyt. Chim. Acta,* 47, 189.
Charpiot, R. (1969). *Cah. Oceanogr.* 22, 773.
Chau, Y. K. and Wong, P-Y. (1969). *Talanta,* 15, 867.
Chau, Y. K. and Riley, J. P. (1965). *Analyt. Chim. Acta,* 33, 36.
Chau, Y. K. and Riley, J. P. (1966). *Deep-Sea Res.* 13, 1115.
Chau, Y. K. and Saitoh, H. (1971). *Effluent Water Treat. J.,* 135.
Chau, Y. K., and Saitoh, H. (1970). *Environ. Sci. Tech.* 4, 839.
Chau, Y. K., Simm, S. S. and Wong, Y. H. (1968). *Analyt. Chim. Acta,* 43, 13.
Cheng, K. L. (1961). *Talanta* 8, 658.
Cheng, K. L. and Cheng, K. (1971). *Amer. Chem. Soc. Div. Water, Air Waste Chem. Gen. Paper* 11, 31.
Chesselet, R. (1970). "Reference Methods for Marine Radioactivity Studies", p. 275. Technical Reports Series, No. 118, International Atomic Energy Agency (Vienna).
Chesselet, R., Lalou, C. and Nordemann, D. (1964). Symposium 'Contamination of the Marine Environment'. Centre d'études recherches de Biologie et d'Océanographie Médicale, Nice.
Chester, R., Gardner, D., Riley, J. P. and Stoner, J. (1973). *Mar. Pollut. Bull.* 4, 28.
Chipman, W. A., Rice, T. R. and Price, T. J. (1958). *Fish Bull. U.S.* 58, 279.
Chow, T. J. (1958). *J. Mar. Res.* 17, 120.
Chow, T. J. (1964). *Analyt. Chim. Acta* 31, 58.
Chow, T. J. (1968). *J. Water Pollut. Control Fed.* 40, 399.
Chow, T. J. and Goldberg, E. D. (1962). *J. Mar. Res.* 20, 163.
Chow, T. J., and Johnson, M. S. (1962). *Analyt. Chim. Acta,* 27, 441.
Chow, T. J. and McKinney, C. R. (1958). *Analyt. Chem.* 30, 1499.

Chow, T. J. and Robinson, R. J. (1952). *J. Mar. Res.* **11**, 124.

Chow, T. J. and Snyder, C. B. (1969). *Earth Planet Sci. Lett.* **7**, 221.

Chow, T. J. and Thompson, T. G. (1955). *Analyt. Chem.* **27**, 18.

Chuecas, L. and Riley, J. P. (1966). *Analyt. Chim. Acta,* **35**, 240.

Chung, Y. C. and Craig, H. (1972). *Earth Planet Sci. Lett.* **14**, 55.

Cigna, A. (1963). *In* "Nuclear Detonations and Marine Radioactivity", p. 95, Norwegian Defence Research Establishment. (Report of a Symposium, Kjeller, Norway. Ed. S. H. Small).

Clark, M. E., Jackson, G. A. and North, W. J. (1972). *Limnol Oceanogr.* **17**, 749.

Clark, R. C., Blumer, M. and Raymond, S. O. (1967). *Deep-Sea Res.* **14**, 125.

Clarke, W. B., Beg, M. A. and Craig, H. (1969). *Earth Planet. Sci. Lett.* **6**, 213.

Clesceri, N. L. and Lee, G. F. (1965). *Air Water Pollution* **9**, 723, 743.

Cline, J. D. (1969). *Limnol Oceanogr.* **14**, 454.

Cline, J. D. and Richards, F. A. (1972). *Limnol Oceanogr.* **17**, 885.

Cohen, S. and Ruchhoft, C. C. (1941). *Industr. Engng Chem. (Anal.)* **13**, 622.

Collier, A. W. and Marvin, K. T. (1953). Fishery *Bull. Fish Wildl. Serv. U.S.* **79**, 71.

Collins, P. and Diehl, H. (1960). *J. Mar. Res.* **18**, 152.

Cooke, R. C. (1973). *Limnol Oceanogr.* **18**, 150.

Cooper, J. A. (1969). Battelle Northwest Pacific Laboratory Annual Report BNWL-1051 part 2, p. 121.

Cooper, J. A. (1973). *In* "Contemporary Activation Analysis." (V. A. Ryan, ed.) Marcel Dekker, New York.

Cooper, J. A. and Perkins, R. W. (1972). *Nucl. Inst. Meth.* **99**, 125.

Cooper, J. A., Wogman, N. A. and Perkins, R. W. (1968). *IEEE Transactions on Nuclear Science,* **15**, 407.

Cooper, L. H. N. (1937a). *J. Mar. Biol. Ass. U.K.* **21**, 673.

Cooper, L. H. N. (1937b). *Proc. Roy. Soc.* **B118**, 419.

Cooper, L. H. N. (1948). *J. Mar. Biol. Ass. U.K.* **27**, 279.

Cooper, L. H. N. (1958). *J. Mar. Res,* **17**, 128.

Coote, A. R., Duedall, I. W. and Hiltz, R. S. (1971). *In* "Advances in Automated Analysis". Vol. 2, p. 347, Thurman, New York.

Copin-Montegut, C. and Copin-Montegut, G. (1973). *Mar. Chem.* **1**, 151.

Corless, J. T. (1965). *J. chem. Educ.* **42**, 421.

Corwin, J. F. (1970). *In* "Symposium on Organic Matter in Natural Waters" (D. W. Hood, ec.), pp. 169–180. University of Alaska.

Cowey, C. B. (1956). *J. Mar. biol. Ass. U.K.* **35**, 609.

Cox, J. L. (1971). *Fish. Bull.,* **69**, 443.

Cox, R. A. (1953). *J. Cons. int. Explor. Mer.* **19**, 297.

Cox, R. A. (1968). UNESCO Technical Papers in Marine Science No. 9.

Coyne, R. V. and Collins, J. A. (1972). *Analyt. Chem.* **44**, 1093.

Craig, H. and Weiss, R. F. (1968). *Earth Planet. Sci. Letters,* **5**, 175.

Craig, H., Weiss, R. F. and Clarke, W. B. (1967). *J. Geophys. Res.,* **72**, 6165.

Cranston, R. E. and Buckley, D. E. (1972). *Environ. Sci. Tech.* **6**, 274.

Cranston, R. E. and Buckley, D. E. (1972a). Unpublished Manuscript, Bedford Institute of Oceanography, Dartmouth, Nova Scotia. Rept. Ser. BI-R-72-7.

Creaser, E. P. (1971). *J. Fish. Res. Bd Can.* **28**, 1049.

Crouch, S. R. and Malmstadt, H. V. (1967a). *Analyt. Chem.* **39**, 1084.

Crouch, S. R. and Malmstadt, H. V. (1967b). *Analyt. Chem.* **39**, 1090.

Crowther, A. B. and Large, R. S. (1956). *Analyst,* **81,** 64.

Culkin, F. (1965). *In* "Chemical Oceanography" (J. P. Riley and G. Skirrow eds.) 1st Ed. Vol. 1. Academic Press, London.

Culkin, F. and Cox, R. A. (1966). *Deep-Sea Res.* **13,** 789.

Cutshall, N., Johnson, V. and Osterberg, C. (1966). *Science,* **152,** 202.

Dagnall, R. M., and Sharp, B. L. (1973). *In* "Analytical Chemistry Part 1, MTP International Review of Science" (T. S. West, ed.), Butterworth's, London.

Dagnall, R. M., Manfield, J. M., Silvester, M. D. and West, T. S. (1972). *Nature,* **235,** 156.

Daisley, K. W. (1958). *J. mar. biol. Ass. UK* **37,** 673.

Dal Pont, G. (1962). *CSIRO Australian Oceanogr. Cruise Rept. No 4* Oceanographic Observations in the Indian Ocean in 1960.

Dal Pont, G. and Newell, B. (1963). *Aust. J. Mar. Freshw. Res.* **14,** 155.

Dal Pont, G., Newell, B. S., and Staniforth, J. (1963). *Aust. J. Mar. Freshw. Res.* **14,** 37.

Dal Pont, G., Klye, J. and Newell, B. (1972). "Laboratory techniques in Marine Chemistry Manual". Commonwealth Scientific and Industrial Research Organisation, Australia. Report 51.

Datsko, V. G. and Datsko, V. E. (1950). *Doklady Akad. Nauk S.S.S.R.,* **73,** 337.

Davankov, A. B., Laufer, V. M., Azhazha, E. G., Gordievskii, A. V. and Kiryushov, V. N. (1962). *Izv. Vysshkh. Uchebn. Zavedenii Tsvetn. Met.* **5,** No. 2, 118.

Dawson, R. and Riley, J. P. (1975). *Mar. Pollution Bull.* (in the press).

De, A. K., Khopkar, S. M. and Chalmers, R. A. (1970). "Solvent extraction of metals", Van Nostrand Reinhold, New York.

Dean, J. A. (1960). "Flame Photometry". McGraw-Hill, New York.

Dearnaley, G. and Northrop, D. C. (1966). "Semiconductor Counters for Nuclear Radiations". John Wiley and Sons, New York.

DeGalan, L. and Samaey, G. F. (1969). *Spectrochim Acta,* **24B,** 679.

Degobbis, D. (1973). *Limnol. Oceanogr.* **18,** 146.

De Manche, J. M., Curl, H. and Coughenower, D. D. (1973). *Limnol Oceanogr.,* **18,** 686.

Degens, E. T., Reuter, J. H. and Shaw, N. F. (1964). *Geochim. Cosmochim. Acta,* **28,** 45.

del Riego, A. F. (1965). *Bol. Inst. Espan. Oceanogr.* No. 120.

Demmitt, T. F. (1965). *In* "Automation in Analytical Chemistry" (L. T. Skeggs, ed.), Mediad Inc., New York.

DeVoe, J. R. and LaFleur, P. D. (1969). "Modern Trends in Activation Analysis". National Bureau of Standards, Washington, D.C.

Diehl, H. and Smith, G. F. (1960). "The iron reagents bathophenanthroline, 2,4,6-tripyridyl-2-triazine and phenyl-2-pyridyl ketoxime". G. F. Smith Chemical Co., Columbus, Ohio, U.S.A.

Dittmar, W. (1884). *In* "The Voyage of H.M.S. *Challenger*" (J. Murray, ed.). Vol. 1. H.M. Stationery Office, London.

Dixon, B. W., Slowey, J. F. and Hood, D. W. (1966a). *Report on U.S. At. Energy Comm. Contract No. AT-(40–1)2799 Ref 66-2F* (Texas A and M University).

Dixon, B. W., Slowey, J. F., and Hood, D. W. (1966b). *U.S. At. Energy. Comm. Rept.* **TID 23295.**

Dobizhanskaya, M. A. and Pshenina, T. I. (1959). *Trav. Sta. biol. Sebastopol,* **11,** 316.

Doerner, H. and Hoskins, W. (1925). *J. Amer. Chem. Soc.* **47,** 662.

Dollman, G. W. (1968). *Environ Sci. Tech.* **2**, 1027.

Donaldson, D. E. (1966). *U.S. Geol. Surv. Prop.* Pap. **550D**, 258.

Doshi, G. R. (1967). *Indian J. Chem.*, **5**, 580.

Douglas, E. (1964). *J. phys. Chem.* **68**, 169.

Douglas, E. (1965). *J. phys. Chem.* **69**, 2608.

Drozhzhin, V. M., Lazarev, K. F. and Nikolaev, D. S. (1965). *Radiokhimiya,* **7**, 374.

Dubravčic, M. (1955). *Analyst,* **80**, 295.

Duce, R. A., Quinn, J. G., Olney, C. E., Piotrowicz, S. R., Ray, B. J. and Wade, T. L. (1972). *Science* **176**, 161.

Durst, R. A. (1968). *Analyt. Chem.* **40**, 931.

Durst, R. A. and Taylor, J. K. (1967). *Analyt. Chem.* **39**, 1483.

Dutton, J. W. R. (1970). Ministry of Agriculture, Fisheries and Food, Technical Report, FRL 4.

Dutton, J. W. R. (In press). *In* "Further Reference Methods for Marine Radioactivity Studies". Technical Reports Series, International Atomic Energy Agency, Vienna.

Duursma, E. K. (1961). *Neth. J. Sea. Res.* **1**, 1.

Duursma, E. K. (1967). *Deep-Sea Res.* **14**, 133.

Duursma, E. K. and Sevenhuysen, W. (1966). *Neth. J. Sea-Res.* **3**, 95.

Dyrssen, D., Jagner, D., and Johansson, H. (1969). "Reports on the Analytical Chemistry of sea water V". Department of Analytical Chemistry, University of Göteborg.

Dyrssen, D. and Hansen, I. (1972). *Mar. Chem.* **1**, 137.

Dyrssen, D. W., Novikov, Y. P. and Uppström, L. F. (1972). *Analyt. Chim. Acta,* **60**, 139.

Eckschlager, K. (1969). "Errors, Measurement and results in chemical analysis". Van Nostrand-Reinhold, London.

Ediger, R. D. (1973). *At. Absn Newsletter* **12**, 151.

Ediger, R. D., Peterson, G. E. and Kerber, J. D. (1974). *At. Absn Newsletter,* **13**, 61.

Edisbury, J. R. (1966). "Practical hints on absorption spectrometry". Adam Hilger, London.

Edwards, G. P., Molof, A. M. and Schneeman, R. W. (1965). *J. Am. Wat. Wks Ass.* **57**, 917.

Ehrhardt, M. (1969). *Deep-Sea Res.* **16**, 393.

Elderfield, H. (1970). *Earth Planet. Sci. Lett.* **9**, 10.

Elgqvist, B. (1970). *J. inorg. Nucl. Chem.* **32**, 937.

Elmer, R. and Robbins, R. C. (1968). *Antarctic J. U.S.* **3**, 194.

Emel'yanov, E. M., Blazhis, I. K., Yuryavichyus, R. Y., Paeda, R. I. Valyukyavichys, C. A. and Yankauskas, I. I. (1971). *Okeanologiya,* **11**, 1116.

Emmet, R. T. (1969). *Analyt. Chem.*, **41**, 1643.

Environmental Protection Agency (1972). "Analyses for mercury in water-a preliminary study". National Environmental Research Center-Office of Research and Monitoring. Report No R4-72-003. U.S. Environment Protection Agency, Cinncinnati, Ohio 45268.

Ernst, T. and Hoermann, H. (1936). *Nachr. Ges Wiss Göttingen Kl* IV [N.S.] **1**, 205.

Evans, R. D. (1935). *Rev. Sci. Instrum.* **6**, 99.

Evans, R. D., Kip, A. F. and Moberg, E. G. (1938). *Amer. J. Sci.* **36**, 241.

Fabricand, B. P., Sawyer, R. R., Ungar, S. G. and Adler, S. (1962). *Geochim. Cosmochim. Acta,* **26**, 1023.

Fabricand, B. P., Imbimbo, E. S., Brey, M. E. and Weston, J. A. (1966). *J. Geophys. Res.,* **71,** 3917.
Fajans, K. and Erdey-Grúz, T. (1932). *Z. phys. Chem.* **A-158,** 97.
Fanning, K. A. and Pilson, M. E. Q. (1973). *Analyt. Chem.* **45,** 136.
Feely, R. A., Sackett, W. M. and Harris, J. E. (1971). *J. Geophys. Res.* **76,** 5893.
Feldman, C. and Rains, T. C. (1964). *Analyt. Chem.* **36,** 405.
Fernandez, A. E. and Okuda, T. (1967). *Bol. Inst. Oceanogr. Univ. Oriente,* **5,** 96.
Fiadeiro, M., Solorzano, L. and Strickland, J. D. H. (1967). *Limnol. Oceanogr.* **12,** 555.
Finucane, J. H. and May, B. Z. (1961). *Limnol. Oceanogr.* **6,** 86.
Fischer, E. (1883). *Chem. Ber.* **26,** 2234.
Fischer, H. (1964). "The Hydrobios Universal series water sampler". Tech. Rept. Hydrobios Apparatebau, Kiel-Holtenau.
Fischer, H. (1968). "The transparent plastic Nansen sampler." Tech. Rept. Hydrobios Apparatebau Gmbh Kiel-Holtenau.
Fishman, M. J. (1970). *Analyt. Chem.* **42,** 1462.
Fishman, M. J. (1972). *At Absn. Newslett.* **11,** 46.
Fitzgerald, G. P. and Faust, S. L. (1967). *Limnol. Oceanogr.* **12,** 332.
Fitzgerald, R. A., Gordon, D. C. and Cranston, R. E. (1974). *Deep-Sea Res,* **21,** 139.
Fitzgerald, W. F. (1970). Ph.D. Thesis, Mass. Inst. of Tech. Cambridge, Mass.
Fitzgerald, W. F. and Lyons, W. B. (1973). *Nature,* **242,** 452.
Fjarlie, R. L. I. (1953) *J. Mar. Res.* **12,** 21.
Fletcher, M. H., May, I., and Anderson, J. W. (1950). *U.S. At. Energy Comm.,* **TEI-133.**
Florence, T. M. (1972). *J. Electroanalyt. Chem.* **35,** 237.
Folkard, A. R. and Jones, P. G. W. (1968). *Cons. int Explor. Mer., Serv Hydrogr. Interlab. Rept. No. 2,* 9.
Folsom, T. R. (1974). *Nature* (in the press).
Folsom, T. R. and Saruhashi, K. (1963). *J. Rad. Res.,* **4,** 39.
Folsom, T. R. and Sreekumaran, C. (1970). *In* "Reference Methods for Marine Radioactivity Studies", p. 129. Technical reports Series, No. 118, International Atomic Energy Agency (Vienna).
Folsom, T. R., Bode, G. W. and Grismore, R. (1970b). *Appl. Spectr.* **24,** 378.
Folsom, T. R., Feldman, C. and Rains, T. C. (1964). *Science,* **144,** 538.
Folsom, T. R., Hansen, N., Parks, G. J. and Weitz, W. E. (1974). *J. Appl. Spectr.* (in the press).
Folsom, T. R., Hansen, N. and Robertson, D. E. (1970a). *Texas A. & M. University Report* TID 25776.
Folsom, T. R., Mohanrao, G. J. and Winchell, P. (1960). *Nature (London),* **187,** 480.
Folsom, T. R., Sreekumaran, C., Weitz, W. E. and Tennant, D. A. (1967). *Scripps. Inst. Oceanogr. Rept* 67–20 TR-922-ERG-A-67.
Fonselius, S. H. (1962). *Fish. Bd. Swed. Ser. Hydrol., Rept* **13,** 31.
Fonselius, S. H. (1970). *Int. Atomic Energy Agency, Radioactivity in the Sea Series,* **No 29** *(Bull. Inst. Océanogr. Monaco 1970,* **69, no. 1407***).*
Fonselius, S. H. and Koroleff, F. (1963). *Bull. Inst. Océanogr. Monaco,* **61,** No 1281.
Forster, W. and Zeitlin, H. (1966). *Analyt. Chim Acta,* **34,** 211.
Forster, W. and Zeitlin, H. (1966a) *Analyt. Chem.* **38,** 649.
Forster, W. and Zeitlin, H. (1966b) *Analyt. Chim. Acta* **35,** 42.
Fossato, V. U. (1968). *Arch. Oceanogr. Limnol.* **16,** 95.

Fossato, V. U. (1968/9). *Atti Ist. Veneto Sci. Lett. Anti, Cl. Sci. Mat. Natur.*, **127**, 135.

Foster, P. and Morris, A. W. (1971). *Deep-Sea Res.* **18**, 231.

Foster, P. and Morris, A. W. (1971a). *Water Res.*, **5**, 19.

Føyn, E. (1968). *Cons int Explor Mer., Serv. Hydrogr. Interlab Rept.* No. **2**, 18.

Føyn, E., Karlick, B., Pettersson, H. and Rona, E. (1939). *Göteborgs. Vetensk-Samh. Handl.* **6**, 44pp.

Franchini, C. (1973). *Arch. Oceanograf. Limnol.* **18**, 39.

Frant, M. S. and Ross, J. W. (1966). *Science,* **154**, 3756.

Frederikson, A. F. and Reynolds, R. C. (1960). *Oil Gas J.* **58**, 154.

Friedman, G. M., Fabricand, B. P., Imbimbo, E. S., Brey, M. E. and Sanders, J. E. (1968). *J. Sediment. Petrol.* **38**, 1313.

Freudenthal, H. D., Blogowski, W. and Stoecker, R. (1968). *Limnol Oceanogr.* **13**, 706.

Fujinaga, T., Kuwamoto, T., Murai, S., Kihara, S., and Nakayama, E. (1971). *Nippon Kagaku Zasshi,* **92**, 339.

Fujinuma, M., Shimada, Y. and Hirano, S. (1971). *Bunseki Kagaku,* **20**, 131.

Fukai, R. (1967). *Nature, Lond.* **213**, 901.

Fukai, R. (1968). *J. Oceanogr. Soc. Japan,* **24**, 265.

Fukai, R. (1969). *Rapp. Comm. int. Mer. Médit.* **19**, 935.

Fukai, R. and Huynh-Ngoc, L. (1967). *Int. At. Energy Ag.* (Vienna) *Radioact. in Sea* Ser. No. 22.

Fukai, R. and Huynh-Ngoc, L. (1971). *Nippon Kaiyo Gakkai-Shi,* **27**, 91.

Fukai, R., and Vas, D. (1967). *J. Oceanogr. Soc. Japan,* **23**, 298.

Fukai, R. and Vas, D. (1969). *J. Oceanogr. Soc. Japan.* **25**, 109.

Fukai, R., Ballestra, S. and Murray, C. N. (1973). *In* "Radioactive Contamination of the Marine Environment", p. 3, International Atomic Energy Agency (Vienna) (Symposium at Seattle, U.S.A.).

Fukai, R., Huynh-Ngoc, L. and Vas, D. (1966). *Nature, Lond.* **211**, 726.

Fukai, R., Huynh-Ngoc, L. and Murray, C. N. (1973). *J. Oceanogr. Soc. Japan,* **29**, 44.

Gales, M. E., Julian, E. C. and Kroner, R. C. (1966). *J. Am. Wat. Wks. Ass.* **58**, 1363.

Gardner, D. and Riley, J. P. (1973). *J. Cons. int. Explor. Mer,* **35**, 202.

Garrett, W. D. (1967). *Deep Sea-Res.* **14**, 221.

Garside, C. and Riley, J. P. (1969). *Analyt. Chim. Acta,* **46**, 179.

Gary, W. W. (1968). U.S. Clearinghouse. Fed. Sci. Tech. Inform. AD 673–426, 61pp.

Gassaway, J. D. (1967). *Int. J. Limnol. Oceanol.* **1**, 85.

Gast, J. A. and Thompson, T. G. (1958). *Analyt. Chem.* **30**, 1549.

Genovese, S. and Magazzu, G. (1969). "Manuale d'analisi per le acque salmastre". La Editrice Universitaria, Messina, Sicily.

Gerard, R. D. (1968). *Mar. Sci. Instr.* **4**, 682.

Gerard, R. and Ewing, M. (1961). *Deep-Sea Res.* **8**, 298.

Giese, A. C. (1967). *Oceanogr. Mar. Biol. Ann. Rev.* 5, 159

Gilbert, T. R. and Caly, A. M. (1973). *Analyt. Chim. Acta,* **67**, 289.

Gilbert, T. R. and Hume, D. N. (1973). *Analyt. Chim. Acta,* **65**, 451.

Gillbricht, M. (1961). *Helgoländ, wiss. Meeresunters,* **8**, 58.

Gillespie, A. S. and Richter, M. G. (1965). *Analyt. Instrum.* 127.

Gilmartin, M. (1967). *Limnol. Oceanogr.* **12**, 325.

Gloss, G. H. (1953). *Chemist Analyst.* **42**, 50.

Goetz, A. and Tsuneishi, N. (1951). *J. Amer. Water Wks. Ass.* **43**, 943.

Gohda, S. (1972). *Bull. Chem. Soc. Japan,* **45**, 1704.

Gold, K. (1964). *Limnol. Oceanogr.* **9**, 343.

Goldberg, E. D. (1952). *Biol. Bull. Woods Hole,* **102**, 243.

Goldberg, E. D., Baker, M. and Fox, D. L. (1952). *J. Mar. Res.* **11**, 194.

Goldberg, E. D., Koide, M., Schmitt, R. A. and Smith, R. H. (1963). *J. Geophys. Res.,* **68**, 4209.

Goldschmidt, V. M. and Strock, L. W. (1935). *Nach. Ges. Wiss Göttingen, Kl. IV* [*N.S.*], **2**, 235.

Good, D. E. (1968). *Mar. Sci. Instr.* **4**, 252.

Gordon, C. M. and Larsen, R. E. (1970). *Radiochem. Radioanalyt. Letters,* **5**, 369.

Gordon, D. C. (1969). *Deep-Sea Res.* **16**, 661.

Gordon, D. C. and Sutcliffe, W. H. (1973). *Mar. Chem.* **1**, 231.

Goulden, P. D. and Afghan, B. K. (1970). *Dept. of Energy Mines and Resources Canada, Inlands Waters Branch Tech Bull.* **No 27**.

Goulden, P. D. and Brooksbank, P. (1974). *Limnol, Oceanogr.* **19**, 705.

Goya. H. A. and Lai, M. G. (1967). "Adsorption of Trace Elements from Seawater by Chelex-100, USNRDL-TR-67-129", U.S. Nava. Radiological Defense Laboratory, San Francisco, California.

Grasshoff (1962a). *Kiel. Meeresforsch.* **18**, 42.

Grasshoff (1962b). *Kiel. Meeresforsch,* **18**, 151.

Grasshoff, K. (1964a). *Deep-Sea Res.* **11**, 597.

Grasshoff, K. (1964b). *Kiel. Meeresforsch.* **20**, 5.

Grasshoff, K. (1964c). *Kiel Meeresforsch.* **20**, 143.

Grasshoff, K. (1966). *Kiel. Meeresforsch.* **22**, 42.

Grasshoff, K. (1966a). *Z. anal. Chem.* **220**, 89.

Grasshoff, K. (1967). *In* "Automation in Analytical Chemistry". Vol. 1, p 573. Mediad, White Plains.

Grasshoff, K. (1968a). *Z. Analyt. Chem.* **234**, 12.

Grasshoff, K. (1968b). *In* "Methoden der Meeresbiologishen Forschung (C. Schlieper, ed.) Gustav Fischer Verlag, Jena, pp 13–31.

Grasshoff, K. (1969). *In* "Chemical Oceanography". (R. Lange, Ed.), Universitets-Forlaget, Oslo.

Grasshoff, K. (1970a). *Technicon Quarterly,* **3**, 7.

Grasshoff, K. (1970b). *In* "Advances in Automated Analysis". Vol. 2, p 133. Mediad, White Plains.

Grasshoff, K. (1970c). *Cons. int. Explor. Mer., Inf. Tech. Methods Sea Water Analysis,* **3**, 5.

Grasshoff, K. and Chan, K. M. (1969). *Adv. Automat. Anal., Proc. Technicon International Congress,* **2**, 147.

Grasshoff, K. and Chan, K. M. (1971). *Analyt. Chim. Acta,* **53**, 442.

Grasshoff, K. and Johannes, H. (1974). *J. Cons. Int. Explor. Mer.* **36**. 90.

Green, E. J. and Carritt, D. E. (1966). *Analyst,* **91**, 207.

Greenfield, L. J. and Kalber, F. A. (1955). *Bull. Mar. Sci. Gulf. Caribb.* **4**, 323.

Grasshoff, K. and Johannsen, H. (1972). *J. Cons. int. Explor. Mer,* **34**, 516.

Greenhalgh, R. and Riley, J. P. (1961). *Analyt. chim. Acta,* **25**, 179.

Greenhalgh, R. and Riley, J. P. (1961a). *J. Mar. Biol. Ass. U.K.* **41**, 175.

Greenhalgh, R. and Riley, J. P. (1962). *Analyst.* **87**, 970.

Greenhalgh, R., Riley, J. P. and Tongudai, M. (1966). *Analyt. Chim. Acta,* **36**, 439.

Griel, J. V. and Robinson, R. J. (1952). *J. Mar. Res.* **11**, 172.

Grimaldi, F. S., May, I., Fletcher, M. H. and Titcomb, J. (1954). *U.S. Geol. Serv. Bull.* **1006**.

Guegueniat, P. (1967). French Atomic Energy Commission Report CEA-R-3284.

Guinn, V. P. and Lukens, Jr., H. R. (1967). *In* "Trace Analysis—Physical Methods" (G. H. Morrison, ed.) Interscience Publishers, New York.

Guiseppe, M. (1965). *J. Electroanalyt. Chem.* **9**, 290.

Gunter, B. D. and Musgrave, B. C. (1966). *J. Gas Chromatog.* **4**, 162.

Guntz, A. A. and Kocher, J. (1952). *C.R. Acad. Sci., Paris,* **67**, 174.

Haber, F. (1928). *Z. Ges Erdk. Berl.* **3**, 3.

Hager, S. W., Atlas, E. L., Gordon, L. I., Mantyla, A. W. and Park, P. K. (1972). *Limnol. Oceanogr.* **17**, 931.

Hahn, J. (1972). *In* "The changing chemistry of the oceans" (D. Dyrssen and D. Jagner, eds.), Wiley-Interscience, New York.

Hahn, O. (1936). "Applied radiochemistry", Cornell University Press, Ithaca, New York.

Hamaguchi, H., Kuroda, R., Hosahara, K. and Shimizu, T. (1963). *Nippon Genshiryoku Gakkaishi,* **5**, 662.

Hamaguchi, H., Kuroda, R., Onuma, N., Kawabuchi, K., Mitsubayashi, T. and Hosohara, K. (1964). *Geochim. Cosmochim Acta,* **28**, 1039.

Hamilton, R. D. and Holm-Hansen O. (1967). *Limnol. Oceanogr.,* **12**, 319.

Hampson, B. L. (1963). *Analyst.* **88**, 529.

Hampson, B. L. (1964). *Analyst,* **89**, 651.

Handa, N. (1966a). *J. Ocean. Soc. Japan,* **22**, 50.

Handa, N. (1966b). *J. Ocean. Soc. Japan,* **22**, 79.

Hansson, I. (1973). *Acta. Chem. Scand.* **27**, 924.

Hansson, I. and Jagner, D. (1973). *Analyt. Chim. Acta,* **65**, 363.

Harsányl, E., Polos, L. and Pungor, E. (1973). *Analyt. Chim. Acta,* **67**, 229.

Harvey, G. R. and Dutton, J. W. R. (1973). *Analyt. Chim. Acta,* **67**, 377.

Harvey, G. R. Steinhauer, W. G. and Teal, J. M. (1973). *Science, N.Y.* **180**, 643.

Harvey, G. R., Steinhauer, W. G. and Milkas, H. P. (1974). *Nature, Lond.* **252**, 387.

Harvey, H. W. (1926). *J. Mar. biol. Ass. U.K.* **14**, 71.

Harvey, H. W. (1948). *J. Mar. biol. Ass. U.K.* **27**, 337.

Harvey, H. W. (1949). *J. Mar. biol. Ass. U.K.* **28**, 155.

Harvey, H. W. (1955). "The chemistry and fertility of sea waters". Cambridge University Press, London.

Harwood, J. E. and Kuhn, D. J. (1970). *Water Res.* **4**, 501.

Harwood, J. E., van Steenderen, R. A. and Kühn, A. L. (1969). *Water Research,* **3**, 425.

Hashimoto, T. (1971). *Analyt. Chim. Acta,* **56**, 347.

Haslam, J. and Moses, G. (1950). *Analyst.* **65**, 343.

Hassenteufel, W., Jagitsch, R. and Koczy, F. F. (1963). *Limnol. Oceanogr.* **8**, 152.

Hayes, D. W. (1969). *Diss. Abstr., Int. B,* **30**, 1782.

Hayes, D. W., Slowey, J. F. and Hood, D. W. (1966). *U.S. Atomic Energy Comm.* Access No. 43087. Rept. No. TID 23295 (see *Nucl. Sci. Abstr.* **20**, 5264).

Hazan, I., Korkisch, J. and Arrhenius, G. (1965). *Z. analyt. Chem.* **213**, 182.

Head, P. C. (1971). *Deep-Sea Res.* **18**, 531.

Head, P. C. (1971a). *J. Mar. Biol. Ass. U.K.* **15**, 891.

Head, P. C. and Burton, J. D. (1970). *J. Mar. Biol. Ass. U.K.* **50**, 439.

Headridge, J. B. (1961). "Photometric titrations", Pergamon Press, Oxford.

Heath, R. L. (1969). *In* "Modern Trends in Activation Analysis". National Bureau of Standards, Washington, D.C.

Hedges, D. H. and Hood, D. W. (1966a). *Texas A and M College, Department of Oceanography Report* AT-(40-1)-2799 Ref. 66-2F.

Hedges, D. H. and Hood, D. W. (1966b). *U.S. Atomic Energy Commission Access. No. 43088* Rept TID-23295.

Hem, J. D. (1967). *Amer. Chem. Soc., Divn Water, Waste, Chem., Preprints*, 7(1), 54.

Henriksen, A. (1965a). *Analyst*, 90, 83.

Henriksen, A. (1965b). *Analyst*, 90, 29.

Henriksen, A. (1970). *Analyst*, 95, 601.

Herdman, H. F. P. (1963). *In* "The Sea", Vol. II, p. 125. Interscience New York.

Hering, R. (1971). *Fortschr. Wasserchem. Ihrer Grenzgebr.* No. 13(3), 175.

Hermann, A. and Suttle, J. F. (1961). *In* "A Treatise on analytical chemistry", (I. M. Kolthoff and P. J. Elving eds.) Part I, Chapter 32. Interscience, New York.

Hermann, F., Kalle, K., Koczy, F. F., Maniece, W. and Tchernia, P. (1959). *J. Cons. int. Explor. Mer*, 24, 429.

Hernegger, F. and Karlick, B. (1935). *Sitzber. Akad. Wiss. Wien, Kl Abt IIa*, 144, 217.

Herring, J. R. and Liss, P. S. (1974). *Deep-Sea Res.* 21, 777.

Heron, J. (1962). *Limnol. Oceanogr.* 7, 316.

Hicks, S. E. and Carey, F. G. (1968). *Limnol. Oceagnogr.* 13, 361.

Higano, R. and Shiozaki, M. (1960). *Contrib. Marine Res. Lab., Hydrogr. Off. Japan*, 1, 137.

Hinkle, M. E. and Learned, R. E. (1969). *U.S. Geol. Surv. Prof. Paper*, 650D, 254.

Hintenberger, M., Konig, H., Schultz, L. and Suets, M. E. (1964). *Z. Naturforsch.* 19, 1227.

Hobbie, J. E. and Wright, R. T. (1965). *Limnol. Oceanogr.* 10, 471.

Høgdahl, O. T. (1965). *Report on NATO Grant* No. 203.

Høgdahl, O. T. (1970). *In* "Reference Methods for Marine Radioactivity Studies", p. 187. Technical Reports Series, No. 118, International Atomic Energy Agency (Vienna).

Høgdahl, O. T., Melsom, S. and Bowen, V. T. (1968). *In* "Trace Inorganics in Water", (R. A. Baker ed.) Amer. Chem. Soc., Washington *(Adv. in Chem. Ser.* 73, 308).

Holm–Hansen, O. (1968). *Limnol. Oceanogr.* 13, 175.

Holm–Hansen, O. (1970). *Plant and Cell Physiol.* 11, 689.

Holm–Hansen, O. C., Lorenzen, C., Holmes, R. W. and Strickland, J. D. H. (1965). *J. Cons. Int. Explor. Mer.* 30, 3.

Holm–Hansen, O. and Booth, C. R. (1966). *Limnol. Oceanogr.* 11, 510.

Holm–Hansen, O., Coombs, J., Volcani, B. E. and Williams, P. M. (1967). *Anal. Biochem.* 19, 561.

Holm–Hansen, O., Sutcliffe, W. H. and Sharp, J. (1968). *Limnol. Oceanogr.* 13, 507.

Holm–Hansen, O., Taylor, F. J. R. and Barsdale, R. J. (1970). *Mar. Biol.* 7, 37.

Hood, D. W. (1963). "Some chemical aspects of the marine environment". *Michigan Univ. Great Lakes Res. Div. Publ.* 10, 91.

Hood, D. W. (1966). *TID 23295, Texas A and M University, Texas.*

Hood, D. W. and Noakes, J. E. (1961). *Deep-Sea Res.* 8, 121.

Hood, D. W. and Slowey, J. F. (1964). *Texas A and M College, Project 276 Ref. 64-27A.*

Hood, D. W., Rona, E., Muse, L. and Buglio, B. (1962). *Amer. Chem. Soc. Div. Water, Waste Chem. Preprints*, 44.

Hosohara, K. (1961). *Nippon Kagaku Zasshi*, **82**, 1107.

Hosohara, K., Kuroda, R. and Hamaguchi, H. (1961). *Nippon Kaguku Zasshi*, **82**, 347.

Hosokawa, I. and Oshima, F. (1969). *Bull. Fukuoka Univ. Educ.* Pt III. *Nat. Sci.* **19**, 23.

Hulthe, P., Uppström, L. and Ostling, G. (1970). *Analyt. Chim. Acta*, **51**, 31.

Hummell, R. W. (1957). *Analyst*, **82**, 483.

Hummel, R. W. and Smales, A. A. (1956). *Analyst*, **81**, 110.

Imai, T. and Sakanoue, M. (1973). *J. Oceanogr. Soc. Japan*, **29**, 76.

International Atomic Energy Agency (1970). "Reference Methods for Marine Radioactivity Studies", Technical Report Series, No. 118, International Atomic Energy Agency (Vienna).

International Atomic Energy Agency (In press). *In* "Further Reference Methods for Marine Radioactivity Studies", Technical Reports Series, International Atomic Energy Agency (Vienna).

Ivanova, L. M. (1967). *Radiokhimiya*, **9**, 622. (Eng. trans. *Soviet Radiochemistry*, 509).

Irving, H. and Cox, J. J. (1963). *J. Chem. Soc. (London)*, 466.

Isaeva, A. B. and Bogoyavlensky, A. N. (1968). *Okeanologiya*, **8**, 539; translation in *Oceanology*, **8**, 433 (1969).

Ishibashi, M. (1953). *Rec. Oceanogr. Wks., Japan, (N.S.)*, **1**, 88.

Ishibashi, M. and Azuma, S. (1949). *Chem. Chem. Ind., Tokyo*, **2**, 14.

Ishibashi, M. and Hara, T. (1955). *Rec. Oceanogr. Wks., Japan, N.S.* **2**, (iii), 45.

Ishibashi, M. and Hara, T. (1959). *Bull. Inst. Chem. Res., Kyoto Univ.* **37**, 179.

Ishibashi, M. and Higashi, S. (1954). *Japan Analyst*, **3**, 213.

Ishibashi, M. and Shigematsu, T. (1950). *Bull. Inst. Chem. Res., Kyoto Univ.* **23**, 59.

Ishibashi, M., Shigematsu, T., Nakagawa, T. and Ishibashi, Y. (1951). *Bull. Inst. Chem. Res., Kyoto Univ.* **24**, 68.

Ishibashi, M., Shigematsu, T. and Nakagawa, Y. (1954). *Bull. Inst. Chem. Res., Kyoto Univ.* **32**, 199.

Ishibashi, M., Shigematsu, T. and Nishikawa, Y. (1956). *Bull. Inst. Chem. Res., Kyoto Univ.* **34**, 210.

Ishibashi, M., Fujinaga, T. and Kuwamoto, T. (1958). *J. Chem. Soc., Japan*, **79**, 1496.

Ishibashi, M., Shigematsu, T. and Nishikawa, Y. (1960a). *Rec. Oceanogr. Wks., Japan*, **5**, 66.

Ishibashi, M., Shigematsu, T., Nishikawa, Y. and Ishibashi, Y. (1960b). *Rec. Oceanogr. Wks., Japan, N.S.* **5**, 63.

Ishibashi, M., Fujinaga, T., Kuwamoto, T., Koyama, T. and Sugibayashi, S. (1960c). *Nippon Kagaku Zasshi*, **81**, 392.

Ishibashi, M., Shigematsu, T., Nishikawa, Y. and Hiraki, K. (1961). *J. Chem. Soc., Japan*, **82**, 1141.

Ishibashi, M., Fujinaga, T., Izutsu, K., Yamamoto, T., and Tamura, H. (1961a). *Rec. Oceanogr. Wks., Japan, N.S.*, **6**, (i), 106.

Ishibashi, M., Shigematsu, T., Tabushi, M., Nishikawa, Y. and Goda, S. (1962). *J. Chem. Soc., Japan*, **83**, 295.

Ishibashi, M., Fujinaga, T. and Kuwamoto, T. (1962a). *Records Oceanogr. Wks., Japan*, **6**, 215.

Ishibashi, M., Fujinaga, T., Kuwamoto, T. and Sawamoto, H. (1964). *Records Oceanogr. Wks., Japan*, **7**, 33.

Ishibashi, M., Fujinaga, T., Kuwamoto, T., Murai, S. (1967). *Nippon Kagaku Zasshi*, **88**, 76.

Ivanoff, A. (1962). *C.R. Acad. Sci., Paris*, **254**, 4493.

Ivanova, L. M. (1967). *Radiokhimiya*, **9**, 622.

Iwantscheff, G. (1958). "Das Dithizon und seine Anwendung in der Mikro- und Spurenanalyse". Verlag Chemie, Weinheim.

Iwasaki, I. and Ishimori, T. (1951). *J. Chem. Soc., Japan* (*Pure Chem.*), **72**, 14.

Jacobi, R. B. (1949). *J. Chem. Soc., Suppl. Vol.* 314.

Jacobsen, J. P. and Knudsen, M. (1940). *Assoc. Ocean. Phys. Publ. Sci.* 7.

Jacobsen, J. P., Robinson, R. J. and Thompson, T. G. (1950). *Assoc. Ocean. Phys. Publ. Sci.* **11**, 5.

Jagner, D. (1970). *Analyt. Chim. Acta*, **52**, 483.

Jagner, D. (1971). Inaugural dissertation, University of Göteborg.

Jagner, D. and Årén, K. (1971). *Analyt. Chim. Acta*, **57**, 185.

Jannasch, H. W. and Maddox, W. S. (1967). *J. Mar. Res.* **25**, 185.

Jarvis, N. L., Garrett, W. D., Schieman, M. A. and Timmons, C. O. (1967). *Limnol. Oceanogr.* **12**, 88.

Jasinski, R. and Trachtenberg, I. (1973). *Analyt. Chem.*, **45**, 1277.

Jeffrey, L. M. (1966). *J. Am. Oil. Chem. Soc.* **43**, 211.

Jeffrey, L. M. and Hood, D. W. (1958). *J. Mar. Res.*, **17**, 247.

Jeffrey, L. M., Fredericks, A. D. and Hillier, E. (1973). *Limnol. Oceanogr.* **18**, 336.

Jenkins, D. (1967). *J. Water Pollut. Control Fed.* **39**, 159.

Jenkins, D. (1968). *In* "Trace inorganics in Water", *Adv. Chem. Ser.* **73**, 265.

Jenkins, D. and Medsker, L. L. (1964). *Analyt. Chem.* **36**, 610.

Jeter, H. W., Føyn, E., King, M. and Gordon, L. I. (1972). *Limnol. Oceanogr.* **17**, 288.

Joeris, L. S. (1964). *Limnol. Oceanogr.* **9**, 595.

Johannesson, J. K. (1958). *Nature, Lond.* **180**, 285.

Johannesson, J. K. (1962). *Analyt. Chem.* **34**, 1111.

Johnson, C. R., McClelland, P. M. and Boster, R. L. (1964). *Analyt. Chem.* **36**, 301.

Johnson, D. L. (1971). *Environ. Sci. Tech.* **5**, 411.

Johnson, D. L. and Pilson, M. E. Q. (1972a). *Analyt. Chim. Acta*, **58**, 289.

Johnson, D. L. and Pilson, M. E. Q. (1972b). *J. Mar. Res.* **30**, 140.

Johnston, R. (1964). *Oceanogr. Mar. Biol. Ann. Rev.* **2**, 97.

Johnston, R. (1966). *Int. Cons. Explor. Mer.*, C.M. 1966/C: 9.

Johnston, R. (1969). *Oceanogr. Mar. Biol., Ann. Rev.* **7**, 31.

Jones, P. G. W. (1963). *J. Cons. int. Explor. Mer.* **28**, 3.

Jones, P. G. W. (1966). *J. mar. biol. Ass. U.K.* **46**, 19.

Jones, P. G. W. (1968). *Cons. Int. Explor. Mer. Inform. Tech. Methods Sea Water Analysis* No. **2**, 7.

Jones, P. G. W. and Folkard, A. R. (1968). *UNESCO Technical Papers in Marine Sciences*, No. 9.

Jones, P. G. W. and Spencer, C. P. (1963). *J. Mar. Biol. Ass. U.K.* **43**, 251.

Jones, R. S. (1970). *U.S. Geol. Surv. Circ.* No. **625**, 15 pp.

Joyce, J. R. (1973). *J. Mar. Biol. Ass., U.K.* **53**, 741.

Joyner, T. (1964). *J. Mar. Res.* **22**, 259.

Joyner, T. and Finley, J. S. (1966). *At. Absn. Newsletter* 5, 4.

Joyner, T., Healy, M. L., Chakravarti, D. and Koyanagi, T. (1966). "Symposium on trace characterization", *National Bureau of Standards*, Washington.

Joyner, T., Healy, M. L., Chakravarti, D. and Koganogi, T. (1967). *Environ. Sci. Tech.* **1**, 417.

Junge, C. and Hahn, J. (1971). *J. Geophys. Res.* **76**, 8143.

Junge, C., Bockholt, B., Schuetz, K. and Beck, R. (1971). *Meteor. Forschungsergeb.* **Reihe, No. 6**.

Junge, C. E., Seiler, W. and Warneck, P. (1971). *J. Geophys. Res.* **76**, 2866.

Kabanova, Yu. G. (1961). *Trud. Inst. Okeand. Akad. Nauk, S.S.S.R.* **47**, 182.

Kahn, H. L. (1973). *Int. J. Environ. Analyt. Chem.* **3**, 121.

Kahn, H. L., Peterson, G. E. and Schallis, J. E. (1968). *At. Absn. Newslett.* **7**, 35.

Kahn, L. and Brezenski, F. T. (1967a). *Env. Sci. Tech.* **1**, 488.

Kahn, L. and Brezenski, F. T. (1967b). *Env. Sci. Tech.* **1**, 492.

Kalle, K. (1939). *Ann. Hydrogr. Berl.* **67**, 267.

Kalle, K. (1966). *Oceanogr. Mar. Biol., Ann. Rev.* **4**, 91.

Kallmann, S. (1961). *In* "Treatise on analytical chemistry". (I. M. Kolthoff and P. J. Elving, eds.). Part II, Vol. 1. Interscience, New York.

Kampitsch, E., Schwarzenbach, G. and Steiner, R. (1945). *Helv. chim .Acta,* **28**, 828.

Kanamori, S. (1971). *In* "The Ocean World". (M. Uda, ed.). Japan Society for the Promotion of Science, Tokyo.

Kappanna, A. N., Gadre, G. T., Bhavnagary, H. M. and Joshi, J. M. (1962). *Current Sci.* **31**, 273.

Kato, K. (1966). *Bolm. Inst. Oceanogr., S. Paulo,* **15**, 25, 29 & 41.

Kato, K. and Kitano, Y. (1966). *J. Earth Sci., Nagoya Univ.* **14**, 151.

Kaufman, A. (1969). *Antarctic J. U.S.* **4**, 187.

Kaufman, S. and Libby, W. F. (1954). *Phys. Rev.* **93**, 1337.

Kautsky, H. and Schmitt, D. E. (1962). *Dtsch. Hydrogr. Z.,* **15**, 199.

Kawabuchi, K. and Kuroda, R. (1969). *Analyt. Chim. Acta,* **46**, 23.

Kawabuchi, K. and Riley (1973). *Analyt. Chim. Acta,* **65**, 271.

Kawasaki, N. and Sugawara, K. (1958). *Rec. oceanogr. Wks., Japan,* **2**, 1.

Keil, G. and Bernt, H. (1972). *Nucl. Instr. Meth.* **101**, 1.

Kennedy, R. H. (1950). *U.S. At. Energy Comm.* **AECD-3187**.

Kentner, E., Armitage, D. B. and Zeitlin, H. (1969). *Analyt. Chim. Acta,* **45**, 343.

Kerfoot, W. B. and Vaccaro, R. F. (1973). *Limnol. Oceanogr.* **18**, 689.

Kerr, J. R. W., Coomber, D. I. and Lewis, D. T. (1962). *Analyst,* **87**, 944.

Kessick, M. A., Vuceta, J. and Morgan, J. J. (1972). *Environ. Sci. Tech.* **6**, 642.

Kester, D. R. (1971). *Deep-Sea Res.* **18**, 1123.

Kholina, Y. B., Zarinskii, V. A., Popov, S. I., Khitrov, L. M. and Khaeva, L. K. (1973). *Dokl. Acad. Nauk SSSR,* **208**, 705.

Kiefer, D. A. (1971). "The *in vivo* measurement of chlorophyll by fluorometry." Symposium on Estuarine Microbiology, Columbia, S. Carolina.

Kiefer, D. A. (1973). *Mar. Biol.* **22**, 263.

Kim, Y. S. and Zeitlin, H. (1968). *Limnol. Oceanogr.* **13**, 534.

Kim, Y. W. and Zeitlin, H. (1969). *Analyt. Chim. Acta,* **46**, 1.

Kim, Y. S. and Zeitlin, H. (1970). *Analyt. Chim. Acta,* **51**, 516.

Kim, Y. S. and Zeitlin, H. (1971a). *Separation Science,* **6**, 505.

Kim, Y. S. and Zeitlin, H. (1971b). *Univ. Hawaii, Univ. Inst. Geophys. Contrib.* 487.

Kim, Y. S. and Zeitlin, H. (1971c). *Analyt. Chem.* **43**, 1390.

Kimura, N., Nagashima, N., Ikeda, N. and Kimura, K. (1955). *Radioisotopes, Japan,* **4**, 31.

Kirijima, T. and Kuroda, R. (1972). *Analyt. Chim. Acta,* **62**, 464.

Kirkbright, G. F. (1971). *Analyst,* **96**, 609.
Kitano, Y. and Furukawa, Y. (1972). *J. Oceanogr. Soc., Japan,* **28**, 121.
Kitano, Y. and Furukawa, Y. (1972). *J. Oceanogr. Soc. Japan,* **28**, 176.
Kiwan, A. M. and Fouda, M. F. (1968). *Analyt. Chim. Acta,* **40**, 517.
Kletsch, R. A. and Richards, F. A. (1970). *Analyt. Chem.* **42**, 1435.
Knapman, F. W. and Robinson, R. J. (1941). *J. Mar. Res.* **4**, 142.
Knauer, G. A. and Martin, J. H. (1973). *Limnol. Oceanogr.* **18**, 597.
Knowles, G. and Lowden, G. F. (1953). *Analyst,* **78**, 159.
Knudsen, M. (1901). "Hydrographical Tables". G.E.C. Gad, Copenhagen.
Knudsen, M. (1923). *Publ. Circ. Cons. Explor. Mer.* **77**, 9.
Knudsen, M. (1929). *J. Cons. int. Explor. Mer.* **4**, 192.
Knudson, E. J. and Christian, G. D. (1974). *At. Absn. Newsletter,* **13**, 74.
Kobayashi, J. (1968). *In* "Chemical environment in the aquatic habitat". (H. Golterman and R. Clymo, eds.). North Holland, Amsterdam.
Koczy, F. F. (1949). *Geol. Fören Stockh. Förh.* **71**, 238.
Koczy, F. F. and Szabo, B. J. (1965). *J. Oceanogr. Soc., Japan (20th Anniv. Vol.),* 590.
Koczy, F. F., Picciotto, E., Poulaert, G. and Wilgain, S. (1957). *Geochim. Cosmochim. Acta,* **11**, 103.
Koczy, G. (1950). *Österr. Akad. Wiss. Math. Natur. Kl. Sitzber. Abt. IIa,* **158**, 113.
Kodama, Y. and Tsubota, H. (1971). *Bunseki Kagaku,* **20**, 1554.
Koide, M. and Goldberg, E. D. (1965). *In* "Progress in Oceanography". (M. Sears, ed.). Vol. 3, p. 173. Pergamon Press, Oxford.
Kolthoff, I. M. and Yutzy, H. (1937). *Industr. Engng. Chem. (Anal.),* **9**, 75.
Kolthoff, I. M. and Chantooni, M. K. (1968). *J. Amer. Chem. Soc.* **90**, 5961.
Konnov, V. A. (1965). *Tr. Inst. Okeanol. Akad. Nauk. S.S.S.R.* **79**, 11.
Kopp, J. F., Longbottom, M. C. and Lobring, L. B. (1972). *J. Amer. Water Wks. Ass.* **64**, 20.
Korkisch, J. (1969). "Modern Methods for the separation of rarer metal ions", Pergamon, Oxford.
Korkisch, J., Thiard, A. and Hecht, F. (1956). *Microchim. Acta,* 1422.
Koroleff, F. (1947). *Acta Chem. Scand.* **1**, 503.
Koroleff, F. (1950). *Merentutkimuslait Julk,* **145**, 69pp.
Koroleff, F. (1965). *UNESCO Technical Papers in Marine Sciences,* p. 9.
Koroleff, F. (1968). *Cons. int. Explor. Mer., Serv. Hydrogr., Interlab Rept.,* **No. 2**, 12.
Koroleff, F. (1970). *Cons. int. Explor. Mer., Serv. Hydrogr. Interlab Rept.* **No. 3**, 19.
Kovaleva, K. N. and Bukser, E. S. (1940). *Rep. Acad. Sci. Ukr.* 35.
Krause, K. A. and Nelson, F. (1956). *Proc. Intern. Conf. on Peaceful Uses of Atomic Energy, United Nations, Geneva,* **7**, 113.
Kraynov, S. R. (1961). *Byul Nauchno-tekh. inform. Ministerstvo geolog i okhrany nedr. SSSR,* **No. 2** (30), 19.
Kremling, K. (1969). *Kiel. Meeresforsch.* **25**, 81.
Kremling, K. (1970). *Kiel. Meeresforsch.* **26**, 1.
Krey, J. (1959). *Proc. Verb. Cons. Perm. Int. Explor. Mer.* **144**, 20.
Krey, J. and Zeitschel, B. (1968). *Kiel. Meeresforsch.* **24**, 38.
Krey, J., Banse, K. and Hagmeir, E. (1957). *Kiel Meeresforsch.* **13**, 35.
Krishnamoorthy, T. M. and Viswanathan, R. (1968). *Indian J. Chem.* **6**, 169.
Krishnamoorthy, T. M., Sastry, V. N. and Sarma, T. P. (1968). *Current Sci.,* **37**, 660.
Krishnaswami, S., Lal, D., Somayajulu, B. L. K., Dixon, F. S., Stonecipher, S. A. and Craig, H. (1972). *Earth Planet. Sci. Lett.* **16**, 84.

Kruger, P. (1971). "Principles of Activation Analysis". Wiley, Interscience, New York.

Kruse, J. M. and Mellon, M. G. (1953). *Analyt. Chem.* **25**, 1188.

Ku, T. L., Li, Y. H., Mathieu, G. G. and Wong, H. K. (1970). *J. Geophys. Res.* **75**, 5286.

Kudo, H. (1964). *Tohoku Kaiku Suisan Kenkyusho Kenkyu Hokoku,* No. **24**, 1.

Kuenzler, E. J., Guillard, R. R. L. and Corwin, N. (1963). *Deep-Sea Res.* **10**, 749.

Kullenberg, B. and Sen Gupta, R. (1973). *Geochim. Cosmochim. Acta,* **37**, 1327.

Kuwamoto, T. (1960). *Nippon Kagaku Zasshi,* **81**, 1669.

Kuwata, K., Hisatomi, K. and Hasegawa, K. (1971). *At. Absn. Newsletter,* **10**, 111.

Kuznetsov, V. I. (1954). *Zhur. Anal. Khim.* **9**, 199.

Kuznetsov, V. I. and Akimova, A. A. (1958). *Zhur. Anal. Khim.* **13**, 79.

Kuznetsov, V. I., Akimova, T. G. and Eliseeva, O. P. (1962). *Radiokhimiya,* **4**, 188.

Kuznetsov, Y. V., Legin, V. K., Lisitsyn, A. P. and Simonyak, Z. N. (1964). *Soviet Radiochemistry,* **6**, 233.

Kuznetsov, Y. V., Simonyak, Z. N., Elizarov, A. N. and Lisitsyn, A. P. (1966a). *Radiokhimiya,* **8**, 455.

Kuznetsov, Y. V., Elizarova, A. N. and Frenklikh, M. S. (1966b). *Radiokhimiya,* **8**, 459.

Kuznetsov, Y. V., Simonyak, Z. N., Elizarova, A. N. and Lisitsyn, A. P. (1966c). *Soviet Radiochemistry,* **8**, 421.

Kwiecinski, B. (1965). *Deep-Sea Res.* **12**, 797.

Laevastu, T. and Thompson, T. G. (1956). *J. Cons. int. Explor. Mer.* **21**, 125.

Lai, M. G. and Goya, H. A. (1966). *U.S. Clearinghouse Fed. Sci. Tech. Inf. AD 648485.*

Lai, M. G. and Weiss, H. V. (1962). *Analyt. Chem.* **34**, 1012.

Lai, M. G. and Goya, H. A. (1965). U.S. Naval Radiological Defense Laboratory Report USNRDL-TR-924.

Lal, D., Arnold, J. R. and Somayajulu, B. L. K. (1964). *Geochim. cosmochim. Acta,* **28**, 1111.

Lambert, R. B., Kester, D. R., Pilson, M. E. Q. and Kenyon, K. E. (1973). *J. Geophys. Res.* **78**, 1479.

Lambert, R. S. and DuBois, R. J. (1971). *Analyt. Chem.* **43**, 955.

Lamontagne, R. A., Swinnerton, J. W. and Linnenbom, V. J. (1971). *J. Geophys. Res.* **76**, 5117.

Landingham, J. W. van (1957). *J. Cons. int. Explor. Mer.* **22**, 174.

Langmyhr, F. J., Klausen, K. S. and Nouri-Nekoui, M. H. (1971). *Analyt. chim. Acta,* **57**, 341.

Lard, E. W. and Horn, R. C. (1960). *Analyt. Chem.* **32**, 878.

Lazarev, K. F., Nikolaev, D. S. and Graschenko, S. M. (1961). *Radiokhimiya,* **3**, 623.

Lazarev, K. F., Nikolaev, D. S., Graschenko, S. M. and Drozhshin, V. M. (1965). *Dokl. Akad. Nauk. SSSR,* **164**, 1151.

Leatherland, T. (1969). M.Sc. Dissertation, University of Southampton.

Lee, M-L. and Burrell, D. C. (1972). *Analyt. Chim. Acta,* **62**, 153.

Lee, M-L. and Burrell, D. C. (1973). *Analyt. Chim. Acta,* **66**, 245.

Lenihan, J. M. and Thompson, S. J. (1969). "Advances in Activation Analysis". Academic Press, New York.

Leroy, M. (1967). *Cah. Oceanogr.* **19**, 53.

Leung, G., Kim, Y. S. and Zeitlin, H. (1972). *Analyt. Chim. Acta,* **60**, 229.

Lewis, G. J. and Goldberg, E. D. (1954). *J. Mar. Res.* **13**, 183.
Lewis, G. J. and Rakestraw, N. W. (1955). *J. Mar. Res.* **14**, 253.
Liam-Ngog-Thu (1967). *Zhur. Anal. Khim.* **22**, 636.
Liddicoat, M. I., Tibbitts, S. and Butler, E. I. (1974). *Limnol. Oceanogr.* (in press).
Lieberman, S. H. and Zirino, A. (1974). *Analyt. Chem.* **46**, 20.
Linnenbom, V. J. and Swinnerton, J. W. (1970). *In* "Symposium on Organic Matter in Natural Waters". (D. W. Hood, ed.) pp. 455–467. University of Alaska.
Lisitsyn, A. P., Belyaev, Y. I., Bogdanov, Y. A. and Bogoyavlenskii (1965). *Chem. Abstr.* **65**, 8561.
Liss, P. S. and Pointon, M. J. (1973). *Geochim. Cosmochim. Acta,* **37**, 1493.
Liss, P. S. and Spencer, C. P. (1969). *J. mar. biol. Ass., U.K.* **49**, 589.
Litchfield, C. D. and Hood, D. W. (1966). *Appl. Microbiol.* **14**, 145.
Litchfield, C. D. and Prescott, J. M. (1970). *Limnol. Oceanogr.* **15**, 250.
Loftus, M. E. and Carpenter, T. H. (1971). *J. mar. Res.* **29**, 319.
Loomis, W. F. (1954). *Analyt. Chem.* **26**, 402.
Lopez-Benito, M. (1970). *Invest. Pesq.* **34**, 385.
Lorenzen, C. J. (1966). *Deep-Sea Res.* **13**, 223.
Lorenzen, C. J. (1967). *Limnol. Oceanogr.* **12**, 343.
Lothian, G. F. (1969). "Absorption Spectrophotometry", 3rd Edition, Adam Hilger, London.
Lotrian, J. and Johannin-Gilles, A. (1969). *Spectrochim. Acta,* **24B**, 479.
Lovelock, J. E., Maggs, R. J. and Rasmussen, R. A. (1972). *Nature,* **237**, 452.
Lovelock, J. E., Maggs, R. J. and Wade, R. J. (1973). *Nature,* **241**, 194.
Loveridge, B. A., Milner, G. W. C., Barnett, G. A., Thomas, A. M. and Henry, W. M. (1960). Atomic Energy Authority Res. Grp. Rept. A.E.R.E., R3323, 6 pp.
Lund, W. and Larsen, B. V. (1974). *Analyt. Chim. Acta.* (in the press).
Lyakhin Yu. I. (1971). *Okeanologiya,* **11**, 635.
Lyle, S. J. (1973). *Select. Rev. Analyt. Sci.* **3**, 1.
Lyakhin (1971). *Okeanologiya,* **11**, 635.
McCarthy, J. J. (1970). *Limnol. Oceanogr.* **15**, 309.
McDaniel, W. H., Hemphill, R. N. and Donaldson, W. T. (1967). *Automation in analytical Chemistry, Proceedings of 3rd Technicon Symposium,* **1**, 363.
McIntyre, W. G. and Platford, R. F. (1964). *J. Fish. Res. Bd. Can.* **21**, 1475.
Macchi, H. (1965). *J. Electroanalyt. Chem.* **9**, 290.
Macchi, G., Cescon, B. and Mameli-d'Errico, D. (1969). *Arch. Oceanogr. Limnol.* **16**, 163.
MacIsaac, J. J. and Olund, R. R. (1971). *Inv. Pesq.* **35**, 221.
Magee, R. J. and Rahman, K. M. (1965). *Talanta,* **12**, 409.
Malavicini, A. and Vido, L. (1961). *Comit. Nazl. Energia Nucleare CNJ,* 1–10.
Malissa, H. and Schoffmann (1955). *Mikrochim. Acta,* **1**, 187.
Mallory, E. C. (1967). *Amer. Chem. Soc., Div. Water, Wastes, Preprints,* **7**, 126.
Mallory, E. C. (1968). *Adv. Chem. Ser.* **73**, 281.
Mameli, D. and Mosetti, F. (1966). *Boll. Geofis. Teor. Appl.* **8**, 294.
Mameli, D. and Mosetti, F. (1967). *Boll. Soc. Adriat. Sci. Trieste,* **55**, 27.
Manabe, T. (1969). *Bull. Jap. Soc. scientif. Fish.* **35**, 897.
Mancy, K. H. and Jaffe, T. (1966). *U.S. Public Health Service Publ.* No. 999-WP-37, 94 pp.
Mangelsdorf, P. C. and Wilson, T. R. S. (1969). *Summary of Progress Report 1968– 1969 on Contract AEC (30–1)-3838 NYO 3838-3,* Woods Hole Oceanogr Inst.

Mangelsdorf, P. C. (1966). *Analyt. Chem.* **38**, 1540.
Mangelsdorf, P. C. and Wilson, T. R. S. (1971). *J. Phys. Chem.* **75**, 1418.
Mangelsdorf, P. C., Wilson, T. R. S. and Daniell, E. (1969). *Science,* **165**, 171.
Manny, B. A., Miller, M. C. and Wetzel, R. G. (1971). *Limnol. Oceanogr.* **16**, 71.
Mantoura, R. F. C. and Riley, J. P. (1975). *Analyt. Chim. Acta,* (in press).
Marchand, M. (1974). *J. Cons. int. Explor. Mer.* **35**, 130.
Marvin, K. T. (1955). *J. mar. Res.* **14**, 79.
Marvin, K. T., Proctor, R. R. and Neal, R. A. (1970). *Limnol. Oceanogr.* **15**, 320.
Marvin, K. T., Proctor, R. R. and Neal, R. A. (1972). *Limnol. Oceanogr.* **17**, 777.
Mascini, M. (1973). *Analyst.* **98**, 325.
Mathews, T. D., Fredericks, A. D. and Sackett, W. M. (1973). *In* "Radioactive Contamination of the Marine Environment", p. 725, International Atomic Energy Agency (Vienna). (Symposium in Seattle, U.S.A. 1972).
Matsudaria, C. and Tsuda, T. (1957). *Tohuku. J. Agr. Res.* **8**, 37.
Matsunaga, K. and Nishimura, M. (1969). *Analyt. Chim. Acta,* **45**, 350.
Matsunaga, K. and Nishimura, M. (1971). *Bunseki Kagaku,* **20**, 993.
Matsunaga, K. and Nishimura, M. (1974). *Analyt. Chim. Acta,* **73**, 704.
Matsunaga, K., Oyama, T. and Nishimura, M. (1972). *Analyt. Chim. Acta,* **58**, 228.
Matthews, A. D. (1969). Ph.D. Thesis, University of Liverpool.
Matthews, A. D. and Riley, J. P. (1969). *Analyt. Chim. Acta,* **48**, 25.
Matthews, A. D. and Riley, J. P. (1970). *Analyt. Chim. Acta,* **51**, 295.
Matthews, A. D. and Riley, J. P. (1970a). *Analyt. Chim. Acta,* **51**, 287.
Matthews, A. D. and Riley, J. P. (1970b). *Nature,* **225**, 1242.
Matthews, A. D. and Riley, J. P. (1970c). *Analyt. Chim. Acta,* **51**, 455.
Matthieu, G. (1969). *Antarct. J.U.S.* **4**, 185.
May, B. Z. (1964). *Quart. J. Florida Acad. Sci.* **27**, 177.
Mazor, E., Wasserburg, G. J. and Craig, H. (1964). *Deep-Sea Res.* **11**, 929.
Mellon, M. G. (1960). "Analytical absorption spectroscopy". Wiley, New York.
Menzel, D. W. (1966). *Deep-Sea Res.* **13**, 963.
Menzel, D. W. and Corwin, N. (1965). *Limnol. Oceanogr.* **10**, 280.
Menzel, D. W. and Ryther, J. H. (1968). *Deep-Sea Res.* **15**, 327.
Menzel, D. W. and Ryther, J. H. (1970). *In* "Symposium on Organic Matter in Natural Waters" (D. W. Hood, ed.) pp. 31–54. University of Alaska.
Menzel, D. W. and Spaeth, J. P. (1962). *Limnol. Oceanogr.* **7**, 155.
Menzel, D. W. and Spaeth, J. P. (1962a). *Limnol. Oceanogr.* **7**, 151.
Menzel, D. W. and Vaccaro, R. F. (1964). *Limnol. Oceanogr.* **9**, 138.
Merrill, J. R., Honda, M. and Arnold, J. R. (1960a). *Analyt. Chem.* **32**, 1420.
Merrill, J. R., Lyden, E. F., Honda, M. and Arnold, J. R. (1960b). *Geochim. Cosmochim. Acta,* **18**, 108.
Merten, J. and Massart, D. L. (1971). *Bull. Soc. chim. Belg.* **80**, 151.
Millero, F. J., Schrager, S. R. and Hansen, L. D. (1974). *Limnol. Oceanogr.* **19**, 711.
Milner, G. W. C., Wilson, J. D., Barnett, G. A. and Smales, A. A. (1961). *J. Electroanalyt. Chem.* **2**, 25.
Minczewski, J. (1967). *In* "Trace Characterization". (W. W. Meinke and B. F. Scribner, eds). pp. 385–414, National Bureau of Standards, Washington, D.C.
Miyake, Y. (1939). *Bull. chem. Soc. Japan,* **14**, 55.
Miyake, Y. and Sakurai, S. (1952). *J. Mar. Met. Soc., Japan,* **30**, 1.
Miyake, Y. and Sugimura, Y. (1965). *Studies in Oceanography, Collected Papers,* 274.
Miyake, Y. and Wada, E. (1967). *Rec. Oceanogr. Wks., Japan,* **N.S. 9**, 37.

Miyake, Y. and Sugimura, Y. (1968). *Pap. Meteorol. Geophys. (Tokyo)*, **19**, 481.
Miyake, Y. and Sugiura, Y. (1955). *Rec. Oceanogr. Wks., Japan*, **2**, 108.
Miyake, Y., Saruhashi, K., Karaourago, Y. and Kanazawa, T. (1961). *J. Rad. Res.* **2**, 25.
Miyake, Y., Saruhashi, K. and Katsuragi, Y. (1960). *Pap. Met. Geophys., Tokyo*, **11**, 188.
Miyake, Y., Saruhashi, K., Katsuragi, Y., Kanazawa, T. and Sugimura, Y. (1964). *In* "Recent Researches in the fields of hydrosphere, atmosphere and nuclear geochemistry". (Y. Miyake and T. Koyama, eds.) pp. 127–141, Maruzen Co., Tokyo.
Miyake, Y., Sugimura, Y. and Uchida, T. (1966). *J. Geophys. Res.* **71**, 3083.
Miyake, Y., Sugimura, Y. and Mayeda, M. (1970a). *J. Oceanogr. Soc., Japan*, **26**, 123.
Miyake, Y., Sugimura, Y. and Yasuyima, T. (1970b). *J. Oceanogr. Soc., Japan*, **26**, 130.
Mizuike, A. (1965). *In* "Trace Analysis—Physical Methods". (G. H. Morrison, ed.). pp. 103–153, Interscience, New York.
Moberg, E. D. and Harding, M. W. (1933). *Science*, **77**, 510.
Mohanrao, G. J. and Folsom, T. R. (1963). *Analyst.* **68**, 105.
Mokievskaya, V. V. (1965). *Tr. Inst. Okeanol. Akad. Nauk. SSSR.* **79**, 3.
Molins, L. R. (1957). *Biol. Inst. Españ. Oceanograf.* **No. 86**, 9 pp.
Montgomery, H. A. C. and Cockburn, A. (1964). *Analyst.* **89**, 679.
Montgomery, H. A. C., Thom, M. S. and Cockburn, A. (1964). *J. appl. Chem.*, **14**, 280.
Moore, W. S. (1969a). *J. Geophys. Res.* **74**, 694.
Moore, W. S. (1969b). *Earth Planet Sci. Lett.* **6**, 437.
Moore, W. S. and Sackett, W. M. (1964). *J. Geophys. Res.* **69**, 5401.
Mor, E. and Beccaria, A. M. (1971). *Ann. Chim. (Roma)*, **61**, 363.
Morcos, S. A. (1968). *Kiel. Meeresforsch.* **24**, 66.
Morcos, S. A. and Riley, J. P. (1966). *Deep-Sea Res.* **13**, 741.
Morita, Y. (1948). *J. Chem. Soc., Japan (Pure Chem. Sect.)*, **69**, 174.
Morita, Y. (1950). *J. Chem. Soc., Japan (Pure Chem. Sect.)*, **71**, 246.
Morita, Y. (1953). *Rec. Oceanogr. Wks., Japan N.S.* **1**, 49.
Morozov, N. P. (1968). *Okeanologiya*, **8**, 216.
Morozov, N. P. (1969). *Okeanologiya*, **9**, 291.
Morris, A. W. (1968). *Analyt. Chim. Acta*, **42**, 397.
Morris, A. W. (1974). *Mar. Pollut. Bull.* (in press).
Morris, A. W. and Riley, J. P. (1963). *Analyt. Chim. Acta*, **29**, 272.
Morris, A. W. and Riley, J. P. (1964). *Deep-Sea Res.* **11**, 899.
Morris, A. W. and Riley, J. P. (1966). *Deep-Sea Res.* **13**, 699.
Morrison, G. H. and Frieser, H. (1957). "Solvent extraction in Analytical Chemistry". John Wiley, New York.
Morrison, G. H. and Potter, N. H. (1972). *Analyt. Chem.* **44**, 839.
Morrison, G. H., Gerard, J. T., Travesi, A., Currie, R. L., Peterson, S. F. and Potier, N. M. (1969). *Analyt. Chem.* **41**, 1633.
Morrison, I. R. and Wilson, A. L. (1967). *Analyst.* **88**, 54.
Motomizu, S. (1973). *Analyt. Chim. Acta*, **64**, 217.
Mulikovskaya, E. P. (1964). *Tr. Vses Nauchn. Issled. Geol. Inst.* **117**, 79.
Mullin, J. B. and Riley, J. P. (1955a). *Analyt. Chim. Acta*, **12**, 464.
Mullin, J. B. and Riley, J. P. (1955b). *Analyt. Chim. Acta*, **12**, 162.

Mullin, J. B. and Riley, J. P. (1955c). *Analyst.* **80**, 83.

Mullin, J. B. and Riley, J. P. (1956). *J. Mar. Res.* **15**, 103.

Munnich, K. O. and Roether, W. (1967). *In* "Symposium on Radioactive Dating and Low-level Counting", p. 93. International Atomic Energy Agency (Vienna).

Murakami, T. (1965). *Bunseki Kagaku,* **14**, 880.

Murakami, T. and Katsuya, U. (1967). *Himeji Kogyo Daigaku Kenkyu Hokoku,* **20A**, 79.

Murakami, T. and Uesugi, K. (1967). *Bunseki Kagaku,* **16**, 781.

Murphy, J. and Riley, J. P. (1956). *Analyt. Chim. Acta,* **14**, 318.

Murphy, J. and Riley, J. P. (1958). *J. mar. biol. Ass. U.K.* **37**, 9.

Murphy, J. and Riley, J. P. (1962). *Analyt. Chim. Acta,* **27**, 31.

Murray, A. J. and Riley, J. P. (1973). *Analyt. Chim. Acta*

Murray, C. N. and Riley, J. P. (1969). *Deep-Sea Res.* **16**, 311.

Murray, C. N. and Riley, J. P. (1970). *Deep-Sea Res.* **17**, 203.

Murray, C. N., Riley, J. P. and Wilson, T. R. S. (1968). *Deep-Sea Res.* **15**, 237.

Murray, C. N., Riley, J. P. and Wilson, T. R. S. (1969). *Deep-Sea Res.* **16**, 297.

Muzzarelli, R. A. A. and Tubertini, O. (1969). *Talanta,* **16**, 1571.

Muzzarelli, R. A. A. and Marinelli, M. (1972). *Inquinamento,* **14**, 27.

Muzzarelli, R. A. A. and Rochetti, R. (1973). *Analyt. Chim. Acta,* **64**, 371.

Muzzarelli, R. R. A. and Rochetti, R. (1974). *Analyt. Chim. Acta,* **69**, 35.

Muzzarelli, R. A. A., Rarth, G. and Tubertini, O. (1970). *J. Chromatog.* **47**, 414.

Muzzarelli, R. A. A. and Marinelli, M. (1972). *Inquinamento,* **14**, 27.

Nagatsuka, S., Suzuki, H. and Nakajima, K. (1971). *Radioisotopes,* **20**, 305.

Nagaya, Y., Nakamura, K. and Saiki, M. (1971). *J. Oceanogr. Soc. Japan,* **27**, 20.

Naito, H. and Sugawara, K. (1957). *Bull. Chem. Soc., Japan,* **30**, 799.

Nakanashi, M. (1952). *Bull. Chem. Soc., Japan,* **24**, 36.

Natarajan, K. V. (1968). *Appl. Microbiol.* **16**, 366.

Natarajan, K. V. (1971). *Limnol. Oceanogr.* **18**, 655.

Natarajan, K. V. and Dugdale, R. C. (1966). *Limnol. Oceanogr.* **11**, 621.

Natterer, K. (1892). *Denkschr. Akad. Wiss. Wien,* **59**, 83.

Nazarenko, V. A. and Shelikhina, O. I. (1967). *Dopov. Akad. Nauk. Ukr. RSR, Ser. B,* **29**, 824.

Nehring, D. (1968). *Cons. int. Explor. Mer, Serv. Hydrogr. Interlab. Rept, No. 2,* 16.

Nelson, K. H., Brown, W. D. and Staruch, S. J. (1972). *Internat. J. Environ. Anal. Chem.* **2**, 45.

Newell, B. S. (1967). *J. mar. biol. Ass. U.K.* **47**, 271.

Newell, B. S. and Dal Pont, G. (1964). *Nature, Lond.* **201**, 36.

Newell, B. S., Morgan, B. and Cundy, J. (1967). *J. Mar. Res.* **25**, 201.

Nicholson, R. A. (1971). *Analyt. chim. Acta,* **56**, 147.

Niskin, S. J. (1962). *Deep-Sea Res.* **9**, 501.

Niskin, S. J. (1968). *Mar. Sci. Instr.* **4**, 19.

Nishikawa, Y., Hiraki, K., Morishige, K., Tsuchiya, A. and Shigematsu, T. (1968). *Bunseki Kagaku,* **17**, 1092.

Nishimura, M., Matsunaga, K., Kudo, T. and Obara, F. (1973). *Analyt. Chim. Acta,* **65**, 466.

Noddack, I. and Noddack, W. (1940). *Ark. Zool.* **32A**, 1.

Nomura, T. (1971). "Preliminary Reports of Hakuhō Maru Cruise KH-70-2". Ocean Research Institute, Tokyo.

Noshkin, V. E. and Mott, N. S. (1967). *Talanta,* **14**, 45.

Novikov, P. D., Miropol'skii, M. U. and Talalaeva, B. M. (1972). *Okeanologiya,* 12, 161.

Novoselov, A. A. and Simolin, A. V. (1965). "Gidrol. i Gidrokhim. Issled. v Tropich Zone Atlant. Okean", *Akad. Nauk. Ukr. SSSR, Mezhvedomsto.*

Nusbaum, I. (1958). *Water and Sewag. Wks.* 105, 469.

Oana, S. (1957). *J. Earth. Sci. Nagoya Univ.* 5, 103.

Odier, M. and Plichon, V. (1971). *Analyt. Chim. Acta,* 55, 209.

Odum, H. T. (1951). *Science,* 114, 211.

Odum, H. T. (1957). *Publ. Inst. Mar. Sci. Univ. Texas,* 4, 22.

Ogata, N. (1968). *Nippon Genshiryoku Gakkaishi,* 10, 672.

Ogata, N. and Inoue, N. (1970). *Nippon Kaisui Gakkaishi,* 23, 148.

Ogata, N. and Kakihana, K. (1969). *Nippon Genshiryoku Gakkaishi,* 11, 82.

Ogata, N., Inoue, N. and Kakihana, M. (1970). *Nippon Kaisui Gakkaishi,* 24, 68.

Ogura, N. and Hanya, T. (1967). *Int. J. Limnol. Oceanogr.* 1, 91.

Ohwada, K. (1972). *Mar. Biol.* 14, 10.

Ohwada, K. and Taga, N. (1972). *Mar. Chem.* 1, 61.

Oka, Y., Kato, T. and Sasaki, M. (1964). *Nippon Kagaku Zasshi,* 85, 643.

Okabe, S. and Moringa, T. (1968). *J. Chem. Soc., Japan,* 89, 284.

Okabe, S. and Moringa, T. (1969). *J. Oceanogr. Soc., Japan,* 25, 223.

Okabe, S. and Toyota, Y. (1967). *Nippon Kagaku Zasshi,* 88, 863.

Okuda, T. (1964). *Bol. Inst. Ocean. Univ. Oriente,* 3, 118.

Olafsson, J. (1974). *Analyt. Chim. Acta,* 68, 207.

Olafsson, J. and Riley, J. P. (1972). *Chem. Geol.* 9, 227.

Olsen, S. (1967). *In* "Chemical environment in the aquatic habitat" (H. L. Golterman and R. S. Clymo eds.). North Holland Publishing Co., Amsterdam. p. 63.

Olsen, V. (1968). *Cons. int. Explor. Mer. Serv. Hydrogr. Interlab. Rept.,* No. 2, 11.

Omang, S. H. (1971). *Analyt. Chim. Acta,* 53, 415.

Oradovskii, S. G. (1966). *Khim. Protsessy Moryakh. Okean. Akad. Nauk. SSSR,* 76.

Orren, M. J. (1971). *J. S. Afr. Chem. Inst.* 24, 96.

Osmolovskaya, E. P. (1964). *Tr. Inst. Okeanol. Akad. Nauk. SSSR.* 75, 141.

Östlund, H. G. (1962). *In* "Tritium in the Physical and Biological Sciences", Volume 1, p. 332. International Atomic Energy Agency (Vienna). (Symposium in Vienna 1961).

Östlund, H. G. and Werner, E. (1962). *In* "Tritium in the Physical and Biological Sciences", Volume 1, p. 95. International Atomic Energy Agency (Vienna). (Symposium in Vienna 1961).

Öström, B. (1973). *Mar. Chem.* 1, 323.

Ostrowski, S. and Czerwinska, A. (1959). *Zeszyty Nauk. Politekh. Gdansk. Chem. No. 3,* 103.

Ostrowski, S. and Nowak, J. (1974). *Gdanskie Towazyst. Nauk., Wydzial Nauk. Mat. Przyrodniczych. Rozprawy Wydzialu,* 3, 107.

Ostrowski, S., Bojanowski, R. and Malewicz, B. (1968). *Golansk. Tow. Nauk. Rozpr. Wydz. z,* 5, 5.

Ostrowski, S., Jasinska, Z. and Zolandziowska, Z. (1965). *Farm. Polska,* 21, 743.

Owa, T., Hiiro, K. and Tanaka, T. (1972). *Bunseki Kagaku,* 21, 878.

Paeda, R., Jurevieius, Blazys, I. and Jankauskas, I. (1968). *Liet. TSR Auskt. Mokyklu Mokslo Dark Chem. Chem. Technol. No. 9,* 61.

Page, J. A. and Lingane, J. J. (1957). *Analyt. Chim. Acta,* 16, 175.

Page, J. O. and Spurlock, W. W. (1965). *Analyt. Chim. Acta.* **32**, 593.

Pakalns, P. and McAllister, B. R. (1972). *J. Mar. Res.* **30**, 305.

Palmork, K. H. (1963a). *Acta. chem. Scand.* **17**, 1456.

Palmork, K. H. (1963b). *Rept. Norweg. Fish. Invest.* **13**, 120.

Palmork, K. H. (1968). UNESCO Technical Papers in Marine Sciences No. 9.

Pamork, K. H. (1969). *In* "Chemical oceanography; an introduction". (R. Lange, ed.). Universitetsforlaget, Oslo.

Park, P. K. (1965). *J. oceanogr. Soc., Japan,* **21**, 28.

Park, P. K. (1968). *Deep-Sea Res.* **15**, 721.

Park, P. K. and Catalfomo, M. (1964). *Deep-Sea Res.* **11**, 917.

Park, P. K., Williams, W. T., Prescott, J. M. and Hood, D. W. (1962). *Science,* **138**, 531.

Park, P. K., Hager, S. W., Pirson, J. E. and Ball, D. S. (1968). *J. Fish Res. Bd. Can.* **25**, 2739.

Parker, C. A. (1968). "Photoluminescence of solutions, with applications to photo-chemistry and analytical chemistry", Elsevier, Amsterdam.

Parker, P. L., Winters, J. K. and Morgan, J. (1972). *In* "Baseline studies of pollutants in the marine environment research recommendations". Working Papers, pp. 555–582. The IDOE Baseline Conference, May 24–26, New York.

Parson, T. R. and Strickland, J. D. H. (1963). *J. Mar. Res.* **21**, 155.

Pascual, J. (1966). "Ion-Exchange Methods for Concentration of Trace Elements from Seawater". USNRDL-TR-67-16, Y.S. Naval Radiological Defense Laboratory, San Francisco, California.

Pate, J. B. and Robinson, R. J. (1958). *J. Mar. Res.* **17**, 390.

Pate, J. B. and Robinson, R. J. (1961). *J. Mar. Res.* **19**, 21.

Paulhamus, J. A. (1972). *In* "Ultrapurity—Methods and Techniques". (M. Zief and R. Speights, eds.) pp. 255–292, Marcel Dekker, New York.

Paus, P. E. (1973). *Z. analyt. Chem.* **264**, 118.

Perez, J. J. and Guillot, P. (1971). Colloque Internationale sur l'Exploitation des Océans, Theme V. Vol. 1.

Perkins, R. W. and Robertson, D. E. (1965). *In* "Proceedings of the International Conference: Modern Trends in Activation Analysis". College Station, Texas, CONF-650405, pp. 48–57, Division of Technical Information Extension, USAEC, Washington, D.C.

Perkins, R. W., Robertson, D. E. and Rieck, H. G. (1966). *In* "Pacific Northwest Laboratory Annual Report for 1965", p. 108, BNWL-235, Part 2. Richland, Washington.

Peshchevitskii, B. I., Anoshin, G. R. and Erenburg, A. M. (1965). *Dokl. Akad. Nauk. SSSR,* **162**, 915.

Petek, M. and Branica, M. (1969). *Thalassia Jugoslavica,* **5**, 257.

Petriconi and Papee, H. M. (1971). *Water, Air Soil Pollut.* **1**, 42.

Petrov, P. S. and Pencheva, Y. (1963). *R. Zhur. Geol.,* No. 3, No. 52.

Pett, L. B. (1933). *Biochem. J.* **27**, 1672.

Pettersson, H. (1955). *Deep-Sea Res.* **5**, *(Suppl.),* 335.

Pillai, K. C., Smith, R. C. and Folsom, T. R. (1964). *Nature (Lond.),* **203**, 568.

Pineda, J. P. (1973). Aspects of the Organic Chemistry of Seawater. M. Phil. Thesis, University of Southampton.

Piper, D. Z. (1971). *Geochim. Cosmochim. Acta,* **35**, 531.

Piper, D. Z. and Goles, G. G. (1969). *Analyt. Chim. Acta,* **47**, 560.

Piro, A. and Rossi, G. (1969). *Publ. Staz. Zool. Napoli,* **37 (Suppl.)**, 290.
Piton, B. and Vorturiez, B. (1968). *Cah. Oceanogr.* **20**, 587.
Pocklington, R. (1971). *Nature,* **230**, 374.
Poluektov, N. S., Viikun, R. A. and Zelyukova, Ya. V. (1964). *Zh. Analit. Khim.* **19**, 937.
Pomeroy, L. R., Smith, E. E. and Grant, C. M. (1965). *Limnol Oceanogr.* **10**, 167.
Pomeroy, R. and Kirschman, H. D. (1945). *Industr. Engng Chem. (Anal.),* **17**, 715.
Portmann, J. E. (1965). Ph.D. Thesis, University of Liverpool.
Portmann, J. E. and Riley, J. P. (1964). *Analyt. Chim. Acta,* **31**, 509.
Portmann, J. E. and Riley, J. P. (1966a). *Analyt. Chim. Acta,* **35**, 35.
Portmann, J. E. and Riley, J. P. (1966b). *Analyt. Chim. Acta,* **34**, 201.
Preston, A. and Dutton, J. W. R. (1970). *In* "Reference Methods for Marine Radio-activity Studies", p. 233. Technical Reports Series, No. 118, International Atomic Energy Agency (Vienna).
Preston, A., Jeffries, D. F., Dutton, J. W. R., Harvey, B. R. and Steele, A. K. (1972). *Environ. Pollution,* **3**, 64.
Price, J. B. and Priddy, R. R. (1961). *Bull. Mar. Sci.* **11**, 198.
Price, W. J. (1972). "Analytical atomic absorption spectrometry". Heyden & Son, London.
Procházková, L. (1964). *Analyt. Chem.* **36**, 865.
Proctor, R. R. (1962). *Limnol. Oceanogr.* **7**, 479.
Putnam, G. L. (1953). *J. Chem. Ed.* **30**, 579.
Quinn, J. G. and Meyers, P. A. (1971). *Limnol. Oceanogr.* **16**, 129.
Quinn, J. G. and Wade, T. L. (1972). *In* "Baseline Studies in the marine environment research recommendations". Working Papers, pp. 633–664. The IDOE Baseline Conference, May 24–26, New York.
Rakestraw, N. W. and Emmel, V. M. (1937). *Industr. Engng. Chem. (Anal.),* **9**, 344.
Rakestraw, N. W. and Lutz, F. B. (1933). *Woods Hole Biol. Bull.* **65**, 397.
Rakestraw, N. W. and Mahncke, H. E. (1935). *Industr. Engng. Chem. (Anal.),* **7**, 425.
Rakevic, M. (1970). "Activation Analysis". CRC Press, Cleveland.
Rama, Koide, M. and Goldberg, E. D. (1961). *Science,* **134**, 98.
Rampon, H. and Cuvelier, R. (1972). *Analyt. Chim. Acta,* **60**, 226.
Rao, S. R., Khan, A. A. and Kamath, P. R. (1964). *Proc. Nucl. Radiation Chem. Symposium, Bombay,* 199.
Ray, B. J. and Johnson, D. L. (1972). *Analyt. Chim. Acta,* **62**, 196.
Reeburgh, W. S. (1967). *Limnol. Oceanogr.* **12**, 163.
Reed, R. K. and Ryan, T. V. (1965). *Deep-Sea Res.* **12**, 699.
Reilly, C. N. and Schmid, R. W. (1959). *Analyt. Chem.* **29**, 264.
Reith, J. F. (1930). *Rec. Trav. Chim. Pays-Bas,* **49**, 142.
Remsen, C. C. (1971). *Limnol. Oceanogr.,* **16**, 732.
Reusmann, G. (1968). *Kiel. Meeresforsch.* **24**, 14.
Revel, J. (1969). *Cah. Océanogr.* **21**, 273.
Rial, J. R. B. and Molins, L. R. (1962). *Bol. Inst. esp. Oceanogr.* No. 111.
Richards, F. A. (1957). *In* "Progress in Physics and Chemistry of the Earth". Vol. 2, p. 77, Pergamon Press, London.
Richards, F. A. and Kletsch, R. A. (1964). "Recent researches in the fields of hydro-sphere atmosphere nuclear geochemistry" (Sugawara Festival Volume) Marrizen, Tokyo.
Richards, F. A. and Thompson, T. G. (1952). *J. Mar. Res.* **11**, 156.

Richter, H. G. and Gillespie, A. S. (1962). *Analyt. Chem.* **34**, 1116.
Rideal, S. and Stewart, C. G. (1901). *Analyst.* **26**, 141.
Rider, B. F. and Mellon, M. G. (1946). *Industr. Engng. Chem. (Anal.)*, **18**, 96.
Riel, G. K., Pedrick, R. A., Attaway, D. H. and Audet, J. J. (1965). *U.S. Dept. Comm.*, **AD 473931**, 47 pp.
Rieman, W. and Walton, H. F. (1970). "Ion exchange in Analytical Chemistry". Pergamon Press, Oxford.
Rigler, F. H. (1964). *Limnol. Oceanogr.* **9**, 511.
Rigler, F. H. (1966). *Verh. int. Verein theor. angew. Limnol.* **16**, 465.
Rigler, F. H. (1968). *Limnol. Oceanogr.* **13**, 7.
Riley, G. A. (1937). *J. Mar. Res.* **1**, 60.
Riley, J. P. (1953). *Analyt. Chim. Acta,* **9**, 575.
Riley, J. P. (1965). *Deep-Sea Res.* **12**, 219.
Riley, J. P. (1965b). *In* "Chemical Oceanography". (J. P. Riley and G. Skirrow, eds.). Vol. II, pp. 295–424. Academic Press, London.
Riley, J. P. and Sunhaseni, P. (1958). *Analyst.* **83**, 299.
Riley, J. P. and Segar, D. A. (1970). *J. Mar. biol. Ass. U.K.* **50**, 713.
Riley, J. P. and Taylor, D. (1968). *Analyt. Chim. Acta,* **40**, 479.
Riley, J. P. and Taylor, D. (1968a). *Analyt. Chim. Acta,* **41**, 175.
Riley, J. P. and Taylor, D. (1968b). *Deep-Sea Res.* **15**, 629.
Riley, J. P. and Taylor, D. (1969). *Anal. Chim. Acta,* **46**, 307.
Riley, J. P. and Tongudai, M. (1964). *Deep-Sea Res.* **11**, 563.
Riley, J. P. and Tongudai, M. (1966a). *Analyt. Chim. Acta,* **36**, 439.
Riley, J. P. and Tongudai, M. (1966b). *Chem. Geol.* **1**, 291.
Riley, J. P. and Tongudai, M. (1967). *Chem. Geol.* **2**, 263.
Riley, J. P. and Topping, G. (1969). *Analyt. Chim. Acta,* **44**, 234.
Riley, J. P. and Williams, H. P. (1959). *Mikrochim. Acta,* 804.
Riley, J. P. and Wilson, T. R. S. (1965). *J. Mar. Biol. Ass. U.K.* **45**, 583.
Ringbom, A., Pensar, G. and Wänninen, E. (1958). *Analyt. Chim. Acta,* **19**, 525.
Riva, B. (1966). *Rend. Seminar Fac. Univ. Cagliari,* **36**, 170.
Riva, N. (1967). *Rend. Seminar Fac. Sci. Univ. Cagliari,* **37**, 595.
Robertson, D. E. (1968a). *Analyt. Chim. Acta,* **42**, 533.
Robertson, D. E. (1968b). *Analyt. Chem.* **40**, 1067.
Robertson, D. E. (1970). *Geochim. Cosmochim. Acta,* **34**, 553.
Robertson, D. E. (1971). Battelle Memorial Institute, Pacific Northwest Laboratories, Report BNWL-1551, Part 2, UC-48.
Robertson, D. E. (1972). *In* "Ultrapurity—Methods and Techniques". (M. Zief and R. Speights, eds.) 207, 254, Marcel Dekker, New York.
Robertson, D. E., Rancitelli, L. A. and Perkins, R. W. (1968). *In* "International Symposium on the Application of Neutron Activation Analysis in Oceanography", p. 143. N.A.T.O., Brussels.
Robinson, J. W. and Sievin, P. J. (1973). *Int. Lab.* 10.
Rocco, G. G. and Broecker, W. S. (1963). *J. Geophys. Res.* **68**, 4501.
Rochford, D. J. (1963). SCOR-UNESCO chemical intercalibration tests, Results of 2nd Series, R. S. Vityaz, August 2–9, 1982 Cronulla.
Rochford, D. J. (1964). SCOR-UNESCO chemical intercalibration tests, Results of 3rd Series, R. R. S. Discovery, May–June, 1964, Cronulla.
Roesmer, J. (1970). "Radiochemistry of Mercury, NAS-NS-3076 (Rev.)). USAEC, Washington. D.C.

Rohde, K. H. (1966). *Beitr. Meeresk.* **19**, 18.
Romanov, V. I. and Eremeeva, L. V. (1970). *Morsk. Gidrofiz. Issled.* **3**, 189.
Rona, E. (1943). *Amer. Phil. Soc. Yearbook,* 136.
Rona, E. and Urry, W. D. (1952). *Amer. J. Sci.* **250**, 241.
Rona, E., Gilpatrick, L. O. and Jeffrey, L. M. (1956). *Trans. Amer. Geophys. Union,* 37, 697.
Rona, E., Akers, L. K., Muse, L. and Hood, D. W. (1960). *Texas A. and M. College Contrib. in Oceanogr. and Meteorol.* **5**, 411.
Rona, E., Hood, D. W. Muse, L. and Buglio, B. (1962). *Limnol. Oceanogr.* **7**, 201.
Rosain, R. M. and Wai, C. M. (1973). *Analyt. Chim. Acta,* **65**, 279.
Rose, J. (1959). *U.K. Atomic Energy Authority,* Industrial Group Headquarters Reprint SCS-R-129, 5 pp.
Rosenbaum, J. P., May, J. T. and Riley, J. M. (1969). *Mines Mag. Golden Colo.* **59**, 14.
Roskam, R. T. and de Langen, D. (1963). *Analyt. Chim. Acta,* **28**, 78.
Roskam, R. T. and de Langen, D. (1964). *Analyt. Chim. Acta,* **30**, 36.
Ross, F. F. (1964). *Water Waste Treatment,* **9**, 528.
Ross, J. W. and Frant, M. S. (1969). *Analyt. Chem.* **41**, 967.
Ross, W. D. and Sievers, R. E. (1970). *Develop. Appl. Spectr.* **8**, 181.
Rotschi, H. (1963). *Cah. Oceanogr.* **15**, 7.
Rottschafer, J. M., Boczkowski, R. J. and Mark, H. B. (1972). *Talanta,* **19**, 163.
Rozhanskaya, L. I. (1964). *Tr. Sevastapol'sk. Biol. Akad. Nauk. SSR,* **15**, 503.
Rozhanskaya, L. I. (1965). *Okeanologiya,* **5**, 983.
Rozhanskaya, L. I. (1965a). *Osnovnye Cherty Geol. Stroeniya Gidrolog. Rezhima i Biol. Sredizemn. Morya Akad. Nauk. SSSR, Okeanog. Kommis.* **146**.
Rozhanskaya, L. I. (1966). *Gidrobiol. Zh. Akad. Nauk. Ukr. SSR,* **2**, 40.
Rozhanskaya, L. I. (1967). *Gidrofiz Gidrokhim Issled. Cherman More,* 60.
Rozhanskaya, L. I. (1970a). *In* "Reference Methods for marine radioactivity studies", *Tech. Rept. Ser. Internat. At. Energy Agency,* **No. 118**, 261.
Rozhanskaya, L. I. (1970b). "Khim Resur. Morei Okeanov" (S. V. Bruevich, ed.) Nauka, Moscow, p. 115.
Ryabinin, A. I. (1972). *Geokhimiya,* 879.
Ryabinin, A. I. and Romanov, A. S. (1970). *Geokhimiya,* 875.
Ryabinin, A. I. and Romanov, A. S. (1973). *Geokhimiya,* 257 (*Geochem. Int.* 1973, **10**, 181).
Ryabinin, A. I., Romanov, A. S. and Khamidova, R. (1971). *Dopov Akad. Nauk Ukr. RSR, Ser. B.* **33**, 834.
Ryabinin, A. I., Romanov, A. S., Khatamov, S. and Khamidova, R. (1972). *Dopov Akad. Nauk. Ukr. RSR, Ser. B,* **34**, 923.
Ryan, P. B. (1966). M.Sc. Thesis, University of Southampton.
Ryan, V. A. (1973). "Contemporary Activation Analysis". Marcel Dekker, New York.
Ryther, J. H. and Guillard, R. R. L. (1962). *Can. J. Microbiol.* **8**, 437.
Ryther, J. H., Menzel, D. W. and Corwen, N. (1967). *J. Mar. Res.* **25**, 69.
Sachdev, S. L. and West, P. W. (1970). *Environ. Sci. Tech.* **4**, 749.
Sackett, W. M. (1960). *Science,* **132**, 1761.
Sackett, W. M. and Arrhenius, G. O. S. (1962). *Geochim. Cosmochim. Acta,* **26**, 955.
Sackett, W. M. and Cook, G. (1969). *Trans. Gulf Coast Assoc. Geol. Soc.* **19**, 233.
Sackett, W. M., Portratz, H. A. and Goldberg, E. D. (1958). *Science,* **128**, 204.
Saeki, A. (1950). *J. Oceanogr. Soc., Japan,* **6**, 39.

Saenger, P. (1972). *Helg. wiss. Meeresunters*, **23**, 32.

Sagi, T. (1966). *Oceanogr. Mag.* **18**, 43.

Sagi, T. (1969). *Oceanogr. Mag.* **21**, 113.

Sakanoue, M., Nomura, T., Imai, T. and Uzuyama, H. (1970). *Proc. International Symposium on Hydrogeochemistry and Biochemistry,* Tokyo H-3.

Sakanoue, M., Nakaura, M. and Imai, T. (1971). *In* "Rapid Methods for Measuring Radioactivity in the Environment", p. 171, International Atomic Energy Agency (Vienna). (Symposium, Nürnberg, Germany, 1971).

Salmon, L. and Creevy, M. G. (1970). *In* "Nuclear Techniques in Environmental Pollution". pp. 47–59, IAEA, Vienna.

Samuelson, O. (1963). "Ion exchange separation in analytical chemistry", John Wiley, New York.

Sandell, E. B. (1959). "Colorimetric determination of trace of metals". Interscience, New York.

Sandell, E. B. and Kolthoff, I. M. (1934). *J. Amer. Chem. Soc.* **56**, 1426.

Sarma, T. P. and Krishnamoorthy, T. M. (1968). *Current Sci.* **37**, 422.

Sastry, V. N., Krishnamoorthy, T. M. and Sarma, T. P. (1969). *Curr. Sci.* **38**, 279.

Scadden, E. M. (1968). *U.S. Naval Radiological Defense Lab., San Francisco, Tech. Rept.* **USNRDL-TR-68-28** 27 pp.

Scadden, E. M. (1969). *Geochim. Cosmochim. Acta* **33**, 633.

Schaefer, H. (1965). *Helgol. Wiss. Meeresunters.* **12**, 253.

Schaefer, M. B. and Bishop, Y. M. M. (1958). *Limnol. Oceanogr.* **3**, 137.

Schall, E. D. (1957). *Analyt. Chem.,* **29**, 1044.

Scheimen, E. W. and Schubel, J. R. (1970). *Limnol. Oceanogr.* **15**, 645.

Schunk, D. R. (1965). *Analyt Chem.* **37**, 764.

Schink, D. R. and Anderson, M. C. (1969). *J. Mar. Tech. Soc.* **3**, 49.

Schink, D. R., Guinasso, N. L., Charnell, R. L. and Sigalove, J. J. (1970). *I.E.E.E. Trans. Nucl. Sci.* **17**, 184.

Schmidt, U. (1974). *Tellus,* **26**, 80.

Schmidt, U. and Seiler, W. (1970). *J. Geophys. Res.* **75**, 1713.

Schmitt, D. E. and Kautsky, H. (1961). *Dtsch. hydrogr. Z.,* **14**, 194.

Schnepfe, M. M. (1972). *Analyt. Chim. Acta,* **58**, 83.

Schoenfeld, I. and Held, S. (1969). *Israel. J. Chem.* **7**, 831.

Schroeder, B. W. and Cherry, R. D. (1962). *Nature, Lond.* **194**, 669.

Schutz, D. E. and Turekian, K. K. (1965a). *J. geophys. Res.* **70**, 5519.

Schutz, D. F. and Turekian, K. K. (1965b). *Geochim. Cosmochim. Acta,* **29**, 259.

Schwarzenbach, G. (1957). "Complexometric titrations", Methuen, London.

Schwarzenbach, G. and Biedermann, W. (1948). *Helv. Chim. Acta,* **31**, 678.

Schwarzenbach, G., Anderegg, G. and Senn, H. (1957). *Helv. chim. Acta.* **40**, 1886.

SCOR/UNESCO (1966). "Determination of photosynthetic pigments in sea-water." *ed.* UNESCO, Paris.

Seba, D. B. and Corcoran, E. F. (1969). *Pesticides Monitoring J.* **3**, 190

Sebba, F. (1962). "Ion flotation." Elsevier, New York.

Sednev, M. P., Starobinets, G. L. and Akulovich, A. M. (1966). *Zh. Analit. Khim.* **21**, 23.

Segar, D. A. (1973). *Int. J. Environ. Analyt. Chem.* **3**, 107.

Segar, D. A. and Gonzalez, J. G. (1972). *Analyt. Chim. Acta,* **58**, 7.

Segar, D. A. (1971). *Joint Conf. Sensing Environ. Pollut. Coll. Tech. Papers,* **2**, 71.

Segar, D. A. and Gonzalez, J. G. (1972). *Analyt. Chim. Acta,* **58**, 7.

Seiler, W. and Schmidt, U. (1973). D.R.P. 2, 224, 408.

Seiler, W. and Schmidt, U. (1974). *In* "The Sea" (E. D. Goldberg, ed.) Vol. 5. Wiley Interscience, New York.

Sen Gupta, R. (1970). FAO Technical Conference on marine pollution and its effects on living resources and fishing. No. FIR/MP/70/E37.

Shah, N. M., Fogg, G. E. (1973). *J. mar. biol. Ass. U.K.,* **53**, 321.

Shah, N. M. and Wright, R. T. (1974). *Mar. Biol.* **24**, 121.

Shah, S. M. and Gogate, S. S. (1969). *Current Sci.* **38**, 94.

Shah, S. M. and Rao, S. R. (1972). *Current Sci.* **41**, 659.

Shannon, L. V. and Orren, M. J. (1970). *Analyt. Chim. Acta,* **52**, 166.

Shannon, L. V., Cherry, R. D. and Orren, M. J. (1970). *Geochim. Cosmochim. Acta,* **34**, 701.

Sharma, N. N. and Parekh, J. M. (1967). *In* "Proceedings of the Symposium on the Indian Ocean, New Delhi, 2–4th March 1967".

Sharma, N. N. and Parekh, J. M. (1968). *Bull. Nat. Inst. Sci. India,* **38**, 236.

Sharma, N. N., Parekh, J. M., Dave, G. N. and Datar, S. S. (1966). *Indian J. Chem.* **4**, 294.

Sharp, J. H. (1973a). *Limnol. Oceanogr.* **18**, 441.

Sharp, J. H. (1973b). *Mar. Chem.* **1**, 211.

Sheldon, R. W. (1972). *Limnol. Oceanogr.* **17**, 494.

Sheldon, R. W. and Sutcliffe, W. H. (1969). *Limnol. Oceanogr.* **14**, 441.

Sheremet'eva, A. I. (1970). *Morsk. Gidrofiz. Issled.* 194.

Shigematsu, T., Nishikawa, Y., Hiraki, K. and Nakagawa, H. (1964). *Nippon Kagaku Zasshi,* **85**, 490.

Shigematsu, T., Tabushi, M., Aoki, T., Fujino, O., Nishikawa, Y. and Goda, S. (1967). *Bull. Chem. Res. Kyoto Univ.* **45**, 307.

Shigematsu, T., Suzuki, T. and Tabushi, M. (1969). *Nippon Kaisui Gakkai-Shi.* **22**, 348.

Shigematsu, T., Nishikawa, Y. Hiraki, K. and Nagano, N. (1970). *Bunseki Kagaku,* **19**, 551.

Shigematsu, T., Nishikawa, Y., Hiraki, K., Goda, S. and Tsujimoto, Y. (1971). **20**, 575.

Shimuzu M. and Ogata, N. (1963). *Japan Analyt.* **12**, 526.

Shimoishi, Y. (1973). *Analyt. Chim. Acta,* **64**, 465.

Shinn, M. B. (1941). *Industr. Engng Chem. (Anal.)* **13**, 33.

Shiozaki, M., Yamagata, N. and Iwashima, K. (In press). *In*: "Further Reference Methods for Marine Radioactivity Studies". Technical Reports Series, International Atomic Energy Agency, Vienna.

Shiozaki, M., Seto, Y. and Higano, R. (1964). *J. Oceanogr. Soc. Japan,* **20**, 31.

Shirasawa, T. H. and Schuert, E. A. (1968). *U.S. Naval Radiological Defense Laboratory Report,* **NRDL-TR-68-93**.

Shiro, H. and Zeitlin, H. (1962). *Limnol. Oceanogr.* **7**, 322.

Shishkina, O. V. and Pavlova, G. A. (1965). *Geokhimiya,* 739.

Sholkovitz, E. R. (1970). *Limnol. Oceanogr.* **15**, 641.

Sieburth, J. McN. (1965). *Trans. Joint Conf. Ocean Sci. Ocean Engng,* MTS-ASLO, Washington, D.C. 1064.

Sieburth, J. McN. and Jensen, A. (1968). *J. exp. mar. Biol. Ecol.,* **2**, 179.

Siegel, A. and Degens, E. T. (1966). *Science,* **151**, 1098.

Siever, R. (1962). *J. Sediment. Petrol.* **32**, 329.

Siever, R., Beck, K. C. and Berner, R. A. (1965). *J. Geol.* **73**, 39.

Sigalove, J. L. and Pearlman, M. D. (1972). *Undersea Technology*, 24.
Silvey, W. D. and Brennan, R. (1962). *Analyt. Chem.* **34**, 784.
Simons, L. H., Monaghan, P. H. and Taggart, M. S. (1953). *Analyt. Chem.* **25**, 989.
Skopintsev, B. A. (1960). *Trudy morsk. gidrofiz. Inst.* **19**, 3; translated in *Soviet Oceanography*, 1963, Series No. 3, pp. 1–14.
Skopintsev, B. A. and Kobanov, V. V. (1958). *Trud. Morsk. Gidiofiz. Inst. U.S.S.R.* **13**, 130.
Skopintsev, B. A. and Mikhailovskaya, L. A. (1933). *Tr. Inst. Oceanogr. Moscow*, **3**, 79.
Skopintsev, B. A., Shtukovskaya, L. A. and Vorob'eva, R. V. (1957). *Gidrokhim. Materialy.* **27**, 146.
Skopintsev, B. A., Timofeyeva, S. N. and Vershinina, O. A. (1966). *Oceanography*, **6**, 201.
Slavin, W. (1968). "Atomic absorption spectroscopy," Wiley-Interscience, New York.
Slawyk, G. and MacIsaac, J. J. (1972). *Deep-Sea Res.* **19**, 521.
Slowey, J. F. and Hood, D. W. (1971). *Geochim. Cosmochim. Acta* **35**, 121.
Slowey, J. F., Hayes, D., Dixon, B. and Hood, D. W. (1965). *In*: 'Symposium on marine geochemistry'', p. 109, Univ. Rhode Island Occasional Publ. No. 3.
Slowey, J. F. Hedges, D. and Hood, D. W. (1962). *TID-22660 Texas A and M University, Texas.*
Slowey, J. F., Jeffrey, L. M. and Hood, D. W. (1962a). *Geochim. Cosmochim. Acta*, **26**, 607.
Slowey, J. F., Jeffrey, L. M. and Hood, D. W. (1967). *Nature, Lond.* **214**, 377.
Smales, A. A. (1967). *In* "Trace Characterization" (W. W. Meinke and B. F. Scribner, eds.), National Bureau of Standards, Washington, D.C.
Smales, A. A. and Pate, B. D. (1952). *Analyst,* **77**, 188.
Smales, A. A. and Salmon, L. (1955). *Analyst,* **80**, 37.
Smales, A. A. and Webster, R. K. (1957). *Geochim. Cosmochim. Acta* **11**, 139.
Smales, A. A. and Webster, R. K. (1958). *Analyt. Chim. Acta,* **18**, 582.
Smirnov, E. V. (1971). *Morsk. Gidrofiz. Issled,* 195.
Smith, A. P. and Grimaldi, F. S. (1954). *U.S. Geol. Surv. Bull.* **1006**, 111.
Smith, G. F., McCurdy, W. H. and Diehl, M. (1952). *Analyst,* **77**, 418.
Smith, J. D. and Burton, J. D. (1972). *Geochim. Cosmochim. Acta,* **36**, 621.
Smith, J. D. and Redmond, J. D. (1971). *J. Electroanalyt. Chem.* **33**, 169.
Smith, J. D., Nicholson, R. A. and Moore, P. J. (1971). *Nature,* **232**, 393.
Smith, K. L. (1971). *Limnol. Oceanogr.* **16**, 675.
Smith, R. C., Pillai, K. C., Chow, T. J. and Folsom, T. R. (1965). *Limnol. Oceanogr.* **10**, 226.
Smith, V. C. (1972). *In* "Ultrapurity—Methods and techniques" (M. Zief and R. Speights, eds), pp. 173–192, Marcel Dekker, New York.
Soares, M. I. V. and Goncalves, E. V. N. (1966). *Rev. Port. Quim.* **8**, 214.
Sodd, V. J., Golding, A. S. and Velten, R. J. (1960). *Analyt. Chem.,* **32**, 25.
Solórzano, L. (1969). *Limnol. Oceanogr.* **14**, 799.
Solórzano, L. and Strickland, J. D. H. (1968). *Limnol. Oceanogr.* **13**, 515.
Solov'ev, A. N. and Doroshenko, G. A. (1970). *Morsk. Gidrofiz. Issled.* 179.
Somayajulu, B. L. K. and Goldberg, E. D. (1966). *Earth Planet. Sci. Letters,* **1**, 102.
Sørensen, C. P. L. (1902). *Kgl. Danske Vidensk. Selsk. Skr. Naturvid. Math. Afd.* **12**, 117.

Sousa, A. de (1954). *Analyt. chim Acta*, **11**, 221.
Soyer, J. (1963). *Vie et Milieu*, **14**, 1.
Spencer, D. W. (1973). *Geosections* No 3.
Spencer, D. W. and Brewer, P. G. (1969). *Geochim. Cosmochim. Acta*, **33**, 325.
Spencer, D. W. and Brewer, P. G. (1970). *CRC Crit. Revs. Solid State Sci.*, **1**, 409.
Spencer, D. W. and Manheim, F. T. (1969). *U.S. Geol. Surv., Prof. Paper*, **650**D, 228.
Spencer, D. W. and Sachs, P. L. (1969). *At. Absn. Newsletter*, **8**, 65.
Spencer, D. W. and Sachs, P. L. (1970). *Mar. Geol.* **9**, 117.
Spencer, D. W., Brewer, P. G., Sachs, P. L. and Smith, C. L. (1970a). *U.S. At. Energy Comm.* NYO-4150–9.
Spencer, D. W., Robertson, D. E., Turekian, K. K. and Folsom, T. R. (1970b). *J. Geophys. Res.* **75**, 7688.
Spencer, D. W., Brewer, P. G. and Sachs, P. L. (1972). *Geochim. Cosmochim. Acta*, **36**, 71.
Spilhaus, A. F. and Miller, A. R. (1948). *J. Mar. Res.* **7**, 370.
Spirn, R. V. (1965). *Ph.D. Thesis*, Massachusetts Institute of Technology.
Sporek, K. F. (1956). *Analyst*, **81**, 540.
Stainton, M. P. (1974). *Limnol. Oceanogr.* **19**, 707.
Stanford, H. M. (1971). *In* "Ecological studies of radioactivity in the Columbia River Estuary and adjacent Pacific Ocean" (J. E. McCauley, ed.), Oregon State University, Corvallis, Oregon.
Starikova, N. D. (1970). *Oceanologiya*, **10**, 786.
Starikova, N. D. and Korzhikova, R. I. (1969). *Oceanologiya*, **9**, 509.
Stark, W. (1943). *Helv. Chim. Acta*, **26**, 424.
Stary, J. (1964). "The solvent extraction of metal chelates", Macmillan, London.
Stauffer, T. B. and MacIntyre, W. G. (1970). *Chesapeake Sci.*, **11**, 216.
Stefanac, Z. and Simon, W. (1967). *Microchem. J.* **12**, 125.
Stefansson, U. and Olafsson, J. (1970). *Cons. int. explor. Mer. Serv. Hydrogr. Interlab. Rept, No. 3*, 23.
Stephen, W. I. (1972). *Proc. Soc. Analyt. Chem.* **9**, 137.
Stephens, K. (1962). *Limnol. Oceanogr.* **7**, 484.
Stephens, K. (1963). *Limnol. Oceanogr.* **8**, 361.
Stewart, D. C. and Bentley, W. C. (1954). *Science*, **120**, 50.
Stock, A. and Cucuel, F. (1934). *Naturwissenshaften*, **22**, 390.
Stoner, J. (1974). *Ph.D. Thesis*, University of Liverpool.
Stoner, J. and Chester, R. (1974). *Mar. Chem.* **2**, 33.
Streeton, R. J. W. (1953). *U.K. At. Energy Res. Rept.* C/M 168.
Strickland, J. D. H. (1952a), *J. Amer. Chem. Soc.* **74**, 862.
Strickland, J. D. H. (1952b). *J. Amer. Chem. Soc.* **74**, 868.
Strickland, J. D. H. (1965). *In* "Chemical Oceanography" (J. P. Riley and G. Skirrow, eds), Vol. I. pp. 478–610. Academic Press, London.
Strickland, J. D. H. (1968). *Limnol. Oceanogr.* **13**, 388.
Strickland, J. D. H. and Austin, K. H. (1959). *J. Cons. int. Explor. Mer.* **2**, 446.
Strickland, J. D. H. and Parsons, T. R. (1966). *Fish. Res. Bd Can., Bull.* No. **125**, 195 pp.
Strickland, J. D. H. and Parsons, T. R. (1965). "A Manual of Seawater Analysis," 203 pp. Fisheries Research Board of Canada, Ottawa.

Strickland, J. D. H. and Parsons, T. R. (1968). "A practical handbook of seawater analysis", *Fish. Res. Bd. Can., Bull.* **167**.

Strickland, J. D. H. and Solórzano, L. (1966). *In* "Some contemporary studies in marine science" (H. Barnes, ed.) Allen and Unwin, London. p. 665.

Sugawara, K. (1955). *Preprint Regional Symposium on Physical Oceanography, Tokyo.*

Sugawara, K. (1969). "On the preparation of CSK standards for marine nutrient analysis". SCOR-UNESCO, ICSU Intergovernmental Oceanographic Commission. Tokyo.

Sugawara, K. and Okabe, S. (1960). *J. Earth Sci. Nagoya Univ.* **8**, 86.

Sugawara, K. and Okabe, S. (1966). *J. Tokyo Univ. Fish. Spec. Ed.* **8**, 165.

Sugawara, K. and Terada, K. (1957). *J. Earth Sci., Nagoya Univ.* **5**, 81.

Sugawara, K., Tanaka, M. and Naito, H. (1953). *Bull. Chem. Soc. Japan*, **26**, 417.

Sugawara, K., Koyama, T. and Terada, K. (1955). *Bull. Chem. Soc. Japan*, **28**, 494.

Sugawara, K., Tanaka, M. and Okabe, S. (1959). *Bull. Chem. Soc. Japan*, **32**, 221.

Sugawara, K., Terada, K., Kanamori, S. and Okabe, S. (1963). *Earth Sci., Nagoya Univ.* **10**, 34.

Sugihara, T. T., Troianello, E. J., James, H. I. and Bowen, V. T. (1959). *Analyt. Chem.* **31**, 44.

Sugimura, Y. and Tsubota, H. (1963). *J. Mar. Res.* **21**, 74.

Sugimura, Y., Torii, T. and Murata, S. (1964). *Nature, Lond.* **204**, 464.

Sutton, D. C. and Kelly, J. J. (1968). *U.S.A.E.C. Health and Safety Lab. Rept. HASL-196.*

Sverdrup, H. U., Johnson, M. W. and Fleming, R. H. (1942). "The Oceans: their physics, chemistry and general biology." Prentice-Hall, New York.

Swinnerton, J. W., Linnenbom, V. J. and Cheek, C. H. (1962a). *Analyt. Chem.* **34**, 483.

Swinnerton, J. W., Linnenbom, V. J. and Cheek, C. H. (1962b). *Analyt. Chem.* **34**, 1509.

Swinnerton, J. W., Linnenbom, V. J. and Cheek, C. H. (1968). *Limnol. Oceanogr.* **13**, 193

Swinnerton, J. W., Linnenbom, V. J. and Cheek, C. H. (1969). *Environ. Sci. Tech.* **3**, 836.

Swinnerton, J. W. and Linnenbom, V. J. (1967a). *J. Gas Chromatogr.* **5**, 570.

Swinnerton, J. W. and Linnenbom, V. J. (1967b). *Science*, **156**, 1119.

Swinnerton, J. W. and Sullivan, J. P. (1962). *U.S. Naval Res. Lab. Rep.* 5806, 13 pp.

Szabo, B. J. (1967). *Bull. mar. Sci.* **17**, 544.

Szabo, B. J. (1967a). *Geochim. Cosmochim. Acta*, **31**, 1321.

Szabo, B. J. (1971). *Bull. Mar. Sci.* **21**, 748.

Szabo, B. J. and Joensuu, O. (1967). *Environ. Sci. Tech.* **1**, 499.

Takai, N. and Yamabe, T. (1971). *Mizi Shori. Gijutsu*, **12**, 3.

Tamont'ev, V. P. and Bruyewicz, S. W. (1964). *Tr. Inst. Okeanol. Akad. Nauk. SSSR.* **67**, 41.

Tanaka, M. and Awata, N. (1967). *Analyt. Chim. Acta*, **39**, 485.

Tanikawa, K. Ochi, H. and Arakawa. K. (1970). *Bunseki Kagaku*, **19**, 1669.

Tatsumoto, M. and Goldberg, E. D. (1959). *Geochim. Cosmochim. Acta*, **17**, 201.

Tatsumoto, M. and Patterson, C. C. (1963). *In* "Earth Science and Meteoritics" (J. Geiss and E. D. Goldberg, eds.) North Holland Publishing Co., Amsterdam.

Tatsumoto, M. T., Williams, W. T., Prescott, J. M. and Hood, D. W. (1961). *J. mar. Res.* **19**, 89.

Taylor, H. J. and Thompson, T. G. (1933). *Industr. Engng. Chem. (Anal.)*, 5, 87.
Telek, G. and Marshall, N. (1974). *Mar. Biol.* 24, 219.
Tera, F., Ruch, C. C. and Morrison, G. H. (1965). *Analyt. Chem.* 37, 358.
Tetlow, J. A. and Wilson, A. L. (1964). *Analyst*, 89, 442, 453.
Thayer, G. W. (1970). *Chesapeake Sci.* 11, 155.
Thiers, R. E. (1957a). *In* "Methods of Biochemical Analysis" (D. Glick, ed.). pp. 274–309. Interscience, New York.
Thiers, R. E. (1957b). *In* "Trace Analysis" (J. H. Yoe and H. J. Kock, Jr., eds.). pp. 637–666, Wiley, New York.
Thompson, E. F. (1940). *J. Mar. Res.* 3, 268.
Thompson, K. C. and Reynolds, G. D. (1971). *Analyst*, 96, 771.
Thompson, T. G. and Korpi, E. (1942). *J. mar. Res.* 5, 28.
Thompson, T. G. and Laevastu, T. (1960). *J. Mar. Res.* 18, 189.
Thompson, T. G. and Robinson, R. J. (1939). *J. Mar. Res.* 3, 268.
Tikhonov, M. K. and Shalimov, G. A. (1965). *Gidrofiz. i Gidrokhim. Issled Akad. Nauk. Uki. SSR*, 133.
Tikhonov, M. K. and Zhavoronkina, V. K. (1960). *Tr. Moskogo Gidrofiz. Inst. Akad. Nauk. S.S.S.R.* 19, 31.
Toelgyessy, J., Braun, T. and Kyrs, M. (1972). "Isotope dilution analysis", Pergamon Press, Oxford.
Topping, G. (1969). *Limnol. Oceanogr.* 14, 798.
Topping, G. and Pirie, J. M. (1972). *Analyt. Chim. Acta*, 62, 200.
Torii, T. (1964). *Japan. Antarctic. Res. Exped.* (*1956–62*). *Sci. Rept. Ser. D. Oceanogr.* No. 1, 1.
Torii, T. and Murata, S. (1964). *In* "Recent Researches in the fields of hydrosphere, atmosphere and nuclear geochemistry" (Y. Miyake and T. Koyama, eds.), pp. 321–334, Maruzen Co., Tokyo.
Traganza, E. D. and Szabo, B. J. (1967). *Limnol. Oceanogr.* 12, 281.
Treguer, P., LeCorre, P. and Courtot, P. (1972). *J. mar. biol. Ass. U.K.* 53, 1045.
Trier, R. M., Broecker, W. S. and Feely, H. W. (1972). *Earth Planet Sci.,* 16, 141.
Trotti, L. and Sacks, D. (1962). *Arch. Oceanogr. Limnol. Roma,* 12, 257.
Truesdale, V. W. (1971). *Analyst,* 96, 584.
Truesdale, V. W. and Smith, C. J. (1975). *Analyst,* 100, 203.
Truesdale, V. W. and Spencer, C. P. (1974). *Mar. Chem.* 2, 33.
Tsalev, D. L., Alimarin, I. P. and Neiman, S. I. (1972). *Zh. Anal. Khim.* 27, 1223.
Tsubota, M. and Kitano, Y. (1960). *Bull. Chem. Soc. Japan.* 33, 770.
Tsunogai, S. (1971a). *Analyt. Chim. Acta*, 55, 444.
Tsunogai, S. (1971b). *Deep-Sea Res.* 18, 913.
Tsunogai, S. and Henmi, T. (1971). *J. Oceanogr. Soc. Japan*, 27, 67.
Tsunogai, S. and Nozaki, Y. (1971). *Geochem. J.* 5, 165.
Tsunogai, S. and Nozaki, Y. (1973). *Analyt. Chim. Acta*, 64, 209.
Tsunogai, S. and Sase, T. (1969). *Deep-Sea Res.* 16, 489.
Tsunogai, S., Nishimura, M. and Nakaya, S. (1968a). *Talanta*, 15, 385.
Tsunogai, S., Nishimura, M. and Nakaya, S. (1968b). *J. Oceanogr. Soc. Japan*, 24, 153.
Turekian, K. K. (1971). *U.S. Atomic Energy Comm. Grant AT(30–1)-2912. Annual Report.*
Turekian, K. K. and Johnson, D. G. (1966). *Geochim. Cosmochim. Acta,* 30, 1153.
Ueno, K. (1969). *Nippon Kaisui Gakkai-Shi,* 22, 337.
Uesugi, K. and Murakami, T. (1966). *Bunseki Kagaku*, 15, 482.

Uesugi, K., Tabushi, M., Marukami, T. and Shigematsu, T. (1964). *Bunseki Kagaku*, **13**, 440.

UNESCO (1966). "Determination of photosynthetic pigments in sea water." UNESCO, Paris.

UNESCO (1973). "International Oceanographic Tables" Vol. 2. National Institute of Oceanography, Wormley, Godalming, Surrey, England.

Uppstrom, L. (1968). *Analyt. Chim. Acta*, **43**, 475.

Uppstrom, L. R. (1974). *Deep-Sea Res.* **21**, 161.

Vaccaro, R. F. and Jannasch, H. W. (1966). *Limnol. Oceanogr.* **11**, 596.

Vaccaro, R. F., Hicks, S. E., Jannasch, H. W. and Carey, F. G. (1968). *Limnol. Oceanogr.* **13**, 356.

Varlet, F. (1958). *Études eburnéennes*, **7**, 249.

Van Dorn, W. G. (1956). *Trans. Amer. Geophys. Union*, **37**, 682.

Van Hall, C. E., Safranko, J. and Stenger, V. A. (1963). *Anal. Chem.* **35**, 315.

Veeh, H. H. (1968). *Geochim. Cosmochim. Acta,* **32**, 117.

Vishniac, H. S. and Riley, G. A. (1961). *Limnol. Oceanogr.* **6**, 36

Viswanathan, R., Sreekumaran, C., Doshi, G. R. and Unni, C. K. (1965). *J. Indian Chem. Soc.* **42**, 35.

Viswanathan, R., Shah, S. M. and Unni, C. K. (1968). *Bull. Nat. Inst. Sci., India*, **38**, 284.

Viswanathan, R., Shah, S. M. and Unni, C. K. (1969). *Bull. Nat. Inst. Sci., India,* **38**, 284.

Vizard, G. S. and Wynne, A. (1959). *Chem. Ind. (London)*, 196.

Vogler, P. (1965). *Fortschr. Wasserchem.* 109.

Vogler, P. (1965a). *Fortschr. Wasserchem.* 100.

Vogler, P. (1966). *Fortschr. Wasserchem.* 211.

Voipio, A. (1959). *Suom. Kemistilehti*, **B32**, 61.

Voipio, A. (1961). *Rapp. Cons. Explor. Mer.* **149**, 38.

Volchok, H. L., Bowen, V. T., Folsom, T. R., Broecker, W. S., Schuert, E. A. and Bien, G. S. (1971). *In* "Radioactivity in the Marine Environment", p. 42. U.S. National Academy of Sciences.

Voyce, D. and Zeitlin, H. (1974). *Analyt. Chim. Acta,* **69**, 27.

Vries, H. de (1959). *In* "Researches in Geochemistry" (P. H. Abelson, ed.), p. 169. Wiley, New York.

Wada, E. and Hattori, A. (1971). *Analyt. Chim. Acta,* **56**, 233.

Wakeel, S. K. El. and Riley, J. P. (1957). *J. Cons. Int. Explor. Mer.* **22**, 180.

Walsh, A. (1955). *Spectrochim. Acta,* **7**, 108.

Walsh, G. E. and Douglas, J. (1966). *Limnol. Oceanogr.,* **11**, 406.

Wangersky, P. J. and Gordon, D. C. (1965). *Limnol. Oceanogr.* **10**, 544.

Wardani, S. A. El. (1958). *Geochim. Cosmochim. Acta,* **15**, 237.

Warner, T. B. (1969a). *U.S. Naval Res. Lab. Rept.* **NRL6905**, 10 pp.

Warner, T. B. (1969b). *Science,* **165**, 178.

Warner, T. B. (1971). *Deep-Sea Res.* **18**, 1255.

Watanuki, K. (1969). *Bunseki Kagaku,* **18**, 1280.

Wattenberg, H. (1943). *Z. anorg. Chem.* **251**, 86.

Webb, D. A. (1939). *J. exp. Biol.* **16**, 178.

Webb, K. L. and Wood, L. (1966). *In* "Automation in Analytical Chemistry," pp. 440–444. Technicon Symposium.

Weimer, W. C. and Lee, G. F. (1971). *Environ. Sci. Tech.* **5**, 1136.

Weiss, H. V. and Crozier, (1972). *Analyt. Chim. Acta*, **58**, 231.
Weiss, H. V. and Lai, M. G. (1960). *Analyt. Chem.* **32**, 475.
Weiss, H. V. and Lai, M. G. (1961). *Talanta*, **8**, 72.
Weiss, H. V. and Lai, M. G. (1963). *Analyt. Chim. Acta*, **28**, 242.
Weiss, H. V. and Reid, J. A. (1960). *J. Mar. Res.* **18**, 185.
Weiss, H. V., Lai, M. G. and Gillespie, A. (1961). *Analyt. Chim. Acta*, **25**, 550.
Weiss, H. V., Yamamoto, S., Crozier, T. E. and Mathewson, J. H. (1972). *Envir. Sci. Tech.* **6**, 644.
Weiss, R. F. (1968). *Deep-Sea Res*, **15**, 695.
Weiss, R. F. (1970). *Deep-Sea Res.* **17**, 721.
Weiss, R. F. (1971a). *Deep-Sea Res.* **18**, 225.
Weiss, R. F. (1971b). *J. Chem. Eng. Data*, **16**, 235.
Werner, A. E. and Waldichuk, M. (1967). *Limnol. Oceanogr.* **12**, 158.
West, T. S. (1961). *Analyt. Chim. Acta*, **25**, 405.
West, T. S. (1971). *Pure Appl. Chem.* **26**, 47.
Wheatland, A. B. and Smith, L. J. (1955). *J. appl. Chem.* **5**, 144.
White, C. E. and Lowe, C. S. (1937). *Industr. Engng Chem.* (*Anal.*), **9**, 430.
Whitfield, M. (1974). *J. Mar. Biol. Ass., U.K.* **54**, 565.
Whitnack, G. C. and Sasselli, R. (1969). *Analyt. Chim. Acta*, **47**, 276.
Wilkniss, P. K. and Linnenbom, V. J. (1968). *Limnol. Oceanogr.* **13**, 530.
Willard, H. H., Merritt, L. L. and Dean, J. A. (1965). "Instrumental Methods of analysis," 4th Ed., Van Nostrand Reinhold, New York.
Williams, P. M. (1961). *Nature*, **189**, 219.
Williams, P. M. (1965). *J. Fish. Res. Bd. Canada*, **22**, 1107.
Williams, P. M. (1969a). *Limnol. Oceanogr.* **14**, 297.
Williams, P. M. (1969). *Limnol. Oceanogr.* **14**, 156.
Williams, P. M. and Chan, K. S. (1966). *J. Fish. Res. Bd. Can.* **23**, 575.
Williams, P. M. and Coote, A. R. (1962). *Limnol. Oceanogr.* **7**, 258.
Williams, P. M. and Strack, P. M. (1966). *Limnol. Oceanogr.* **11**, 401.
Williams, P. M. and Weiss, H. V. (1973). *J. Fish Res. Bd Can.* **30**, 293.
Williams, P. M. and Zirino, A. (1964). *Nature*, **204**, 462.
Williams, R. T. and Bainbridge, A. E. (1973). *J. Geophys. Res.* **78**, 2691.
van Willigen, J. H. H. G. and Schonebaum, R. C. (1966). *Rec. Trav. chim. Pays-Bas*, **85**, 35.
Wilson, A. L. (1965). *Analyst.* **90**, 270.
Wilson, J. D., Webster, R. K., Milner, G. W. C., Barnett, G. A. and Smales, A. A. (1960). *Analyt. chim. Acta*, **23**, 505.
Wilson, R. F. (1961). *Limnol. Oceanogr.* **6**, 259.
Wilson, T. R. S., Sayles, F. L. and Mangelsdorf, P. C. (1972). *Trans. Amer. Geophys. Union*, **53**, 529.
Windham, R. L. (1972). *Analyt. Chem.* **44**, 1334.
Windom, H. L. (1971). *Limnol. Oceanogr.* **16**, 806.
Windom, H. L. and Smith, R. G. (1972). *Deep-Sea Res.* **19**, 727.
Winkler, L. W. (1888). *Ber. dtsch. chem. Ges.* **21**, 2843.
Wittig, G. (1950). *Z. angew. Chem.* **62**, 231.
Wittig, G., Keiches, G., Rückert, A. and Raff, P. (1949). *Ann. Chem.* **573**, 195.
Wogman, N. A. (1970). *Nuclear Instruments and Methods*, **83**, 277.
Wogman, N. A., Robertson, D. C. and Perkins, R. W. (1967). *Health Physics*, **13**, 767.
Wolgemuth, K. (1970). *J. geophys. Res.* **75**, 7686.

514   J. P. RILEY

Wolgemuth, K. and Broecker, W. S. (1970). *Earth Planet. Sci. Lett.* **8**, 372.
Wong, G. T. F. and Brewer, P. G. (1974). *J. Mar. Res.* **32**, 25.
Wong, K. M. (1971). *Analyt. Chim. Acta*, **56**, 355.
Wood, E. D. (1971). University of Alaska, Dept of Oceanography Rept No. R71-17.
Wood, E. D., Armstrong, F. A. J. and Richards, F. A. (1968). *J. mar. biol. Ass., U.K.* **47**, 25.
Woodson, T. T. (1939). *Rev. Sci. Instr.* **10**, 308.
Wooster, W. S., Lee, A. J. and Dietrich, G. (1969). *Deep-Sea Res.* **16**, 321.
World Health Organization (1966). "Methods of Radiochemical Analysis", p. 84. Geneva.
Wunderlich, F. (1969). *Senckenberg Mar.* **1**, 147.
Wüst, G. (1932). *Wiss. Ergebn. dtsch. Atlant. Exped. Meteor. 1925–27*, **4(A)**, 21.
Yamabe, T. and Nobuhara, T. (1969). *Seisan Kenkyu*, **21**, 530.
Yamabe, T. and Takai, N. (1970). *Nippon Kaisui Gakkai-Shi*, **24**, 16.
Yamagata, N. (1957). *J. Chem. Soc. Japan*, **78**, 513.
Yamagata, N. and Yamagata, T. (1960). *Analyst.* **85**, 282.
Yamagata, N. and Iwashima, K. (1963). *Nature, Lond.*, **200**, 52.
Yamagata, N. and Iwashima, K. (1965). *Bull. Inst. Publ. Health*, **14**, 183.
Yamaguchi, R., Machida, T. and Ueki, M. (1969). *Yakugaku Zasshi*, **89**, 1534.
Yatsimirski, K. B., Emel'yanov, E. M., Pavtova, V. K. and Savichenko, Ya. S. (1970). *Okeanologiya*, **10**, 1111.
Yatsimirski, K. B., Emel'yanov, E. M., Pavlova, V. K. and Savichenko, Y. S. (1971). *Okeanologiya*, **11**, 730.
Yentsch, C. S. (1967). *In* "Chemical Environment in the Aquatic Habitat" (H. L. Golterman and R. S. Clymo, eds. pp. 255–270. N.V. Noord-Hollansche Vitgevers Maatschappy, Amsterdam.
Yentsch, C. S. and Menzel, D. W. (1963). *Deep-Sea Res.*, **10**, 221.
Yonehara, N. (1964). *Bull. Chem. Soc. Japan*, **37**, 1107.
Yoshida, Y. (1967a). *Bull. Jap. Soc. sci. Fish.* **33**, 343.
Yoshida, Y. (1967b). *Bull. Jap. Soc. sci. Fish.* **33**, 348.
Youden, W. J. (1959). *In* "Treatise on analytical chemistry" (I. M. Kolthoff and P. J. Elving, eds), Part I, Vol. 1. p. 47, Interscience, New York.
Youden, W. J. (1967). "Statistical techniques for collaborative tests," Association of Official Analytical Chemists, Washington, D.C.
Young, A. W., Buddemeier, R. W. and Fairhall, A. W. (1969). *Limnol. Oceanogr.* **14**, 634.
Young, E. G., Smith, D. G. and Langille, W. M. (1959). *J. Fish. Res. Bd., Can.* **16**, 7.
Yoza, N. (1973). *J. Chromatogr.* **86**, 325.
Yuen, S. H. (1958). *Analyst*, **83**, 350.
Zeller, H. D. (1955). *Analyst*, **80**, 632.
Zharikov, V. F. (1970). *Tr. Gos. Okeanogr. Inst.* Nö. **101**, 128.
Zhavoronkina, V. K. (1960). *Morsk. Gidiofiz. Inst. Akad. Nauk. S.S.S.R.* **19**, 38.
Zirino, A. and Healy, M. L. (1970). *Limnol. Oceanogr.* **15**, 956.
Zirino, A. and Healy, M. L. (1971). *Limnol. Oceanogr.* **16**, 773.
Zirino, A. and Healy, M. L. (1972). *Environ. Sci. Tech.* **6**, 243.
Zobell, C. E. (1941). *J. Mar. Res.* **4**, 173.
Zsolnay, A. (1971). *Kieler Meer.* **27**, 129.

Appendix

Tables of physical and chemical constants relevant to marine chemistry

TABLE 1

Some physical properties of pure water (after Dorsey, 1940)

Molecular weight	18·0153
Heat of formation	285·89 kJmol^{-1} (at 25°C and 1 atm)
Ionic dissociation constant	10^{-4} M^{-1} (at 25°C and 1 atm)
Heat of ionization	55·71 kJmol^{-1} (at 25°C and 1 atm)
Viscosity	8·949 mP (at 25°C and 1 atm)
Velocity of sound	1496·3 ms^{-1} (at 25°C and 1 atm)
Density	0·9979751 g cm^{-3} (at 25°C and 1 atm)
Freezing point	0°C (at 1 atm)
Boiling point	100°C (at 1 atm)
Isothermal compressibility	45·6 × 10^{-6} atm^{-1} (at 25°C over the range 1–10 atm)
Specific heat at constant volume	4·1786 int.J (g°C)$^{-1}$ (at 25°C and 1 atm)
Thermal conductivity	0·00598 W cm^{-1} °C^{-1} (at 20°C and 1 atm)
Temperature of maximum density	3·98°C (at 1 atm)
Dielectric constant	81·0 (at 1 atm, 17°C, and 60 MHz)
Electrical conductivity	Less than 10^{-8} Ω^{-1} cm^{-1} (at 25°C and 1 atm)

TABLE 2

Concentrations of the major ions in sea water of various salinities $(g\ kg^{-1})$*

Salinity (‰)	Na^+	Mg^{2+}	Ca^{2+}	K^+	Sr^{2+}	B	Cl^-	SO_4^{2-}	Br^-	F^-	HCO_3^-
5	1·539	0·185	0·058	0·057	0·001	0·001	2·763	0·387	0·010	0·0002	0·020
10	3·078	0·370	0·118	0·114	0·002	0·001	5·527	0·775	0·019	0·0004	0·041
15	4·617	0·555	0·177	0·171	0·003	0·002	8·290	1·162	0·029	0·0005	0·061
20	6·156	0·739	0·235	0·228	0·005	0·003	11·054	1·550	0·038	0·0007	0·081
25	7·695	0·924	0·294	0·285	0·006	0·003	13·817	1·937	0·048	0·0009	0·101
30	9·234	1·109	0·353	0·342	0·007	0·004	16·581	2·325	0·058	0·0011	0·122
31	9·542	1·146	0·365	0·353	0·007	0·004	17·133	2·402	0·059	0·0011	0·126
32	9·850	1·183	0·377	0·365	0·007	0·004	17·685	2·480	0·062	0·0012	0·130
33	10·157	1·220	0·388	0·376	0·008	0·004	18·239	2·557	0·063	0·0012	0·134
34	10·465	1·257	0·400	0·388	0·008	0·004	18·791	2·635	0·065	0·0012	0·137
35	10·773	1·294	0·412	0·399	0·008	0·004	19·344	2·712	0·067	0·0013	0·142
36	11·081	1·331	0·424	0·410	0·008	0·005	19·897	2·789	0·069	0·0013	0·146
37	11·389	1·368	0·435	0·422	0·008	0·005	20·449	2·867	0·071	0·0013	0·150
38	11·696	1·405	0·447	0·433	0·009	0·005	21·002	2·944	0·073	0·0014	0·154
39	12·004	1·442	0·459	0·445	0·009	0·005	21·555	3·022	0·075	0·0014	0·158
40	12·312	1·479	0·471	0·456	0·009	0·005	22·107	3·099	0·077	0·0015	0·162
41	12·620	1·516	0·482	0·467	0·009	0·005	22·660	3·177	0·079	0·0015	0·166
42	12·928	1·553	0·494	0·479	0·009	0·005	23·213	3·254	0·081	0·0015	0·170

* Cations concentrations; averages of mean results of Cox and Culkin (1967) and Riley and Tongudai (1967). Sulphate and bromide concentration based on mean values from Morris and Riley (1966).

TABLE 3

Preparation of artificial sea water (S = 35·00‰)

Lyman and Fleming (1940) (g.)		Kalle (1945) (g.)	
NaCl	23·939	NaCl	28·566
$MgCl_2$	5·079	$MgCl_2$	3·887
Na_2SO_4	3·994	$MgSO_4$	1·787
$CaCl_2$	1·123	$CaSO_4$	1·308
KCl	0·667	K_2SO_4	0·832
$NaHCO_3$	0·196	$CaCO_3$	0·124
KBr	0·098	KBr	0·103
H_3BO_3	0·027	$SrSO_4$	0·0288
$SrCl_2$	0·024	H_3BO_3	0·0282
NaF	0·003		
Water to	1 kg	Water to	1 kg

Kester *et al.* (1967)
A. Gravimetric salts $g\,kg^{-1}$

NaCl	23·926
Na_2SO_4	4·008
KCl	0·667
$NaHCO_3$	0·196
KBr	0·098
H_3BO_3	0·026
NaF	0·003

B. Volumetric salts (standardized by Mohr method)

	Approx. molarity	Use volume equivalent to
$MgCl_2 6H_2O$	1·0 M	$1·297\,g\,Mg\,kg^{-1}$
$CaCl_2 2H_2O$	1·0 M	$0·406\,g\,Ca\,kg^{-1}$
$SrCl_2 6H_2O$	0·1 M	$0·0133g\,Sr\,kg^{-1}$

C Water to 1 kg

Note: (i) Allowance must be made for water of crystallization of any of the salts used.
(ii) After aeration the pH should lie between 7·9 and 8·3.

TABLE 4

Collected conversion factors

Conversion	Factor	Reciprocal
μg NO_3^- $\longrightarrow \mu$g N	0·2259	4·427
μg NO_2^- $\longrightarrow \mu$g N	0·3045	3·286
μg NH_3 $\longrightarrow \mu$g N	0·8225	1·216
μg NH_4^+ $\longrightarrow \mu$g N	0·7764	1·287
μg PO_4^{3-} $\longrightarrow \mu$g P	0·3261	3·066
μg P_2O_5 $\longrightarrow \mu$g P	0·4364	2·291
μg SiO_2 $\longrightarrow \mu$g Si	0·4675	2·139
μg SiO_4^{4-} $\longrightarrow \mu$g Si	0·3050	3·278
μg N $\longrightarrow \mu$g-at N	0·07138	14·008
μg P $\longrightarrow \mu$g-at. P	0·03228	30·975
μg Si $\longrightarrow \mu$g-at. Si	0·03560	28·09

TABLE 5

Table for conversion of weights of nitrogen, phosphorus and silicon expressed in terms of μg into μg-at.

μg N, P, or Si^{-1}	μg-at Nl^{-1}	μg-at. P l^{-1}	μg-at. Si l^{-1}
1	0·071	0·032	0·036
2	0·143	0·065	0·071
3	0·214	0·097	0·107
4	0·286	0·129	0·142
5	0·357	0·161	0·178
6	0·428	0·194	0·214
7	0·500	0·226	0·249
8	0·571	0·258	0·284
9	0·643	0·291	0·320
10	0·714	0·323	0·356
20	1·428	0·646	0·712
30	2·142	0·968	1·068
40	2·856	1·291	1·424
50	3·569	1·614	1·780
60	4·283	1·937	2·136
70	4·997	2·260	2·492
80	5·711	2·582	2·848
90	6·425	2·905	3·204
100	7·139	3·228	3·560

TABLE 6

Solubility of oxygen (C) in sea water ($cm^3 dm^{-3}$) with respect to an atmosphere of 20·95% oxygen and 100% relative humidity at a total atmospheric pressure of 760 mm Hg. (UNESCO, 1973)*

T (°C)	Salinity (‰)														
	0	5	10	15	20	25	30	31	32	33	34	35	36	37	38
0	10·22	9·87	9·54	9·22	8·91	8·61	8·32	8·27	8·21	8·16	8·10	8·05	7·99	7·94	7·88
1	9·94	9·60	9·28	8·97	8·68	8·39	8·11	8·05	8·00	7·94	7·89	7·84	7·78	7·73	7·68
2	9·67	9·35	9·04	8·74	8·45	8·17	7·90	7·85	7·79	7·74	7·69	7·64	7·59	7·53	7·48
3	9·41	9·10	8·80	8·51	8·23	7·96	7·70	7·65	7·60	7·55	7·50	7·45	7·40	7·35	7·30
4	9·16	8·86	8·57	8·29	8·02	7·76	7·51	7·46	7·41	7·36	7·31	7·26	7·22	7·17	7·12
5	8·93	8·64	8·36	8·09	7·83	7·57	7·33	7·28	7·23	7·18	7·14	7·09	7·04	7·00	6·95
6	8·70	8·42	8·15	7·89	7·64	7·39	7·15	7·11	7·06	7·01	6·97	6·92	6·88	6·83	6·79
7	8·49	8·22	7·95	7·70	7·45	7·22	6·98	6·94	6·89	6·85	6·81	6·76	6·72	6·67	6·63
8	8·28	8·02	7·76	7·52	7·28	7·05	6·82	6·78	6·74	6·69	6·65	6·61	6·57	6·52	6·48
9	8·08	7·83	7·58	7·34	7·11	6·89	6·67	6·63	6·59	6·54	6·50	6·46	6·42	6·38	6·34
10	7·89	7·64	7·41	7·17	6·95	6·73	6·52	6·48	6·44	6·40	6·36	6·32	6·28	6·24	6·20
11	7·71	7·47	7·24	7·01	6·80	6·58	6·38	6·34	6·30	6·26	6·22	6·18	6·14	6·10	6·07
12	7·53	7·30	7·08	6·86	6·65	6·44	6·24	6·21	6·17	6·13	6·09	6·05	6·01	5·98	5·94
13	7·37	7·14	6·92	6·71	6·50	6·31	6·11	6·07	6·04	6·00	5·96	5·93	5·89	5·85	5·82
14	7·20	6·98	6·77	6·57	6·37	6·17	5·99	5·95	5·91	5·88	5·84	5·80	5·77	5·73	5·70
15	7·05	6·84	6·63	6·43	6·24	6·05	5·87	5·83	5·79	5·76	5·72	5·69	5·65	5·62	5·58
16	6·90	6·69	6·49	6·30	6·11	5·93	5·75	5·71	5·68	5·64	5·61	5·58	5·54	5·51	5·48
17	6·75	6·55	6·36	6·17	5·99	5·81	5·64	5·60	5·57	5·53	5·50	5·47	5·43	5·40	5·37
18	6·61	6·42	6·23	6·05	5·87	5·69	5·53	5·49	5·46	5·43	5·40	5·36	5·33	5·30	5·27
19	6·48	6·29	6·11	5·93	5·75	5·59	5·42	5·39	5·36	5·33	5·29	5·26	5·23	5·20	5·17
20	6·35	6·17	5·99	5·81	5·64	5·48	5·32	5·29	5·26	5·23	5·20	5·17	5·14	5·10	5·07
21	6·23	6·05	5·87	5·70	5·54	5·38	5·22	5·19	5·16	5·13	5·10	5·07	5·04	5·01	4·98
22	6·11	5·93	5·76	5·60	5·44	5·28	5·13	5·10	5·07	5·04	5·01	4·98	4·95	4·92	4·89
23	5·99	5·82	5·65	5·49	5·34	5·18	5·04	5·01	4·98	4·95	4·92	4·89	4·87	4·84	4·81

TABLE 6 cont.

Solubility of oxygen (C) in sea water (cm³ dm⁻³) with respect to an atmosphere of 20·95% oxygen and 100% relative humidity at a total atmospheric pressure of 860 mm Hg. (UNESCO, 1973)*

T (°C)	Salinity (‰)														
	0	5	10	15	20	25	30	31	32	33	34	35	36	37	38
24	5·88	5·71	5·55	5·39	5·24	5·09	4·95	4·92	4·89	4·86	4·84	4·81	4·78	4·75	4·73
25	5·77	5·61	5·45	5·30	5·15	5·00	4·86	4·84	4·81	4·78	4·75	4·73	4·70	4·67	4·65
26	5·66	5·51	5·35	5·20	5·06	4·92	4·78	4·75	4·73	4·70	4·67	4·65	4·62	4·59	4·57
27	5·56	5·41	5·26	5·11	4·97	4·83	4·70	4·67	4·65	4·62	4·60	4·57	4·54	4·52	4·49
28	5·46	5·31	5·17	5·03	4·89	4·75	4·62	4·60	4·57	4·55	4·52	4·50	4·47	4·45	4·42
29	5·37	5·22	5·08	4·94	4·81	4·67	4·55	4·52	4·50	4·47	4·45	4·42	4·40	4·37	4·35
30	5·28	5·13	4·99	4·86	4·73	4·60	4·47	4·45	4·43	4·40	4·38	4·35	4·33	4·31	4·28
31	5·19	5·05	4·91	4·78	4·65	4·53	4·40	4·38	4·36	4·33	4·31	4·28	4·26	4·24	4·22
32	5·10	4·96	4·83	4·70	4·58	4·45	4·33	4·31	4·29	4·26	4·24	4·22	4·20	4·17	4·15

* Based on measurements by Carpenter (1966) and Murray and Riley (1969a) fitted by Weiss (1970) to the thermodynamically consistent equation:

$$\ln C = A_1 + A_2(100/T) + A_3 \ln(T/100) + A_4(T/100) + S‰[B_1 + B_2(T/100) + B_3(T/100)^2]$$

where

A_1	A_2	A_3	A_4	B_1	B_2	B_3
−173·4292	249·6339	143·3483	−21·8492	−0·033096	0·014259	−0·017000

and T and $S‰$ are the absolute temperature (K) and salinity in parts per mille respectively.

TABLE 7

Solubility of nitrogen in sea water ($cm^3 dm^{-3}$) with respect to an atmosphere of 78·084 % nitrogen and 100 % relative humidity at a total pressure of 760 mm Hg (Weiss (1970) from data by Murray and Riley (1969b)).

	Salinity ‰								
T(°C)	0	10	20	30	34	35	36	38	40
−1	—	—	16·28	15·10	14·65	14·54	14·44	14·22	14·01
0	18·42	17·10	15·87	14·73	14·30	14·19	14·09	13·88	13·67
1	17·95	16·67	15·48	14·38	13·96	13·86	13·75	13·55	13·35
2	17·50	16·26	15·11	14·04	13·64	13·54	13·44	13·24	13·05
3	17·07	15·87	14·75	13·72	13·32	13·23	13·13	12·94	12·76
4	16·65	15·49	14·41	13·41	13·03	12·93	12·84	12·66	12·47
5	16·26	15·13	14·09	13·11	12·74	12·65	12·56	12·38	12·21
6	15·88	14·79	13·77	12·83	12·47	12·38	12·29	12·12	11·95
8	15·16	14·14	13·18	12·29	11·95	11·87	11·79	11·62	11·46
10	14·51	13·54	12·64	11·80	11·48	11·40	11·32	11·17	11·01
12	13·90	12·99	12·14	11·34	11·04	10·96	10·89	10·74	10·60
14	13·34	12·48	11·67	10·92	10·63	10·56	10·49	10·35	10·21
16	12·83	12·01	11·24	10·53	10·25	10·19	10·12	9·99	9·86
18	12·35	11·57	10·84	10·16	9·90	9·84	9·77	9·65	9·52
20	11·90	11·16	10·47	9·82	9·57	9·51	9·45	9·33	9·21
22	11·48	10·78	10·12	9·50	9·26	9·21	9·15	9·03	8·92
24	11·09	10·42	9·79	9·20	8·98	8·92	8·87	8·76	8·65
26	10·73	10·09	9·49	8·92	8·71	8·65	8·60	8·50	8·39
28	10·38	9·77	9·20	8·66	8·45	8·40	8·35	8·25	8·15
30	10·06	9·48	8·93	8·41	8·21	8·16	8·12	8·02	7·92
32	9·76	9·20	8·67	8·18	7·99	7·94	7·89	7·80	7·71
34	9·48	8·94	8·43	7·96	7·77	7·73	7·68	7·59	7·51
36	9·21	8·69	8·20	7·75	7·57	7·53	7·48	7·40	7·31
38	8·95	8·46	7·99	7·55	7·38	7·33	7·29	7·21	7·13
40	8·71	8·23	7·78	7·36	7·19	7·15	7·11	7·03	6·95

The solubility at any value of salinity and temperature in the above range can be calculated if the following constants are substituted in the equation below (Table 6).

A_1	A_2	A_3	A_4	B_1	B_2	B_3
−172·4965	248·4262	143·0738	−21·7120	−0·049781	0·025018	−0·003486

APPENDIX

TABLE 8

*Solubility of argon in sea water ($cm^{-3} dm^{-3}$) with respect to an atmosphere of 0·934%
argon and 100% relative humidity at a total atmosphere pressure of 760 mm Hg (Weiss
(1970) from data by Douglas (1964, 1965)).*

T (°C)	0	10	20	30	Salinity ‰ 34	35	36	38	40
−1	——	——	0·4456	0·4156	0·4042	0·4014	0·3986	0·3931	0·3877
0	0·4980	0·4647	0·4337	0·4048	0·3937	0·3910	0·3883	0·3830	0·3777
1	0·4845	0·4524	0·4224	0·3944	0·3837	0·3811	0·3785	0·3733	0·3682
2	0·4715	0·4405	0·4115	0·3845	0·3741	0·3716	0·3691	0·3641	0·3592
3	0·4592	0·4292	0·4012	0·3750	0·3650	0·3625	0·3601	0·3552	0·3505
4	0·4474	0·4184	0·3912	0·3659	0·3562	0·3538	0·3515	0·3468	0·3422
5	0·4360	0·4080	0·3817	0·3572	0·3478	0·3455	0·3432	0·3387	0·3342
6	0·4252	0·3980	0·3726	0·3488	0·3397	0·3375	0·3353	0·3309	0·3265
8	0·4049	0·3794	0·3555	0·3331	0·3246	0·3225	0·3204	0·3162	0·3121
10	0·3861	0·3622	0·3397	0·3186	0·3106	0·3086	0·3066	0·3027	0·2989
12	0·3688	0·3463	0·3251	0·3053	0·2977	0·2958	0·2939	0·2902	0·2866
14	0·3528	0·3316	0·3116	0·2929	0·2857	0·2839	0·2822	0·2787	0·2752
16	0·3380	0·3180	0·2991	0·2814	0·2746	0·2729	0·2712	0·2679	0·2647
18	0·3242	0·3053	0·2875	0·2707	0·2642	0·2626	0·2610	0·2579	0·2548
20	0·3114	0·2935	0·2766	0·2607	0·2546	0·2531	0·2516	0·2486	0·2457
22	0·2995	0·2825	0·2665	0·2514	0·2455	0·2441	0·2427	0·2399	0·2371
24	0·2883	0·2722	0·2570	0·2426	0·2371	0·2357	0·2344	0·2317	0·2291
26	0·2779	0·2626	0·2481	0·2344	0·2292	0·2279	0·2266	0·2241	0·2215
28	0·2681	0·2535	0·2398	0·2268	0·2217	0·2205	0·2193	0·2169	0·2144
30	0·2588	0·2450	0·2319	0·2195	0·2147	0·2136	0·2124	0·2101	0·2078
32	0·2502	0·2370	0·2245	0·2127	0·2081	0·2070	0·2059	0·2037	0·2015
34	0·2420	0·2294	0·2175	0·2062	0·2019	0·2008	0·1997	0·1976	0·1955
36	0·2342	0·2222	0·2109	0·2001	0·1959	0·1949	0·1939	0·1919	0·1899
38	0·2269	0·2154	0·2046	0·1943	0·1903	0·1893	0·1883	0·1864	0·1845
40	0·2199	0·2090	0·1986	0·1888	0·1849	0·1840	0·1831	0·1812	0·1794

The solubility at any value of salinity and temperature in the above range can be calculated if
the following constants are substituted in the equation below (Table 6).

A_1	A_2	A_3	A_4	B_1	B_2	B_3
−173·5146	245·4510	141·8222	−21·8020	−0·034474	0·014934	−0·0017729

TABLE 9

Literature citations for solubilities of other gases in sea water

Gas	Reference
Carbon dioxide	Murray and Riley (1971); see also Chapter 9, Table 9.
Helium	Weiss (1971); see also Chapter 8, Table A8.5.
Neon	Weiss (1971); see also Chapter 8, Table A8.4
Krypton	Wood and Caputi (1966); see also Chapter 8, Table 8.5.
Xenon	Wood and Caputi (1966); see also Chapter 8, Table 8.5.
Carbon monoxide	Douglas (1967); see also Chapter 8, Table 8.12.
Hydrogen	Crozier and Yamamoto (1974).

TABLE 10

The density of artificial sea water as a function of temperature and chlorinity*
(Millero and Lepple, 1973)

Cl (‰)	0°C	5°C	10°C	15°C	20°C	25°C	30°C	35°C	40°C
0	0·999868	0·999992	0·999728	0·999129	0·998234	0·997075	0·995678	0·994063	0·992247
$3·42_6$	1·004944	1·004959	1·004599	1·003921	1·002962	1·001744	1·000295	0·998643	0·996783
$6·05_5$	1·008665	1·008705	1·008292	1·007566	1·006575	1·005335	1·003868	1·002190	1·000307
$8·17_4$	1·011851	1·011731	1·011265	1·010502	1·009472	1·008201	1·006707	1·005013	1·003113
$11·69_5$	1·016982	1·016758	1·016208	1·015368	1·014275	1·012949	1·011407	1·009669	1·007745
$13·67_3$	1·019835	1·019564	1·018970	1·018102	1·016986	1·015641	1·014087	1·012346	1·010406
$16·33_3$	1·023703	1·023352	1·022695	1·021772	1·020611	1·019229	1·017642	1·015866	1·013920
$19·05_6$	1·027648	1·027227	1·026511	1·025538	1·024335	1·022921	1·021311	1·019528	1·017564
$21·53_7$	1·031240	1·030774	1·029989	1·028941	1·027731	1·026307	1·024658	1·022890	1·020925

* These densities are relative to those tabulated by Kell (1967) for pure water assuming the density of pure water is $1·000000 \text{ g ml}^{-1}$ at 3·98°C.

TABLE 11

The expansibility of artificial sea water as a function of temperature and chlorinity, $\alpha \times 10^6 \; (deg.^{-1})$ *(Millero and Lepple, 1973)*

Cl‰	0°C	5°C	10°C	15°C	20°C	25°C	30°C	35°C	40°C
0·000	−68·1	16·0	87·9	150·7	206·6	257·0	303·1	345·7	385·4
3·426	−46·9	35·5	105·1	165·4	218·7	266·7	310·7	351·8	391·0
6·055	−28·0	49·4	115·2	172·7	224·1	271·0	314·8	356·4	396·7
8·174	−14·8	60·4	124·4	180·5	230·7	276·6	319·4	359·9	398·8
11·695	8·2	79·2	140·2	194·1	242·7	287·2	328·3	366·7	402·6
13·673	18·4	88·1	147·5	199·6	246·6	289·8	330·1	368·6	405·7
16·333	36·1	102·3	159·2	209·4	254·8	296·8	335·8	372·8	407·9
19·056	51·0	115·2	170·2	218·5	262·2	302·4	339·9	375·2	408·8
21·537	61·9	127·6	181·6	227·5	267·9	304·9	340·1	375·0	410·7

$\alpha = -1/d(\partial d/\partial t)$ where d is the density of the sea water.

TABLE 12

The isothermal compressibility of sea water at 1 atm as a function of salinity and temperature (Lepple and Millero, 1971)

S(‰)	$\beta \times 10^{-6} \; (bar^{-1})$								
	0°C	5°C	10°C	15°C	20°C	25°C	30°C	35°C	40°C
0·00	50·886	49·171	47·811	46·736	45·895	45·250	44·774	44·444	44·243
6·14	50·07	48·42	47·10	46·09	45·31	44·71	44·26	43·92	43·75
11·80	49·25	47·70	46·43	45·41	44·66	44·13	43·68	43·34	43·19
14·75	48·84	47·30	46·11	45·15	44·38	43·83	43·43	43·13	43·01
21·01	48·14	46·71	45·59	44·63	43·92	43·45	42·96	42·71	42·63
24·52	47·63	46·25	45·17	44·29	43·61	42·98	42·68	42·33	42·23
29·38	47·01	45·62	44·62	43·74	43·17	42·56	42·24	41·96	41·86
34·25	46·49	45·17	44·15	43·32	42·69	42·18	41·88	41·69	41·55
35·00	46·32	45·03	44·02	43·19	42·58	42·11	41·78	41·49	41·48
39·00	45·84	44·62	43·63	42·80	42·30	41·73	41·53	41·23	41·15

TABLE 13

Observed values for the change in the specific volume of sea water from 0° to T°C at various pressures and salinities. Unit of specific volume = $10^{-6}\ cm^3\ g^{-1}$. (Cox et al., 1970)

S = 35.00‰

P, bars absolute →	8.3	201.3	401.2	601.0	800.9	1000.8
S, ‰ →	35.000	35.004	35.005	35.002	35.002	35.002
pH (1 bar, 25°C) →	7.91	7.95	7.94	7.94	8.00	7.96
T(°C) ↓						
-2.000	—	—	-277.1	-356.9	-424.3	-480.5
-1.000	—	-97.5	—	—	—	—
0.000	0	0	0	0	0	0
2.000	132.2	224.9	310.03	383.2	444.9	497.6
4.000	311.2	489.5	652.0	791.7	910.6	1012.5
6.000	535.0	791.0	1023.0	1225.3	1396.8	1544.7
8.000	801.0	1127.3	1424.4	1683.3	1902.4	2094.3
10.000	1107.1	1498.0	1854.4	2163.3	2427.7	2660.1
12.000	1452.7	1901.6	2312.4	2668.5	2971.6	3243.6
14.000	1836.3	2336.9	2796.2	3198.3	3535.9	3827.4
16.000	2255.5	2804.0	3306.4	3745.3	4119.2	4448.9
18.000	2709.3	3299.8	3840.9	4315.0	4721.5	5075.4
20.000	3196.3	3823.9	4400.2	4906.4	5341.0	5719.9
22.000	3717.0	4376.5	4984.1	5516.1	5975.8	6378.6
24.000	4268.2	4957.5	5591.4	6151.3	6630.1	7051.0
26.000	4850.1	5564.4	6223.6	6803.5	7303.4	7738.2
28.000	5461.6	6197.0	6877.1	7472.2	7990.8	8439.8
30.000	6102.8	6855.3	7554.3	8165.9	8693.8	9159.8

S = 30.50‰

P, bars absolute →	8.3	201.3	601.0	1000.9
S, ‰ →	30.502	30.504	30.506	30.510
pH (1 bar, 25°C) →	8.06	7.98	8.03	8.00
T(°C) ↓				
-2.000	—	—	-341.9	-472.3
-1.000	—	-86.9	—	—
0.000	0	0	0	0
2.000	106.9	204.7	368.6	489.0
4.000	262.7	450.5	766.4	998.5
6.000	464.7	734.7	1189.8	1523.0
8.000	712.1	1055.7	1637.2	2064.2
10.000	1000.9	1412.3	2106.9	2623.2
12.000	1330.2	1802.6	2603.7	3197.3
14.000	1698.8	2226.3	3123.4	3790.3
16.000	2104.7	2682.5	3665.3	4398.0
18.000	2547.1	3169.6	4228.6	5021.7
20.000	3022.8	3685.2	4815.6	5661.0
22.000	3533.0	4230.0	5421.9	6315.4
24.000	4075.6	4804.1	6049.9	6985.9
26.000	4649.9	5403.9	6698.9	7673.4
28.000	5255.2	6032.2	7367.9	8374.8
30.000	5889.4	6687.1	8056.6	9091.8

S = 39.50‰

P, bars absolute →	8.3	201.3	601.0	1000.8
S, ‰ →	39.503	39.502	39.504	39.507
pH (1 bar, 25°C) →	8.22	8.18	8.13	8.16
T(°C) ↓				
-2.000	—	—	-370.6	-489.4
-1.000	—	-107.1	—	—
0.000	0	0	0	0
2.000	155.8	245.2	394.4	504.3
4.000	355.8	527.9	815.7	1026.8
6.000	599.1	846.2	1259.1	1556.1
8.000	883.0	1198.6	1726.2	2117.5
10.000	1205.8	1582.7	2216.4	2686.7
12.000	1566.5	1999.4	2729.5	3274.6
14.000	1962.6	2446.4	3264.0	3872.9
16.000	2394.3	2923.1	3818.4	4489.2
18.000	2858.7	3428.0	4394.0	5119.5
20.000	3355.8	3961.8	4994.4	5764.6
22.000	3883.4	4522.1	5610.2	6423.2
24.000	4442.8	5108.6	6246.4	7100.4
26.000	5031.4	5721.8	6901.7	7789.7
28.000	5648.6	6358.8	7576.4	8493.4
30.000	6294.8	7021.5	8269.4	9211.1

TABLE 14

Specific gravity and percentage volume reduction of sea water* under pressure (amended from Cox, 1965)

Pressure (db)	Specific gravity	% decrease in volume
0	1·02813	0·000
100	1·02860	0·046
200	1·02908	0·093
500	1·03050	0·231
1,000	1·03285	0·460
2,000	1·03747	0·909
3,000	1·04199	1·349
4,000	1·04640	1·778
5,000	1·05071	2·197
6,000	1·05494	2·609
7,000	1·05908	3·011
8,000	1·06314	3·406
9,000	1·06713	3·794
10,000	1·07104	4·175

* Salinity, 35·00‰; Temperature 0°C.

TABLE 15

Percentage reduction in volume of sea water under a pressure of 1,000 db at various temperatures and salinities. (After Cox, 1965).

S‰	Temperature (°C)			
	0	10	20	30
0	0·500	0·470	0·451	0·440
10	0·486	0·459	0·442	0·432
20	0·474	0·448	0·432	0·423
30	0·462	0·438	0·424	0·415
35	0·457	0·433	0·419	0·411
40	0·450	0·428	0·415	0·407

TABLE 16

Thermal expansion of sea water under pressure $(10^{-6} cm^3 (°C)^{-1})$. (*Bradshaw and Schleicher, 1970*)

Pressure (bars)	Temperature (°C)			
	0	10	20	30
		S = 30·50‰		
1	39	155	246	324
500	158	229	290	346
1000	240	284	323	362
		S = 35·00‰		
1	52	162	251	327
500	166	234	293	347
1000	244	286	325	363
		S = 39·50‰		
1	65	170	256	329
500	174	239	296	348
1000	248	289	326	363

TABLE 17a*

Velocity of sound in sea water†

Pressure (db)	Temperature (°C)						
	0	5	10	15	20	25	30
0	1449·3	1471·0	1490·4	1507·4	1522·1	1534·8	1545·8
1000	1465·8	1487·4	1506·7	1523·7	1538·5	1551·3	1562·5
2000	1482·4	1504·0	1523·2	1540·2	1555·0	1567·9	1579·2
3000	1499·4	1520·7	1538·6	1555·6			
4000	1516·5	1537·7	1555·2	1572·2			
5000	1533·9	1554·8	1571·9	1588·9			
6000	1551·5	1572·1					
7000	1569·3						
8000	1587·3						
9000	1605·4						
10000	1623·5						

* Reproduced by permission of U.S. Navy Oceanographic Office.
† Velocities in $m s^{-1}$; pressures in decibars above atmosphere. Salinity 35‰. For other salinities see Table 17b.
For detailed tables of the velocity of sound in sea water, see U.S. Naval Oceanographic Office (1962) and Bark *et al.* (1964).

TABLE 17b*

Effect of salinity on sound velocity†

$S\%_{\circ}$	Temperature (°C)						
	0	5	10	15	20	25	30
30	−7·0	−6·7	−6·5	−6·2	−5·9	−5·6	−5·3
32	−4·2	−4·0	−3·9	−3·7	−3·5	−3·4	−3·2
33	−2·8	−2·7	−2·6	−2·5	−2·4	−2·2	−2·1
34	−1·4	−1·3	−1·3	−1·2	−1·2	−1·1	−1·1
35	0	0	0	0	0	0	0
36	1·4	1·3	1·3	1·2	1·2	1·1	1·1
37	2·8	2·7	2·6	2·5	2·4	2·3	2·1
38	4·2	4·1	3·9	3·7	3·6	3·4	3·2
40	7·0	6·8	6·5	6·2	6·0	5·7	5·3

* Reproduced by permission of U.S. Navy Oceanographic Office.

Corrections to be applied to the values in Table 17a for salinities other than 35‰.

TABLE 18

Specific heat of sea water at constant pressure ($J g^{-1}\,^{\circ}C^{-1}$) at various salinities and temperatures (Millero et al., 1973).

Salinity, ‰	0°C	5°C	10°C	15°C	20°C	25°C	30°C	35°C	40°C
0	4·2174	4·2019	4·1919	4·1855	4·1816	4·1793	4·1782	4·1779	4·1783
5	4·1812	4·1679	4·1599	4·1553	4·1526	4·1513	4·1510	4·1511	4·1515
10	4·1466	4·1354	4·1292	4·1263	4·1247	4·1242	4·1248	4·1252	4·1256
15	4·1130	4·1038	4·0994	4·0982	4·0975	4·0977	4·0992	4·0999	4·1003
20	4·0804	4·0730	4·0702	4·0706	4·0709	4·0717	4·0740	4·0751	4·0754
25	4·0484	4·0428	4·0417	4·0437	4·0448	4·0462	4·0494	4·0508	4·0509
30	4·0172	4·0132	4·0136	4·0172	4·0190	4·0210	4·0251	4·0268	4·0268
35	3·9865	3·9842	3·9861	3·9912	3·9937	3·9962	4·0011	4·0031	4·0030
40	3·9564	3·9556	3·9590	3·9655	3·9688	3·9718	3·9775	3·9797	3·9795

TABLE 19

The relative partial equivalent heat capacity of sea salt (cal(eq deg)$^{-1}$) (Millero et al., 1973a)

Salinity	Temperature (°C)						
(‰)	0	5	10	15	20	25	30
0	0	0	0	0	0	0	0
5	2·8	3·1	3·4	3·7	4·0	4·3	4·6
10	4·7	5·0	5·4	5·7	5·9	6·3	6·6
15	6·9	7·1	7·3	7·5	7·7	7·9	8·1
20	9·5	9·4	9·4	9·4	9·3	9·3	9·2
25	12·2	11·9	11·5	11·2	10·8	10·5	10·2
30	15·3	14·6	13·8	13·1	12·3	11·6	10·8
35	18·6	17·4	16·2	15·0	13·8	12·6	11·3
40	22·2	20·5	18·7	17·0	15·2	13·4	11·7

TABLE 20

Thermal conductivity (K in 10^{-5} W cm deg^{-1}) of sea water (S = 34·994‰) as a function of temperature and pressure. (After Castelli et al., 1974)*

Pressure (p)	Temperature (t°C)			
(bars)	1·82	10	20	30
200	563	578	594	605
400	570	585	601	613
600	578	592	609	619
800	585	599	615	627
1000	591	606	622	634
1200	596	613	628	641
1400	602	618	634	647

$K = 5·5286 \times 10^{-3} + 3·4025 \times 10^{-7}P + 1·8364 \times 10^{-7}t - 3·3058 \times 10^{-9}t$

* Other data have been published by Caldwell (1974).

TABLE 21

Freezing point of sea water (T_f) at atmospheric pressure based on the data of Doherty and Kester (1974).

$S‰$	T_f (°C)	$S‰$	T_f (°C)	$S‰$	T_f (°C)
5	−0·275	17	−0·918	29	−1·582
6	−0·328	18	−0·973	30	−1·638
7	−0·381	19	−1·028	31	−1·695
8	−0·434	20	−1·082	32	−1·751
9	−0·487	21	−1·137	33	−1·808
10	−0·541	22	−1·192	34	−1·865
11	−0·594	23	−1·248	35	−1·922
12	−0·648	24	−1·303	36	−1·979
13	−0·702	25	−1·359	37	−2·036
14	−0·756	26	−1·414	38	−2·094
15	−0·810	27	−1·470	39	−2·151
16	−0·864	28	−1·526	40	−2·209

The freezing point at *in situ* pressure is given by
$T_f(°C) = -0·0137 - 0·051990\ S‰ - 0·00007225\ (S‰)^2 - 0·000758z$ where z is the depth in metres.

TABLE 22

Boiling point elevation of sea water ($S = 35·00‰$) at various temperatures, (Stoughton and Lietzke, 1967)

Temp. (°C)	30	40	50	60	70	80	90	100
Vap. press. (atm)	0·042	0·073	0·122	0·197	0·309	0·469	0·694	1·003
Elevation of B.P (°C)	0·325	0·350	0·377	0·405	0·433	0·463	0·493	0·524
Temp. (°C)	120	140	160	180	200	220	240	260
Vap. press. (atm)	1·965	3·577	6·119	9·931	15·407	22·99	33·18	46·52
Elevation of B.P (°C)	0·590	0·660	0·735	0·817	0·906	1·003	1·111	1·232

TABLE 23

Osmotic pressure and vapour depression of sea water at 25°C (Robinson, 1954)

					Chlorinity					
	12	13	14	15	16	17	18	19	20	21
Osmotic pressure (atm)	15·51	16·85	18·19	19·55	20·91	22·28	23·366	25·06	26·47	27·89
Vap. press. lowering* $\times 10^2$	1·139	1·237	1·334	1·433	1·532	1·631	1·732	1·832	1·936	2·039

$(p^0 - p)/p^0$ where p and p^0 are the vapour pressures of sea water and pure water respectively ($p^0 = 23.75$ mm at 25°C).

TABLE 24

Surface tension of clean sea water (in N m⁻¹) at various salinities and temperatures (from data by Krümmel (1900)* and others (After Fleming and Revelle, 1939)

S‰	Temperature (°C)			
	0	10	20	30
0	$75{\cdot}64 \times 10^{-3}$	$74{\cdot}20 \times 10^{-3}$	$72{\cdot}76 \times 10^{-3}$	$71{\cdot}32 \times 10^{-3}$
10	75·86	74·42	72·98	71·54
20	76·08	74·64	73·20	71·76
30	76·30	74·86	73·42	71·98
35	76·41	74·97	73·53	72·09
40	76·52	75·08	73·64	72·20

Surface tension (N m⁻¹) = 10^3 (75·64–0·144t + 0·0221 S‰)

* Measurements made on bubbles below the surface, they therefore take no account of the effects of surface contamination which may be very considerable (e.g. see Lumby and Folkard, 1956 and Vol. 2, pp. 233–4.

TABLE 25

The viscosity of sea water (η) at various salinities and temperatures (in centipoises) computed from values for distilled water (η₀) by Korson et al. (1969) using equations developed by Millero (1974)

Salinity ‰	Temperature °C															
	0	2	4	5	8	10	12	14	16	18	20	22	24	26	28	30
0	1·7916	1·6739	1·5681	1·4725	1·3857	1·3069	1·2349	1·1691	1·1087	1·0532	1·0020	0·9547	0·9109	0·8703	0·8326	0·7975
5	1·8049	1·6868	1·5808	1·4849	1·3979	1·3189	1·2466	1·1807	1·1200	1·0644	1·0129	0·9655	0·9215	0·8807	0·8428	0·8076
10	1·8180	1·6995	1·5930	1·4968	1·4095	1·3302	1·2576	1·1913	1·1304	1·0745	1·0228	0·9751	0·9309	0·8900	0·8519	0·8165
15	1·8312	1·7122	1·6054	1·5087	1·4210	1·3412	1·2685	1·2018	1·1407	1·0845	1·0327	0·9847	0·9402	0·8991	0·8608	0·8252
20	1·8445	1·7251	1·6178	1·5208	1·4325	1·3525	1·2794	1·2125	1·1513	1·0945	1·0424	0·9942	0·9495	0·9082	0·8697	0·8339
25	1·8579	1·7380	1·6302	1·5327	1·4442	1·3638	1·2903	1·2231	1·1614	1·1046	1·0522	1·0036	0·9588	0·9172	0·8786	0·8426
30	1·8713	1·7509	1·6427	1·5448	1·4560	1·3751	1·3012	1·2338	1·1717	1·1146	1·0619	1·0132	0·9682	0·9263	0·8875	0·8513
32	1·8767	1·7563	1·6478	1·5497	1·4607	1·3797	1·3057	1·2379	1·1758	1·1186	1·0658	1·0171	0·9719	0·9300	0·8910	0·8547
34	1·8823	1·7643	1·6528	1·5545	1·4652	1·3843	1·3101	1·2423	1·1800	1·1227	1·0698	1·0210	0·9757	0·9336	0·8945	0·8582
36	1·8876	1·7696	1·6578	1·5594	1·4701	1·3888	1·3146	1·2465	1·1841	1·1267	1·0737	1·0248	0·9793	0·9372	0·8981	0·8617
38	1·8932	1·7752	1·6630	1·5644	1·4748	1·3934	1·3189	1·2508	1·1883	1·1308	1·0778	1·0286	0·9831	0·9409	0·9017	0·8651
40	1·8986	1·7805	1·6680	1·5692	1·4795	1·3980	1·3233	1·2551	1·1925	1·1348	1·0817	1·0325	0·9869	0·9446	0·9053	0·8686
42	1·9041	1·7861	1·6732	1·5741	1·4842	1·4026	1·3278	1·2595	1·1967	1·1389	1·0857	1·0363	0·9906	0·9483	0·9089	0·8721

Viscosity of pure water η_t at temperature $t\,°C$ is given by $\log \dfrac{\eta_t}{\eta_{20}} = \dfrac{1\cdot1709(20 - t) - 0\cdot001827(t - 20)^2}{t + 89\cdot93}$ where η_{20} is the viscosity at 20°C.

Viscosity of sea water calculated from ratio $\dfrac{\eta}{\eta_0} = 1 + A\mathrm{Cl}_t^{\frac{1}{2}} + B\mathrm{Cl}_t$

Where Cl_t is the volume chlorinity ($\mathrm{Cl}_t = \mathrm{Cl}‰ \times$ density) and A = 0·000366, 0·001403 and B = 0·002756, 0·003416 at 5° and 25°C; constants at other temperatures obtained by linear interpolation or extrapolation. According to Matthäus (1972) the change in dynamic viscosity ($\Delta\eta_p$, centipoises) produced by increase in pressure (P, kg cm^{-2}) at temperature, $T\,°C$) can be calculated from the expression

$$\Delta\eta_p = -1\cdot7913 \times 10^{-4}\,P + 9\cdot5182 \times 10^{-8}\,P^2 + P(1\cdot3550 \times 10^{-5}\,T - 2\cdot5853 \times 10^{-7}\,T^2 - P^2(6\cdot0833 \times 10^{-9}\,T - 1\cdot1652 \times 10^{-10}\,T^2)$$

* The assistance of Miss J. Wolfe with the computations is gratefully acknowledged.

Table 26

Relative viscosity of Standard Sea Water ($S = 35.00‰$) at various temperatures and pressures. (Stanley and Batten, 1969)

Pressure, kg cm^{-2}	η_p/η_1 at −0.024°C	η_p/η_1 at 2.219°C	η_p/η_1 at 6.003°C	η_p/η_1 at 10.013°C	η_p/η_1 at 15.018°C	η_p/η_1 at 20.013°C	η_p/η_1 at 29.953°C
176	0·9828	0·9852	0·9891	0·9914	0·9949	0·9977	0·9997
352	0·9709	0·9742	0·9814	0·9876	0·9926	0·9972	0·0001
527	0·9620	0·9670	0·9766	0·9843	0·9900	0·9978	1·0031
703	0·9560	0·9626	0·9735	0·9821	0·9915	0·9998	1·0071
878	0·9533	0·9598	0·9733	0·9836	0·9932	1·0040	1·0131
1055	0·9526	0·9600	0·9750	0·9874	0·9964	1·0070	1·0179
1230	0·9533	0·9637	0·9767	0·9902	1·0014	1·0110	1·0244
1406	0·9559	0·9673	0·9821	0·9961	1·0073	1·0166	1·0313

Where η_p/η_1 is the ratio of the viscosity at pressure p (kg cm^{-2}) relative to that at 1 atm.

T

TABLE 27

Specific conductivity of sea water* (Weyl (1964). From data by Thomas et al., (1934)

$S\text{‰}$	Temperature (°C)					
	25	20	15	10	5	0
10	17·345	15·628	13·967	12·361	10·816	9·341
20	32·188	29·027	25·967	23·010	20·166	17·456
30	46·213	41·713	37·351	33·137	29·090	25·238
31	47·584	42·954	38·467	34·131	29·968	26·005
32	48·951	44·192	39·579	35·122	30·843	26·771
33	50·314	45·426	40·688	36·110	31·716	27·535
34	51·671	46·656	41·794	37·096	32·588	28·298
35	53·025	47·882	42·896	38·080	33·457	29·060
36	54·374	49·105	43·996	39·061	34·325	29·820
37	55·719	50·325	45·093	40·039	35·190	30·579
38	57·061	51·541	46·187	41·016	36·055	31·337
39	58·398	52·754	47·278	41·990	36·917	32·094

* Conductivity in millimho cm^{-1}.

TABLE 28

Effect of pressure on the conductivity of sea water* (after Bradshaw and Schleicher, 1965)

Temp.	Pressure (db)	S‰			Temp.	S‰		
		31	35	39		31	35	39
0°C	1,000	1·599	1·556	1·512	15°C	1·032	1·008	0·985
	2,000	3·089	3·006	2·922		1·996	1·951	1·906
	3,000	4·475	4·345	4·233		2·895	2·830	2·764
	4,000	5·759	5·603	5·448		3·731	3·646	3·562
	5,000	6·944	6·757	6·569		4·506	4·403	4·301
	6,000	8·034	7·817	7·599		5·221	5·102	4·984
	7,000	9·031	8·787	8·543		5·879	5·745	5·612
	8,000	9·939	9·670	9·401		6·481	6·334	6·187
	9,000	10·761	10·469	10·178		7·031	6·871	6·711
	10,000	11·499	11·188	10·877		7·529	7·358	7·187
5°C	1,000	1·368	1·333	1·298	20°C	0·907	0·888	0·868
	2,000	2·646	2·578	2·510		1·755	1·718	1·680
	3,000	3·835	3·737	3·639		2·546	2·492	2·438
	4,000	4·939	4·813	4·686		3·282	3·212	3·142
	5,000	5·960	5·807	5·655		3·964	3·879	3·795
	6,000	6·901	6·724	6·547		4·594	4·496	4·399
	7,000	7·764	7·565	7·366		5·174	5·064	4·954
	8,000	8·552	8·333	8·114		5·706	5·585	5·464
	9,000	9·269	9·031	8·794		6·192	6·060	5·929
	10,000	9·915	9·661	9·408		6·633	6·492	6·351
10°C	1,000	1·183	1·154	1·125	25°C	0·799	0·783	0·767
	2,000	2·287	2·232	2·177		1·547	1·516	1·485
	3,000	3·317	3·237	3·157		2·245	2·200	2·156
	4,000	4·273	4·170	4·067		2·895	2·837	2·780
	5,000	5·159	5·034	4·910		3·498	3·429	3·359
	6,000	5·976	5·832	5·688		4·056	3·976	3·896
	7,000	6·728	6·565	6·402		4·571	4·481	4·390
	8,000	7·415	7·236	7·057		5·045	4·945	4·845
	9,000	8·041	7·847	7·652		5·478	5·369	5·261
	10,000	8·608	8·400	8·192		5·872	5·756	5·640

* Percentage increase compared with the conductivity at one atmosphere.

TABLE 29

Conductivity ratio of sea water at 15°C (R_{15}) and 20°C (R_{20}) relative to sea water of salinity 35·000‰. (From data in UNESCO, 1966).

S‰	15°C	20°C	S‰	15°C	20°C
29·50	0·85795	0·8583	36·00	1·02545	1·0254
30·00	0·87101	0·8714	36·50	1·03814	1·0380
30·50	0·88404	0·8844	37·00	1·05079	1·0506
31·00	0·89705	0·8973	37·50	1·06341	1·0632
31·50	0·91002	0·9103	38·00	1·07601	1·0758
32·00	0·92296	0·9232	38·50	1·08858	1·0883
32·50	0·93588	0·9361	39·00	1·10112	1·1008
33·00	0·94876	0·9489	39·50	1·11364	1·1133
33·50	0·96160	0·9617	40·00	1·12613	1·1257
34·00	0·97444	0·9745	40·50	1·13849	1·1381
34·50	0·98724	0·9873	41·00	1·15103	1·1505
35·00	1·00000	1·0000	41·50	1·16344	1·1629
35·50	1·01275	1·0127	42·00	1·17583	1·1752

For 15°C $S‰ = -0.08996 + 28.29720 R_{15} + 12.80832 R_{15}^2 - 10.678969 R_{15}^3 + 5.98624 R_{15}^4 - 1.32311 R_{15}^5$.

TABLE 30

Correction values ($\times 10^4$) to be applied to conductivity ratios measured at temperatures differing from 20°C to correct them to ratios at 20°C (to be used only in conjunction with 20°C ratios in Table 29). (After UNESCO, 1966)

Measured ratio	Temperature (°C)								
	10	12	14	16	18	20	22	24	26
0·85	80	62	45	29	14	0	−14	−26	−38
0·90	56	44	32	21	10	0	−9	−18	−27
0·95	29	23	17	11	5	0	−5	−10	−14
1·00	0	0	0	0	0	0	0	0	0
1·05	−33	−25	−19	−12	−6	0	5	11	15
1·10	−69	−54	−39	−25	−12	0	11	22	32
1·15	−109	−85	−62	−40	−19	0	18	35	50

TABLE 31

Light absorption of typical sea waters. Extinction for 10 cm path length. (After Clarke and James, 1939)

Sample	Wavelength Å							
	3600	4000	5000	5200	6000	7000	7500	8000
Pure water	0·001	0·001	0·002	0·002	0·010	0·025	0·115	0·086
Artificial sea water	0·011	0·003	0·005	0·007	0·010	0·025	0·115	0·086
Ocean water, unfiltered	0·012	0·009	0·007	0·008	0·011	0·025	0·115	0·086
Continental slope waters, unfiltered	0·052	0·030	0·011	0·010	0·012	0·035	0·130	0·088
Continental slope waters, filtered	0·016	0·010	0·005	0·005	0·012	0·030	0·115	0·086
Inshore water unfiltered	0·055	0·042	0·028	0·026	0·035	0·052	0·140	0·100
Inshore water, filtered	0·015	0·010	0·005	0·005	0·010	0·025	0·110	0·086

TABLE 32

Differences between the extinctions of sea waters and pure water*. (From data by Clarke and James, 1939)

Sample	Wavelength Å							
	3600	4000	5000	5200	6000	7000	7500	8000
Artificial sea water	0·010	0·002	0·003	0·005	nil	nil	nil	nil
Ocean water unfiltered	0·011	0·008	0·005	0·006	0·001	nil	nil	nil
Continental slope water, unfiltered	0·051	0·029	0·009	0·008	0·002	0·010	0·015	0·002
Continental slope water, filtered	0·015	0·009	0·003	0·003	0·002	0·005	nil	nil
Inshore water, unfiltered	0·054	0·041	0·026	0·024	0·025	0·027	0·025	0·015
Inshore water, filtered	0·014	0·009	0·003	0·003	nil	nil	nil	nil

* $E_{SW(10\,cm)} - E_{PW(10\,cm)}$

Note: The values given in Tables 31 and 32 for unfiltered inshore waters should be taken as no more than a rough indication, since actual values vary widely with time and location.

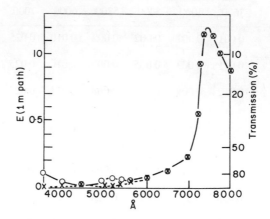

FIG. 1. Extinction (1 m path) against wavelength. Solid line—filtered ocean water. Broken line—pure water. (After Clarke and James, 1939).

TABLE 33

Refractive index differences (Δn) for sea water at a wavelength of 589.3 nm at various temperatures and salinities. (Δn = (n−1.30000) · 10⁵). (Matthäus, 1974)

T[°C]	0	2	4	6	8	10	12	14	16	18	20	22	24	26	28	30	32	34	36	38	40
											S [‰]										
0	3402	3441	3481	3520	3559	3598	3637	3677	3716	3755	3794	3833	3873	3912	3951	3990	4029	4069	4108	4147	4186
1	3400	3439	3478	3517	3556	3595	3634	3674	3713	3752	3791	3830	3869	3908	3947	3986	4025	4064	4103	4142	4181
2	3398	3437	3476	3515	3553	3592	3631	3670	3709	3748	3787	3826	3865	3904	3942	3981	4020	4059	4098	4137	4176
3	3395	3434	3473	3511	3550	3589	3628	3666	3705	3744	3783	3821	3860	3899	3938	3976	4015	4054	4093	4131	4170
4	3392	3431	3469	3508	3547	3585	3624	3662	3701	3740	3778	3817	3855	3894	3933	3971	4010	4048	4087	4126	4164
5	3389	3427	3466	3504	3543	3581	3620	3658	3697	3735	3773	3812	3850	3889	3927	3966	4004	4043	4081	4120	4158
6	3385	3424	3462	3500	3538	3577	3615	3653	3692	3730	3768	3807	3845	3883	3922	3960	3998	4037	4075	4113	4152
7	3381	3419	3458	3496	3534	3572	3610	3648	3687	3725	3763	3801	3839	3878	3916	3954	3992	4030	4068	4107	4145
8	3377	3415	3453	3491	3529	3567	3605	3643	3681	3719	3757	3795	3833	3871	3909	3948	3986	4024	4062	4100	4138
9	3372	3410	3448	3486	3524	3562	3600	3638	3675	3713	3751	3789	3827	3865	3903	3941	3979	4017	4055	4093	4130
10	3367	3405	3443	3481	3518	3556	3594	3632	3669	3707	3745	3783	3821	3858	3896	3934	3972	4010	4047	4085	4123
11	3362	3399	3437	3475	3512	3550	3588	3625	3663	3701	3738	3776	3814	3851	3889	3927	3964	4002	4040	4077	4115
12	3356	3394	3431	3469	3506	3544	3581	3619	3656	3694	3732	3769	3807	3844	3882	3919	3957	3994	4032	4069	4107
13	3350	3387	3425	3462	3500	3537	3575	3612	3649	3687	3724	3762	3799	3837	3874	3911	3949	3986	4024	4061	4098
14	3344	3381	3418	3456	3493	3530	3568	3605	3642	3679	3717	3754	3791	3829	3866	3903	3941	3978	4015	4053	4090
15	3337	3374	3411	3449	3486	3523	3560	3597	3635	3672	3709	3746	3783	3821	3858	3895	3932	3969	4006	4044	4081
16	3330	3367	3404	3441	3478	3515	3552	3590	3627	3664	3701	3738	3775	3812	3849	3886	3923	3960	3997	4035	4072
17	3323	3360	3397	3434	3470	3507	3544	3581	3618	3655	3692	3729	3766	3803	3840	3877	3914	3951	3988	4025	4062
18	3315	3352	3389	3425	3462	3499	3536	3573	3610	3647	3684	3720	3757	3794	3831	3868	3905	3942	3979	4016	4052
19	3307	3344	3380	3417	3454	3491	3527	3564	3601	3638	3675	3711	3748	3785	3822	3858	3895	3932	3969	4006	4042
20	3298	3335	3372	3408	3445	3482	3518	3555	3592	3629	3665	3702	3739	3775	3812	3849	3885	3922	3959	3995	4032
21	3290	3326	3363	3399	3436	3473	3509	3546	3582	3619	3656	3692	3729	3765	3802	3838	3875	3912	3948	3985	4021
22	3281	3317	3354	3390	3427	3463	3500	3536	3573	3609	3646	3682	3719	3755	3792	3828	3865	3901	3938	3974	4011
23	3271	3308	3344	3380	3417	3453	3490	3526	3562	3599	3635	3672	3708	3745	3781	3817	3854	3890	3927	3963	3999
24	3261	3298	3334	3370	3407	3443	3479	3516	3552	3588	3625	3661	3697	3734	3770	3806	3843	3879	3915	3952	3988
25	3251	3288	3324	3360	3396	3433	3469	3505	3541	3578	3614	3650	3686	3723	3759	3795	3831	3868	3904	3940	3976
26	3241	3277	3313	3349	3386	3422	3458	3494	3530	3566	3603	3639	3675	3711	3747	3783	3820	3856	3892	3928	3964
27	3230	3266	3302	3338	3375	3411	3447	3483	3519	3555	3591	3627	3663	3699	3736	3772	3808	3844	3880	3916	3952
28	3219	3255	3291	3327	3363	3399	3435	3471	3507	3543	3579	3615	3651	3687	3723	3759	3796	3832	3868	3904	3940
29	3208	3244	3279	3315	3351	3387	3423	3459	3495	3531	3567	3603	3639	3675	3711	3747	3783	3819	3855	3891	3927
30	3196	3232	3268	3303	3339	3375	3411	3447	3483	3519	3555	3591	3627	3662	3698	3734	3770	3806	3842	3878	3914

TABLE 34

Refractive index differences (Δn) for sea water of salinity 35·00‰ at various temperatures and wavelengths ($\Delta n = (n - 1\cdot30000) \cdot 10^5$). (After Matthäus, 1974)

T[°C]	Wavelength (nm)														
	404·7	435·8	457·9	467·8	480·0	488·0	501·7	508·5	514·5	546·1	577·0	579·1	589·3	632·8	643·8
0	5099	4840	4684	4621	4549	4504	4433	4400	4372	4240	4130	4124	4091	3961	3929
1	5094	4835	4679	4616	4544	4500	4428	4395	4367	4235	4126	4119	4086	3956	3925
2	5089	4830	4674	4611	4539	4495	4423	4390	4362	4230	4121	4114	4081	3951	3920
3	5084	4825	4669	4606	4534	4489	4418	4385	4357	4225	4115	4109	4076	3946	3914
4	5078	4819	4664	4601	4528	4484	4412	4379	4351	4219	4110	4103	4070	3941	3909
5	5072	4814	4658	4595	4522	4478	4407	4374	4345	4213	4104	4097	4065	3935	3903
6	5066	4807	4652	4589	4516	4472	4400	4367	4339	4207	4098	4091	4052	3929	3897
7	5060	4801	4645	4582	4510	4465	4394	4361	4333	4201	4091	4085	4052	3922	3890
8	5053	4794	4639	4576	4503	4459	4387	4354	4326	4194	4085	4078	4045	3916	3884
9	5046	4787	4632	4569	4496	4452	4380	4347	4319	4187	4078	4071	4038	3909	3877
10	5039	4780	4624	4561	4489	4444	4373	4340	4312	4180	4071	4064	4031	3901	3869
11	5031	4773	4617	4554	4481	4437	4366	4332	4304	4172	4063	4056	4023	3894	3862
12	5023	4765	4609	4546	4473	4429	4358	4325	4297	4164	4055	4048	4016	3886	3854
13	5015	4757	4601	4538	4465	4421	4350	4317	4288	4156	4047	4040	4008	3878	3846
14	5007	4748	4592	4529	4457	4412	4341	4308	4280	4148	4039	4032	3999	3869	3837
15	4998	4740	4584	4521	4448	4404	4333	4300	4271	4139	4030	4023	3991	3861	3829
16	4989	4731	4575	4512	4439	4395	4324	4291	4262	4130	4021	4014	3982	3852	3820
17	4980	4721	4566	4503	4430	4386	4314	4281	4253	4121	4012	4005	3972	3843	3811
18	4971	4712	4556	4493	4421	4376	4305	4272	4244	4111	4002	3995	3963	3833	3801
19	4961	4702	4546	4483	4411	4366	4295	4262	4234	4102	3993	3986	3953	3823	3791
20	4951	4692	4536	4473	4401	4356	4285	4252	4224	4092	3982	3976	3943	3813	3781

TABLE 34 cont.

T[°C]	0.4047	0.4358	0.4579	0.4678	0.4800	0.4880	0.5017	0.5085	0.5145	0.5461	0.5770	0.5791	0.5893	0.6328	0.6438
21	4940	4682	4526	4463	4390	4346	4275	4242	4214	4081	3972	3965	3933	3803	3771
22	4930	4671	4515	4452	4380	4335	4264	4231	4203	4071	3961	3955	3922	3792	3760
23	4919	4660	4504	4441	4369	4324	4253	4220	4192	4060	3951	3944	3911	3781	3749
24	4908	4649	4493	4430	4358	4313	4242	4209	4181	4048	3939	3932	3900	3770	3738
25	4896	4637	4482	4419	4346	4302	4230	4197	4169	4037	3928	3921	3888	3759	3727
26	4884	4626	4470	4407	4334	4290	4219	4186	4157	4025	3916	3909	3877	3747	3715
27	4872	4614	4458	4395	4322	4278	4207	4174	4145	4013	3904	3897	3865	3735	3703
28	4860	4601	4445	4382	4310	4265	4194	4161	4133	4001	3892	3885	3852	3722	3690
29	4847	4589	4433	4370	4297	4253	4182	4149	4120	3988	3879	3872	3840	3710	3678
30	4834	4576	4420	4357	4284	4240	4169	4136	4108	3975	3866	3859	3827	3697	3665

λ[nm]

TABLE 35

Absolute refractive index of sea water ($S = 35·00‰$) as a function of temperature, pressure and wavelength. (Stanley, 1971)

Pressure	\multicolumn{7}{c}{Temperature (°C)}
	0·03	5·03	10·03	15·02	20·00	24·99	29·98
			6328 Å				
Atm.	1·34015	1·33977	1·33935	1·33899	1·33850	1·33795	1·33737
352 kg cm²	1·34539	1·34487	1·34431	1·34388	1·34331	1·34270	1·34207
703 kg cm⁻²	1·35025	1·34962	1·34896	1·34844	1·34780	1·34713	1·34647
1055 kg cm⁻²	1·35481	1·35403	1·35380	1·35269	1·35200	1·35129	1·35059
1406 kg cm⁻²	—	1·35813	1·35738	1·35668	1·35592	1·35519	1·35443
			5017 Å				
Atm	1·34455	1·34455	1·34422	1·34379	1·34327	1·34272	1·34215
352 kg cm⁻²	1·35008	1·34969	1·34924	1·34873	1·34813	1·34757	1·34694
703 kg cm⁻²	1·35507	1·35450	1·35394	1·35333	1·35269	1·35208	1·35137
1055 kg cm⁻²	1·35953	1·35891	1·35834	1·35764	1·35695	1·35632	1·35561
1406 kg cm⁻²	—	1·36314	1·36241	1·36166	1·36095	1·36019	1·35946

TABLE 36

Velocity of light ($\lambda = 589.3$ nm) in sea water at 1 atm (km s^{-1}) (Sager, 1974)

$S‰$	Temperature (°C)								
	0	5	10	15	20	25	30	35	40
0	224.732	224.749	224.785	224.837	224.904	224.985	225.080	225.185	225.305
2.5	224.650	224.668	224.705	224.759	224.827	224.909	225.004	225.110	225.230
5.0	224.567	224.588	224.626	224.681	224.749	224.832	224.928	225.035	225.156
7.5	224.485	224.507	224.547	224.603	224.672	224.756	224.852	224.960	225.081
10.0	224.402	224.426	224.468	224.524	224.595	224.679	224.776	224.885	225.006
12.5	224.319	224.346	224.388	224.446	224.518	224.603	224.700	224.810	224.931
15.0	224.236	224.265	224.309	224.368	224.441	224.527	224.625	224.735	224.857
17.5	224.154	224.185	224.230	224.290	224.364	224.450	224.549	224.660	224.782
20.0	224.072	224.104	224.151	224.212	224.287	224.374	224.473	224.585	224.707
22.5	223.990	224.024	224.072	224.134	224.210	224.297	224.398	224.510	224.633
25.0	223.907	223.943	223.994	224.057	224.133	224.221	224.322	224.435	224.559
27.5	223.825	223.863	223.915	223.979	224.056	224.145	224.247	224.360	224.485
30.0	223.743	223.783	223.836	223.901	223.979	224.069	224.171	224.285	224.411
32.5	223.661	223.703	223.758	223.823	223.903	223.993	224.096	224.211	224.336
35.0	223.579	223.623	223.679	223.746	223.826	223.917	224.020	224.136	224.262
37.5	223.498	223.543	223.600	223.669	223.749	223.841	223.945	224.061	224.188
40.0	223.416	223.463	223.521	223.591	223.673	223.765	223.870	223.986	224.114

References

Bark, L. S., Ganson, P. P. and Meister, N. A. (1964), "Tables of the Velocity of Sound in Sea Water" Pergamon, Oxford.

Bradshaw, A. and Schleicher, K. E. (1965). *Deep-Sea Res.* **12**, 151.

Bradshaw, A. and Schleicher, K. E. (1970). *Deep-Sea Res.* **16**, 691.

Caldwell, D. T. (1974). *Deep-Sea Res.* **21**, 131.

Carpenter, J. H. (1966). *Limnol. Oceanogr.* **11**, 264.

Castelli, V. J., Stanley, E. M. and Fischer, E. C. (1974). *Deep-Sea Res.* **21**, 311.

Clarke, G. L. and James, H. R. (1939). *J. Opt. Soc. Amer.* **29**, 43.

Cox, R. A. (1965). *In* "Chemical Oceanography" (J. P. Riley and G. Skirrow, eds), Vol. I. Academic Press, London.

Cox R. A. and Culkin, F. (1967). *Deep-Sea Res.* **13**, 789.

Cox, R. A., McCartney, M. J. and Culkin, F. (1970). *Deep-Sea Res.* **17**, 679.

Crozier, T. E. and Yamamoto, S. (1974). *J. Chem. Eng. Data,* **19**, 242.

Doherty, B. T. and Kester, D. R. (1974), *J. Mar. Res.* **32**, 285

Dorsey, N. E. (1940). "Properties of Ordinary Water-substance". Reinhold, New York.

Douglas, E. (1964). *J. Phys. Chem.* **68**, 169.

Douglas, E. (1965). *J. Phys. Chem.* **69**, 2608.

Douglas, E. (1967). *J. Phys. Chem.* **71**, 1931.

Fleming, R. H. and Revelle, R. R. (1939). "Recent Marine Sediments" (N. Trask ed.). Amer. Soc. Petrol. Geol., Tulsa, Oklahoma.

Kalle, K. (1945). In "Probleme der Kosmischen Physik" 2nd Edn., Vol. 23. Leipzig.

Korson, L., Drost-Hansen, W. and Millero, F. J. (1969). *J. Phys. Chem.* **73**, 34.

Kester, D. R., Duedall, I. W., Connor, D. N. and Pytkowicz, R. M. (1967). *Limnol. Oceanogr.* **12**, 176.

Krümmel, O. (1900). *Wiss. Meeresuntersuch.* **5**, 9.

Lepple, F. K. and Millero, F. J. (1971). *Deep-Sea Res.* **18**, 1233.

Lumby, J. R. and Folkard, A. R. (1956). *Bull. Inst. Océanogr. Monaco,* **1080**, 1.

Lyman, J. and Fleming, R. H. (1940). *J. Mar. Res.* **3**, 134.

Matthäus, W. (1974). *Beitr. Meeresk.* **29**, 93.

Matthäus, W. (1974). *Beitr. Meeresk.* **33**, 73.

Millero, F. J. (1974). *In* "The Sea" (E. D. Goldberg ed.), Vol. 5, Interscience, New York.

Millero, F. J. and Lepple, F. K. (1973). *Mar. Chem.* **1**, 89.

Millero, F. J., Hansen, L. D. and Hoff, E. V. (1973b). *J. Mar. Res.* **31**, 21.

Millero, F. J., Perron G. and Desnoyers, J. E. (1973). *J. Geophys. Res.* **78**, 4499.

Morris, A. W. and Riley, J. P. (1966). *Deep-Sea Res.* **13**, 689.

Murray, C. N. and Riley, J. P. (1969a). *Deep-Sea Res.* **16**, 311.

Murray, C. N. and Riley, J. P. (1969b). *Deep-Sea Res.* **16**, 297.

Murray, C. N. and Riley, J. P. (1971). *Deep-Sea Res.* **18**, 533.

Riley, J. P. and Tongudai, M. (1967). *Chem. Geol.* **2**, 263.

Robinson, R. A. (1954). *J. Mar. Biol. Ass. U.K.* **33**, 449.

Sager, G. (1974). *Beitr. Meeresk.* **33**, 68.

Stanley, E. M. (1971). *Deep-Sea Res.* **18**, 833.

Stanley, E. M. and Batten, R. C. (1969). *J. Geophys. Res.* **74**, 3415.

Stoughton, R. W. and Lietzke, M. H. (1967). *J. Chem. Eng. Data,* **12**, 101.

Thomas, B. D., Thompson, T. G. and Utterback, C. L. (1934). *J. Cons. Int. Explor. Mer,* **9**, 28.

UNESCO (1966). International Oceanographic Tables, Vol. 1. National Institute of Oceanography, Wormley, Surrey, England.

UNESCO (1973). International Oceanographic Tables. Vol. 2. National Institute of Oceanography, Wormley, Surrey, England.

U.S. Naval Oceanogr. Office (1962) Tables of sound speed in sea water. Publ. SP58 U.S. Naval Oceanographic Office, Washington, U.S.A.

U.S. Navy (1961). Tables for the velocity of sound in sea water. Bureau of ships reference NObsr 81564 S-7001-0307, Washington, U.S.A.

Weiss, R. F. (1970). *Deep-Sea Res.* **17**, 721.

Weiss, R. F. (1971). *J. Chem. Eng. Data,* **16**, 235.

Weyl, P. (1964). *Limnol. Oceanogr.* **9**, 75.

Wilson, W. D. (1960). *J. Acoust. Soc. Amer.* **32**, 1357.

Wood, D. and Caputi, R. (1966). "Technical Report No. 988" U.S. Naval Radiological Defense Laboratory, San Francisco.

Subject Index

(Numbers in bold type indicate the page on which a subject is treated most fully.)

A

Absorption spectrometry, 290–298
  automatic, 294–298
  blanks in, 293
  errors in, 292, 293
  hidden blanks, 293, 294
  salt error in, 293
  sensitivity, 294
  standardization of, 293
  effect of turbidity, 293
Acetone, determination of, 474
Achromobacter, 475
Actino-uranium series, 97
Adenosine triphosphate, 420, 457
  determination, 457
ADP, 470
Adsorption of organic compounds, 468
Airborne contamination, 273, 325
Albacore, 174
Aldehydes, determination of, 470
Aldrin, 71
Algae, thorium in, 119
Alizarin complexone, 248
Alkalinity, changes during sampling, 202
n-alkanes, persistence of, 58
Alpha-spectrometry, 432, 434, 436, 437
Aluminium, determination by AAS, 337
  fluometry, 298, 336
  gas chromatography, 311, 336
  X-ray fluorescence, 337
  particulate, 336, 337
  preconcentration, 280
  speciation, 336
Aluminium-26, 139, **140**, 154, 155
Amino acids, bioassay of, 464
  determination of, 472, 473
  preconcentration of, 466, 468

Aminophosphonic acids, 420
Ammonia, contamination with, 414
  determination of, automatic, 412, 414
  determination of electrometric, 414
    by indophenol blue, 410, 411, 412
    by oxidation, 412, 413
    by rubazoic method, 413
  intercomparison, 414
  relationship to nitrite, 8
  storage for, 222
Ammonium—see ammonia
Amphidinium carteri, 475
Analytical error, 197–198
Analytical methods, development of, 278
Anion exchange, 284, 343, 358, 359, 377
  392, 394, 395, 403, 405
Anodic stripping voltammetric methods,
  337, 343, 347, 356, 359, 371, 372, 385,
  407
Anoxic basins, water-renewal in, 22–29
Anoxic conditions, development of, 2, 10,
  11, 12, 16–22, 30, 31
  in open ocean, 2–11
Anoxic marine basins list of, 13–15
Antarctic Convergence, Ra near, 121,
  123
Antarctic Ocean, tritium in, 152, 153
Antimony, determination by ASV, 337
  colorimetry, 337
  spectrometry, 337
  NAA 338
  particulate, 338
  preconcentration, 280, 286, 337
  speciation, 337
Antimony-124, determination of, 449
Antimony-125 determination of, 449
Antimony-129m, 160
Apparent oxygen utilization, 7, 8, 21

549

Arabian Sea, anoxic conditions in, 10
  hydrogen sulphide in, 10
  nitrite in, 10
Argon, determination by isotope dilution, 264
  gas chromatography, 261, 262, 264
  separation from oxygen, 261, 262
  solubility of, 253
  solubility tables for, 522
Argon-40, 93–95
Aroclor see polychlorinated biphenyls
Aromatic amines (chlorinated), 48
Arsenic, determination by colorimetry, 339
  NAA, 340
  spectroscopy, 340
  organic forms, 338, 340
  particulate, 341
  preconcentration, 280, 283, 338, 339
  speciation of, 338
Arsenomolybdic acid, 339
Arsine, 338, 340
Artificial sea water formulae, 517
Atlantic, anoxic conditions in, 2, 3
  ^{14}C in, 146
  ^{137}Cs in, 166, 168, 169
  ^{210}Pb in, 130, 131
  phosphate in, 2
  ^{211}Po in, 130, 131
  Pu in, 176
  Ra in, 121, 122, 125
  ^{90}Sr in, 166, 167, 168, 169, 170, 171
  tritium in, 151, 152
Atmospheric metal pollution, 68
Atomic absorption spectrophotometers, 301, 303, 306
Atomic absorption spectrophotometry, 300–309
  analytical applications, 306, 307
  atomization, 304, 307
  background correction, 307
  carbon tube furnace, 305, 308
  cold vapour technique, 305
  flame burner, 304
  instrumentation, 303–306
  interferences, 306, 307
  non-atomic attenuation, 307
  principles, 301, 302
  sensitivity, 307, 310
  tantalum boat, 305

Atomic absorption spectrometry application to trace metals, 336, 340, 341, 345, 348, 351, 355, 366, 371, 372, 374, 375, 379, 380, 383, 385, 393, 406, 407
Atomic absorption spectrometry for calcium, 230, 237, 307
  magnesium, 230, 237, 307
  potassium, 230, 307
  strontium, 230, 238, 307
Atomic fluorescence spectrometry, 309
Atmosphere carbon-14 in, 141
  tritium in, 151
Atomic bombs production of tritium by, 151
Atomic weapons, 143
AutoAnalyzer, 294–298
  for determination of ammonia, 297
  boron, 252, 298
  fluoride, 249, 298
  hydrogen sulphide, 266, 298
  nitrate, 297
  nitrite, 297
  organic P, 297
  phosphate, 297
  silicate, 297
  filtration in, 296
  profiling with, 296
Automatic photometric analysis, 294–298, 412, 413, 417, 419, 424, 430

B

Bacteria, degradation of oil by, 57
  nitrate-reducing, 5
  sulphate reducing, 7
Bahama Banks, 171
Baltic, 12, 13, 16–22, 24, 25
  oil pollution of, 54
  uranium in sediments of, 107
Barium, determination by AAS, 341
  flame photometry, 341
  isotope dilution, 342
  NAA, 341, 342
  particulate, 342
  profile in sea, 122
  preconcentration, 341
  speciation, 341
Barium-140, determination of, 451
Bathophenanthroline, 287, 364, 365
Bathythermograph, 208

Beer's Law, 291
Beryllium, determination of, 298, 343
  preconcentration of, 280, 281, 343
  speciation, 342
Beryllium-7, 139, **140**, 153
  determination of, 444
Beryllium-10, 139, **140**, 141, 153, 154
  determination of, 444
Beta-counting, 318, 319, 441, 442
Bilges, oil from, 55
Bioassay of organic compounds, 464, 465
Biological hydrocarbons, differentiation
  of, 59
Biomass, measurement of, 455, 457
Biotin, determination of, 465, 475
Bipyridyl, 364
Biquinolyl, 287, 353, 354, 355
Bismuth, determination of, 343
  preconcentration of, 280, 284, 285,
    288, 343
  speciation, 343
Black Sea, 12, 13, 16, 21, 31, 33
  radium in, 123, 125
  uranium in sediments of, 107
Boiling point elevation of sea water, 615
Bolstadtfjord, 13
Boron, determination of, 250–252
  colorimetric, 251, 252
  fluorometric, 252
  titrimetric, 250, 251
Box models, 142, 147–149
Bristle Cone pine, 142
Bromide, determination of, 246, 247
Bromine-82, 330
Brucine, 416
Butane, determination of, 268
Butyraldehyde, determination of, 474

C

Cadmium, anthropogenic, 69
  determination by AAS, 308, 309, 344,
    345
    ASV, 345
    colorimetry, 344
    polarography, 345
    particulate, 345
    preconcentration, 280, 284, 285, 288,
      289, 308, 344
    speciation, 343
Cadmium-113m, 175

Caesium determination of, 346
  intercalibration for, 335
  preconcentration, 345, 346
  speciation, 345
Caesium-137, 49, 50, 52, 53, 159, 160,
  164, **165–171**
  determination of, 450, 451
Calcein, 235
Calcium, determination of, 230–237
  difference chromatography, 237,
    242, 243
  gravimetric, 230
  spectrometric, 237, 300
  volumetric, 231–237
  ion exchange separation of, 226, 227,
    235, 236
Calcon, 235
Calred, 235
Calver II, 235
Carbohydrates, determination of, 470,
  472
Carbon, determination in POM, 456, 458
Carbon dioxide annual flux of, 40
  box models for, 142, 147–149
  residence time of, 142
  (total) determination, 265
Carbon monoxide, determination of,
  267, 268
Carbon, organic oxidation of, 2, 5, 6, 10,
  11
Carbon tetrachloride, 81, 82
  determination of, 474
Carbon-13 in Black Sea sediments, 16
Carbon-14, 139 ff.
  atmospheric activity, 141
  bomb derived, 52, 143, 145, 160
  data on, 140
  data presentation, 144
  dating by, 145
  determination of, 442, 444
  in the sea, 145–150
  transport of, 150
  variations of, 142–143
Carbonates, uranium in, 109–110
Cariaco, Gulf of, 14
Cariaco Trench, 2, 13, 21, **24, 25**, 26, 30,
  31
Caribbean Sea, radium profile in, 123
  ^{90}Si in, 168, 169
Cation exchange, 284, 372

CDTA, 237
Centrifugation, 212
Cerium, see also lanthanides
  determination, 347
  preconcentration, 283
  speciation, 347
Cerium-141, 52, 160, 161
Cerium-144, 52, 160, 161, 163, 172, 440
  determination of, 451
Chelates in sea water, 270
Chelating ion exchange, 285, 286, 344,
  348, 352, 359, 370, 374, 377, 382, 392,
  403
Chelating resins for recovery of amino
  acids, 468, 471
Chelatometric titration, 231–235
Chemical wastes, dumping of, 48
Chemical yield, measurement of, 432,
  441
Chitosan, 286, 344, 352, 405
Chlorinated hydrocarbons determina-
  tion, 269
Chlorine-36, 165
Chlorinity, determination of, 225
Chloroform, 81
Chlorophyll, determination of, 457, 459,
  460, in situ, 299
CHN Analyzer, 456
Choline containing lipids, 34
Chromium, adsorption, 162
  anthropogenic, 69
  determination by colorimetry, 348
    AAS, 308, 348
    NAA, 348
    X-ray fluorescence, 348
  particulate, 348
  preconcentration, 280, 347
  speciation, 347
Chromium-51, 46, 47, 52, 160, 173, 440
  determination of, 445
Clams, 175
Cobalt determination by AAS, 308, 309,
  351
    colorimetry, 350
    NAA, 351
    polarography, 351
    particulate, 351
    preconcentration, 280, 283, 285, 289,
      308, 349, 350
    speciation, 348

Cobalt-60, 47, 175, 440
  determination of, 445, 446
Co-crystallization, 282–283
Cod, 176
Colloid flotation, 377
Colloidal species, 271
Colorimetric methods application of,
  337, 339, 343, 344, 347, 348, 350, 351,
  353, 355, 357, 358, 361, 364, 365, 366,
  374, 378, 383, 385, 386, 391, 393, 395,
  396, 397, 400, 406, 408
Columbia River, 46, 47, 48, 171, 173, 174
Compressibility of sea water, 524
Conductivity ratio of sea water, 538
Conductivity, changes during sampling,
  202
Contamination of samples, 201, 202, 203,
  271
Conversion factors for micronutrients,
  518
Copper anthropogenic, 69
  determination by AAS, 308, 309, 355
    ASV, 356
    colorimetry, 353–355
    NAA, 356
    polarography, 355, 356
    X-ray fluorescence, 311, 353
    particulate, 356
    preconcentration, 285–288, 352, 353
    speciation, 352
    total, determination, 356
Coprecipitation, 278–281
Coprecipitation of organic compounds,
  468
Corals, U in, 109
Cosmic radiation, nuclides production
  by, 139–157
Critical pathway analysis, **84**, **85**, 162,
  163, 164, 165
Cryptococcus albidus, 465, 475
Curcumin, 251, 252
Cyclotella nana, 465, 475

D

Darwin Bay, 13
Data processing in NAA, 319, 320
Dating of deep-water, 26–29
DDT, 41, 70–72, 76–78, 80, 81
  determination of, 474
  preconcentration of, 467

Deep water, age of, 150
Denitrification in anoxic waters, 4, 5, 6 8, **9**, 11
Density of sea water as a function of t and chlorinity, 523
Detrital aluminosilicates, 96
Deuterium background corrector, 307
Dianthrimide, 252
Diethyldithiocarbamate (sodium), 289, 290, 353, 354, 365, 375, 384, 403, 405
Difference chromatography, 228, 237, **240–243**
Dimethylarsinic acid, 338, 340
Dimethylglyoxime, 384, 385
Dimethyl mercury, 62
Dimethyl sulphide, determination of, 473
Dinoflagellates, Pu in, 177
Diphenylbenzidine, 416
Diphenylphenanthroline, 287, 364, 365
Dissolved organic matter, determination of amino acids, 467, 468, 471
    carbohydrates, 470
    carbon, 461–462
    lipophilic compounds, 471, 472
    of nitrogen, 461, 462, 463
    of phosphorus, 462, 463
    various compounds, 472–475
    vitamins, 471
  preconcentration of, 466–470
Dissolved organic carbon, determination of, 461, 462
Dissolved organic nitrogen, determination, 461, 462, 463
Dissolved organic phosphorus, determination, 462, 463
Dithiocarbamic acid, 288
Dithizone, 287, 288, 344, 352, 353, 354, 370, 377, 392, 405
DNA, 457
Doerner-Hoskins distribution law, 282, 283
Drammensfjord, 13
Dredge spoils, discharge of, 47
Dysprosium, see lanthanides

E

East Pacific Rise sediments, U in, 105

EDC tar, 48, 82
EDTA, 231–236
EGTA, 236
Electroanalytical methods, 313
Emission spectroscopy, 299
Emission spectroscopy application of, 337, 340, 343, 358, 368, 385, 389, 395, 396
Endosulfan, 48, 83
Enzymatic determination of organic compounds, 464
Erbium, see lanthanides
Eriochrome Black T, 231, 233, 234, 235
Error, analytical, 197–199
Ethane in anoxic waters, 31
  determination of, 268
Ethylene in anoxic waters, 31
Euglena gracilis, 465
Euphausia pacifica, 174
Europium, see lanthanides
Expansibility of sea water, 524

F

Fallout, form of, 161
  penetration into deep water, 169, 170
Färo Deep, 13
Fatty acids, removal by filtration
Fatty acids, determination of, 473
  preconcentration of, 467
Feldspars, dating of, 95
Fermentation in anoxic waters, 31, 32
Ferromanganese concretions, ^{26}Al in, 155
  dating of, 95
Filtration, apparatus for, 218, 329
  of samples, 211–218, 454, 455
  changes caused by, 213, 215–217, 271 272, 273
  of particulate organics, 454, 455
  retention during, 215
Fish, pelagic, ^{210}Po in, 131
Fission products, 158, 159, 160
  in oceans, 171–177
Fjarlie sampler, 204
Fjellangervåg, 13
Flame photometry, 299–300
Flame photometry applications of, 341, 346

Flame photometry for calcium, 237, 300
  magnesium, 237, 300
  potassium, 230, 300
  strontium, 238, 300
Flame photometry for major ions, 230, 237, 238, 239, 300
Fluoride determination of, 247–250
  photometric, 247–249
  potentiometric, 249–250
Fluoride specific electrodes, 249, 250
Fluorimetry, 298
Fluorimetry, applications of, 336, 347, 357, 399
Fluorimetric determination of chlorophyll, 459, 460
Fluxes into sedimentary cycle, 40
Framvaren, 13
Freezing point of sea water, 615
Freons, 81
  determination of, 474
Frierfjord, 13
Freundlich isotherm, 279
Fulvic acids, 61

G

Gadolinium, see lanthanides
Galactose, determination of, 472
Gallium, determination of, 298, 357
  preconcentration of, 280, 283, 357
  speciation, 357
Gamma ray spectrometry, 317, 318, 319, 328, 330, 333, 441, 442
Gas chromatography application of, 311, 337, 473, 474
Gas chromatography for dissolved gases, 261, 262, 264, 267, 268, 269
Gases, dissolved storage of samples for, 219–220
Geochronology application to sediments, 131–139
  use of $^{230}Th/^{232}Th$, 133–135
  decay of ^{234}U, 135–136
  use of ^{230}growth, 136–138
  use of short-lived nuclides, 138
  use of ^{32}Si, 156, 157
Geosecs, 152
Germanium determination of, 357, 358
  preconcentration of, 280, 357
  speciation, 357

Ge(Li) detectors, 317, 318, 328
Glacial debris, flux of, 40
Glass apparatus, use of, 273–275, 324
Glasses, dating of, 95
Glucose, determination of, 470, 472
  by bioassay, 464
Glucose-6-phosphate, 470
Glycollic acid, determination of, 474
Gold, determination of, 358, 359
  preconcentration of, 280, 283, 284, 285, 286, 358
  speciation, 358
Golfo, 14
Gotland Deep, 14, 25
Greenland ice, ^{26}Al in, 154
Gross errors, 197
Gross National Product, 43, 44

H

Hafnium, determination of, 359
  preconcentration of, 283
Hanford Reactor, 46, 47, 48
Heavy metals, atmospheric pollution by, 67–68
  baselines in oceans, 67–68
  pollution of sea with, 60–72
Helium, determination of, 265
Hellefjord, 14
Helvigfjord, 14
Holmium, see lanthanides
Hot brines of Red Sea, 105
Hydrocarbons, biological, differentiation, 49
  determination of, 268, 269, 474
  emission of, 56
  formation in anoxic waters, 31, 32
  preconcentration of, 467
  in sediments, 66
Hydrogen, determination of, 267
Hydrogen sulphide in Arabian Sea, 10
  Baltic, 19
  Black Sea, 27, 28, 31
  determination, 265, 266
  dissociation of, 265
  effect on oxygen determination, 259
  oxidation of, 265
  relation to DOC, 31
  storage of samples for, 219
Hydrowire, contamination from, 201

Hydroxylamine, determination of, 414
8-hydroxyquinoline, 353, 365, 370, 375, 384, 398, 405

I

Indian Ocean, anoxic conditions in, 2, 3, 10
^{14}C in, 147
uranium in sediments of, 105
Intercalibration of methods, 199, 313, 314, 335, 435, 443
Indicators for titration of Ca and Mg, 235
Indium, determination of, 359, 360
preconcentration of, 283, 285, 359
speciation of, 359
Iodate, determination of, 361, 362, 363
Iodine determination by catalytic method, 362
electrochemistry, 362, 363
photometry, 361, 362
titrimetry, 361, 362
preconcentration, 360, 361
speciation, 360
storage for, 363
Iodine-129, 159, 173
Iodine-131, 159, 172, 441
determination of, 449
Iodomethane, 81
Ion exchange for preconcentration, 283–286, see also under individual elements
Ion exchange, separation of major ions by, 226, 227, 228, 235, 236, 240–243
Ionium, see thorium-230
Iridium, determination of, 363
preconcentration, 283
Irish Sea, 163
Iron, colloidal, 363
determination by AAS, 308, 309, 366
colorimetric, 364–366
NAA, 366
X-ray fluorescence, 311, 366
oxide(-hydrous) as coprecipitant, 280, 281, 468
particulate, 364, 367, 368
preconcentration, 286, 287, 289, 364, 365
speciation, 363, 364
total dissolved, 366

Iron-55, 160, 175, 176
determination of, 445
Iron-59, 160, 175
determination of, 445
Isefjaefjord, 14
Isotope dilution, 312, 313
Isotope dilution, application of, 239, 240, 342, 360, 361, 389, 400
Isotopic equilibration, 440

K

Karlsö Deep, 14
Ketones, determination of, 470
Knudsen sampler, 204
Krypton, determination of, 265

L

Lactobacillus Leichmanii, 465
Lambert-Beer's Law, 290, 291
Laminar flow hood, 271, 325, 326
Landsort Deep, 15
Lanthanides, determination of, 368, 369
preconcentration of, 280, 281, 283, 285, 368
speciation of, 368
Lanthanum, see lanthanides
Lanthanum alizarin complexone, 248
Lead aerosols of, 63, 67
annual budget, 63
anthropogenic, 65, 66, 69
determination by anodic stripping,

isotope dilution, 312, 371
polarography, 371
intercalibration for, 370
isotopic composition, 66
as marine pollutant, 63, 67
particulate, 372
preconcentration, 280, 281, 285, 289, 369, 370
profiles in sea, 64
in sediments, 63, 65
speciation of, 369
Lead-210, 98, 128–130, 200, 281
determination of, 433
in lobsters, 131
Lenefjord, 15
Light absorption of sea water, 539

Lithium, determination, 307, 308, 373
  preconcentration of, 284, 372
  speciation of, 371
Litter, marine, 82–83
Lobsters, 58, 131
London, discharge of sewage from, 48
Lutetium, see lanthanides

M

Magnesium, determination of, 230–237
  difference chromatography, 237, 242, 243
  gravimetric, 230
  spectrometric, 237
  volumetric, 231–237
  ion exchange separation of, 226, 227, 235, 236
Major anions, determination of, 243–252
  boric acid, 250–252
  bromide, 246–247
  carbonate, 247
  fluoride, 247–250
  sulphate, 243–246
Major cations, determination of, 225–252
  difference chromatography of, 228, 237, 240–243
  ion exchange separation of, 226–227, 228, 235, 236
  precision required for, 226
  determination of calcium, 230–237
    sodium, 227, 228
    magnesium, 230–237
    potassium, 228–230
    strontium, 237–240
Major ions definition of, 225
Major ions, concentrations of at various S‰, 516
Manganese, determination by AAS, 308, 309, 374, 376
  colorimetry, 374, 375
  NAA, 375
  polarography, 375
  X-ray fluorescence, 311, 375
  dioxide for preconcentration, 281
  particulate, 376
  preconcentration, 283, 285, 373, 374
  speciation, 373
  total, 375, 376

Manganese-54, 161, 174, 175, 440
  determination of, 445
Mass spectrometry for noble gases, 264, 265
Mediterranean, radium in, 123
Membrane salinometer, 241
Mercaptans, 34
2-Mercaptobenzimidazole, 283, 358
Mercury, desorption of, 60
  determination by AAS, 379, 380
    colorimetry, 380
    NAA, 380
    spectroscopy, 380
  environmental flux of, 61
  filtration for, 218, 376
  in fish, 61, 62
  organic compounds of, 376, 381
  poisoning by, 85, 86
  pollution by, 60–62
  preconcentration of, 280, 285, 288, 377, 378
  speciation of, 376
Mercury-203, 175
Metals, pollution of sea by, 60–70
Methane in anoxic waters, 31, 32
  determination of, 268, 269, 474
  hydrate, 32
Methylarsonic acid, 338, 340
Methyl mercuric chloride, 62
  poisoning by, 85, 86
Micronutrients, storage for, 220–223
Minimata Bay, 61, 62
  Disease, 62, 85, 86
Model of mixing of surface waters, 169
Model, advection diffusion, 34
Molar absorptivity, 294
Molybdenum determination of, 383
  particulate, 383
  preconcentration, 280, 285, 286, 382
  speciation, 382
Monochrysis lutheri, 465, 475,
Monsoons, 42, 43
Murexide, 235
Mustard gas, 83
Mytilus edulis, 174

N

Nansen sampler, 203, 204
Na(Tl) detectors, 318

Neodymium-147, 159
Neon, determination of, 264, 265
Neptunium, 49, 159
Neutron activation analysis, 313–334
  accuracy, 320–322
  advantages of, 315
  data handling, 319
  freedom from interference, 315
  irradiation in, 316, 317
  interferences, 323, 324
  precision, 320–322
  principles, 314, 317
  rabbit irradiations, 316
  scope of, 320–324
  sensitivity, 315, 323, 334
  spectrometry, 317–320
Neutron activation analysis of sea water
  cooling period in, 327, 330, 332
  instrumental, 330–334
  irradiation step, 326
  manipulations, 325
  preconcentration for, 328, 331, 332
  radiochemical separations in, 333, 334
  radiological hazards of, 327, 330
  standardization of, 329, 330
Neutron activation applications of, 338,
  340, 341, 342, 347, 348, 351, 356, 358,
  359, 360, 366, 368, 369, 375, 380, 385,
  388, 389, 390, 391, 394, 401, 402, 407
Neutron activation products in ocean,
  173–177
New York, discharges from, 48
Nickel, determination by, AAS, 308,
  309, 385
    ASV, 385
    colorimetry, 384, 385
    NAA, 385
    X-ray fluorescence, 384
  particulate, 386
  preconcentration, 280, 285, 286, 289,
    384
  speciation, 383, 384
Nickel-63, 175
  determination of, 446
Nile Blue A, 252
Nicotinic acid, determination of, 471
Niobium, determination of, 386
  preconcentration of, 280, 386
Niobium-95, determination of, 448
Niskin sampler, 205, 208

Nitinat, Lake, 14, 31, 34
Nitrate, contamination with by filtra-
    tion, 217
  determination of, 415–419
  reduction of, 2, 4, 5, 6, 7, 8, 30
  relation to AOU, 7, 8
  storage for, 222
Nitrite, determination of, 414, 415
  formation of, 4, 5, 6, 7, 8, 10
  maxima, 4, 5, 6, 7
  storage for, 222
Nitrogen, budget of, 9, 11
  determination of, 261–263
    in POM, 456–458
  formation of, 253
  solubility of, 253
  solubility tables, 520
Nitropiazoselenol, 391
1-Nitroso-2-naphthol, 283
Nitrous oxide, 5
  determination of, 266, 267
Noble gases determination of, 264, 265
  storage of samples for, 219, 220
Nordåsvatn, 15
North Sea, dumping into, 48
Nuclear detonations, source of pollu-
    tants, 49, 53
Nuclear waste, discharge of, 158
Nutrients automatic determination of,
    297, 412, 413, 417, 419, 424, 430
  storage for, 220–223

O

Ochromonas malhamensis, 465
Oil, fluxes, to the sea, 55, 56
Organic carbon, anoxic decomposition
    of, 29–34
  oxidation rate of, 30
  relation to H_2S, 331
Organic chemicals, synthetic, dumping
    of, 83
  production of, 70
Organic matter in anoxic sediments, 30
Organic phosphorus, presentation of,
    221, 222
Organisms concentration of radionu-
    clides by 174, 175, 177
Orthophosphate, see phosphate

Osmium, determination of, 386, 387
  preconcentration of, 283, 386
Osmotic pressure of sea water, 616
Oxygen in Baltic, 17, 19, 20, 21
  in Cariaco Trench, 25
  consumption rate of, 30
  depletion in oceans, 2–11
  determination of, 7, 253 ff
    by chemical methods, 260
    gasometry, 260
    gas chromatography, 261, 262
    radiochemistry, 260, 261
    by Winkler method, 254–260
    accuracy, 259, 260
    end point in, 257, 258
    errors in, 255, 256
    interferences, 258, 259
    at low levels, 7, 258
    photometric, 258
    standardization, 256
  sampling errors in, 203, 253
  solubility tables for, 520
  storage of samples for, 219
Oxalyldihydrazide, 355
Outfalls as sources of pollutants, 47
Oysters, 164

P

P_{CO_2}, determination of, 265
Pacific, anoxic conditions in, 2, 3, 6, 7, 8
  caesium-137 in, 166, 168, 169
  carbon-14 in, 146, 147
  denitrification in, 4, 5, 8, 9
  lead-210 in, 130
  nitrite maxima in, 4, 5, 6, 7, 8
  plutonium in, 176
  polonium in, 130
  radium in, 121, 123
  silicon-32 in, 155, 156
  strontium-90 in, 166, 167, 168, 169
  tritium in, 152
Palladium, preconcentration of, 285
Panama, Gulf of, 2
Particulates analysis of, 329 see also
  under individual elements
  determination of, 215, 216, 329
Particulate organic matter, analysis of,
  455–460
  carbon determination, 456, 458
  nitrogen determination, 456, 457, 458

Particulate material collection of, 211,
  212
PCBs, see also polychlorinated biphenyls
Pelagophyscus porra, 177
Perchlorate, determination of, 387
Peru current, 4
Petroleum products, baseline of, 59, 60
  decomposition in sea, 57
  effect on organisms, 58–60
  persistence of, 58
  as pollutants, 53–60
  solubility of, 57
pH in anoxic basins, 21
  changes during sampling, 222
Phaeophytin, 459
Phenylphosphonic acid, 427
Pheromones, 58
Phillipsite, dating of, 95
Phosphate, in Baltic, 19, 21
  condensed, 427
  determination of, 421–426
    automatic, 424
    calibration, 423
    comparative studies, 425
    composition of reagents for, 423
    interference in, 424, 425
    at low concentrations, 424, 425
    mechanism of, 422
    species determined in, 426
  minerals association with U, 108
  release from sediments, 21
  storage for, 220, 221
Phosphomolybdic acid, 421, 422
Phosphorus, fractionation of, 420
  organic, 420
    breakdown of, 420, 421, 426, 427
    particulate forms of, 420
    species in sea water, 419, 420
Phosphorus-32, 160
  determination of, 445
Photo-oxidation of organic matter, 463
Pinus aristata, 142
Plankton, oil in, 59
Plant pigment analysis, 218, 459, 460
Plastics in the sea, 83
Platinum, determination of, 387
Plutonium, 48, 49, 53, 85, 157, 159, 161,
  176, 177
  determination of, 452, 453
  preconcentration of, 283

Polar Easterlies, 42
Polarography, applications of, 344, 351, 355, 356, 375, 400, 401, 407
Pollutants, control of, 84–86
  transport path, 41–48
Polonium-210, 98, 129–131
  determination of, 433
Polychlorinated biphenyls, 41
  alteration of, 75, 76
  atmospheric transport, 74, 80
  determination, 474
  dumping of, 76
  in fish, 79
  fluxes of, 73, 74, 75
  losses to environment, 76
  in phytoplankton, 79
  production rate, 71, 73
  ratio to DDT, 76
  routes into ocean, 73, 74
  preconcentration of, 467
  sales of, 76
  in sea water, 80
  toxicity of, 80
  use of, 73
  in zooplankton, 77–79
Polyethylene apparatus, use of, 273–275, 324
Polyphosphate, hydrolysis of, 221
Polystyrene spherules in the sea, 82–83
Polytetrafluoroethylene, use of for apparatus, 273–275, 325
Porphyra, 41, 53, 163
Potassium-argon dating, 94, 95
Potassium, determination of, 228–230
  difference chromatography, 228, 242, 243
  gravimetric, 228–229
  photometric, 230, 300
  titrimetric, 229–230
  tetraphenyl boron for, 228
Potassium rhodizonate, 283
Potassium-40, 93, 94
  determination of, 433, 439
  oceanic inventory of, 49
Praseodymium, see lanthanides
Praseodymium-143, 159
Preconcentration techniques, 278–289
Preconcentration of organic compounds, 466–470
Primordial nuclides, 92

Probes, in situ, 200, 209
Promethium-147, 160, 164, 165, 172
  determination of, 451
Propane in anoxic waters, 31, 33
  determination of, 268, 269
Protactinium, behaviour in sea water, 119–120
  concentration in sea water, 119
  determination of, 434
  in ferromanganese nodules, 120
  residence time of, 119
  in sediments, 119, 120
Pumping, sampling by 200, 209–211
Purines, bioassay of, 464
Pyrimidines, bioassay of, 464
Pyridine-2-aldehyde-2-quinolyl hydrazone, 385
Pyrrolidine dithiocarbamate (ammonium), 288, 289, 344, 353, 365, 375, 384, 392, 405, 406, 433

                    Q

Quinoline-2-aldehyde-2 quinolyl hydrazone, 353, 355
Quinoxaline-2,3-dithiol, 385

                    R

Radiation, ecological effects of, 162–165
Radioactive waste disposal, 162–165
Radioactivity artificially produced, 157–174
Radioactivity, discharge into sea, 48–53
Radiocarbon dating of deep waters, 26, 28, 29
Radiochemical separations, 432, 433, 440–443
Radionuclides (artificial) determination of, 437–453
  direct determination, 439
  distribution of, 438
  intercalibration of, 443
  oceanic inventory, 49
Radionuclides (natural), determination of, 431–437
Radionuclides, preconcentration of, 432, 440, 441

Radium-226, behaviour in the sea, 120
  in coastal waters, 124
  determination of, 434, 435
  for geochronology, 124
  intercalibration for, 435
  models for, 121–122, 125
  in mollusc shells, 124
  profiles in sea, 121–124, 130
  in river water, 120
  in sea water, 120–124
  in sediments, 124
  use as tracer, 121
Radium-228, 98, 113, 120, **124**, *125*
  determination of, 435
Radon-222, concentration gradients, 125
  decay of, 98
  determination, 434, 435, 436
  model of, 126, 128
  profiles of near sea floor, 126, 127
  in surface waters, 128, 129
  use in gas exchange studies, 128
  use in mixing studies, 128
Random errors, 198
Rare earth elements, see lanthanides
Reactors, 316
Reagents, purification of, 276, 325
  trace elements in, 276, 277, 325
Refineries, losses to sea, 55, 56
Refractive index of sea water, 541, 542
Relative partial equivalent heat capacity
  sea salt, 530
Residence time of water in Black Sea, 27–29
  Cariaco Trench, 26, 33
Rhenium, determination of, 285, 388
  preconcentration, 388
  speciation, 387
Rhodamine B, 337, 357
Rhodium, determination of, 388
Rhodium-102, 159
Rivers, transport of pollutants by, 44–47
Rosocyanin, 251, 252
Rubber, trace elements in, 275
Rubidium determination of, 307, 309, 312, 388, 389
  preconcentration, 388
  speciation, 388
Rubidium-strontium dating, 95–96
Rubidium-87, 95
Rubrocurcumin, 251

Ruthenium, determination of, 390
  preconcentration of, 280, 283
Ruthenium-103, 160, 172, 440
Ruthenium-106, 41, 52, 53, 160, 161, 163, 164, 165, 441
  determination of, 448, 449

S

Saanich Inlet, 15, 23, 24, 107
Salinity, determination of, 223, 224
  relation to chlorinity, 224
  storage of samples for, 219
Salmon, 175
Salpha, 174
Samarium-151, 159
Samplers, Ekman, 203, 204
  Fjarlie, 204
  for gases, 206, 207
  general purpose, 203–205
  IOS, 204
  Knudsen, 204
  Nansen, 203, 204
  Large volume, 205, 206, 209
  Niskin, 205, 206, 208
  for organic compounds, 206
  sterile, 208
  Van Dorn, 204, 205
  for pore waters, 208, 209
  pumping, 209–211
  under way, 208
Sampling, 199–211
  alteration of sample during, 202
  contamination in, 201
  criteria for samplers, 199, 202–3
  depths for, 200, 201
  effect on alkalinity, 202
    conductivity, 202
    pH, 202
    trace metals, 202
  for organic compounds, sterile, 208
  near sea floor, 207
  profiling in, 200, 209, 210
  by pumping, 200, 209–211
  standard depths for, 200, 201
Santa Barbara Basin, 23
  metal pollution of, 67, 69
  radon in, 128
Santa Barbara blowout, 55
Santa Monica Basin, 69

Sargasso Sea, plutonium in, 176, 177
Sargassum weed, 54, 59
Scandium determination by AAS, 308, 391
    NAA, 391
    particulate, 391
    preconcentration, 280
Scandium-46, 160, 175
    determination of, 440, 445
Sea salt flux, 40
Sea water, boiling point elevation of, 531
    compressibility of, 524
    conductivity ratio of, 538
    density of, 523
    expansibility of, 527
    freezing point of, 531
    light absorption of, 539
    major ion concentrations in, 516
    osmotic pressure of, 532
    refractive index of, 541, 542
    specific conductivity of, 536
        under pressure of, 537
    specific heat of, 529
    specific volume of, 525
    surface tension of, 533
    thermal conductivity of, 530
    thermal expansion of, 527
    vapour pressure of, 532
    velocity of light in, 545
    velocity of sound in, 527
    viscosity of, 534
    volume reduction under pressure, 535
Secular equilibrium, 97
Sediments, adsorption of nuclides by, 164, 165
    ^{26}Al in, 155
    ^{14}C in, 145
    lead in, 63, 66
    uranium content of, 104–112
Selenium, determination by colorimetry, 391
    gas chromatography, 391
    NAA, 391, 392
    preconcentration, 280, 391
    speciation, 391
Sensitivity of atomic spectroscopic methods, 310
Separation processes, 289, 29
Serratia marinorubra, 464, 465
Seston, oil contamination of, 59

Shipboard analysis, 196
Ships, discharges from, 47, 48, 55
    nuclear powered, 158
Silica apparatus, trace elements in, 274, 275, 276
Silicate, see silicon
Silicic acid, see silicon
Silicomolybdic acid, 428, 429
Silicon, determination of, 427–431
    automatic, 430
    comparative studies, 431
    molybdenum blue methods, 429, 430
    silicomolybdic acid method, 428, 429
    particulate, 427
    preconcentration of, 280
    speciation of, 428
    standard solution of, 430
    storage for, 222, 223
    total, determination of, 431
Silicon-32, 139, 155–157, 200, 281
    data on, 140
    determination of, 444
Silver, anthropogenic, 69
    determination by AAS, 309, 393
        colorimetry, 393
        NAA, 392
    particulate, 393
    preconcentration, 280, 283, 285, 288, 289, 392
    speciation, 392
Silver-110m, 175, 440
    determination of, 449
Skoldafjord, 15
Sodium, determination of, 227, 228
Sodium-24, 330
Soledad Basin, 69
Solvent extraction for preconcentration, 286–289, 467
Søndeledspoll, 15
Speciation of trace elements, 269
Specific conductivity of sea water under pressure, 537
Specific heat of sea water, 529
Specific volume of sea water change with t and P, 526
Spectrophotometers, 291
    calibration of, 293
Sponges, silicon-32 in, 155

Squid, 175
STD probes, 209
Standard addition technique, 300, 301, 308
Standards for nutrient determination, 199
Sterols, 34
Storage of samples, 218–223, 271, 273
Strontium, determination of 237–240
  colorimetric, 240
  isotope dilution, 239–240, 312
  neutron activation, 239, 321
  spectrochemical, 238, 300
Strontium-85, 441
Strontium-87 : strontium-86 ratio, 96
Strontium-89, determination of, 446
Strontium-90, 49–52, 159, 160, 164, **165–171**
  determination of, 447
  intercomparison of methods, 443
Strychnidine, 416
Suess effect, 143
Suigetsu, Lake, 14
Sulphate, determination, 243–246
    gravimetric, 243
    volumetric, 244, 245
  as a measure of pollution, 44, 46
  reduction, absence in Pacific, 7
  transport by rivers, 44–46
Sulphite, formation of, 20
Sulphur, elemental, 20
Sulphur-35, 160
Surface-tension of sea water, 533
Suspended river solids flux of, 40
Systematic errors, 197

T

Tantalum determination of, 393
Tar-balls, 54
Tellurium, determination of, 394
Tellurium-129m, 160, 161
Terbium, see lanthanides
Tetrachloroethylene, 81
Thallium, determination of, 394
  preconcentration of, 283, 284, 285
Thermal conductivity of sea water, 530
Thermal expansion of sea water, 527
Thiamine, determination of 465, 475
Thin layer chromatography, 459, 466

Thionalide as a coprecipitant, 283
Thiosulphate, formation of, 20
  determination of, 394, 395
Thiosulphate solution, standardization of, 256, 259
Thorium-228, 98, 113, 114, 116, 117, 440
Thorium-230, 98, 113, 116, 118, 119, 120
Thorium-232, 98, 113, 116, 117, 118, 119
Thorium-234, 98, 113, 116
Thorium, in barites, 117
  in coral, 118
  decay series, 98
  in ferromanganese concretions, 117, 120
  in marine particulates, 116
  removal from sea, 116
  in river water, 116
  in sea water 113–115
  in sediments, 117–118
  in tissues of organisms, 118, 119
  in zooplankton, 119
Thorium, determination of, 436, 437
Thorium, preconcentration of, 280, 281
Thorium-230/thorium-232 dating, 133–135
Thraustochytrium globosum, 465
Thulium, see lanthanides
Tin, determination of, 395, 396
  preconcentration of, 283, 395
Titanium, determination of, 396
  preconcentration of, 280, 396
Tofino Inlet, 15
"Torrey Canyon", 55
Total cations, determination of, 225
Total trace metals, determination of, 270
Trace elements, contamination with, 223, 271, 272, 273, 276
    from glass, 273, 275, 324
    from plastics, 273, 274, 275
  filtration for, 271
Trace elements in glass, 273, 275, 324
  in plastics, 273, 274, 275
Trace elements, storage for, 223, 271, 273
Trace metal, contamination from apparatus, 271–277
    from reagents, 276, 277
  organic complexes of, 270
  total, determination of, 270
Trade winds, 42
Transport paths of pollutants, 42–48

Trichloroethylene, 81
Tritium, 139, 140, 150–153, 160
  determination of, 442, 444
  inventory of, 49, 52
Tungsten, determination of, 397
  preconcentration of, 280, 283, 396, 397
  speciation of, 396
Tungsten-181, 159
Tungsten-185, 159

U

Ulva, 161
Uranium, in anoxic muds, 106, 107
  behaviour in anoxic waters, 107
  in carbonate sediments, 105, 109
  complex formation, 99
  decay series of, 98
  determination by colorimetry, 400
    fluorimetry, 398, 399
    isotope dilution, 312, 400
    NAA, 401, 402
    polarography, 400, 401
    α-spectrometry, 401
  determination of particulate, 402
  geochemical balance of, 111
  geological variation in sea, 112
  intercalibration of analyses, 335
  particulate, 99
  in phillipsite, 108
  in phosphate minerals, 107–108
  preconcentration of, 280, 283, 400
  in river water, 99–102
  relation to DOC, 107
  in sea water, 98, 100–101, 102
  in sediments, 104–111
  speciation of, 397
  in tissues, 112–113
Uranium-helium dating, 94
Uranium-234, use for dating, 135–136
Uranium-238, fission of, 157, 159
Urea, determination of, 464, 465, 474

V

Vanadium, anthropogenic, 69
  determination by AAS, 309
  colorimetry, 403, 404
  polarography, 404
  preconcentration, 280, 285, 402, 403
  speciation, 402

Vapour pressure depression of sea water, 526
Vanna, Lake, 15
Velocity of light in sea water, 545
Viscosity of sea water under pressure, 535
Velocity of sound in sea water, 527
Vestihusfjord, 15
Vinyl chloride, 82
Vinyl esters, 48
Vitamin B_{12}, determination of 465, 466, 467, 475
Volatilization of organic compounds, 468, 470
Volcanic debris, flux of, 40
Volume reduction of sea water under pressure, 525

W

Water, purification of, 325
Water, properties of, 515
Water renewal in anoxic basins, 22–29
Westerlies, 42, 43
Wind systems, transport of pollutants by, 42–44
Winkler method, see oxygen determination

X

Xenon, determination of, 265
X-ray fluorescence spectrometry, 311
X-ray fluorescence spectrometry applications of, 337, 348, 353, 366, 375, 385

Y

Ytterbium, see lanthanides
Yttrium, see lanthanides
Yttrium-90, 447
Yttrium-91, 160, 161

Z

Zinc, anthropogenic, 69
  determination by AAS, 308, 309, 406, 407
  ASV, 407
  colorimetry, 406
  NAA, 407
  polarography, 406

intercalibration for, 335
particulate, 407, 408
preconcentration, 283, 285, 286, 288,
  289, 404–406
speciation, 404
Zinc-65, 47, 84, 85, 164, 173, 174, 175,
  440
determination of, 446
Zirconium, determination of, 408
  preconcentration of, 283

Zirconium alizarin red S for fluoride, 247
Zirconium-95, 52, 159, 160, 161, 163, 172
  440, 441
determination of, 447, 448
Zooplankton, lead in, 131
  plutonium in, 177
  thorium in, 119